Relativity *in Curved* ~~me~~

Relativity *in Curved Spacetime*
Life without special relativity

Eric Baird

Chocolate Tree Books

Copyright © 2007, Eric Baird.

First printing: September 2007

All rights reserved. No reproduction, copy or transmission of this publication may be made in any format or medium, including but not limited to photographic, electronic, holographic or involving intricate arrangements of ornamental carved fish, without prior written permission.

The right of Eric Baird to be identified as the author of this work has been asserted by him in accordance with the Copyright, Designs and Patents Act 1988.

Published by Chocolate Tree Books, UK   www.chocolatetreebooks.com

Printed and bound by CPI Antony Rowe, Eastbourne

Cataloguing data

Baird, Eric

    Relativity in curved spacetime: Life without special relativity / Eric Baird.

    First edition.

    394 pp. (i-xvi, 1-378) size: 234 mm × 156 mm

    Includes bibliographic references and index, and ~200 figures and illustrations

    ISBN 978–0–9557068–0–6 (paperback)

    1. General Relativity, Special Relativity, Quantum Mechanics

5-69-10-162

> I do not see any reason to assume that ... the principle of general relativity is restricted to gravitation and that the rest of physics can be dealt with separately on the basis of special relativity ... I do not believe that it is justifiable to ask: what would physics look like without gravitation?
>
> Albert Einstein, 1950

RELATIVITY IN CURVED SPACETIME

OVERVIEW

00 [BOOK INTRODUCTION]
The usual things that one finds at the start of a book, abbreviations, and a table of contents

01 BACKGROUND PHYSICS
The speed of light, local and global ... Gravity, light and time, space and spacetime curvature ... Inertial and gravitational mass, energy has gravity, $E=mc^2$ ... Different principles of relativity, Mach's principle ... Wave-particle duality and the Newtonian crisis

02 EFFECTS DUE TO RELATIVE MOTION
Doppler effects on signal frequencies and wavelengths ... Analogous effects on the apparent lengths of rulers ... Aberration: associated distortion of angles ... Moving and accelerated bodies drag light, light-dragging as a fundamental aspect of the General Principle

03 LIMITS TO OBSERVATION
Quantum Mechanics and observability, observerspace, "Copenhagen" and "Hidden Variable" interpretations ... Dark Stars and Black Holes, the "classical" and "quantum" versions of Hawking Radiation, the Black Hole Information Paradox and some associated black hole paradoxes and problems

04 UPDATING STANDARD THEORY
Problems with the current implementation of general relativity, should GR ever have incorporated Einstein's Special Theory? ... Zooming in on a single solution, Special Relativity's relationships seem to be incompatible with fully curvature-based models

05 THE FLAT SPACETIME PARADIGM
The build-up to Einstein's special theory – Lorentz, Einstein and Minkowski, Einstein's "special" theory in brief ... Some shortcomings of the Special Theory, cumulative redshift issues ... Experimental evidence for Special Relativity, the way it has been presented, and some unsettling procedural discrepancies

06 FUTURE PHYSICS
A few interesting universes, the Hartle-Hawking Bubble ... Wormholes, antiwormholes and pseudo-wormholes ... Warp drive theory

07 THE HUMAN FACTOR
Logic and language, limitations of sequential logic, "thinkable" vs. "unthinkable", quality control and filtering, belief systems, quantum mechanics applied to the behaviour of societies (including physicists) ... Seeing what we want to see, what do experiments really tell us?, exploration is a dangerous business ... Two possible histories, signs associated with systemic failure, coming to terms with the potential loss of special relativity

08 [BOOK LEADOUT]
Calculations too tedious to include in the main body of the book (but demanded by physicists) ... Lists of some relevant reference papers and books (about 400 entries), a few significant people, and the final, inevitable, keyword index

TABLE OF CONTENTS

OVERVIEW VI
TABLE OF CONTENTS VII
WELCOME XIII
ABOUT THE AUTHOR XIV
NOTES TO THE FIRST EDITION XIV
ACKNOWLEDGEMENTS AND CREDITS XV
    ABBREVIATIONS XVI

PART I BACKGROUND PHYSICS — 1

1 THE SPEED OF LIGHT 3
    1.1 : LIGHT IS PRETTY FAST 5
    1.2 : LIGHTSPEED VARIES 6
    1.3 : LIGHTSPEED IS NOT JUST THE SPEED OF LIGHT 7
    1.4 : LIGHTSPEED AFFECTS INERTIA 8
    1.5 : LIGHTSPEED CONTROLS TIMEFLOW 8
    1.6 : LIGHTSPEED IS LOCALLY CONSTANT 8
    1.7 : LIGHTSPEED IS NOW DEFINED AS CONSTANT 9
    1.8 : THE GRAVITY WELL 9
    1.9 : LIGHT TRAVELS IN STRAIGHT LINES. EXCEPT WHEN IT DOESN'T ... 11
    1.10 : LIGHT USED TO DEFINE A STRAIGHT LINE 12

2 GRAVITY, ENERGY AND MASS 13
    2.1 : WHAT IS MASS? 15
    2.2 : DOES LIGHT HAVE MASS? 16
    2.3 : GENIE IN A BOTTLE: THOUGHT-EXPERIMENTS WITH BOTTLED LIGHT ... 17
    2.4 : DIFFICULTY OF DETECTING THE EFFECT 19
    2.5 : MASS-TO-ENERGY CONVERSION 20
    2.6 : HISTORY OF $E=mc^2$ 21
    2.7 : ENERGY HAS MASS, PERIOD 22

3 CURVED SPACE AND TIME 23
    3.1 : GRAVITY IS ... WHAT, EXACTLY? 25
    3.2 : GRAVITY BENDS LIGHT 25
    3.3 : GRAVITY WARPS GEOMETRY 26
    3.4 : GRAVITY AS A VARIATION IN INERTIA 28
    3.5 : ENERGY-CHANGE IN LIGHT DUE TO GRAVITY 28
    3.6 : GRAVITATIONAL REDSHIFTS AND BLUESHIFTS 29
    3.7 : GRAVITATIONAL TIME DILATION 30
    3.8 : NOT JUST CURVED SPACE, BUT CURVED SPACETIME 32

4 RELATIVITY 33
    4.1 : RELATIVITY OF SPACE 35
    4.2 : RELATIVITY OF TIME 35
    4.3 : RELATIVITY OF VELOCITY 36
    4.4 : ISAAC NEWTON'S "PRINCIPIA" 37
    4.5 : MACH AND RELATIVITY 38
    4.6 : PRACTICAL ADVANTAGES OF "RELATIVISTIC" ARGUMENTS 39
    4.7 : APPLYING OCCAM'S RAZOR 39
    4.8 : DIFFERENT "PRINCIPLES OF RELATIVITY" 40
    4.9 : CAUSES OF CONFUSION 41
    4.10 : RELATIVITY OF ACCELERATION 42
    4.11 : RELATIVE ACCELERATION VS. ABSOLUTE ACCELERATION 43
    4.12 : RELATIVITY OF ROTATION 45

4.13 : "Centrifugal" and "Coriolis" fields	45
4.14 : Rotational dragging	47
4.15 : Experimental verification	49
4.16 : Equivalence principles	50

5 The Newtonian Catastrophe ... 51

5.1 : Newton's unification scheme:	53
5.2 : The lightspeed mistake	54
5.3 : The "space-density" mistake	54
5.4 : The light-energy mistake	55
5.5 : Loss of wave-particle duality	55
5.6 : Newton vs. Huyghens	56
5.7 : The lightspeed trap	57
5.8 : Consequences for physics	59

PART II EFFECTS DUE TO RELATIVE MOTION   61

6 Doppler Shifts ... 63

6.1 : "Stationary observer" Doppler effect	65
6.2 : "Stationary source" Doppler effect	66
6.3 : Comparisons	66
6.4 : Transverse Doppler effects (audio)	67
6.5 : Optical Doppler effects	69
6.6 : Longitudinal Doppler effect under Special Relativity	69
6.7 : Transverse Doppler effect under Special Relativity	70

7 Apparent Length-Changes in Moving Objects ... 71

7.1 : Apparent changes in length	73
7.2 : Approaching objects appear elongated	73
7.3 : Receding objects appear contracted	73
7.4 : Degree of contraction or elongation	74
7.5 : Special relativity and length-changes	75
7.6 : Rulers and gravitation	76

8 Aberration of Angles ... 77

8.1 : Aberration of Angles	79
8.2 : Relativistic aberration at 90 degrees	80
8.3 : The Relativistic Ellipse	81
8.4 : Putting it all together	83
8.5 : Relativistic ellipse: Newtonian theory	83
8.6 : Relativistic ellipse: Special relativity	84

9 Moving bodies drag light ... 85

9.1 : Generality of dragging effects	87
9.2 : Naming conventions: Gravitomagnetism, frame-dragging	87
9.3 : Argument #1: Linear GM as a gravitational timelag effect	87
9.4 : Argument #2: "Effective gravitational potential" depends on relative velocity	88
9.5 : Argument #3: Gravitational smudging	89
9.6 : Argument #4: The slingshot effect	89
9.7 : Argument #5: Rotational GM and gravitational timelag	90
9.8 : Argument #6: QM and "probabilistic" smudging	91
9.9 : Argument #7: Experiment: The Fizeau effect	91
9.10 : Inconsistencies in our approach to velocity	92
9.11 : Cancellation and unification?	94
9.12 : Implementation – the tilted gravity-well	95
9.13 : Zeno revisited: the "impossibility" of motion	96
9.14 : Worldlines and curvature	97
9.15 : Uh-oh	98
9.16 : The score chart	99
9.17 : "Relativistic" implementations of lightspeed constancy	100

PART III LIMITS TO OBSERVATION — 101

10 Quantum Mechanics and Observability — 103
10.1 : The origin of quantum mechanics — 105
10.2 : Is quantum mechanics a theory? — 106
10.3 : The "Copenhagen" and "Hidden Variable" interpretations — 107
10.4 : The two-slit experiment — 108
10.5 : Quantum mechanics and everyday experience — 111
10.6 : Illusion and reality — 112
10.7 : Pair Production — 114
10.8 : Virtual particles — 114
10.9 : Pseudo- pair production — 115

11 Dark Stars and Black Holes — 117
11.1 : John Michell's dark stars — 119
11.2 : Properties of a compact gravitational object — 120
11.3 : Escape velocity calculations and the gravitational horizon — 121
11.4 : Tidal forces — 121
11.5 : "Visiting" particles around a dark star — 122
11.6 : Dark stars and "acoustic" metrics — 123
11.7 : Acoustic metrics and nonlinearity — 124
11.8 : Black holes under GR1915 — 125
11.9 : The Kerr black hole — 129
11.10 : The expansion problem — 130
11.11 : The acceleration problem — 130
11.12 : Black holes according to Quantum Mechanics — 131
11.13 : Hawking radiation — 132
11.14 : Pair-production and pseudo-pair-production — 133
11.15 : Attempts to eliminate the "dark star" explanation — 134
11.16 : Acoustic metrics, once again — 135
11.17 : "Acceleration radiation" — 136
11.18 : The Black Hole Information Paradox — 137
11.19 : The BHIP and Microcausality — 138
11.20 : "Observerspace" arguments — 139
11.21 : The Membrane Paradigm — 140
11.22 : Holographic arguments — 141
11.23 : The Holographic Principle in action — 142
11.24 : The "no-signal" problem — 143
11.25 : The verdict — 144

PART IV UPDATING STANDARD THEORY — 145

12 What's wrong with General Relativity? — 147
12.1 : "Core" experimental tests of general relativity — 149
12.2 : Experimental significance — 152
12.3 : Incompatibility with quantum mechanics — 153
12.4 : Fudge factor?: The Cosmological Constant — 154
12.5 : Possible breaking of conservation laws — 155
12.6 : Possible incompatibility with Mach's principle — 155
12.7 : Fudge factor?: Galactic curves and Dark Matter — 156
12.8 : Arbitrary suspension of the Equivalence Principle — 157
12.9 : Invoking reduction to flat spacetime — 159
12.10 : Use of tailor-made definitions — 160
12.11 : Do cosmological horizons count as "acoustic"? — 163
12.12 : Doppler effects and the Black Hole Information Paradox — 164
12.13 : Grand unification? — 165
12.14 : Gravitomagnetic incompatibility? — 167
12.15 : Complexity — 168
12.16 : Is GR1915 scientifically falsifiable? — 169

12.17 : Blaming special relativity 170
13 Horrible Nasty Mathematics 171
13.1 : A family of relativistic theories 173
13.2 : Selecting a reference theory 174
13.3 : Defining the range 174
13.4 : Ellipses 175
13.5 : Special relativity as a special solution 175
13.6 : Positive values of © and positive curvature 176
13.7 : Rejecting negative solutions for © 176
13.8 : Gravitomagnetism suggests positive © 177
13.9 : Graphed Doppler responses 177
13.10 : Setting "one" as a higher limit for © 178
13.11 : Using the BHIP to set a minimum of "one" for © 178
13.12 : Oops? 179
13.13 : Preliminary conclusions 180

PART V FLAT SPACETIME AND SPECIAL RELATIVITY 181

14 Einstein's "special" theory of relativity 183
14.1 : The birth of special relativity 185
14.2 : Failure of earlier theories 185
14.3 : ... "Draggable" aethers 185
14.4 : ... Absolute aether 186
14.5 : Aether, either, neither neither 187
14.6 : Lorentz Ether Theory (LET), → 1904 188
14.7 : Special relativity, 1905 189
14.8 : Additional interpretational overhead 190
14.9 : Minkowski Spacetime 192
14.10 : Implications of Minkowski spacetime 194

15 So, what's wrong with the special theory? 195
15.1 : SR and Observerspace 197
15.2 : Is the special theory "robust"? 198
15.3 : Minkowski spacetime as an argument against SR 199
15.4 : The "stratification" problem 200
15.5 : Does SR "do" acceleration? 201
15.6 : Extensibility 203
15.7 : Cumulative redshift effects 203
15.8 : ... Thermal redshifts 205
15.9 : ... Cosmological redshifts 206
15.10 : Round-trip effects in general 208

16 Experimental Evidence for Special Relativity .. 209
16.1 : Commonly-cited evidence for special relativity 211
16.2 : ... $E=mc^2$ 211
16.3 : "Classical Theory" vs. Special Relativity 212
16.4 : ... "Transverse" redshifts 213
16.5 : ... "Longitudinal" Doppler shifts 214
16.6 : ... The lightspeed upper limit in particle accelerators 215
16.7 : The "searchlight" effect 216
16.8 : Velocity-addition 216
16.9 : Particle tracklengths 216
16.10 : Muon showers 217
16.11 : ... Particle storage rings and centrifugal time dilation 218
16.12 : deSitter / Brecher disproof of simple emission theory 219
16.13 : "Domain of applicability" issues 220
16.14 : Conclusions 222

PART VI FUTURE PHYSICS   223

17 COSMOLOGIES 225
- 17.1 : THE EXPANDING UNIVERSE 227
- 17.2 : THE "BIG BANG" 227
- 17.3 : SPATIAL CLOSURE 228
- 17.4 : EXPANSION CURVES 231
- 17.5 : COSMOLOGICAL TIME COORDINATES 232
- 17.6 : THE HARTLE-HAWKING "BUBBLE UNIVERSE" 234
- 17.7 : ENTROPY, ARROWS OF TIME, AND THE BIG CRUNCH 235
- 17.8 : EXTENDING THE "BUBBLE" MODEL 236
- 17.9 : VARIABLE DIMENSIONALITY? 237
- 17.10 : "MIRROR" AND "KALEIDOSCOPE" UNIVERSES 237
- 17.11 : ORANGES AND RASPBERRIES 239
- 17.12 : A FEW MULTIVERSES 240
- 17.13 : FRACTAL UNIVERSE ARGUMENTS 243
- 17.14 : WHY IS THE UNIVERSE RATIONAL? 246
- 17.15 : THE DRAKE EQUATION 248
- 17.16 : BEFORE EVENT ZERO 249

18 TROUBLE WITH WORMHOLES 251
- 18.1 : WHAT IS A WORMHOLE? 253
- 18.2 : "SPACETIME SURGERY" AND SIMPLE OPTICS 253
- 18.3 : WORMHOLE INSTABILITY? 254
- 18.4 : THE DISTANCE PROBLEM 256
- 18.5 : THE AGEING PROBLEM 256
- 18.6 : THE "ANTIHORIZON" PROBLEM 256
- 18.7 : "ANTI-WORMHOLES" AND SPATIAL REVERSAL 257
- 18.8 : THE KERR WORMHOLE 260
- 18.9 : THE FIELDLINE PROBLEM 261
- 18.10 : THE GRAVITY PROBLEM 261
- 18.11 : WORMHOLE POLITICS 262
- 18.12 : THE TIME-CONNECTION PROBLEM 262
- 18.13 : WORMHOLE TIME TRAVEL? 263
- 18.14 : MISTAKEN TIME MACHINE BEHAVIOUR 264
- 18.15 : QUANTUM FOAM 265
- 18.16 : SCALE-DEPENDENT TOPOLOGY 266
- 18.17 : PSEUDOWORMHOLES 266
- 18.18 : DOES QUANTUM FOAM CONTAIN ONLY PSEUDOWORMHOLES? 268
- 18.19 : DO WORMHOLES EXIST AT ALL? 268

19 METRIC ENGINEERING AND WARP DRIVES 269
- 19.1 : "SPACE BUNGEES" AND REGENERATIVE BRAKING 271
- 19.2 : BOOMERANGING 272
- 19.3 : EXOTIC-MATTER DRIVES 273
- 19.4 : THE NEGATIVE-FIELD PROBLEM 274
- 19.5 : ULTRAFAST TRAVEL USING SIMPLE GRAVITY 275
- 19.6 : THE "CRESTING" PROBLEM 276
- 19.7 : THE KRASNIKOV TUBE 277
- 19.8 : WARPFIELD HAWKING RADIATION? 278
- 19.9 : THE "ACOUSTICS" ANALOGUE 279
- 19.10 : SIMPLE WARPFIELD GENERATORS 281
- 19.11 : TOROIDAL CONFIGURATIONS 282
- 19.12 : CANCELLATION AND NON-CANCELLATION 283
- 19.13 : THE 2-SPIN TORUS 284
- 19.14 : SELF-REFRACTION AND CROSS-REFRACTION 286
- 19.15 : GENERAL FIELD-REFRACTION ISSUES 287
- 19.16 : MOMENTUM CONVERSION 288
- 19.17 : "REACTIONLESS" DRIVES AND DEFERRED MOMENTUM 289
- 19.18 : CAN WE BUILD A WORKING WARP DRIVE? 289

PART VII THE HUMAN FACTOR — 291

20 Limitations of Language and Procedure ... 293
- 20.1 : The order in which things are written ... 295
- 20.2 : Lightspeed, velocity, and language traps ... 295
- 20.3 : Fractured logic ... 297
- 20.4 : Logic traps and logical black holes ... 298
- 20.5 : More examples of circular thinking ... 300
- 20.6 : Is consistency all it's cracked up to be? ... 302
- 20.7 : "First answer" syndrome ... 303
- 20.8 : Life, Death, and the Square Root of Two ... 304
- 20.9 : The story of Pi ... 305
- 20.10 : Pi and global extinction ... 307
- 20.11 : Naming rituals, binary logic and Giant Pandas ... 308
- 20.12 : Intransitive logics ... 309
- 20.13 : Complex logical spaces ... 310
- 20.14 : Intransitive ordering and gravitation ... 312
- 20.15 : "Certainty" parameters ... 314
- 20.16 : Living with uncertainty ... 315
- 20.17 : Conclusions ... 316

21 The Perils of Experimentation ... 317
- 21.1 : Evaluating science neutrality ... 319
- 21.2 : Perception filters ... 321
- 21.3 : System bias and "v1.0" syndrome ... 323
- 21.4 : Safety in numbers ... 325
- 21.5 : Accident reporting ... 326
- 21.6 : Quantum sociology? ... 327
- 21.7 : Pattern Recognition and group decisionmaking ... 328
- 21.8 : Market Forces ... 331
- 21.9 : Physics nightmares ... 333

22 Conclusions ... 335
- 22.1 : SR-based or NM-based physics? ... 337
- 22.2 : The fork in the road ... 339
- 22.3 : Warning signs ... 339
- 22.4 : Mathematical "truth" vs. relevance ... 344
- 22.5 : Alternative alternatives ... 346
- 22.6 : Life after special relativity ... 346

PART VIII CALCULATIONS, REFERENCES AND INDEX — 347

- Calculations 1 : Doppler shifts ... 348
- Calculations 2 : $E=mc^2$ from Newtonian mechanics ... 350
- Calculations 3 : Non-SR transverse Doppler effect / "Aberration redshift" ... 352
- Calculations 4 : The "Box of Frogs" depiction of classical Hawking radiation ... 354
- Calculations 5 : Comparison table ... 357

Major Players ... 358
Topic References ... 359
General References: ... 365
Index ... 371

WELCOME

The development of modern relativity theory is one of the major achievements of the Twentieth Century, and there seems to be a general appetite for understanding of what "the theory" actually says. So it's a shame that most of the books written on the subject either insist on repeating the standard explanations of *special* relativity, or are aimed at people with university-level mathematics. The reader typically has to choose between accepting Minkowski spacetime and twin paradoxes and the like as away of earning access to the wider world of general relativity, with its even more obscure mathematical conventions and technical details, or giving up and walking away.

But the properties of the *general* principle of relativity are arguably more intuitive, easier to understand, and more accessible to everyday experience than those of the special theory. To say "gravity slows time" requires just three words and no mathematics – a childlike imagination that loves Lewis Carroll and fairy tales should have no problem in looking at the Moon and saying: " Time there runs faster than it does here". This information is part of humanity's scientific and cultural heritage, and we are failing if we reserve these nuggets only for people with a technical "need to know".

The book attempts to explain some of the "best bits" of modern theory, with only occasional equations to placate those aggrieved scientifically-trained readers who may demand such things. Scary Numbers, where they do appear, can be skipped by those not interested, and the same goes for any technical details about special relativity that may have crept in by the back door. The book also concentrates, quite deliberately, on aspects of modern theory that are at the frontiers, and tackles black holes, wormholes, and at the far limits, warpdrive theory.

It's also a kind of detective story.

Many books on physics are written to convince the reader that a current theory is right, or to help them follow a syllabus, and scraps of "unhelpful" information that might lead someone to question whether a thing is *actually* true tend to be left out. Which is unfortunate, because it's sometimes the things that might lead us to suspect that a theory *doesn't* work that are the most interesting things about it. You can't always find the things that *don't* work mentioned in textbooks, and there's also a certain amount of information in teaching texts that doesn't seem to be historically, geometrically or mathematically true. How does a reader sift scientific fact from fantasy? The book doesn't claim to give a complete listing of *everything* that seems less than perfect with current theory, but it does try to be cheerfully honest about where things may be going wrong.

The last aspect of the book is its attempt to reassemble all these pieces into some sort of order. The book's attitude to special relativity can be summed up the opinion that Einstein stated towards the end of his life, that the use of SR as a foundation for more advanced physics may be a historical leftover that can't be justified with hindsight. Taking that position as a mission statement, we look at the heretical idea that special relativity, one of the cornerstones of Twentieth-Century science, might conceivably not be correct.

This is not meant to be a resource for special relativity. There are already plenty of books explaining why SR is considered "true" – *this* book tries to skip past Einstein's special theory with its flat-spacetime logic, and where SR really *has* to be mentioned, attempts to explain why the "special" theory might not be correct after all. Some people will hate that idea. I'd advise those people not to buy the book.

EB 2007

About the Author

Eric Baird has lived in London, UK, for most of his life. In the latter part of the Twentieth Century he produced a free "helpfile" hypertext version of Einstein's "Relativity..." book, and wrote and maintained what (in its time) was probably the most popular relativity website on the internet.

His phone number is unlisted, and he doesn't answer correspondence or emails. He also doesn't help with homework problems, career advice or demands from theoretical physicists to be "convinced".

He is, in short, Not In.

Notes to the First Edition

It is traditional for descriptions of unfamiliar systems of physics to make at least one Truly Awful Mistake that could be used to justify writing off the whole scheme. For the original version of Newtonian mechanics it was Newton's mistaken idea that longer wavelengths carried greater energies than shorter ones, for Einstein's 1905 paper it was the prediction that equatorial clocks should run slower than polar clocks, and in general relativity it was the use of the cosmological constant to "explain" why the universe was supposed to be static rather than expanding. In all three cases, we now regard these mistakes as "user error" rather than as fundamental problems with the models in question.

If Einstein and Newton were unable to avoid the inevitable "Truly Awful Mistake" in their writing, it would be naive to think that something similarly bad hasn't crept into this book, and while the background material used here has been extensively researched, the nature of the book has made independent proofreading problematic. I apologise in advance for any errors, typographic or otherwise, that may have slipped through the net.

EB 2007

ACKNOWLEDGEMENTS AND CREDITS

Image of the Hubble Space Telescope: Courtesy of NASA, which makes a range of copyright-free images available, as long as the source is credited. The image reproduced here has been cropped. More images are available from **www.hubblesite.org/**

Research papers (preprints and e-prints) with LANL numbers can be downloaded free of charge from the **Los Alamos Nuclear Laboratory** e-print repository, at **lanl.arxiv.org/**

Photograph of the Pleiades (backing picture for section 1) comes from a licensed Corel clipart collection.

Leonardo DaVinci sketch (backing picture for Part V) has been degraded to emphasise lineart features. Leonardo (1452-1519)'s original work is assumed to be out of copyright.

"Tweedle Dum and Tweedle Dee" illustration from "Through the Looking Glass" is by John Tenniel (1820-1914) and is assumed to be out of copyright.

All other original images, diagrams, photographs and graphics are copyright by the author.

Most of the historical research behind this book was done over a period of years at the London-based collections of the **Science and Patent Office Library** (now merged with the larger central collection of the **British Library**, and accessible at the newer **St. Pancras** site), and at the libraries of **Imperial College, London** (now accessible, with the **Science Museum Library**, in the **Central Library** block). Without this deep, shelf-level access, much of this background research would not have been feasible.

Abbreviations

NM   **Newtonian Mechanics** – Newton's relationships for the behaviour of moving bodies, or (in this book) the matching relationships that light would have to have to use to be compatible with Newtonian theory. Textbook accounts of how "Newtonian theory" compares with modern relativity theory often don't correspond to the historical record, so a degree of redefinition has been necessary.

The historical application of NM to light in flat spacetime ("emission theory") wasn't consistent with experiment or with the idea of wave-particle duality. This suggested that at high velocities, either the NM relationships were wrong, or the assumption of flat spacetime was wrong. This is a significant decision-fork.

SR   **Special Relativity** – Einstein's "special" or "restricted" theory of relativity models moving-body problems by assuming that spacetime is completely flat, and that the motion of bodies has no effect on the propagation of light. SR has to change the NM relationships (by a "Lorentz factor") to get this to work. The special theory only applies the principle of relativity to the case of observers with simple inertial motion in flat spacetime.

CT   **Classical Theory** – In the context of special relativity, "Classical Theory" refers to a hybrid of two earlier sets of predictions, with the energy and momentum relationships for matter derived using NM, and those for light derived by assuming a fixed "absolute" aether stationary in the observer's frame, or lightspeed globally fixed for the observer. This combination is not consistent.

Special relativity removes the conflict by creating a third, intermediate set of predictions.

GR   **General Relativity** – Can refer either to a very broad subject (the general application of relativity to physics), or can refer more specifically to Einstein's *implementation* of a general theory in 1915 (with minor updates).

GR1915   To reduce confusion, this book usually refers to Einstein's specific 1915 theory as "**GR1915**". GR1915 tends to be regarded as the "default" general theory.

PoR   **Principle of Relativity,**

GPoR   **General Principle of Relativity**

QM   **Quantum Mechanics** – the beginning of the Twentieth Century it was suggested that we could produce a better agreement between theory and experiment if the absorption and emission of light were treated as *quantised* events. From this foundation, quantum mechanics has grown into a major subject impacting on both engineering and philosophy.

PART I

BACKGROUND PHYSICS

RELATIVITY IN CURVED SPACETIME

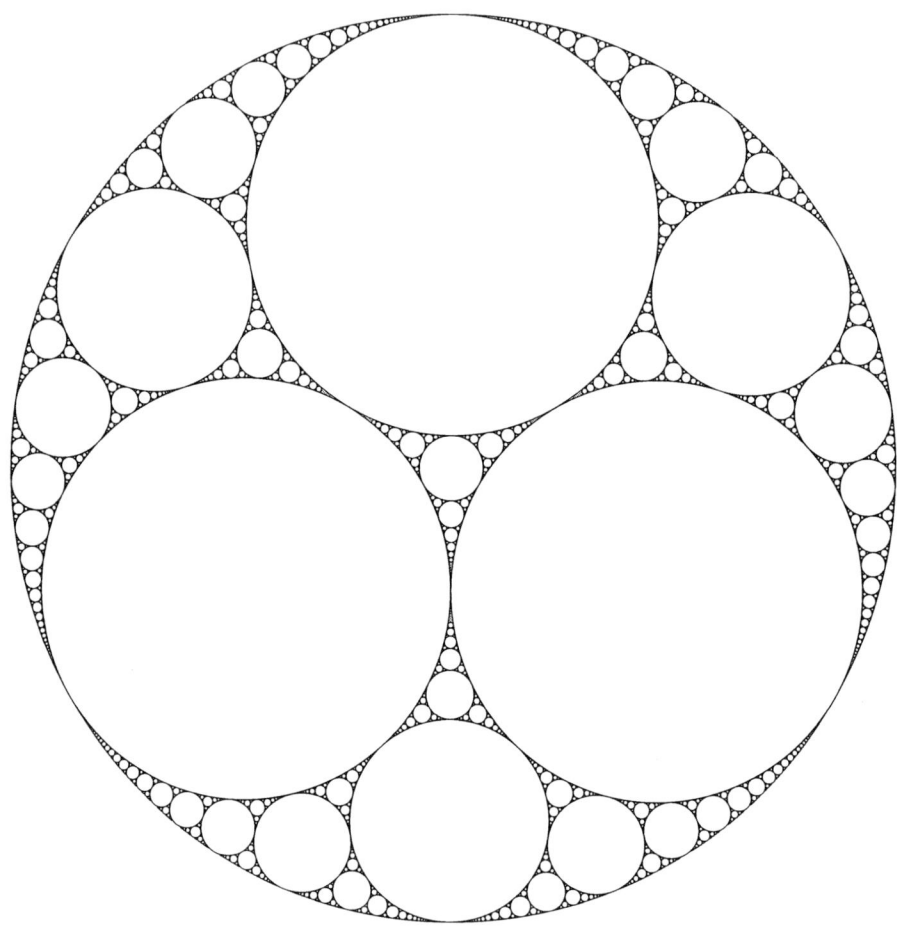

1
The Speed of Light

> "... it is now certain from the phenomena of Jupiter's satellites, confirmed by the observations of different astronomers, that light is propagated in succession, and requires about seven or eight minutes to travel from the sun to the earth."
>
> **Isaac Newton, "Principia..." (Scholium to Proposition XCVI)**

> "Where lighter gases, circumfused on high,
> Form the vast concave of exterior sky;
> With airy lens the scatter'd rays assault,
> And bend the twilight round the dusky vault;"
>
> **Erasmus Darwin on atmospheric lensing, "The Botanic Garden", 1791**

> "Every child at school knows, or believes he knows, that this propagation takes place in straight lines with a velocity $c = 300,000$ km./sec."
>
> **Albert Einstein, "Relativity ..."**

> "... the opposite of a profound truth may well be another profound truth."
>
> **Niels Bohr**

> "The principle of the constancy of the velocity of light holds good according to this theory in a different form from that which usually underlies the ordinary theory of relativity."
>
> **Albert Einstein, "The influence of gravitation on the propagation of light", (section 3), 1911**

The Pleiades star cluster:
Its central "question mark" of young bright, blue stars
spans a distance of about ten lightyears, and is thought to
be approximately ~400 lightyears away (estimates vary)

1.1: Light is pretty fast

a sense of scale

The **speed of light** is approximately 300,000 kilometres per second, or a little over 186,000 miles per second in old measures. The official speed is 299,792,458 metres per second.

To get some idea of just how fast this is, the **Earth** is about 12,756 km across at the equator, and its equatorial circumference is just over 40,000 km, so a lightbeam bouncing around a series of mirrors could circle the Earth seven-and-a-bit times in one second. Or, a lightsignal created on the Earth's surface and aimed at the **Moon** would manage, in that one second, to get about four-fifths of the way there.

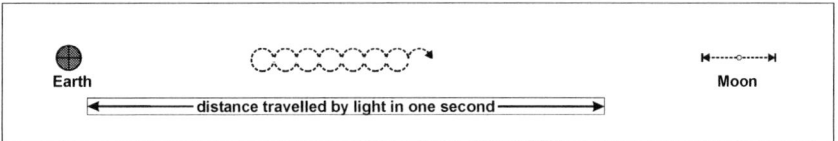

Figure 1-1: The distance covered by light in a second

Light is *so* fast that it makes a very useful ruler for describing astronomical distances, where these distances expressed in kilometres or miles would start to boggle the mind. We can say that the Earth's equator is roughly 0.135 **lightseconds (ls)** around, the Moon's average distance from us is about 1.2 lightseconds, the **Sun's** radius is about two-and-a-third lightseconds (so that the Moon's entire orbit would fit comfortably inside the Sun), and the distance from the Sun to the Earth is a shade over 500 lightseconds (light from the Sun takes just over eight minutes to reach us).

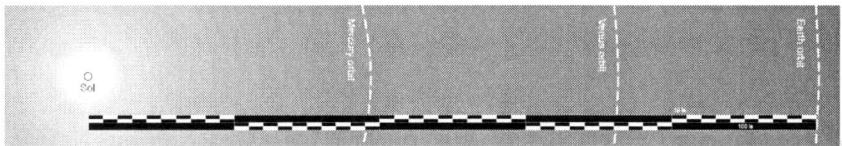

Figure 1-2: Sun→Earth, ~500 seconds

Pluto is nearly 40 times further away from the Sun, at a distance of just under 20,000 ls.

Figure 1-3: Sun→Pluto, ~20,000 seconds, approx 5½ hours

For interstellar distances the numbers get a bit big even when they're counted in lightseconds, so we use **light*years*** ("**ly**") instead, with one lightyear being, logically enough, the distance that light moves in one Earth year (nearly 32 million times bigger than our lightsecond). The nearest star to our solar system (**Alpha Centauri**) is about 4.3 ly away, and **Sirius**, the brightest star in our sky, is 8.7 ly away.

Our ten nearest stars just fit into a sphere 10 ly across.

Figure 1-4: Milky Way: over 100,000 ly across

Our own galaxy is over a hundred thousand lightyears across, and the nearest galaxy of a similar size, **Andromeda** (also known as "**M31**"), is about two million lightyears away. The observable universe extends over a distance of billions of lightyears.

1.2: Lightspeed varies

We all know that the speed of light is constant, right? Well, that's not entirely true, because, in a very real physical sense, lightspeed can be said to appear to vary from place to place and from time to time, if it's provided with a good enough reason. Lightspeed "over there" can seem to be different to lightspeed "over here". However, for reasons we'll get into shortly, lightspeed "here-and-now", measured in a vacuum, always gives the same value ("lightspeed is *locally* constant"). The "local" part of that phrase is quite important.

lightspeed is affected by matter

Although light travels more slowly through regions containing matter, the speed of light in air at sea level is still quite high, with its speed only reduced by about three parts in a thousand. For light in glass or water we get down to values around two-thirds or three-quarters of the "official" speed of light, and electromagnetic signals passing down copper wires are supposed to travel at about two-thirds of standard lightspeed. These rough figures change with density: the more matter we squeeze into a region, the longer it takes light to travel through it.

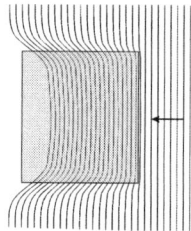

Figure 1-5: Light moves more slowly though solid objects

This trend doesn't just apply to the spaces between atoms: it also applies to the spaces between planets and stars. **Gravitational fields** affect lightspeeds, and the more stars we squeeze into a region, the longer it takes light to cross that region.

Modern gravitational theory also tells us that objects should *drag* nearby light when they accelerate or rotate. The (reduced) speed of light in a region containing particulate matter can also be *biased* by the velocity of the particles: light within a moving block of glass should find it easier to move in the same direction as the glass than in the opposite direction.

It seems that just about anything that we might do in a region can affect the speed of light. To exclude these complicating effects we say that the speed of light is only effectively constant **in a vacuum** ("*in vacuo*"), when these complicating factors aren't reckoned to be significant.

lightspeed and gravity

Gravity affects the behaviour of light. If we aim a beam of light across a region that contains a gravity-source, the signal should take longer to reach the other side than if the gravity-source wasn't there – this is known as the **Shapiro effect**.

Because of this gravitational light-slowing, if we build two identical "light-clocks" whose "ticks" are pulses of light bouncing back and forth between two parallel mirrors, and put one of these clocks into deep space, it should "tick" faster (for a given "official" distance between its plates) than a matching clock back on the laboratory bench in Earth's gravity.

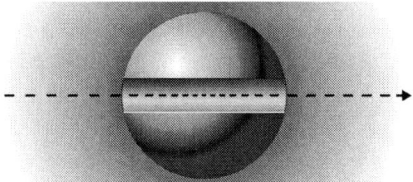

Figure 1-6: Shapiro effect : Light takes longer to cross a region where the gravitational field is stronger

This gravitational light-slowing effect doesn't depend on anyone necessarily being able to feel any sort of gravitational "pull". If we put an observer into a scooped-out hollow at the centre of the Earth, they may insist that their region *seems* to be "gravity-free", because there won't be any overall force pulling them in any particular direction. They'll feel weightless. But *we'll* still insist that their background gravitational field is stronger then ours, that their region is "downhill" from ours, and that their speed of light should be correspondingly lower.

For a particularly extreme example, we might imagine a lump of transparent material being compacted down to form a black hole. As the material compacts, light should take progressively longer to move through the region and through the lump, and once it's collapsed to less than its critical **event horizon** radius, simple calculations for how long it takes light to cross the region become *infinite*, because the light never manages to get out at all.

1.3: Lightspeed is not just the speed of light

Light is now generally understood to be an **electromagnetic wave**: the motion of an **electric field** is supposed to cause **magnetic** side-effects, and the motion of a *magnetic* field is supposed to cause *electrical* side-effects. The disturbances in the electrical and magnetic fields are supposed to move at the same speed – c – (for "*c*elerity", meaning "speediness") and when we create an electromagnetic disturbance or fluctuation, the resulting electromagnetic ripples scoot away from the source, hand-in-hand, as a stable electromagnetic ("**EM**") wave.

In the last half of the Nineteenth Century, **James Clerk Maxwell** calculated the speed that these hypothetical electromagnetic waves should have, and found that they had the same speed as light – light *was* an electromagnetic wave that just happened to be within the range that our eyes can see.

The spread of EM **wavelengths** that can be seen by the human eye is quite small – about an octave – but this corresponds to a range that is particularly useful. Within this range, human eyes can only properly distinguish between three overlapping colour bands centred on "**red**", "**green**" and "**blue**", and computer monitors and TV sets try to mimic the full range of visible colours by adjusting the balance of red, green and blue light for each pixel. Longer wavelengths make up the **infra-red, microwave** and **radio** parts of the **EM spectrum**, and shorter wavelengths are classed as **ultraviolet** ("**UV**") light, **x-rays** and **gamma rays**.

1.4: Lightspeed affects inertia

But *c* isn't just the speed of electromagnetism, it's the default speed at which electric and magnetic disturbances carry *information* across space – light just happens to be the handiest (and cleanest) example available. We can think of *c* as being the speed of *communication* without having to know anything terribly much about the properties of light or about electromagnetism in general. This "communication speed" or "reaction speed" seems to be fundamentally tied into the rate at which everyday processes occur, and affects other basic properties such as the apparent **inertial mass** of objects.

A ball's **inertia** can be defined according to how easy it is to get it to move (for instance, by hitting it with a baseball bat with a known amount of force, and seeing how quickly it leaves the bat after the impact). But when atoms in materials jostle against each other, the things that interact and bump together are the atoms' external electric fields, so the speed of light also controls the rate at which everyday forces are transmitted though solid objects.

If we reduce a region's speed of light, the forces from our bat will be (electromagnetically) transferred *to* the ball and *through* the ball's structure more slowly, and the ball will take longer to react to the impact. If the amount of energy in the moving ball is expressed as a measure of the ball's velocity compared to the speed of light, then for a fixed amount of energy, the speed of the ball leaving our bat should also be lower. Our ball seems to have an increased resistance to the applied force, and seems to have greater inertial mass.

1.5: Lightspeed controls timeflow

But there's more to it than this. Reducing *c* means that all collisions between molecules inside the bat and inside the batsman's body should *also* now take place at a reduced speed, and so should the chemical reactions that result from those collisions. The batter thinks more slowly because the nerve impulses travelling through their body and brain move more slowly. If they wear a mechanical wristwatch, the watch ticks at a rate determined by the time that a flywheel takes to rock from side to side under the influence of a spring, and increasing the little wheel's inertia makes it more resistant to acceleration: the watch ticks more slowly. Increasing the inertia of an *electronic* watch makes the quartz timing crystal resonate at a lower frequency, and if we happen to have an atomic clock handy, the frequencies generated by *that* clock will be slowed, too. If all this is taking place at Stagg Field, University of Chicago, above the site used by the **Manhattan Project**, increasing the inertia of any radioactive residues still under the ground will enhance their stability, and give them correspondingly longer decay times.

If a reduction in the speed of light makes *all* these different sorts of clocks run more slowly, by the same amount, then perhaps the simplest description of what is happening is that *time itself* is passing more slowly in the region of slower lightspeed.

1.6: Lightspeed is *locally* constant

But what about the local observer?

A far as our batter is concerned, *nothing seems to have changed*. Since their brain is slowed by the same rate as all the other processes around them, they're entitled to believe that the bat and ball are behaving normally. All their clocks seem to be keeping perfect time with each other. Unless they start comparing their rate of timeflow with that *outside* the pitch, or comparing gravitational field strengths, they don't have any obvious way of telling that the speed of light is any different. Even if there happens to be a University of Chicago lightspeed-testing experiment going on at the side of the pitch, bouncing light between reflectors carefully placed around the stand and measuring the time taken for the light to perform a complete circuit, the equipment won't be able to show that anything has changed.

We can argue that light *ought* to be taking longer to travel across the stadium, but since the experiment's local reference clocks should be taking longer between tick by the same proportion, the equipment should be obstinately displaying the same numerical value for the speed of light that it showed before we made our change.

This is what we mean when we say that lightspeed is **globally variable but locally constant**: an outsider beaming light across the time-slowed region would be able to tell that it was taking longer than normal to reach the other side, but every observer making strictly *local* observations of the speed of light in a vacuum ought to always get exactly the same figure. Any uniform gravitational effects that modify *local lightspeeds* also modify the rest of the *local physics* in such a way as to make the change invisible to any *locally-calibrated* detection equipment. At least, that's the theory.

1.7: Lightspeed is now <u>defined</u> as constant

The speed of light in highly controlled situations can now be measured to exquisite accuracy, and these measurements are exquisitely repeatable. As the accuracy with which we were able to measure the speed of light started to approach the accuracy with which we could define distances and times, we needed a new approach. The new atomic clocks gave us a better reference "time-base", but we needed an equally accurate and replicable measure of distance.

In the end we took a pragmatic approach: we now *define* lengths by the distance that light travels in a known time, and the metre is now *defined* as the distance that light moves in exactly 1/ 299 792 458 of a second.

1.8: The gravity well

If we follow these definitions and use light-signals to measure distances, we can end up saying that, in effect, a gravitational field's light-slowing abilities seem to modify *the amount of space that we measure* inside a region:

Let's take a cube-shaped box and a stopwatch and measure how long it takes light to travel from one side of the box to the other. If we now place a microscopic black hole at the centre of the box and repeat the exercise, rays that pass closer to the box centre should take longer to reach the other side. Using these light rays to map out distances, we can then say that the middle of the box seems to "contain more distance" that we'd expect from its external dimensions: it seems to be bigger on the inside than the outside.

We can express this by drawing an **embedding diagram** to express these modified light-distances. If we take a cross-section through the box when it's empty and map out its insides, we might end up with a map something like this:

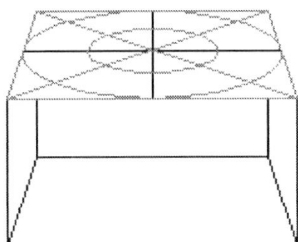

Figure 1-7: Cross section through an empty box:
flat spacetime

This is what we'd normally refer to as an example of **flat geometry**.

RELATIVITY IN CURVED SPACETIME

Now we'll repeat the exercise when the box contains its tiny central gravity-source.

In the new map, the lightbeams will take progressively longer to cross the region depending on how close they get to the centre of the box, so the new map will describe space as seeming to contain more "space per unit volume" towards its middle, with the "spatial density" appearing to be higher towards the middle of the box. We might use darker shading and/or contour lines to describe the apparent change in density, like this:

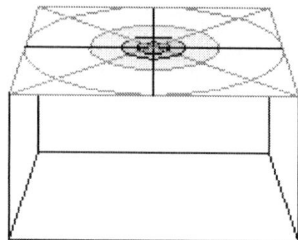

Figure 1-8: Contour map of apparent spatial density around a gravity-source

Finally, we might decide that we don't like this approach of using contour lines and shadings: we'd prefer the map's distances to be an accurate "scale" representation of the light-distances involved. The problem with this is approach that the extra distances in the centre of the box won't fit politely into a "flat" map – in order to cram them in, we have to make our map bulge in the middle: we have to **extrude** our two-dimensional map of the slice through the box into a third dimension, in order to get everything to fit in.

This gives us a sort of curved "funnel" shape, something like this:

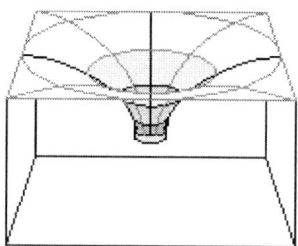

Figure 1-9: Gravity-well

We refer to this distinctive shape as a **gravitational well**. We usually see it drawn using a horizontal plane and with the well's throat pointing "down", but that's mostly because the diagram looks more natural to us that way. We could equally well have drawn the diagram with the funnel bulge pointing upwards, like an odd-looking mountain – what's important here is the way that distances are packed *inside* the shape, and the resulting angles.

Shapes drawn on this curved surface no longer obey the ancient rules of conventional classical Greek geometry as described by **Euclid**, and we refer to geometry carried out in these sorts of curved surfaces (or in curved spaces) as **non-Euclidean geometry**. The rules concerning non-Euclidean surfaces and volumes weren't properly worked out until the middle of the Nineteenth Century, by pioneers **Karl Friedrich Gauss** and **Georg Friedrich Riemann**, and we'll be dealing with some of the implications of this curvature in section 3.

1.9: Light travels in straight lines. Except when it doesn't

Another thing that we learn in school is that light travels in straight lines. Except that, of course, we know that it often doesn't. Lightrays bend when they move from air to water or glass to air, and light also *seems* to bend even when it's only skimming the surfaces of objects.

A transparent body designed with a special shape to exploit this first effect and make lightrays converge or diverge in a particular way is called a **lens**, and "lensing" effects are quite common: if we look above a hot radiator in winter, objects behind it seem to "ripple" as the warmer and less dense air rising from the radiator creates variations in the way that light moves through the region, and similar lightspeed-lensing effects in air above hot sand are supposed to be to blame for the phenomena of desert **mirages**.

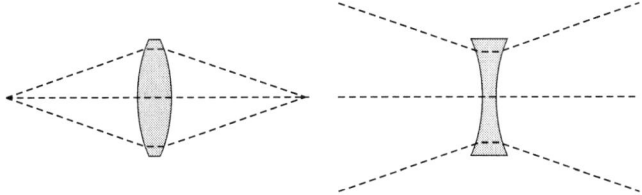

Figure 1-10: Conventional lensing effects

Newton and his contemporaries recognised that these deflections could be explained by variations in the speed that light as it encountered materials with different optical properties (section 5), and Newton also went on to explore how conventional gravitational effects might be explained by similar variations in lightspeed, caused by the regional variations in density of a gravitational medium.

If conventional lenses deflect light by changing lightspeeds in a region, and gravitational fields also affect lightspeeds, does *gravity* deflect lightbeams, too?

The short answer is: yes. Suppose that we repeat our "box" experiment and rig things so that one face of the box illuminates all-at-once. In the case of the empty box, we expect the resulting electromagnetic shockwave to move away from the box wall with a nice flat leading edge and travel as a **plane wave** to the opposite side of the box, hitting it face-on. Because, really, it has no obvious excuse to do anything else. We can then draw in lines at right angles to this plane wave, and describe how these **rays** describe the paths that light moves along when travelling from one side of the box to the other. They are nice straight lines.

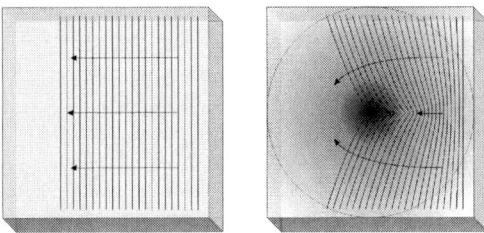

Figure 1-11: Gravitational deflection of light

But things change when we put our tiny gravity-source back into the box. The rays moving through the centre of the box now take longer to reach the far end than the rays further away from the gravity-source, so the centre of our wavefront progresses more slowly than its edges, and the result of "holding back" the middle of the wave is that the wavefront ends up distorted, and parts of its surface begin to tilt more towards the direction of the gravity-source.

We refer to the deflection of lightbeams around a gravity-source as **gravitational lensing**, and astronomers have collected some nice examples of images that seem to show this effect in action. When light bends around the gravitational field of an extremely massive, distant object (such as a galaxy) it produces quite a distinctive-looking effect.

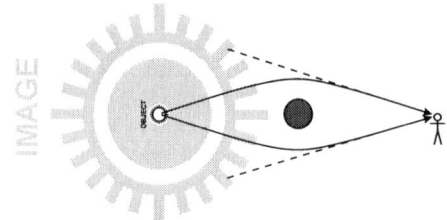

Figure 1-12: Gravitational lensing

These images tend to be referred to as **Einstein rings** (although since **Orest Chwolson** suggested the effect earlier, it might be more proper to refer to them as **Chwolson rings**).

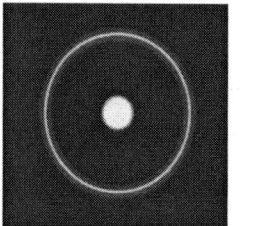 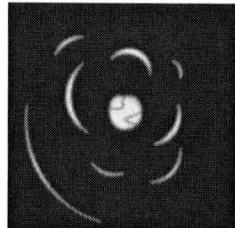

Figure 1-13: Results of "perfect" and "irregular" gravitational lensing

Although we've only recently had the technology required to *see* the effects, the *idea* of gravitational lensing isn't especially novel, and when Einstein submitted his paper on the subject, he apologised to the editor for sending in something so unimportant (explaining that he'd only written it to stop someone hassling him). The general problem – of calculating lensing effects caused by a medium whose density changes with distance – is quite old, and Newton is known to have carried out similar calculations for the more conventional lensing effects caused by a planet's atmosphere (which become weaker with distance in a broadly similar way). Gravitational lensing would have been a natural result of Newton's physics.

1.10: Light used to <u>define</u> a straight line

The gravitationally-deflected rays that we've drawn look suspiciously like the sort of lines that we'd end up with it we tried to draw a set of parallel straight lines across our gravity-well shape in Figure 1-9 using a flexible ruler, without realising that the surface was curved. The curvature could fool us into drawing lines that followed the curve of the surface ("**geodesics**") so that they ended up crossing each other at strange angles, and this suggested to some clever mathematicians in the Nineteenth Century that perhaps our innocent-looking curved-space map might have more powerful properties: it might describe and map *all of* the properties of light in a gravitationally-distorted region, and might be able to generate a full *geometrical* description of how gravitational fields influenced nearby matter.

The idea of modelling gravitational physics as the result of **spatial curvature** was brilliant, but the mathematicians involved couldn't seem to get it work. Some vital component was still missing. But before we can look at what it was, we need to look again at how gravity interferes with some of our older ideas about light and matter.

2
Gravity, Energy and Mass

> *"Are not gross Bodies and Light convertible into one another, and may not Bodies receive much of their Activity from the Particles of Light which enter their Composition? For all fix'd Bodies being heated emit Light so long as they continue sufficiently hot, and Light mutually stops in Bodies as often as its Rays strike upon their Parts, as we shew'd above. ...*
>
> *The changing of Bodies into Light, and Light into Bodies, is very conformable to the Course of Nature, which seems delighted with Transmutations. ... And among such various and strange Transmutations, why may not Nature change Bodies into Light, and Light into Bodies?"*
>
> Isaac Newton, "Opticks" 3:1:Qu30:

> *"If the theory corresponds to the facts, radiation conveys inertia between the emitting and absorbing bodies."*
>
> Albert Einstein,
"Does the inertia of a body depend upon its energy-content?", 1905

> *" 'And we know now that the atom, that once we thought hard and impenetrable, and indivisible and final and — lifeless, is really a reservoir of immense energy. ... A little while ago we thought of the atoms as we thought of bricks, as solid building material, as substantial matter, as unit masses of lifeless stuff, and behold! these bricks are boxes, treasure boxes, boxes full of the intensest force.'*
>
> *'Why does not all the uranium change to radium and all the radium change to the next lowest thing at once? Why this decay by driblets; why not a decay en masse? ... Suppose presently we find it is possible to quicken that decay?'*
>
> *'We cannot pick that lock at present, but — ... — we will.' "*
>
> H.G. Wells, "The World Set Free", 1915

2.1: What is mass?

The "**mass**" of an object is essentially a measure of how substantial it is. We could try to talk about mass being a measure of the amount of "tangible stuff" that a thing contains, but it is more satisfyingly technical to refer to "the quantity of *matter*". The words "matter" and "stuff" are fairly interchangeable, except that "matter" is recognised as the standard term in physics. We can judge how much there is to an object using two methods, by measuring its **inertial mass** or its **gravitational mass**. Both methods measure the "obstinacy" of the object – the more "mass" an object has, the more difficult it is to wrangle it into changing its behaviour.

gravitational mass

If a one-tonne iron weight is placed on your foot, your foot hurts because the object has **gravitational mass**, and is trying to freefall downwards under the Earth's gravity, pinning your foot against the floor ... the more material there is in the object, the more strongly the combined attraction pulls the object downwards, and the more difficult it is to stop it falling where it wants, or to drag your foot out from underneath it.

Figure 2-1: Gravitational mass: the enthusiasm of an object for following the laws of gravitational attraction

inertial mass

Alternatively, if the one-tonne weight is sitting on a frictionless set of rollers, and you run up and kick it, *this* time the reason your foot hurts is because of the object's **inertial mass**. If you had kicked a much smaller lump, it could have more easily yielded to the power of your foot and hurtled away after the kick, but since our large chunk has a lot more "stuff" to it, persuading it *all* to move out of the way of your foot *quickly* is a lot more difficult. The larger object will react to a given applied force more slowly, and once it's moving (on its rollers), it will then tend to *keep* moving unless you can apply a similar amount of force in the opposite direction to stop it.

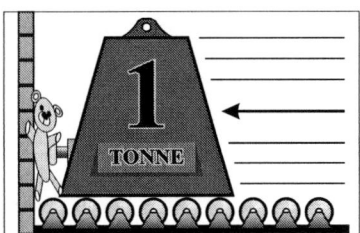

Figure 2-2: Inertial mass: The determination of an object to keep doing exactly what it was doing a moment ago

Once a large inertial mass is in motion, if you stand between it and a wall, you are liable to get squished.

2.2: Does light have mass?

One of the more perplexing things in physics is the idea that although light *also* has a certain amount of "stuff" to it (it has both **energy** and the ability to push, or **momentum**), we still insist on saying that light it doesn't have any mass. Matter has mass, light doesn't.

This doesn't sound very intuitive, and in fact there's some "small print" here that we need to look at.

"mass" suggests *locatability*

To appreciate what's happening here, we have to remember that our *concept* of "mass" was originally developed and designed around the behaviour of solid objects, and around the measurement protocols that we typically use to measure how substantial these objects are. We can put an object onto a weighing-scale in a known gravitational field to measure its gravitational mass, or we can dangle it on a piece of string and throw objects at it to see how easy it is to get it to flinch, or twirl it around on the end of the string, to measure its inertial mass. But neither of these approaches works with light, because light doesn't tend to stay put of its own accord. You can't place some light on a scale to weigh it, or put it on a table and expect it to still be there when you return with a hammer to test it. Light moves at ... well ... the speed of light, and since this is as fast or faster than any of the forces that we would usually be trying to use to measure it, our usual functional definitions don't hold for it.

Since the behaviour of "free" light bypasses our usual definitions and understandings of what mass "is" (which were formulated for measuring more solid things), we commonly say that light is "massless". "Mass" can be assigned to a **persistent position** or to an "owning" object that we can easily track – we point to the object and say, "inside that bounding surface is a particular mass" – but a lightbeam will tend to zip away from the region we are pointing at before we can get the words out.

"trapped" light contributes mass

However, as we've said, light *does* have energy and momentum, and if we slow light down or trap it by passing it through an intense gravitational field, or through some physical medium in which lightspeed is slower, we can start to assign that energy and momentum to a persistent *region*, and can then start to say that the presence of the light contributes some additional mass to the region. If we look at a vast expanse of interstellar space, and the scale of the region is large compared to the speed of light, we'll be able to argue that the energy of the light moving within that expanse contributes to its total gravitational effect. If we aim two laser-beams parallel to each other in opposite directions, we might expect the energy tied up in the two constant, "antiparallel" beams to cause them to attract one another, and it's even been argued that if we could create a sufficiently intense momentary concentration of light, the associated gravitational effect might (theoretically) be strong enough to create a black hole made entirely of self-trapped light, sometimes referred to as a "**kugelblitz**" (Wheeler 1955).

So ... although "*free*" light is said not to have mass, if we were to imprison some light-energy inside a sealed container (in order to hold it in one place long enough to allow us to make our measurements), then the "full" light-container should weigh more and be more resistant to applied force than when it is "empty".

Let's see how this idea works ...

2.3: Genie in a bottle: Thought-experiments with bottled light

To illustrate how "trapped" light appears to contribute inertial and gravitational mass, we'll imagine taking an impossibly-perfect reflecting container and putting some light inside it.

Because the speed of light is so unreasonably fast, it should only take a moment for the light-pressures acting on the container's internal surfaces, due to the light bouncing around, to settle down and reach a nice state of equilibrium.

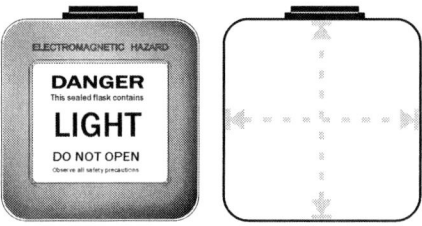

Figure 2-3: A sealed flask containing light

We'll now use this imaginary flask to perform some basic thought-experiments.

captive light has inertial mass

Lets suppose that we give one side of the flask a mighty thump. The flask's own inertial mass resists the forced acceleration of the flask, but if the flask contains light, there's an additional resistance. Although the speed of light is very, very fast, it isn't *infinite*, and while the flask is being accelerated, the contained light will be impacting harder on the wall that's being pushed towards it than the one that's being pushed out of its way. The "light in flight" hits the approaching wall with a slight **Doppler blueshift** (increased energy), and hits the receding wall with a slight **Doppler redshift** (reduced energy), and as a result, this fleeting imbalance in light-pressures makes the "full" flask seem to push back against our hand more strongly than if it was empty.

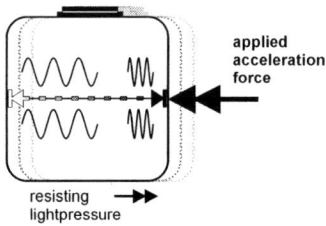

Figure 2-4: Captive light contributes inertial mass

When we stop pushing the flask, the internal light quickly reaches a new equilibrium state in the flask at its new speed.

If we then try to halt the moving flask by pushing against the *other* side, the light now slams harder into the *other* side of the decelerating flask, and the higher internal light-pressure on *that* side now resists our attempt to slow the flask down.

Since the "full" flask has a greater resistance to being accelerated than an empty one, the trapped light seems, as far as our external measurements are concerned, to be contributing some inertial mass to its container.

captive light "falls" in a gravitational field

Introductory school physics tells us that all objects will fall at the same rate in a gravitational field. This can seem like a confusing idea when we hear it for the first time, because it *seems* to conflict with our everyday experience, that heavier, denser objects *do* tend to fall faster: If we drop a bag of flour it hurtles to the floor, but if we empty out the contents of the bag, the resulting cloud of flour wafts down towards the floor much more slowly. A balloon filled with air will fall more slowly than a similar-sized cannonball, and a balloon filled with helium will "fall" upwards!

Although we can congratulate ourselves on having noticed these behaviours, they are caused by the complicating presence of air, and when the effects of **buoyancy** and **air resistance** are removed (for instance, by dropping the items in a sealed chamber with all the air pumped out of it), our objects no longer have an obvious excuse to fall at different rates. A hammer and a feather (and a helium balloon), dropped together by an astronaut standing on the Moon, should all accelerate towards the lunar surface at identical speeds.

The idea that gravity should affect all objects in the same way independently of their composition is known is known as **Eötvös' Principle**. If a given amount of *inertial* mass is always associated with a certain amount of *gravitational* mass, this makes it easier for us to appreciate how planets and moons can have stable, well-behaved orbits, and this **principle of equivalence** suggests that if we can't force independent changes in inertial and gravitational mass, they may be different manifestations of a single underlying property.

But there's a complication: objects can contain electromagnetic energy, and if we expect all objects to fall in a gravitational field at the same rate irrespective of their internal composition, then it seems that a gravitational field needs to be able to deflect EM energy as well as crude matter, suggesting that lightbeams should be deflected by a gravitational field.

We met a "geometrical" argument for this in section 1.9, here's the "flask" version:

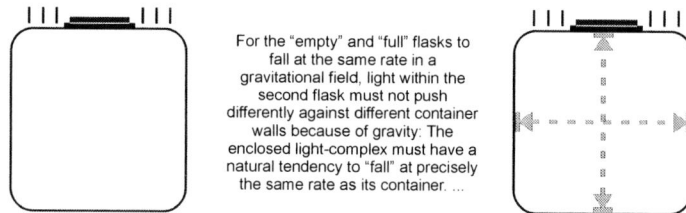

Figure 2-5: Eötvös' principle, applied to light

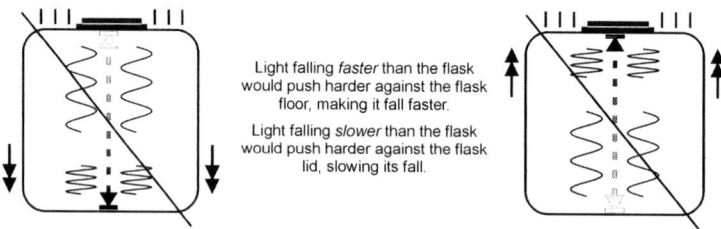

Figure 2-6: ... if light was *not* deflected by gravity

Suppose that we take two of our light-flasks, one full and one empty, and drop them both from a great height (in a vacuum). For Eötvös' Principle to hold, both flasks must fall at exactly the same rate.

We'll now look at the case of light bouncing about inside the "full" flask. If this light *wasn't* deflected downwards by a gravitational field, then the downward acceleration of the flask would force it down against the contained light, which would then be impacting against the upper part of the flask's insides more strongly than against the base of the flask – the internal imbalance in radiation-pressures due to the trapped light would then buoy the flask up slightly against the gravitational field and make it fall more slowly.

On the other hand, if the light inside the flask was deflected *more strongly* by the gravitational field than the matter making up the flask, then the light would be deflected towards the bottom surface of the slower-falling flask and would press down against it more strongly, so that the flask was pushed towards the gravity-source and would fall more quickly.

The only way that the flask could fall at *precisely* the same rate regardless of its contents would be if a lightbeam aimed at right angles to a gravity-source was attracted towards it at *precisely* the same rate as a conventional lump of matter.

2.4: Difficulty of detecting the effect

What do we mean by light "falling"? For light-signals aimed directly upward or downward, local variations in definitions of distance and time can make it difficult to produce straightforward statements about any change in speed (*see:* Newton's problems in section 5). However, if light is aimed *horizontally*, parallel to the Earth's surface, we can expect it to undergo a natural deflection towards the Earth at the same rate as a dropped ball.

Suppose that we drop our ball from a first-floor window, and fire a horizontal beam of light from a laser pointer at the same moment, and from the same position. We expect the ball to fall towards the ground in a second or so, so if light *really* falls at the same speed, shouldn't we expect the light ray to be bent by gravity so that the ray hits the ground at the same moment as the ball? Why don't we notice this happening?

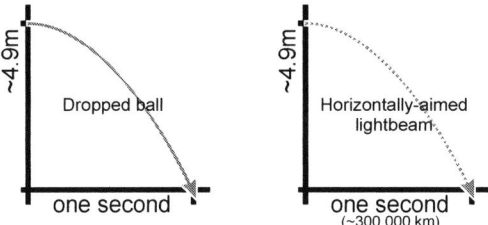

Figure 2-7: Falling light

The explanation is that light moves so ridiculously fast that it's already out of sight before the ball hits the floor. If the ball takes a second to drop to the ground, then in that same second, the light will have moved away from our position by a distance of about 300,000 kilometres, which we've already established is about 80% of the distance from here to the Moon.

If we have a line 300,000,000 metres long, and the position of one end of the line is deflected by less than *five* metres, then a deflection this small is very difficult to measure. And of course in real life the deflection would be even less, because most of our line is significantly further away from the Earth, and the attraction at those distances is correspondingly weaker. And this doesn't even take into account the additional time needed for the signal to get back to us. If we use a much shorter distance (to make it easier for us to contain the light-ray inside our laboratory), the amount of time that the light takes to reach the far end of our apparatus is so small that neither the ball nor the lightbeam have very much time to fall.

Curvature this weak is difficult to notice in everyday life.

"captive" light provides gravitational mass.

For our third thought-experiment, we'll take our full light-flask and place it on a set of kitchen scales. Since we've already decided that the light inside the flask ought to be deflected downwards by gravity, the fact that the flask is being *suspended* and isn't being *allowed* to free-fall means that its contained light is constantly being deflected by gravity towards the floor of the flask, and should be hitting it more strongly than it hits the flask's upper part.

The resulting downward radiation pressure should push the flask against the scales more strongly, and should produce a fractionally larger reading for the weight of the flask. When we lift the container off the scales, it should feel slightly heavier than its empty counterpart.

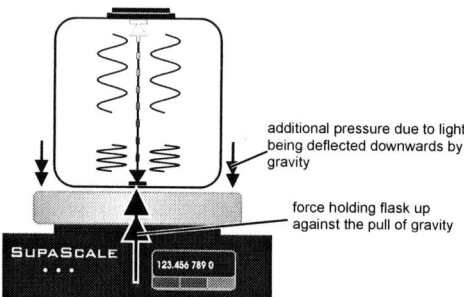

Figure 2-8: Captive light contributes gravitational mass

At this point we've run out of obvious ways to tell whether the "full" flask is heavier because it contains electromagnetic energy or because it contains a tiny piece of conventional matter. Since the attraction between the flask and the Earth has to be mutual, the contained light ought to make the flask's own gravitational field stronger, too.

2.5: Mass-to-energy conversion

Having established that trapped EM energy appears to contribute mass to a container, we may be a bit impatient to know exactly how much.

Well, we can work it out ...

momentum

As we've already mentioned, moving objects and light both carry **momentum**.

Momentum is a measure of how hard something pushes against obstructions placed in its path. In Newtonian mechanics, the momentum "p" of a moving object is just its mass "m", multiplied by its velocity "v", giving us $p=mv$ ... the object's momentum is the quantity of material, times its direction and speed. But as we've seen, adding lightenergy to a flask increases its effective inertial mass, and also the momentum if the flask is seen to be moving.

If we knew how much additional momentum is due to the trapped light, we could calculate how much additional mass the light seems to contribute to its container. The momentum of an individual lightbeam is proportional to its energy, and **Newtonian Mechanics** gives us a prediction for how that energy should change with relative speed (Doppler shifts, section 6), so we can go on to work out the momentum associated with a *complex* of trapped light with a given rest energy, **E**, when its container is seen to be moving. If this momentum varies with velocity in the same way as the momentum of conventional matter, we can use these momentum relationships to derive a conversion factor between the amount of energy placed inside the box, **E**, and the amount of "effective" mass that the energy contributes to the box, **m**.

"Newtonian" calculations

Taking a box whose lightrays are perfectly aligned with the direction of the box's motion, we can calculate how much these forward-pointing and rearward-pointing rays will be **Doppler-shifted** by the motion of the box. Using the **NM** Doppler relationship (section 6.2), we find that the frequency, and energy, and momentum of a ray all increase or decrease linearly with the box's speed. Summing the effects of light hitting the opposite sides of the box, we find that the *overall* momentum of a "complex" of trapped light is then proportional to its velocity, too, just as we'd expect if we'd replaced the trapped light with a piece of something more substantial. Where the "Newtonian" prediction for the momentum of matter is $p=mv$, the momentum of our lightcomplex with an quantity of energy **E**, will be $p=E[something]\ v$.

When we do this exercise using the "Newtonian" relationships (Appendix, Calculations 2), we find that almost everything in the two calculations cancels out, apart from "m" (the change in mass), "E" (the light-energy in the flask), and a couple of spare "c" lightspeed terms left over from the "light-momentum" equations, which didn't have anything in the "mass-momentum" equations to cancel with. We then find ourselves staring at a surprisingly short equation for how much energy we have to put into or take out of a stationary box to change its "effective mass" by a given amount …

$$E = mc^2$$

2.6: History of E=mc²

Albert Einstein's famous paper on $E=mc^2$ was published as a small supplementary piece to his much larger 1905 paper on the **special theory of relativity**. Einstein said that if his earlier arguments were correct, then a body should lose this amount of mass when it radiates, and that this might explain how it was that radium was mysteriously able to continuously give off energy without undergoing any chemical reactions.

A few other researchers had also been tentatively moving towards this equation (or variations on it) to try to describe, say, the momentum of a cavity containing trapped light, but Einstein is given the credit because he seems to be the first person who got the relationship exactly right, recognised its general significance, and came out and said so, unambiguously, in print.

generality of E=mc²

Einstein didn't use our Newtonian calculations – according to the newer "SR" relationships presented in his 1905 "electrodynamics" paper, the energy (and momentum) of emitted light should be greater than it is under NM by something called a "**Lorentz factor**" (section 14.7), and the momentum of a body should *also* be greater than in the NM account, by another Lorentz factor – but since these two changes to our calculations cancel out, they don't affect the final answer. It still comes out as $E=mc^2$, either way.

So, while **E=mc²** is sometimes presented as compelling evidence for the correctness of Einstein's **special theory of relativity**, and while it's probably right for Einstein to get the credit for it, the credit properly belongs to *him* rather than to the particular theory that he happened to use to work it out. The same logic would have worked (with slightly simpler math) under old-fashioned Newtonian **emission theory**, and in fact, $E=mc^2$ would seem to be a result of just about *any* theory that reduces to NM, good or bad, provided that any modifications to the Newtonian momentum laws for matter are accompanied by a matching modification to those for light (which is pretty much essential if one want a theory to agree with our general arguments about flasks). We can even derive *a continuous range* of possible relativistic theories that do this (section 13.1), based on different Doppler shift laws, and provided that we construct these consistently and assume basic conservation laws, we should always get $E=mc^2$ as the amount of energy that the stationary flask sitting on the scales has to give off in order to become lighter by a given amount.

brevity

The fact that the $E=mc^2$ equation is *so* short can sometimes makes non-physics people suspicious. Isn't it rather fishy that the energy (in Joules) is *exactly* the same as the mass times the velocity of light squared, in kilogrammes and metres per second? What if we decided to use miles-per-hour for speed and pound-weight for mass, shouldn't there be a scaling factor? Well, yes, if we'd used Imperial weights and measures (or a mix of any other sorts of weights and measures), then we'd have to include some sort of conversion factor, "k", and write the equation as $E = k \times mc^2$. Our factor "k" would then have different values depending on whether we chose to define "velocity" in kilometres per second or miles per hour or cubits per lunar month. But it just so happens that the relative magnitudes of the various units used by the international **metric system** ("Système Internationale", or "**S.I. units**") are cleverly devised to interlock with each other so that these sorts of type conversions are unnecessary.

If we stick with S.I. units, all we have to write down is "$E=mc^2$".

2.7: *Energy has mass, period*

What if the escaping light *hadn't* been contained inside the box as EM radiation? What if the box was fitted with a pair of flashlights pointing in opposite directions, powered by conventional chemical batteries? When we switch the flashlights on, and the chemical potential energy stored up inside the batteries turns into electrical potential, which then drives the two oppositely-directed bulbs, does the effective inertial mass of the equipment shrink as it sheds lightenergy? Well, yes, exactly the same considerations apply. A passing observer could say that since the "moving" box has forward momentum, and the pair of opposing Doppler-shifted lightbeams *also* carries overall "forward" momentum, then if the escaping light's total momentum isn't appearing "from nowhere", and the equipment isn't changing speed as the light is given out, then the hardware (in this case, the batteries) must be losing mass.

This is quite a serious change to the way that we used to think about the world. *In theory*, a hot object ought to weigh more than a cold one because of the kinetic energy bound up in its jiggling atoms. If we choose to power our flashlight with clockwork mechanism that drives a dynamo, the steel spring should have a greater gravitational and inertial mass when it is wound up and its inter-atomic bonds are stressed, than when it's run down. Winding up an elastic band *should* make it weigh more.

But because c is *so* big, and *c squared* is so *absurdly* big, the additional mass contributed by everyday amounts of energy is normally far too small for us to have any hope of measuring it. Although chemical compounds should *in theory* weigh slightly more or less depending on the energetic configuration of their components, we can't expect to make practical use of this by popping a battery onto a set of kitchen scales to see how much "oomph" it has left in it.

But the "flipside" is that a tiny change in mass corresponds to a *huge* change in energy. We *do* find noticeable discrepancies when we compare the **atomic weights** of different elements against their **atomic numbers**, and this suggested that there might be huge differences in energy associated with the different configurations that could exist inside the nucleus of an atom – **nuclear reactions** involve vastly more energy than chemical ones, and the significance of $E=mc^2$ is therefore usually considered to be its relevance to nuclear physics.

3
Curved Space and Time

> *"I am at least of opinion that gravity is nothing more than a natural tendency implanted in particles ... by virtue of which, they, collecting together in the shape of a sphere, do form their own proper unity and integrity. And it is to be assumed that this propensity is inherent also in the sun, the moon, and the other planets."*
>
> — Copernicus on the idea of universal gravitation, 1543 (quoted by Mach)

> *"Therefore because of the analogy there is between the propagation of the rays of light and the motion of bodies, I thought it not amiss to add the following Propositions for optical uses; not at all considering the nature of the rays of light, or inquiring whether they are bodies or not; but only determining the trajectories of bodies which are extremely like the trajectories of the rays. ..."*
>
> — Isaac Newton on the generality of geometry, and the expected gravitational deflection of light, "Principia"

> *"Query 1: Do not Bodies act upon Light at a distance, and by their action bend its rays, and is not this action (cateris paribus) strongest at the least distance?"*
>
> *"Qu.4. Do not the Rays of Light which fall upon Bodies, and are reflected or refracted, begin to bend before they arrive at the Bodies ... ?"*
>
> — Isaac Newton, "Opticks" (1704 edition onwards)

> *"By no means, sir. Time travels in divers paces with divers persons. I'll tell you who Time ambles withal, who Time trots withal, who Time gallops withal and who he stands still withal."*
>
> — Rosalind, "As You Like It" act 3 scene 2, William Shakespeare

> *"Space is precisely analogous with time."*
>
> — "Time and Space", Edgar Allen Poe, 1844

> *"I have no objections to Budde's conception of space as a sort of medium ... although I think that the properties of this medium should be demonstrable physically in some other manner, and that they should not be assumed **ad hoc**. If all apparent actions at a distance, all accelerations, turned out to be effected through the agency of a medium, then the question would appear in a different light ..."*
>
> — Ernst Mach, "The Science of Mechanics" (2nd ed.) 1902

> *"An atom absorbs or emits light of a frequency which is dependent on the potential of the gravitational field in which it is situated"*
>
> — Einstein, "Relativity ...", Appendix 3c

RELATIVITY IN CURVED SPACETIME

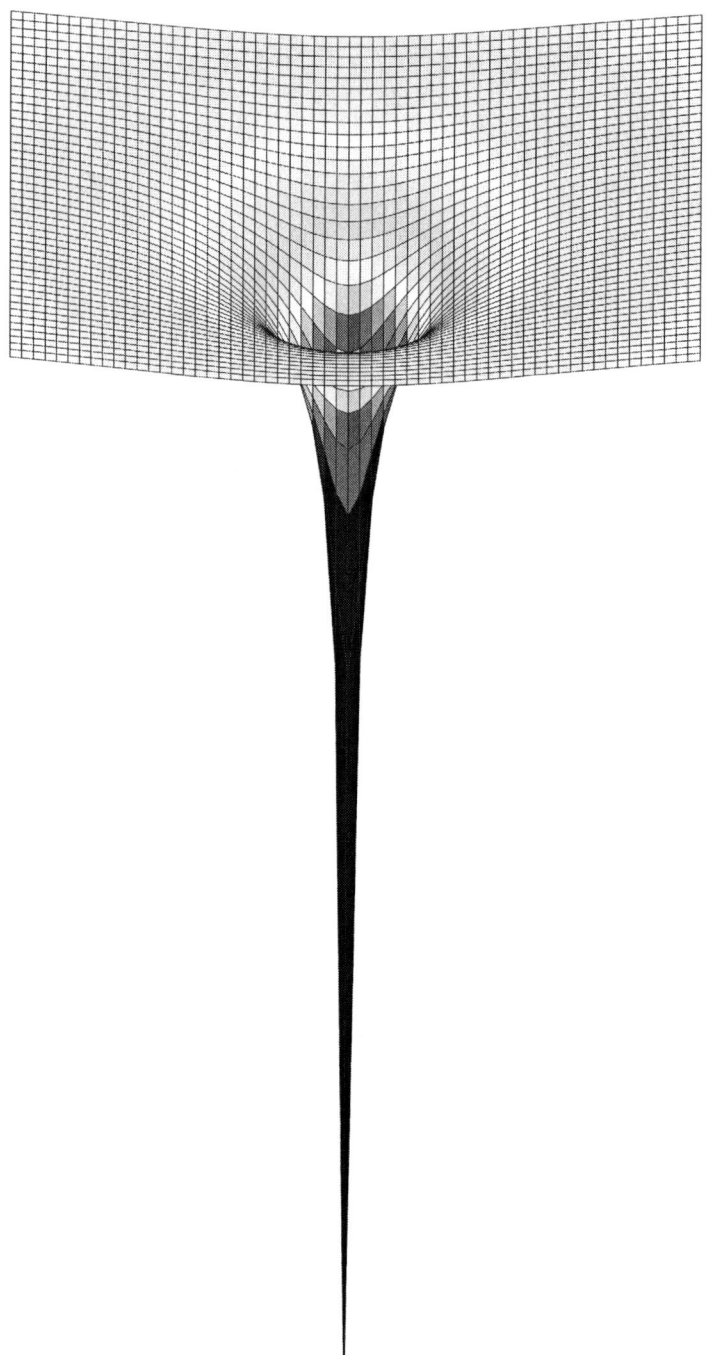

Extreme gravitational well

3.1: Gravity is ... what, exactly?

So, what *is* gravity? There are many possible answers: we might say, gravity is "a variation in the speed of light", or "a variation in the rate of timeflow or inertia", or "spacetime curvature". We might try to define it in all sorts of ways.

Gravity is pretty fundamental. It existed long before we did, and long before language, so instead of trying to give a single prescriptive definition of what gravity *is* (by comparing it to something else), we'll try to describe what gravity *does*, and the way that it seems to do it, and hopefully that'll give us a good enough answer.

3.2: Gravity bends light

Let's introduce a strong gravity-source to a region, We'll suppose that light propagates more slowly nearer the source, either because lightspeed is slower there, or because space there is "more dense" (we won't worry ourselves with which).

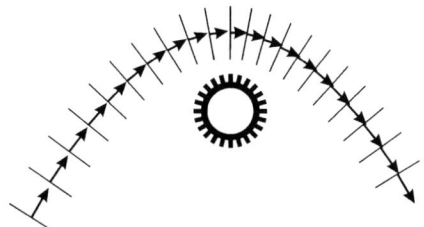

Figure 3-1: Light-bending as a "wave-dragging" effect

A set of plane-waves will advance more slowly when passing through the stronger-gravity region, and this lightspeed difference results in the "slower" side of the waves being held back, so that the whole wavefront ends up being steered around to point more towards the region of slowest lightspeed (section 1.9).

A gravity-source causes lightbeams to deflect towards it.

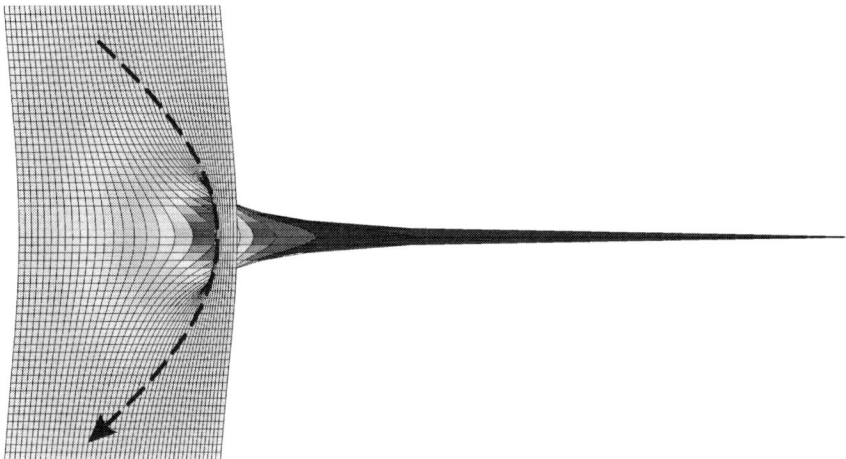

Figure 3-2: Light-bending as a "curvature" effect

3.3: *Gravity warps geometry*

Take three equidistant observers, **A**, **B** and **C**, forming an equilateral triangle.

School geometry tells us that the angle at each corner of the triangle will be 60 degrees, and this can be verified if observer **A** carries a laser-pointer and the others carry mirrors. If the beam bounces from **A** to **B** to **C** and back to **A** again, each observer ought to be able to use a **protractor** to verify that the angle between the incoming and outgoing beams is 60 degrees. Even if **A**, **B** and **C** *aren't* equally spaced, school geometry tells us no matter what sort of triangle is formed, the *sum* of the three protractor measurements will still be 180 degrees.

If **A**, **B** and **C** are placed around a gravity-source, things are different: this geometry no longer works. In order to aim their laser-pointer at **B**, **A** now has to aim slightly *away* from the gravity-source in order to compensate for the gravitational bending of the beam. Since any light coming back from **B** is being bent by the same angle, **A** sees **B**'s position to be again, slightly further away from the gravity-source than the "true" position. In practice this doesn't matter: **A** simply aims their sights at where **B** *appears* to be, switches on their laser, and sees the red laser dot show up on the target exactly where it would be expected to appear. Nothing particularly strange seems to be happening, and **A** might feel justified in saying that the path of the beam is a straight line, after all, neither **A** or **B** can see any deflection from what seems to them to be a simple, straight, line-of-sight. However, when all three observers have adjusted their mirrors to get the laser beam to form what seems to them to be a conventional triangle, something strange happens: the equal corner angles now add up to more than 180 degrees.

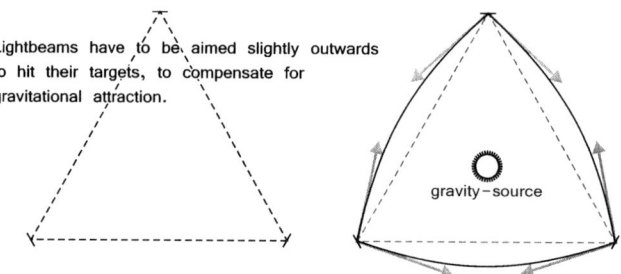

Figure 3-3: Sum >180 degrees!

If this visible distortion affects *every* method that we might use to try to measure angles and distances, then, if *everything* in a region gets distorted in the same way, we may as well say that, to all practical intents and purposes, *space itself* is distorted in the region.

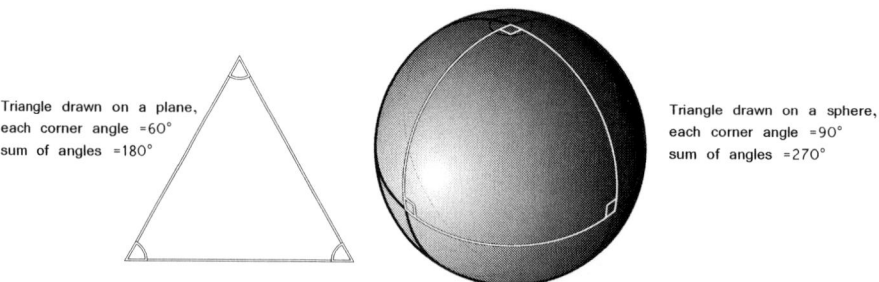

Figure 3-4: Triangle drawn on a positively-curved surface

3: Curved Space and Time

Some people become unhappy at talk of space and time being distorted, and complain that the things we're talking about here are merely *apparent* quantities – what clocks and rulers *experience* and *measure* rather than a "*real*" underlying space and time. But physics is often *about* what can be experienced: we can invent a set of *undistorted* reference-coordinates and then overlay fields on it to explain how our actual readings deviate from the "perfect" values, or we can deal with the readings directly, and talk about spacetime itself being curved.

To prefix every mention of "time" or "space" with the word "apparent", and to always talk about spacetime having a warped *lightbeam* geometry, quickly becomes tedious ... so it seems reasonable to just talk about "warped spacetime", and to hope that people who dislike these phrases will understand what we mean, even if they disapprove of our choice of words.

positive and negative curvature

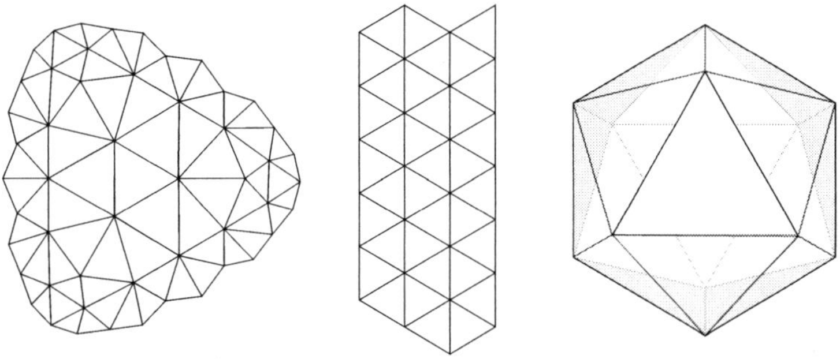

Figure 3-5: Sample grid-maps of negatively-curved, "flat", and positively-curved surfaces

When a plane passing through **flat space** is mapped with a triangular grid, six equilateral triangles will meet at each point. In the left-hand map, we've used *seven*. This is referred to as space with **negative curvature**, and as we move through this sort of region, we find that we're continually being confronted with more alternative directions than we're used to seeing – negatively-curved space is very easy to get lost in! If we measure the perimeter of a region with negative curvature, we'll find that there's less space inside that perimeter than normal.

In the map on the right, we have only *five* triangles meeting at each point instead of six. With *this* configuration, our map eventually closes completely in on itself (in this case, it turns into a twenty-sided icosahedron). If we mark out a region of **positively-curved space**, then the perimeter is *smaller* than we'd expect for a given radius – the region "bulges", and contains *more* space than we might think from taking measurements around its perimeter.

We can also express this uniform curvature with the embedding diagrams that we met earlier.

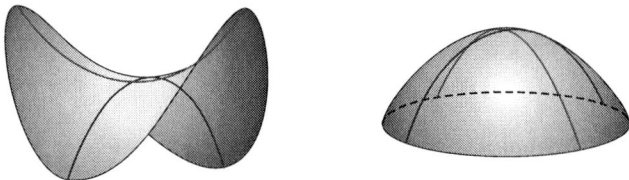

Figure 3-6: Extrusions of negatively and positively curved surfaces

Again, on the left we have a section of negatively curved space, and on the right, a region with positive curvature. The left-hand shape is sometimes referred to as a "saddle" shape.

3.4: Gravity as a variation in inertia

variable timeflow → variable inertia

If a lightspeed difference deflects light, how does this affect matter?

If the forces holding an atom together in equilibrium propagate at "lightspeed" then by altering those speeds we throw the system out of balance. An overall deflection of those lightspeed-regulated forces towards a "slow" region means that the atom, staying nicely in equilibrium, should then find itself naturally moving progressively faster towards the "slow" region, too. When a larger object made from these atoms finds itself deflected, then since all of its atoms are influenced in the same way, there should be no obvious geeforces or stresses or strains in the object's structure to reveal the acceleration (apart from perhaps tidal forces). The object is entitled to believe that it is in fact following an *unaccelerated* **inertial path**, but that this path has somehow been warped by the distorted properties of spacetime: the variation in lightspeed can be reconsidered as an ***inertial* field**.

inertial field → gravitational field

What happens when we strengthen the inertial properties of a region of spacetime? Time appears to slow down. In our "inertia-based" description, clocks now run slower at the bottom of a gravity-well because increasing the inertia of their structures makes them react to forces more slowly. Our earlier arguments (section 1.5) about what happens to timeflow when we change the speed of light cut in once again, so all true **inertial clocks** run more slowly in the more intense regions of the field. Mechanical pocket-watches tick more slowly, the speed of molecules with a fixed energy is reduced, chemical reactions occur more slowly for a given energy, and observers' brains, quartz-based digital watches and atomic clocks all run more slowly. If all these effects are slowed by the same amount (which they should be), then an observer in the region should be unable to tell from local observations that anything untoward is happening.

So our inertial field makes time run more slowly, the effect of this field slowing the propagation of light through a region (Figure 3-1) is to make lightbeams deflect towards it, and Eötvös principle then requires the field to also attract matter. By allowing *inertia* to vary from place to place, we end up reinventing *gravity*.

The terms "inertial field" and "gravitational field" seem to be functionally identical. Reconsidering the conventional gravitational field as an "inertial" field helps to explain why we can't seem to isolate the inertial mass of an object from its gravitational mass – the two properties seem to be interchangeable and inseparable.

3.5: Energy-change in light due to gravity

A body falling downhill increases its speed, momentum, and kinetic energy, and for a complete implementation of **Eötvös' Principle**, we need the energy and momentum of trapped light to change in the same way (because otherwise we'd need different gravitational equations of motion for different sorts of materials). This tells us that light falling downhill across a gravitational gradient needs to gain energy.

There are only two obvious parameters that we can adjust to change the energy-content of one cycle of a "simple" signal: its **frequency** (or **wavelength**), which can describe the separation in time (or space) between two successive signal peaks, and its **amplitude**, which is a measure of the "height" of the signal peaks, and is analogous to the "loudness" of the signal. Our signal must increase its energy by changing its *character* or its *intensity*, and since there's no obvious mechanism for the *quantity* of light to change as it falls, we're left with the idea that the signal arrives at its final destination showing a different frequency and wavelength.

If we were using a "particle-compatible" description of light, or trying to construct a theory of gravity that works with either particles or waves, an *amplitude* change would also tend to mean that the *number of particles* was changing in mid-experiment, which wouldn't be a very nice feature.

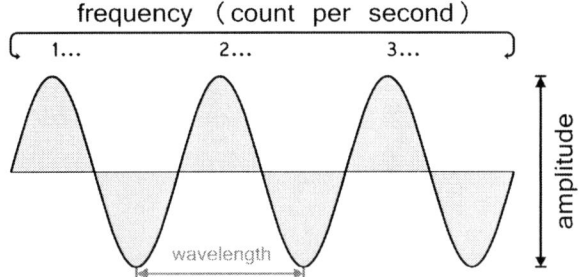

Figure 3-7: Amplitude and frequency

Worse things can sometimes happen under quantum mechanics, but back in 1783, **John Michell** decided on the more straightforward option of saying that it's not the quantity of light that is changed by gravity, but its character: light reaching us from high-gravity stars should be *weakened* by the uphill climb, and should then be deflected differently by a prism – when the "gravitationally-weakened" starlight was fed through the device, the emerging light should be seen to be offset or displaced towards the weaker end of the spectrum.

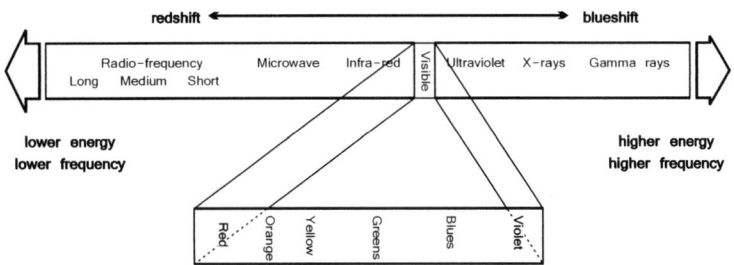

Figure 3-8: The electromagnetic spectrum
(not to scale)

3.6: Gravitational redshifts and blueshifts

Since we now know that short-wavelength-high-frequency light carries more energy, this gives us a simple rule:

> When light has passed *uphill* across a gravitational gradient, from a region of stronger gravity to one where the background field intensity is more dilute, it loses energy in the climb and arrives with a reduced frequency and a longer wavelength, and we say that it's showing a **gravitational redshift**.

> When a lightsignal has travelled *downhill* across a gravitational gradient, from a weak-gravity-region to a strong-gravity region, it gains energy and arrives at its destination with an increased frequency and a shorter wavelength, and we say that it's showing a **gravitational blueshift**.

RELATIVITY IN CURVED SPACETIME

3.7: Gravitational time dilation

Reading the previous pages on the gravitational shifting of light, we may feel a queasy uneasiness stirring at the back of our minds. Certainly, the idea that light gains or loses energy when it moves into or out of a gravitational field *sounds* reasonable, but the next part about the signal changing *frequency* sounds a bit disturbing: How can a signal possibly be received at a different rate to the one it started out with, in cases where there's no explicit relative motion between the source and the observer? More specifically, how can we possibly be receiving a blueshifted signal *faster* than it's being generated by the source?

At first sight the idea doesn't seem to make any sense, and this may help to explain why the subject of gravitational shifts seemed to disappear for a century ... until Einstein resurrected the problem in 1911, and cheerfully insisted that the argument *must* be right, and the discrepancy *can't* be explained – **unless the weaker-gravity observer is also aging faster**. Michell's predicted gravitational shifts don't just change objects' *appearances*, they also tell us something about the underlying rates at which observers in different gravitational environments will experience time.

Einstein's breakthrough (which was so simple that it seemed to have eluded everyone else before him), was to connect the gravitational field to the rate of timeflow: according to our "gravity-shift" arguments, when we see a gravitationally-blueshifted signal, this can *only* be explained if our own reference clocks are running slower than those of the signal source – a stronger gravitational field makes time run slower.

gravitational time dilation in practice

Let's suppose that we decide to build and launch a satellite fitted with a very accurate onboard atomic clock, programmed to broadcast pictures and time-checks on a particular radio frequency. Once the satellite is in orbit, we notice something strange: the broadcast signal is gravitationally blueshifted as described in the previous section, and arrives back on Earth with a slightly higher energy and frequency than we intended.

Figure 3-9: Objects in weaker gravitational fields age more quickly

Perhaps the frequency change is so small that we don't care. But there's another problem with the signal that is more disturbing. On broadcast, it takes a fixed number of carrier-wave pulses to carry each image, and this ratio has to be the same when the images are received. No extra pulses have been sneakily inserted into the signal stream en route, and so the rate at which we receive pictures from the satellite is slightly fast. But if the satellite is engineered to take and transmit one picture per second, how can we be receiving pictures faster than this?

The satellite's broadcasts, based on its own internal clocks, are gradually getting out of sync with our ground-based atomic clocks, and eventually, if we receive the incoming pictures at the higher rate for long enough, we should be getting pictures today that are timestamped with tomorrow's date, or next week's. If this is Monday, then how can we already be seeing satellite broadcasts scheduled for *Tuesday*?

We decide that the launched satellite's clocks are faulty, and are running at a faster rate than they're supposed to. We grab some recalibration tools, hop onto a space shuttle and visit the satellite, to retune it to the "right" frequency. This is where we are hit by the next surprise: When we reach the satellite and hook up our equipment we find that the problem seems to have mysteriously fixed itself, and to make things even more confusing, it now seems that it's instead the *Earth-based* radio signals that have the wrong rate ("slow" due to gravitational redshift). When we return to Earth, scratching our heads, and recheck out our calibration equipment, we find that it once again runs perfectly in synch with the other Earth-based atomic clocks, but that the satellite clock seems to have decided to start running too fast again.

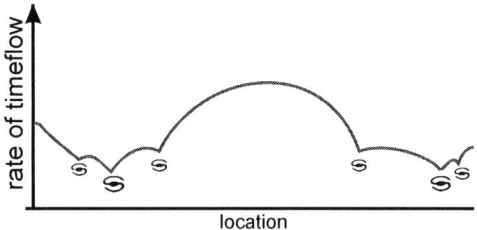

Figure 3-10: Time can run at different rates in different parts of the universe

What's happening is that the timing references that run our various pieces of equipment are operating at different rates as a function of location: clocks in the region of space away from the Earth are running at a faster rate than those at sea level, and as we take our portable calibration equipment from place to place, it naturally finds itself running at the same rate as the other hardware at those locations (section 1.5).

avoiding acceleration effects

Eagle-eyed readers will notice that we haven't *quite* proved our case, because the "Earth" clocks are feeling stresses and strains due to resisting Earth's gravity, while the satellite clocks are drifting freely in space. What if the effect of these *physical* accelerations on the clocks (which we ignored) have some additional effect that invalidates our earlier argument?

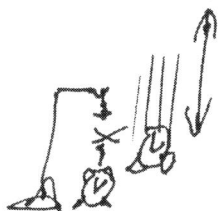

Figure 3-11: "Dropped clocks"

But we can get rid of this objection by saying that the "Earth" clock could be in freefall, too: it could be initially dangling at the end of a piece of string or sitting on a trap-door, and "dropped" the moment before it takes its readings of the deep-space clock's signals: when it takes its readings, the "dropped" clock is then *also* freefalling, but hasn't had time to pick up enough speed to significantly change the readings – the blueshift that it should see can then be considered as a simple function of the gravitational difference between the two locations.

If we want to go further and insist that the condition that everyone is in freefall has to be *sustainable*, we can imagine hollowing out a cavity at the centre of the Earth where our "strong-gravity" reference clocks can float weightlessly (section 1.2) – now *both* sets of clocks appear to have equivalent locally-measured conditions, but the existence of the gravitational differential between them is still responsible for an asymmetrical gravitational shift in signals passed between them, and a corresponding difference in ageing rate.

Or, we might like to consider the case of two observers with no obvious relative motion, one drifting weightlessly between stars at the centre of a galaxy, and the other out deep in an intergalactic void – although both observers might occupy local gravitational plateaus, and most of their local physics might appear equivalent, signals passed between them should show an asymmetrical change in energy, and the "strong-gravity" observer should age more slowly than their distant, weak-gravity counterpart.

There are some interesting subjects that we could branch off to here, such as the difficulty of establishing which object is "really" downhill when spacetime is twisted up around a rotating body, but we'll ignore them for now and press on.

pendulum clocks are not "good" clocks

When we talk about "clocks" in this context, we have to avoid using devices driven by gravitational *differences*, such as such as hour-glasses and pendulum clocks. These measure a hybrid of *two different* effects – the passage of time and the gravitational gradient acting across the device – and even in Newton's time it was known that pendulum clocks placed at different latitudes ran at different rates for reasons other than the ones we are interested in here. A pendulum clock in freefall won't run at all: this doesn't mean that time "stops" in freefall, it just means that these gadgets aren't much good at measuring timeflow in the sorts of situations that we're interested in. In order to "fix" a pendulum clock to work in freefall, we could specify that the pendulum be rigid, use a spring or set of springs to accelerate it instead of gravity, and use extravagant counterbalancing to allow it to operate in any orientation. When we do these things we essentially convert our swung pendulum into a spring-loaded flywheel, and we then have the same sort of mechanism as a mechanical pocket watch. This should then be slowed by a stronger gravitational-field level like our other clocks.

3.8: Not just curved space, but curved space<u>time</u>

Einstein's 1911 argument tells us that not only does gravity warp apparent *spatial* distances and times as measured by lightbeams, it also affects the underlying rate at which clocks run. Instead of just the effective *spatial* density varying from place to place, we also have the effective density of *time coordinates* changing with gravitational field strength. In our "box" example in section 1.8, the perimeter of the box containing a gravity-source doesn't just enclose *more space* than we might expect from its perimeter, the centre of the box also contains *fewer clock-ticks* per second than the perimeter.

onward to general relativity

It was this critical 1911 realisation that gravity affected clocks as well as rulers that allowed Einstein to go on to produce his general theory of relativity in 1915. Some very clever mathematicians had already been attempting to produce mathematical models that applied the theory of curved surfaces to *three*-dimensional curvature of *space*, but Einstein's breakthrough was to realise that *the problem itself* was wrong, and that **Gauss'** geometry should instead be applied to *four*-dimensional curvature of *spacetime*.

At last, all the pieces seemed to be starting to fit together properly.

Unfortunately, Einstein's general theory introduced a new assumption of its own …

4
Relativity

> *"A substance offers as much resistance to the air as the air to the substance"*
> Leonardo

> *"... whether heavn move or earth*
> *Imports not, if thou reckon right; ... "*
> – "Paradise Lost", Book VIII (~1667), John Milton

> *Wherefore relative quantities are not the quantities themselves, whose names they bear, but those sensible measures of them (either accurate or inaccurate), which are commonly used instead of the measured quantities themselves. And if the meaning of words is to be determined by their use, then by the names time, space, place, and motion, their [sensible] measures are properly to be understood; and the expression will be unusual, and purely mathematical, if the measured quantities themselves are meant. On this account, those violate the accuracy of language, which ought to be kept precise, who interpret these words for the measured quantities. Nor do those less defile the purity of mathematical and philosophical truths, who confound real quantities with their relations and sensible measures.*
> Newton, "Principia" (1687)

> *".. or, in other words, these two propositions, "the earth turns round," and, "it is more convenient to suppose that the earth turns round," have one and the same meaning. There is nothing more in one than in the other."*
> Henri Poincaré, "Science and Hypothesis" 1901
(Chapter 7: Relative and Absolute Motion)

> *" But if we take our stand on the basis of facts, we shall find we have knowledge only of **relative** spaces and motions. **Relatively**, not considering the unknown and neglected medium of space, the motions of the universe are the same whether we adopt the Ptolemaic or the Copernican point of view. Both views are, indeed, equally **correct**; only the latter is more simple and more **practical**. The universe is not **twice** given, with an earth at rest and an earth in motion; but only **once**, with its **relative** motions, alone determinable."*
>
> *"It is not necessary to refer the law of inertia to a special absolute space."*
>
> *"Nature behaves like a machine. The individual parts reciprocally determine one another."*
>
> *"[re: Streintz] I cannot share this view. For me, only relative motions exist ... , and I can see, in this regard, no distinction between rotation and translation. ... Can we fix Newton's bucket of water, rotate the fixed stars, and **then** prove the absence of centrifugal forces? ... The experiment is impossible, the idea is meaningless, for the two cases are not, in sense-perception, distinguishable from each other. I accordingly regard these two cases as the **same** case ... "*
> Ernst Mach, "The Science of Mechanics" (2nd ed), 1902

"Yin-Yang", symbolising balance, mutuality and completeness.
A suitable symbol for the principle of relativity.

4.1: Relativity of space

We're already used to the idea of the **relativity of position**: space doesn't come stamped, tagged or engraved with ready-made **coordinate-system** labels, and when we assign, say, a **GPS** location to a building, it's understood that that location tells us where the building is *relative* to the other features on the Earth's surface. We choose to use particular agreed references and coordinate systems, not because we believe that they correspond to some deeper, absolute, *fundamental* concept of location, but for convenience.

Ships used to be in the habit of using the astronomical observatory at **Greenwich, London** to synchronise their clocks, and since these clocks were used to calculate position, the Greenwich Meridian became our international reference, "zero longitude". We've kept it partly because it puts the international dateline at the other side of the planet to Greenwich, at a position that doesn't pass through the middle of any major landmasses where lots of people live. We also find it useful to use gridlines that terminate at the Earth's rotational axis, because this gives us two well-defined points, the **North and South Poles**, and because we can use the relative rotation of the Sun and stars to help find our position on this grid. It's very handy for astronomy and calculating daylight time zones. But this system also has the very *political* advantage that the two places where this coordinate system goes crazy (the Earth's poles) are places where, again, almost nobody lives. If we lived on a warmer planet where the poles were inhabited and the equator was a barren desert, we might use a different system with locations defined by surface distances from important cities, or with the surface divided into polygons that approximated political territories. We might use projections in which the polar regions were nicely defined, but coordinate numbers went crazy at the equator.

We don't believe in absolute locations, only in distances between persistent *things* that we consider important to us, or which are easy to describe and agree on. Even stars and galaxies drift with time.

We've also learnt (section 1.8) that even the *scaling* of space, which we might once have considered absolute, seems to vary with the strength of the background gravitational field. Even apparent distances between agreed locations depend on how we measure them, and have to be defined locally, or defined globally using a particular reference-length and a given projection method.

4.2: Relativity of time

We're also familiar with the idea that the date and time values that we assign to an event are relative – we compare the temporal location of our event with a well-known reference event, and use the difference between them as our agreed time location. If we talk about the year being "2007", it means that we're referring the current time to a **reference event** just over two thousand years ago. But different cultures use different calendars and references, for instance, after the French Revolution of 1792, the country is supposed to have adopted a "Republican Calendar" that started again at year zero, and used ten-hour days and hundred-minute hours. These things are, to a great extent, arbitrary.

We might feel that at least the *rate* of time is absolute and agreed ("a second is a second is a second"), but the concept of gravitational time dilation (sections 1.5, 3.7) spoils this idea, too. A second on the Earth and a second on the Moon, timed with identical local atomic clocks, are not going to agree. For global measurements, the time measured between two events can depend on the route that we use for the measurement. When our coordinate systems become inconsistent, we usually strike some sort of convenient compromise, and we tend to fall back on using averages of different atomic clocks, or we decide to use a particular timebase as a system of reference. Again, it's a matter of convenience.

4.3: Relativity of velocity

Newtonian mechanics is founded on the idea that **velocity** and/or **speed** are strictly relative properties, too.

Zeno's paradox against absolute motion

Zeno of Elea (*b.* ~450 BC) was a Greek philosopher and student of **Parmenides** who churned out a series of philosophical problems relating to the divisibility of time, space and motion.

Zeno argued that motion couldn't exist as a *fundamental* property of a single object, since the object's motion didn't exist for all observers. Two runners jogging around a racetrack, side by side, *could* be said to have a certain speed as seen by observers sitting in the stadium, but they'd also be stationary with respect to each other. If the runners' supposed speed didn't exist for *everyone*, then "speed" couldn't be an *intrinsic* aspect of a body, and could only be defined as a *relationship* between a selected body and other things around it.

We can say that the runners are stationary *relative to one another*, are moving *relative to the stadium*, or are moving somewhat faster relative to the rest of the solar system, and to our galaxy, and to background galaxies.

The choice of reference object doesn't seem to have any physical consequences. For objects moving in simple straight lines at constant speed, Nature doesn't seem to have a preferred state of motion, and seems happy to allow bodies to continue coasting along "inertially" with their initial speeds and directions, until something else intervenes.

relativity in practice

If two ice-pucks strike each other, and if the **principle of relativity** is valid for the laws of physics that describe their motion, we should be able to calculate the resulting paths of the pucks *either* by saying that puck #1 was stationary when it was hit by puck #2, *or* by saying that puck #2 was stationary when it was hit by puck #1 ... or by using some other reference system in which both pucks are said to be moving.

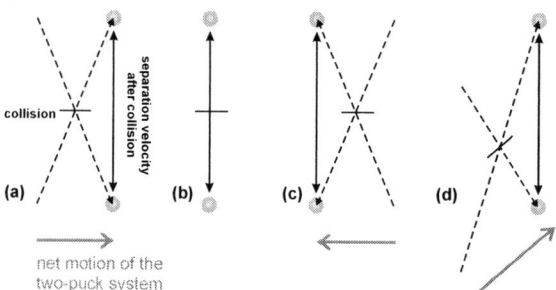

Figure 4-1: Collision between two pucks, as seen by different observers

In diagram **(a)**, the colliding pucks are seen to be moving from left to right at the same speed when they bounce off each other.

In **(b)**, the observer is moving along with the pucks, and sees no sideways motion. For them, the two pucks are aimed directly at each other and when they bounce away they retrace their original paths in reverse.

In **(c)**, the observer could be said by **(a)** to be moving rightwards twice as fast as the pucks, and overtaking them. For onlooker **(c)**, the pucks are both moving leftwards.

In **(d)**, the observer has a sideways motion that matches **(a)**, but is also moving down the page relative to **(a)**, **(b)** and **(c)**. The observer in **(d)** sees the pucks to be moving to the right, and also moving *up* the page. For *this* observer, the two pucks have different speeds. Before the collision, the lower puck is described as moving upwards at a faster speed than its companion, but momentum is then transferred *between* the pucks during the collision, and afterwards it's the upper puck that moves the fastest, away from the collision site.

What **Galileo** and **Isaac Newton** did next was to try to work out what the common laws of physics would have to be if all these reference observers were to be able to predict the same outcome for an experiment by applying the same rules to their different observations. If all onlookers were to agree that some fundamental laws were being obeyed – for instance that the energy and momentum of the two-puck system should be *conserved* during the collision (no energy appearing from nowhere or disappearing to nowhere, for no reason, for any observer), then this was enough for Newton to start to distil out a simple set of definitions and laws for How Things Must Behave When They Move.

Newton's "relativistic" rules of mechanics would let us calculate how long it takes a ball to fall to the floor from a height of ten metres or a hundred metres assuming standard Earth gravity, and could tell us the exact trajectory of a cannonball (ignoring air resistance). Although we now tend to think of some of these rules as being self-evident, at the time they were revolutionary. If an object could hurtle through space and feel as if it was *stationary*, then it was slightly less outrageous to imagine that the Earth could be circling the Sun at nearly 30 km/s without us feeling any obvious side-effects. Galileo's work (building on that of the executed **Giordano Bruno**) helped us to step away from the idea of the Earth as being a special body fixed solidly at the centre of the universe and allowed us to think of it as being just another planet, obeying the same laws as its neighbours. This breakthrough helped us to derive more general (and more realistic) laws of gravitation and motion.

4.4: Isaac Newton's "Principia"

To those taught that Newton said that space and time were absolute, reading the "Definitions" section of Principia can be a surprise. Newton started his book by laying out his definitions of what he meant by absolute *and relative* spaces and times. Locations and velocities (said Newton) were purely relative and their values could be "zeroed" with a purely mathematical change of reference coordinate system.

Apparent times of events were also relative because of the finite transmission time of light. Astronomical events could be *deduced* to happen at different times to when they were *seen* to happen, and astronomical technique was already good enough for astronomers to have seen apparent timing discrepancies in the eclipses of Jupiter's moons due to the variations in the distance between Earth and Jupiter, and to have used this to calculate a decent value for the speed of light. Newton *did* also define "*absolute time ... which flows equably ...*", but didn't claim that this necessarily corresponded to the rate at which any Earthly or celestial clock would actually run, or would be seen to run. We could *deduce* absolute time by calculating and compensating for factors that would be thought to alter clockrates, but even celestial machinery was irregular, and a "mathematical", artificially-imposed time coordinate system might not correspond to any actual clock in the universe.

Newton's scheme did, however, say that *physical accelerations and rotations* had real physical consequences that couldn't be eliminated with mathematical sleight of hand, and could therefore be said to be "real" or "absolute". His system imposed the principle of relativity onto the physics of simple **inertial** motion, but allowed asymmetrical, absolute differences to exist between the physics of these objects and the physics of objects that were accelerated. The idea that all viewpoints were equally valid seemed to apply to simple inertial motion, but didn't seem to work in situations involving geeforces.

Newton's bucket

"Principia" gives the example of a rotating bucketful of water as an example of "absolute" motion – the water's surface would be flat for a stationary bucket, or for a bucket moving smoothly with constant speed ... but it has a *curved* surface in the *spinning* bucket, as the water, trying to move at constant speed along what inertial observers would say were straight lines, ends up riding up the sides of the bucket wall. This (said Newton) showed that a body's rotation was a real, physical, "absolute" property. The curvature of the water's surface was there for *all* observers, and showed unambiguously that the water really *was* rotating.

Figure 4-2: Swirling water in a bucket climbs the walls.

Newton's contemporary **Bishop Berkeley** objected to this characterisation of noninertial motion as "absolute" on the grounds that (according to Berkeley) *nothing* ought to be truly absolute and fundamental apart from God. Stripping away the religious part of the argument, Berkeley's objection can be seen as a valid invocation of **William of Ockham**'s idea of minimalism, known as **Occam's Razor** – that a theory should have as few independent parts as possible. The more "special cases" we need to include, the weaker the theory is.

4.5: Mach and relativity

In the late Nineteenth Century, **Ernst Mach** championed the idea that *all* motion, of *every* type, could be considered as being purely relative. Mach took the view that our view of reality should, as far as possible, be based on things that we can actually verify first-hand. The "absolute space" in Newton's Principia that supposedly defined a body's state of rotation *could not be seen*, and when we talked of a "rotating object" feeling geeforces, what we *meant* by this was an object that could be seen to be rotating relative to a persistent or reasonably-static distribution of material arranged around it (such as the Earth or the background stars). So all we were entitled to claim from our *direct experience* was that these rotational geeforces always seemed to be associated with a *relative* rotation between bodies.

In Newton's "bucket" example, parts of a rotating body of water push harder against the walls of the container because their *inertial* mass tells them to try to move along what would be straight lines in a non-rotating frame. But according to Mach, we should be able to say with equal validity that the bucket is stationary and the *outside universe* is rotating, in which case the outward attraction of the "stationary" water ends up being described as a *gravitational* side-effect of this shell of distant matter rotating around it. Instead of having two separate things, inertia and gravity, we then have just *one* thing, described in two different ways.

Mach's ideas didn't fit textbook Newtonian mechanics, but Mach countered that if we rotate *with* the bucket, the rotation of the universe around us and the associated gravitational field are both "real" to us, in the sense that they describe tangible experiences. We *see* the stars circling around us and we *feel* an outward force, and the two things are associated – this was our reality. This view was just as real (and just as legitimate) as that of an inertial observer, and if existing theory couldn't describe these relationships, then that was the theory's fault, not Nature's.

4.6: Practical advantages of "relativistic" arguments

From a purely practical point of view, the **principle of relativity** is powerful for four reasons:

1: It allows extreme minimalism in the design of a theory

The principle of relativity allows us to explain more effects using fewer different laws. For instance, instead of blaming rotational effects on an arbitrary distinction between inertial and noninertial motion, Mach's arguments let us blame them on the cumulative effects of matter that we already know is out there.

2: It places strict limits on the possible laws that a universe might obey

If we want the laws of physics to be the same for any observer, then many types of law simply won't work, and can be discarded.

3: It allows us to derive new laws from existing effects

Relativity theory lets us recycle existing relationships in order to derive new effects.

Suppose that we already have a good understanding of the workings of the solar system, derived from Newtonian laws, and that we know that these predictions are pretty accurate. Now, let's pretend for a moment that the Earth is really stationary, and that the Sun and outside universe rotate around it in a concentric set of crazy loop-the-loop orbits. In this (slightly wacky) description, the "loopy" orbits have to be balanced by some very odd gravitational effects associated with the moving background stars and Sun to predict the same final outcome. If the exercise is valid, we have derived a new set of gravitational effects for what *has* to happen when huge masses are hurled around in peculiar and unfamiliar ways, without our having to put together an experiment to make large masses actually *do* that. The ability of relativity to "recycle" known situations and turn them into "new" experiments is very cost-effective.

4: Sanity-checking

Relativity theory allows us to perform the same calculations in two or more different ways and check whether the answers agree – if they *don't* agree, then we know that we've screwed up somewhere. The ability to carry out these sorts of independent checks can be important when we're working in unfamiliar areas of research where solid supporting experimental data isn't available. A calculation may *seem* to be telling us that a particular outcome is correct, but if we don't have an independent way to check the answer, it isn't always obvious whether we've used a method correctly, or whether we've tried to apply it to a situation where it's not really valid. With the principle of relativity we can construct alternative descriptions of the same situation, and check whether the different descriptions can consistently predict the same agreed outcome. If they can, our confidence in that prediction becomes stronger.

4.7: Applying Occam's Razor

Occam's Razor frowns on "the multiplication of entities": in practice we use it to argue for a minimum number of *different types* of things, even if this increases their *quantity*. In astronomy, we find it convenient to name and classify stars as separate objects, but when we try to predict the behaviour of those stars, we assume that all stars are examples of a more generic sort of object, and that they all follow the same underlying sets of rules and laws. We assume that stars are assembled from atoms just like "normal" objects, and even though this dramatically increases the number of atoms in the universe, it also makes the universe a conceptually simpler place. We assume that the atoms in any star obey the same general laws as the atoms in our Sun, and the same laws as the atoms in our laboratories.

4.8: Different "Principles of Relativity"

The phrase "*the* principle of relativity" can be slightly misleading, as there are a few different versions of "the principle" in circulation, not all of which are completely interchangeable. In modern physics we usually deal with the **"special "** or **"restricted" principle ("SPoR")** that tackles the "special case" of simple inertial motion according to the conventions of Einstein's special theory, and the **general principle (GPoR)**, which applies the idea to all motion. We also seem to have ended up with two slightly different versions of the general principle.

"Newtonian relativity"   **(as used by Newtonian mechanics):**
" The same laws of physics apply for all "inertial", unaccelerated bodies moving in straight lines at constant speed. The same laws can be applied using any of these bodies as our reference, and the physical results will be precisely the same. "

"Special" (or "Restricted") Principle of Relativity   **(as used by SR):**
" The same laws of physics apply in all inertial **reference frames***, describing motion in straight lines at constant speed. These laws include those concerning the behaviour of light, whose speed is assumed to be* **globally constant** *for all valid inertial observers. "*

General Principle of Relativity (Mach):
" A single set of physical laws applies for all bodies and observers alike, regardless of how they move. "

General Principle of Relativity (Einstein):
" A single set of physical laws applies within all **reference frames***, regardless of whether these frames are nominally inertial or noninertial. "*

comparisons

Newtonian Mechanics was "relativistic" in broadly the same sense as Einstein's later "special" theory, but after the recognition of Newton's "optical" mistakes (in section 5), it became increasingly common to think of Newtonian relativity as only applying to matter, and as not having so much to say about lightspeed issues. Newtonian **"Ballistic Emission Theory"** *did* apply the principle of relativity to light (by treating it as little particles obeying Newtonian rules), but wasn't obviously compatible with a light-metric or flat spacetime.

Special relativity made some very specific assumptions about light. It's often presented as a high-velocity "extension" to NM, that successfully applies the PoR to the propagation of light and to interactions between bodies moving at a significant fraction of lightspeed. It does this by creating an idealised mathematical world in which the motion of bodies has *no* effect on light, and in which each observer's local sense of the speed of light is extended and extrapolated outwards to create a global **reference frame** that can be used to label the locations and times of distant events in the region for that observer, according to the belief that the speed of light is wholly, globally constant in that frame. We can then "abstract away" the observers, and use just the properties of these derived reference frames to isolate a single set of relationships that can work within them. The resulting coordinate-mesh then requires **Lorentz** distortions and rotations to make it compatible for different observers.

Special relativity's notion of global lightspeed constancy isn't really true, but it lets the "special" theory produce a unique and unambiguous mathematical solution, quickly. Some people find SR counterintuitive, and there are some problems integrating SR with a *fully* relativistic theory, so we'll leave further discussion of this theory for later.

Special relativity's approach of defining a principle using "frames" rather than "bodies" lives on in **Einstein's general theory** ("**GR1915**"), where the relationships between mathematical *frames of reference* are used to derive the way that objects are expected to interact. GR1915 assumes that a "frame-based" geometry must reduce to the geometry of special relativity over small regions, so again, we're dealing with idealisations that don't quite correspond to the known physics of real objects, but which *do* seem to make the problem easier to solve.

Current textbook discussions of "the general principle" tend to fall in line with the way that GR1915 *implements* relativity theory, by including frames as part of the definition and treating a reduction to SR as essential. In this book we'll be using the "broader" version of the general principle, and won't be imposing these additional restrictions.

4.9: Causes of confusion

" the theory of relativity says ... "

Of course, the principle of relativity can be used to make different predictions, and can "say" different things under different theories, depending on the other assumptions and approaches we've decided to use with it.

There are some quite authoritative statements in print that "the principle of relativity says ... [*insert theory-specific prediction here*]". These statements can easily be mistaken for general *definitions* of the principle of relativity, when they're more properly considered as descriptions of what the principle *can be used to predict* in a given context. The difference between *what something can say* and *what it is* can be confusing: it might be reasonable for me to state that the law of gravity says that my mug of coffee will fall onto the carpet if I drop it, but this isn't the same as *defining* the law of gravity according to a particular configuration of carpets and coffee (if I nail my carpet to the ceiling, the law of gravity doesn't say that my coffee will fall upwards). Relativity theory is especially prone to this sort of definitional quicksand because the subject's nature is so slippery and iterative – it makes so many effects mutually interdependent that we can sometimes find ourselves longing for the old-fashioned linear logics that started with "absolute" first principles and work upward, and when we see something in a book that *looks* like an absolute, certain "*fact*", it's tempting to pounce on it and cling onto it tightly. Unfortunately, the subject has a habit of destroying clear definitions and distinctions.

The word "the" can also be troublesome in physics. When we talk of "*the* principle of relativity", or "*the* theory of relativity", it's not always obvious from context whether we are talking about the most *general* form of the principle, it most *popular* implementation, or a broad subject that could include many possible theories and variations. When a source refers to "relativity theory", it might refer to "*a*" specific theory, or it might refer to "theory in general", and disputes about who gets to define and police the usage of these words can lead to some long-winded and ultimately fruitless arguments. The principle of relativity is such an open-ended idea that perhaps to define it too strictly is to damage it.

"Mach's Principle says ... "

Similar confusions exist over what "**Mach's Principle**" (capital "M", capital "P") "is" and what it "says", partly because the phrase seems to have been originally coined by Einstein rather than Mach. MP is associated with certain ideas ("inertia is gravity", "gravity *there* defines inertia *here*"), but "the principle" itself doesn't seem to be officially defined anywhere. Perhaps Mach's Principle should just be the general principle of relativity, in its most general form, but the idea of a "general" principle that might *not* be supported by Einstein's general theory is politically sensitive, and different people have arrived at different conclusions as to what "the principle" is (or what they think it ought to be). These arguments are notoriously difficult to resolve.

4.10: Relativity of acceleration

One of the easiest ways of distinguishing between "inertial" and "non-inertial" motion is that in the first case an object will "coast", while in the second, some sort of force normally needs to be applied to the object to compel it to move in a particular way, and the object itself usually feels squashed or stretched as this force is transmitted through its structure.

If you're in a train that's moving at constant speed along a straight track, then it can sometimes be difficult to tell whether the train is moving without looking out of the window, and even then, you could try to claim that the train is still "really" stationary and that it's the outside scenery that's moving, sweeping past your window-seat. But if the train suddenly *accelerates*, and you're thrown back into your chair, then this acceleration feels like a real, physical force, and these **geeforces** will suggest that the train has just changed its state of motion.

We may think that there is no way to explain this "lurching" without admitting that the train really *is* physically changing speed, but if we are consistent we can produce an alternative explanation for the forces pushing us back against our seat and knocking over our coffee. The force *feels* like the effect of a gravitational field (although we don't usually think of gravity as something that can suddenly switch on and off), and from the point of view of a camera fixed to the train, it certainly *looks* as if objects that are rolling down the aisle of the accelerating train are being pulled backwards by a gravitational field.

If we look outside the train in an attempt to find an explanation for why this field might exist, we find that it is associated with the relative acceleration of the world outside our window, in the same direction as the field. Perhaps the sudden *relative* acceleration of all that background mass is creating a *real* gravitational field inside the train?

Einstein's chest

How seriously should we take this argument? Einstein's thought-experiment on the subject involved an observer in a sealed room or chest, suspended by a rope. We'll use a standard shipping container dangling from a loading-crane:

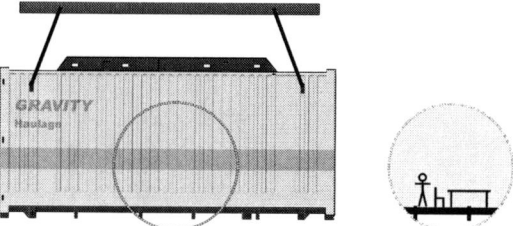

Figure 4-3: Suspended room

The inside of our container has been fitted out as a study, with furniture and modern conveniences. It also contains a volunteer experimenter, who has been locked inside the room while unconscious. If the suspended container is undisturbed, the observer (when they have woken up) will be able to walk around inside and measure the force holding the container's contents against the floor. This will be one standard Earth gravity, "**one gee**", equivalent to a natural acceleration of about ~9.81 meters per second, each second. Everything seems normal.

But now our experimenter has a nasty thought: The cables outside *might* indeed be suspending the airtight container near the Earth's surface, but everything seen inside the box would also be compatible with the container being in deep space, with the cables attached to a rocket ship smoothly accelerating at ~9.81 m/s^2 to *simulate* one Earth gravity. What if they have been the victim of a cruel hoax? How would they be able to tell the difference?

By jumping up and down, the observer can feel the vibrations of the suspension cables, but the tension in those cables would be the same regardless of whether they were pulling against the container's *gravitational* mass suspended in the Earth's gravitational field, or pulling against its *inertial* mass in deep space. A ball thrown at the far wall of the container *seems* to deflect downwards, but it's not clear whether this is due to the pull of gravity on the ball, or due to the container accelerating upwards to meet it. Walking, sitting and lying down inside the container would feel the same, and when the experimenter pours themself a cup of tea, the tea will pour the same and sit in the cup the same way, in either case.

What if they pressed a button on the room's remote control panel to disconnect the cables?

If they were in deep space, the room's "artificial gravity" would suddenly disappear, and its contents would then be floating about, weightless.

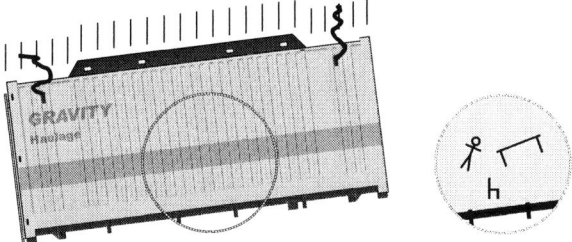

Figure 4-4: A freefalling observer doesn't feel gravity

But if the container was experiencing *real* Earth gravity, cutting the cables would produce the same effect: everything in the container would now be in freefall, and it would seem as if the Earth's gravity had ceased to exist (until the container hit the ground). Until then, ignoring the puny resistance of the air against the outside of the falling container, there would still be no obvious way to tell whether the original gravitational field had been "real" or "simulated".

4.11: Relative acceleration vs. absolute acceleration

For the sake of argument, we'll say that our experimenter finally plucks up enough courage to drill a hole in the side of the container wall. They then quickly clap a small sheet of glass over the hole to seal it and make a window to let them peek outside at their surroundings. It is just as well that they did this, because it turns out that the room actually *is* in deep space. When they put their eye to their new window they see blackness dotted with stars, and a rocketship pulling them along. Surely this means that their "gravitational field" is a fake after all?

According to the general principle of relativity things aren't so straightforward. It's perfectly possible for the experimenter to claim that the space outside appears to be permeated by a *genuine* gravitational field. After all, they can see freefalling dust and planets and stars around them, all accelerating "downwards" in the same direction as if the entire outside universe was "falling" in the direction of the alleged gravitational field, and it *looks* as if the only reason why their room *isn't* falling in the field is because it is supported by the rocketship above.

This argument sounds slightly insane, but it is consistent. According to Machian ideas, if accelerational effects only exist because of *relative* accelerations, then if the entire universe and its contents were nominally accelerating *together*, there would be nothing external for it to have to push against, no reaction forces and no required energy expenditure. The supposed acceleration would have no consequences. We could *claim* that the entire universe is jumping from side to side in any way that we want, and as long as *everything* is said to be jumping in the same way, it should make absolutely no difference to the internal physics. We can also suggest that perhaps the only reason that the experimenter was able to sense the effects of the field was because they were resisting it. To physically produce this *observation* did require some energy, but the amount of energy needed to accelerate the ship against the outside universe, and the amount needed to accelerate the outside universe against the ship (according to Mach), should be exactly the same. Neither the rocketship or the outside universe are anchored to anything else other than each other.

Aha, says the sceptic, But this *can't* be true! If the rocketship's engines were really able to accelerate the *entire outside universe*, where is the timelag? How can engines "here" move stars and galaxies billions of lightyears away, almost instantaneously? How can we see the entire outside universe moving *as a block*, as if the forces were being transmitted to all parts of it instantly? What happened to the idea that transmission is limited to the speed of light?

Well, here we have to admit some new concepts. While introductions to general relativity may say that we can talk with equal validity of one object or another being "really" accelerated or moved, it's closer to the truth to say that neither object is "really" being accelerated or moved in the old "absolute" sense: If we were to "really" move the distant stars, then this *might* require faster-than light transmission of forces, but producing *purely relative* motion is a different thing.

To visualise this, divide the universe into two regions, the region containing the spaceship and the region containing Everything Else. Now, in order to produce relative motion between these two *regions*, we *don't* have to physically affect anything *within* a region, or out at the distant stars: we only have to produce a physical effect at the *boundary* between the two regions, to allow them to move past each other. Suppose that we fire up our rocket engine, accelerate for a while, and then cut the engines and coast. When the ship starts accelerating it produces a **gravitational wave** around the ship that distorts lightbeams, and this distortion spreads out into the region surrounding the ship. When the spaceship stops accelerating, no more acceleration distortion is generated, and the region within this spreading distortion-wave can return to a more normal state. But the consequences of the *relative acceleration* of the two regions still has a physical existence in the spreading, distorted, transitional boundary layer.

physical consequences

It might seem as if all we've done here is to take a simple effect that we already understood very well and to turn it into a more unfamiliar and unnecessarily-complicated form, but along the way we've come up with some new physical predictions:

In our exercise, the relative acceleration of a shell of matter surrounding a test object produced a "real" gravitational field acting in the direction of acceleration. But since the principle of relativity requires that these effects ought to be *mutual*, we can now go on to predict that not only should the relative acceleration of the background produce a gravitational field in the *object*, the relative acceleration of the *object* should also produce a gravitational field in the *background*. The physically-accelerating spaceship ship tugs at us as it passes by. Our re-examination of standard accelerational effects using the general principle of relativity gives us a piece of "new physics" that we didn't have before: accelerate a mass – any mass – and we create a distortion in spacetime.

When we compel an object to move in a direction that it doesn't like, our actions produce a *genuine* gravitational field around the forcibly-moved body, as the protesting object desperately tries to cling to the surrounding region of spacetime and ends up dragging some of it along. The forcible acceleration of a mass should warp the surrounding region – when we "shove" a coffeemug and feel its inertial resistance to acceleration pressing back against our hand, we're deforming spacetime. What we're feeling when the mug shows an inertial resistance, creating pressure against our hand, is spacetime acting *through* the mug, trying to avoid the creation of this additional distortion effect.

4.12: Relativity of rotation

We can use similar arguments to claim that there's really no such thing as "absolute" rotation.

As inhabitants of the Earth's surface, we see the Sun rising in the East and setting in the West, and if we were to look down on the Earth from above its North Pole, we should see the planet rotating anticlockwise relative to the background stars.

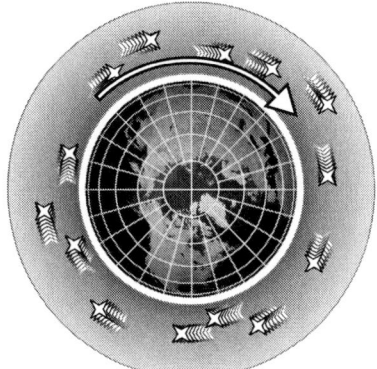 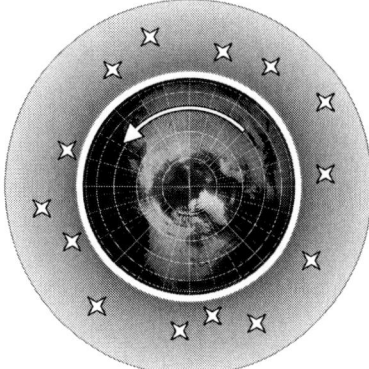

**Figure 4-5: Stars circle relative to stationary Earth
... or Earth rotates relative to background stars**

Does the Earth "really" rotate anticlockwise relative to the background stars, or do the background stars "really" circle the Earth clockwise? According to the general principle of relativity, it shouldn't matter. For Earth-dwellers it's easier to assume that the Earth is fixed, while for planetary calculations it's easier to assume that the Earth rotates, but ... according to the GPoR, as long as we model the situation in sufficient detail, either explanation should be equally valid and should work equally well. What's *really* important here is the *relative* rotation between the two things.

4.13: "Centrifugal" and "Coriolis" fields

How can these two descriptions give us the same physics?

The first objection that springs to mind is that we can normally tell if we are on a body that is *really* rotating because we feel ourselves thrown outward by rotational effects, away from the rotation axis. We could argue that the parts of a rotating body should have a natural tendency to try to move inertially, in a straight line, and that for particles near the rim of a rotating body, this means that they should be trying to fly off at a tangent to the body, with only the object's internal tensions (and/or its gravitational field) holding everything together.

In the case of the Earth, these outward forces give the Earth's surface a noticeable outward bulge at the equator. Isn't this evidence of an "absolute" rotation?

But according to Mach's principle, we can choose to say that the Earth *isn't* rotating, and that the outward pull must therefore be something to do with the observed rotation of the outside universe around it. If we treat the universe's background matter as a hollow shell of material surrounding the Earth, we notice that when this external shell rotates relative to the Earth (and to us), there seems to exist an outward gravitational field pointing away from the rotation axis: In the first description we can say that Earth's equatorial bulge was caused by the *inertial* mass of its material, in the second version, we say that the bulge is due to the material's *gravitational* mass responding to this outward gravitational pull caused by the rotating shell.

RELATIVITY IN CURVED SPACETIME

Another way to visualise this field-effect is to imagine a ball bouncing around the walls of a perfectly circular room, as seen from above. Experience tells us to expect to see the ball moving in a (fairly) straight line between bounces.

Figure 4-6: "Rotating room" / centrifuge

But if the room rotates *with* the ball, so that the ball hits the same point on the wall *every time*, then a video camera fixed to the rotating wall will record images of a ball apparently bouncing up and down on the spot, under the influence of a gravitational field.

back-reaction

What other conclusions can we draw? Well, if the outside universe is pulling preferentially on the Earth's equatorial plane, then perhaps the Earth also ought to be pulling preferentially on the outside universe's rotation plane, too ... so does the Earth's equator attract outside matter more strongly than its poles? A number of arguments say that it should:

Topology: If we said that the universe was **spatially closed**, and that it "wrapped around" on itself, then we could treat the Earth and the outside universe as if they were two spheres with parallel faces. Topologically, it shouldn't matter too much which of these two spheres we say is "on the outside facing in", and which is "on the inside facing out". If the rotation of the Universe-sphere produces an additional attraction to *its* equator, then so should the rotation of the Earth-sphere.

Kinetic energy: in a non-Machian description we could say that the rotating Earth carries a certain amount of rotational kinetic energy, and since the equator moves fastest, it has more energy. We've already said that energy is associated with gravitation, so perhaps a rotating body's equator should have an enhanced gravitational pull (although in a *fully* relativistic description, it isn't the isolated rotation of a body that contains this energy but the relative rotation of two parts of a larger system).

Fieldlines: We can also say that, in general, a concentration of **field lines** between two bodies is associated with an increased gravitational attraction between them. When one body rotates relative to another, the finite speed of signal transmission means that fieldlines connecting the two objects should show appear slightly "wound up", and the region where this effect is strongest is around the equator.

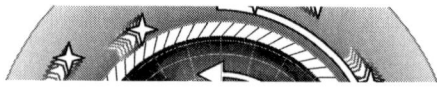

Figure 4-7: Twisted fieldlines

The "fieldline" argument also suggests another effect: Since the *apparent* concentration of these twisted fieldlines depends on the direction that we are looking in, perhaps the *apparent* gravitational attraction of a piece of equator depends on whether we are watching it rotate away from us or towards us. This would suggest that a body rotating with respect to its environment should have a tendency to drag objects and light around with it.

4.14: Rotational dragging

The general principle of relativity agrees with our previous fieldline argument, and says, yes, light (and matter) *should* be pulled around by a rotating body.

Let's place a laser-pointer at the Earth's equator, aimed straight upwards, and switch it on. As a first approximation, we'd probably expect the lightbeam to be travelling away from the Earth at a fixed speed and in a straight line as far as outside inertial observers are concerned. After twenty-four hours, the front of the beam should have travelled 86,400 lightseconds, over four times the radius of Pluto's orbit, and should be out of what we'd normally think of as the main body of our solar system. During this time, the Earth will have completed one full revolution.

If we're drifting in space somewhere above the Earth's North Pole, the left-hand part of the following sketch should be a pretty good approximation of what we expect to see. The light travels (up the page) in pretty much a straight line, and the Earth rotates beneath it.

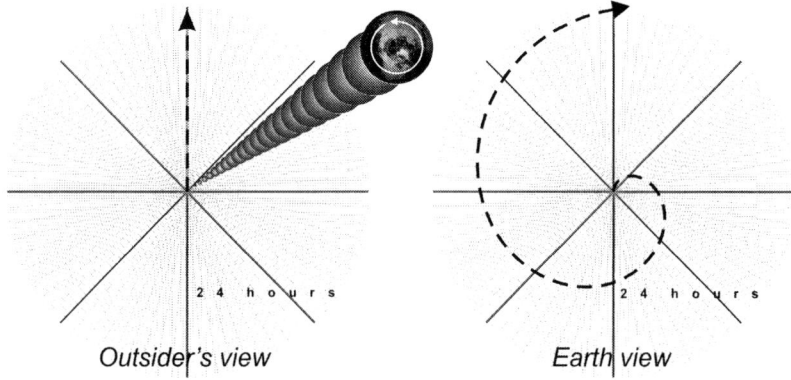

Figure 4-8: Signal progress over 24 hours as seen by observers nearby, and on the rotating Earth

But for an Earth observer standing next to the laser source and looking up, things look a bit weird. If the light is *really* travelling in a straightish line according to proper inertial observers, then for *this* observer, the beam should seem to be veering Westwards and performing one complete circuit around the Earth.

In Earthly coordinates, the beam appears to deflect Westward and disappears out of sight, then swings into view again from the East to line up with the pointer again after 24 hours (or 48 hours if we take into account the time that signals take to get back to the Earth). It seems to be marking out a giant expanding spiral in space, circling the Earth in lock-step with the similarly-rotating starfield.

We may be tempted to say that because light is supposed to travel in straight lines, and because we are seeing it doing something very different, this seems like good evidence that the Earth is "really" rotating, and that our Earth-observer's account ought to be set aside in favour of the description given by the first observer. Since the rotating observer doesn't see light travelling in straight lines, their data isn't any good.

But the general principle of relativity insists that the Earth-observations are valid. Since a curved lightpath is normally evidence of a gravitational field, we can say that the Earth observer sees the rotating shell of background stars generating a "sideways" gravitational field that pulls the lightbeam around with it – the relative rotation of the Earth and the background stars creates a gravitational twist in spacetime between them that deflects light and matter.

RELATIVITY IN CURVED SPACETIME

rotational dragging by a single object

Once again, by deliberately describing some familiar behaviour from an unfamiliar viewpoint and applying the principle of relativity, we end up with a new effect. Having established the idea that a rotating *shell* drags light and matter around with it, the fact that relativistic effects need to be *mutual* means that as well as the revolving starfield warping spacetime as seen from the *Earth*, the rotating *Earth* should also be warping spacetime as seen by a background observer stationary with respect to the starfield. If we're floating in space somewhere above the Earth's equator, the Earth's rotation should be trying to pull us around with it, at least a little bit.

The rotational dragging influence of the Earth is small and difficult to measure, but this is part of what makes the principle of relativity so powerful: it lets us take old, "obvious", "trivial" observations (such as the tendency of light to move in apparently straight lines as seen by inertial observers), apply a relativistic "twist", and use them to predict new, *interesting* effects that aren't nearly so obvious, and which and were missed by earlier generations of physicists. And this process often doesn't even need any mathematics – we've just successfully used the principle to work out that a spinning body should create a rotational gravitational drag around it ... a sort of rotational gravitational field ... without having to make a single calculation.

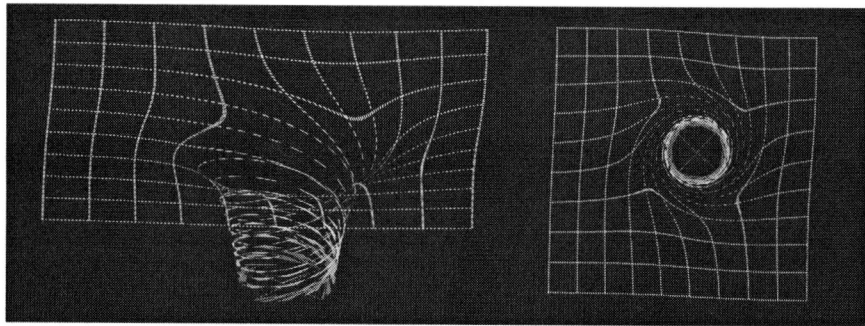

Figure 4-9: Distortion due to a gravity-well's rotation

Admittedly, to calculate the exact *strength* of the twisting of lightbeam geometry around a rotating body will require a bit more theoretical know-how, but to predict that the effect should *exist* doesn't need fancy hardware, geometry or math. It just needs the GPoR, an open mind, and a little basic reasoning.

Figure 4-10: Gravitational waves given off by a pair of orbiting stars

4.15: Experimental verification

Foucault's Pendulum

One of the more famous experiments in gravitational physics was carried out by **Jean Bernard Leon Foucault** in 1851. Foucault reasoned that although we don't normally *notice* any local side-effects due to the Earth's rotation, we could use the tendency of a perfectly-swung pendulum to keep swinging in the same plane to demonstrate that these effects really existed.

If we allowed a pendulum to swing freely at the Earth's North Pole, then if the fixed stars define our sense of rotation, the pendulum's swing should stay in alignment with the starfield, with the Earth rotating beneath it. An Earth observer should see the pendulum's "swing plane" slowly rotating to stay aligned with the background stars, making a complete 360-degree rotation every 24 hours. Foucault set up his experiment in Paris, where things where a little more complicated because of the angle of the Earth's surface, so his pendulum showed one complete cycle every ~31¾ hours. Because it's such a nice simple experiment, science museums all over the world now display their own working copies of Foucault's pendulum.

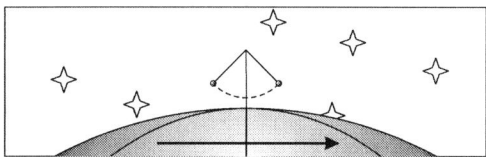

Figure 4-11: Foucault pendulum

Gravity Probe B

In order to show that rotational dragging effects are *mutual* we need to show that a Foucault-type experiment doesn't track the starfield *perfectly*, and that the Earth rotating beneath the starfield also has some influence on it. The experiment designed to show this, **Gravity Probe B**, was launched by NASA into a polar orbit in 2004, with its results expected in 2007.

GP-B contains a set of four exquisitely-engineered spinning quartz spheres that act as gyroscopes, and whose natural states of rotation are monitored over long periods of time. GP-B's results, when they're published, are expected to show that the natural state of rotation at the satellite's position doesn't track the Earth's rotation rate *or* the starfield's, but is somewhere in between, dragged around by both the Earth and the stars.

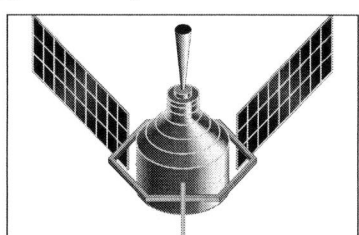

Figure 4-12: Gravity Probe B's satellite

The gyroscopes will identify what *seems to them* to be a particular "absolute" state of non-rotation, but their identification of that special non-rotating frame will only be valid **locally**, representing a local compromise between the influences of the Earth's rotating gravitational field "below" it and the outside universe's rotating gravitational field "above" it.

According to the GPoR, when we find a state of motion to be "rotating" or "nonrotating", these aren't *global* or *universal* states, but only *local averages* of the competing influences of background material with different states of motion. **John Wheeler** has usefully referred to this as the **democratic principle** – all surrounding matter gets to "vote" on what happens in any given region.

4.16: Equivalence principles

According to Mach's way of looking at things, inertia isn't an intrinsic property of a body, but exists as an *interaction* or *coupling* between the body and its surrounding universe. If we could totally shield a baseball from the outside world, it would have zero inertia. Alternatively, if we were to increase an objects' coupling with outside matter, by placing it into a region where the local mass-density was higher, then we'd expect its inertia to increase. Einstein referred to this inertia-increase effect in his 1921 Princeton lectures as being the third major consequence of Mach's principle that was supported by GR1915: it's our old friend "gravitational time dilation" (section 1.5) dressed up in fancy field theory language.

This gives us a good general rule: Objects in denser regions age more slowly, and "slower" regions attract. "Slower" means "down", and "down" means "slower".

inertial fields

The inertial coupling between objects, described as a field effect, gives us the concept of the *inertial field*. Increase the local density of the inertial field and a passing test mass will experience an imbalance in its own local inertia that will steer it towards the region of highest field strength. We don't usually talk about "inertial fields", because we already have a perfectly good name for them: "gravitational fields". Inertia produces gravitation. Or gravitation produces inertia. It doesn't seem to matter which.

This is an example of how relativity theory sometimes lets us calculate the same results from radically different starting assumptions. If we were asteroid-belt-dwelling creatures living in a microgravity environment, we could take what we'd learnt about inertial physics by throwing rocks at each other, and by applying the GPoR we could *predict* the existence of gravitational effects, even if we'd previously had no idea that gravity existed. In our case, this prediction is rather spoilt by the fact that, as creatures living on a planetary surface, we're already quite familiar with conventional gravity. But there are other situations that we have less personal experience of where the GPoR can help us out. For instance: Can we build things like **warp drives** (section 19)?

Everyday experience would suggest that we can't. If a spaceship's "default" gravitational field is almost imperceptibly weak, we'd be forgiven for thinking that even a *very significant* distortion in this miniscule field could only have vanishingly weak consequences. We could think that *any* modification of the field would have to have effects that were even weaker than those of the object's conventional, default, gravitational effect.

But Mach's viewpoint and the most general principle of relativity (and the concept of the inertial field) allow us to think about these problems differently. In a Machian model, it isn't so much that the strength of gravitational fields is weak compared to inertia, it's more that the combined effects of the universe's background masses produce a balanced gravitational field pulling us in all directions (responsible for inertia) that is *already so strong* that the additional field contribution produced by our spaceship represents a very small *proportional* increase in the total field. We can sense the existence of the background field by deliberately moving along an accelerated path to spoil its state of equilibrium, and feeling geeforces as a result. But if our spaceship has a warp drive (section 19), the drive isn't just distorting the ship's own field, but also the background field that the ship inhabits. We aren't then distorting a field that is *insignificant* compared to the ship's inertia, we're distorting a field that *is* the ship's inertia.

5
The Newtonian Catastrophe

> *"Therefore, because of the analogy there is between the propagation of the rays of light and the motion of bodies, I thought it not amiss to add the following Propositions for optical uses; not at all considering the nature of the rays of light, or inquiring whether they are bodies or not; but only determining the curves of bodies which are extremely like rays."*

Isaac Newton, Principia vol.1, Prop XCVI, Scholium

> *"Qu. 21 ... I see no reason why the* ~~increase~~ **[decrease]** *of density should stop anywhere, and not rather be continued through all distances for the Sun to Saturn and beyond. And although this* ~~increase~~ **[decrease]** *in density be exceeding slow, yet if the elastick force of this Medium be exceeding great, it may suffice to impel Bodies from the* ~~denser~~ **[rarer]** *parts of the Medium towards the* ~~rarer~~ **[denser]**, *with all that power which we call Gravity ... "*

Isaac Newton, on gravity considered as a variation in the density of an optical medium [words in square brackets changed]
"Opticks", 1721 edition

> *"As stones, by falling upon Water put the Water into an undulatory Motion and all Bodies by percussion excite vibrations in the Air; so the Rays of Light, by impinging on any refracting or reflecting Surface, excite vibrations in the ... Medium... and, by exciting them agitate the ... Body; ... the vibrations thus excited are propagated in the ... Medium ... and move faster than the Rays so as to overtake them."*

Newton's anticipation of the concept of a pilot wave under QM
("Opticks", Query 17)

> *"And so if any one should suppose that* **Aether** *(like our Air) may contain Particles which endeavour to recede from one another (for I do not know what this* **Aether** *is) and that its Particles are exceeding smaller than thofe of Air, or even than those of Light: The exceeding smallness of its Particles may contribute to the greatness of the force by which those Particles may recede from one another, and thereby make that Medium exceedingly more rare and elastick than Air ..."*

Newton, "Opticks"

> *"Recapitulating, we may say that according to the general theory of relativity space is endowed with physical qualities; in this sense, therefore, there exists an ether.*
>
> *According to the general theory of relativity space without ether is unthinkable; for in such space there not only would be no propagation of light, but also no possibility of existence for standards of space and time (measuring-rods and clocks), nor therefore any space-time intervals in the physical sense. But this ether may not be thought of as endowed with the quality characteristic of ponderable media, as consisting of parts which may be tracked through time. The idea of motion may not be applied to it."*

Albert Einstein "Ether and the general theory of relativity", 1920

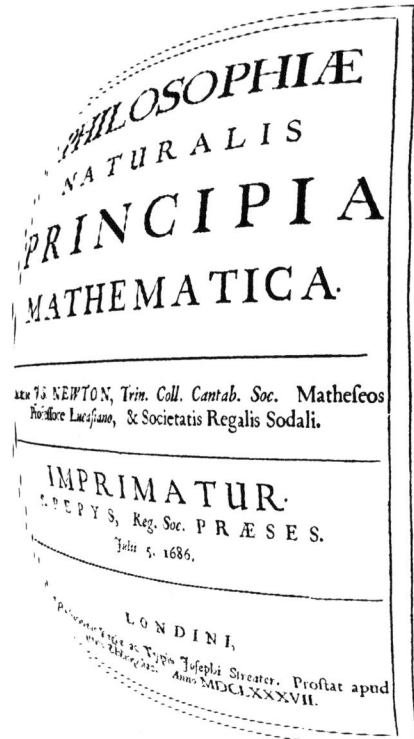

Title pages from Newton's two major works,
"Principia Mathematica" and "Opticks"

5.1: Newton's unification scheme:

gravity and optics

One of **Isaac Newton**'s more brilliant ideas was to try to unify gravity with optics by taking the behaviour of light as the common unifying factor. With *optics*, the amount of deflection of light at an air/glass surface seemed to depend on the relative densities of the two media, suggesting that the deflection might be due to a lightspeed differential, while with *gravity*, we had an attraction between bodies, which, if it also applied to light, might be expected to *produce* a lightspeed differential. This apparent attraction of both light and matter to denser regions could (reckoned Newton) be explained by a variation in the density of some underlying medium, producing variations in the speed of light.

Since Newton's system made gravitation the natural result of a variable-density medium, we can think of it as a forerunner to what would now be considered a **curved-space** model of gravity. Particles were either *attracted* by gravitational force, or their associated waves were *refracted* by the change in density of the medium. Waves and particles could either "fall" or be steered by lightspeed gradients, with both calculations producing suspiciously similar results.

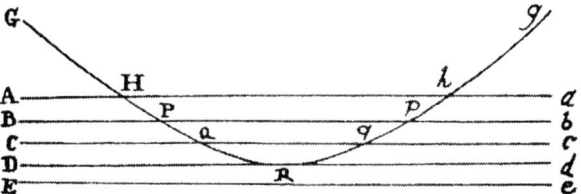

Figure 5-1: Newton's Principia: Path of a body associated with gradients in a gravitational field, and in refractive index (Prop XCVI)

waves and particles

Newton's approach suggested a form of **wave-particle duality**. Light might be *considered* as a stream of particles, but was also associated with a form of wave (but not a conventional compression wave, since this would make light tend to repel other light and disperse). In "**Opticks**", Newton suggested that any particles or **corpuscles** of light might be surrounded by waves that he likened to the waves surrounding a stone hitting the surface of a pond. The propagation and interaction characteristics of these waves ("early fits") were then supposed to dictate where the particle moved, in a manner reminiscent of the **pilot wave** idea that later appeared under modern quantum mechanics.

Newton tried to use this "wavelike" feature to explain the wavelengths associated with light and the phenomenon of **total internal reflection**. If light inside a denser medium (such as glass or water) hits a less-dense region at a sufficiently awkward angle, it'll be reflected back inwards, as if the surface between the two regions is mirrored – this is the effect that makes small air bubbles in water or glass look "silvery". "Total internal reflection" posed philosophical problems for the "particle" description of light: if a light-corpuscle gets deflected by a different speed of light at the other side of an glass-air boundary, then, if the light never actually *passes out* of the glass, how does it *know* that the speed outside the glass is any different? How can it be deflected by the properties of a region that it never actually enters? The behaviour was arguably even more awkward because, according to Newton's "gravitational" analogy, the transitional nature of the boundary should be "smudged" and smoothed, so that the light should start to deflect *before* it reached the surface.

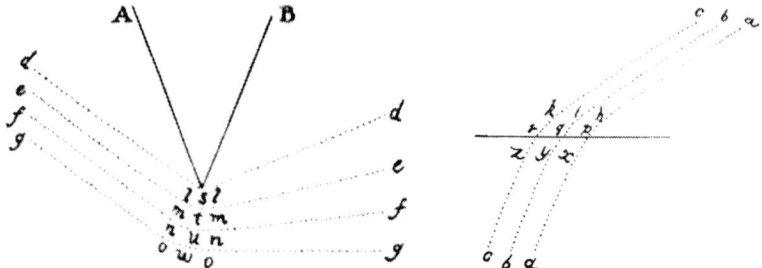

Figure 5-2: Newton's Principia: "The refraction ... is made not in the point of incidence, but gradually, by a continual inflection of the rays; which is done partly in the air before they touch the glass, partly ... within the glass, after they have entered it." (Prop XCVI)

For Newton, the wavelike aspect of light was the answer: a light-corpuscle's associated wave (moving ahead of the particle), could pass through the surface and the resulting interference pattern could then steer the particle away from the surface before it actually reached it: the wave allowed the particle to tentatively "feel" what was on the other side of the glass, like a bather dipping their toe into a bath to check the temperature of the water before jumping in.

Newton's approach was desperately ambitious: without knowing some of the critical relationships, he was attempting to produce a unified, geometrically-compatible description of the interplay between light, gravity and particulate matter. His approach was arguably centuries ahead of its time: even in the Twentieth Century we didn't manage to produce a theory that successfully combined particulate behaviour (quantum mechanics) with gravitation (general relativity). But Newton's advanced model shared an unfortunate characteristic in common with many great inspirational works. It was badly wrong.

5.2: The lightspeed mistake ...

Newton took inspiration from the idea that what a gravitational field did to the paths of particles ... changing their speed and deflecting their paths ... looked rather like what happens to light as it moves between different media. If we throw a ball in the air, it slows as it fights against gravity, and then speeds up again on its way down. If light and matter obeyed a single set of rules for the energy and momentum, this meant that light ought to lose and gain energy as it passed uphill and downhill, too, and this energy-increase (said Newton) related to a change in speed. Thus: the natural speed of light ought to be faster in a stronger gravitational field, and by analogy, the speed of light ought to be faster in glass than in air.

Both of these relationships were *exactly* the wrong way round.

5.3: The "space-density" mistake

Although Newton was just trying to work some reasonable-seeming ideas though to their logical conclusions, the inverted relationships, applied consistently, wrecked his model. In **"Opticks"**, Newton went on to convert these speed-variations into the equivalent of a curved geometrical space: he described a new light-medium whose density was affected by the presence of matter, and whose density-variations generated the effects that we know as gravity. This *should* have given him an efficient gravitational medium similar to that of general relativity, that doesn't exist *in* space, but which *is* space ("the medium is the metric") but the inverted wavelength/energy relationships forced Newton to assume a more complex system in which his new medium had to be *displaced* by objects and their fields.

5.4: The light-energy mistake

Newton's next natural and logical mistake was to say (quite reasonably) that if "fast" light had higher energy, then its increase in speed should be associated with a "stretching" of the waveform and an increase in its wavelength: Newton reasoned that longer-wavelength light should therefore be more energetic, whereas we now understand that forcing an electromagnetic disturbance into a smaller region creates a stronger distortion that requires *more* energy, not less – *compacting* a waveform represents an energy-increase.

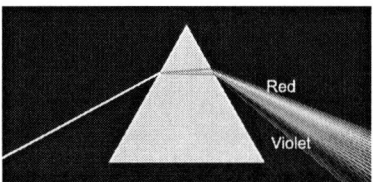

Colour-dependent speed in glass: Red light is slowed less by glass than blue light, and its refraction by a prism is correspondingly weaker. We can suggest that perhaps the shorter-wavelength blue light interacts more strongly with fine detail within the glass, and therefore takes longer to cross the distance: Red light is affected less because it interacts with this detail more weakly and "sees" less distance. Newton reasoned instead that red light was deflected less than blue because it was more powerful.

These wavelength-dependencies are a feature of *particulate* media, and aren't supposed to happen with geometrically "smooth" gravitational fields and features that don't have this additional fine detail.

Figure 5-3: Interactions between light and matter

The principle that Newton was missing was the "wavelike" idea of local lightspeed constancy. But to Newton, the behaviour of prisms, which suggested that lightsignals with different colours and energies could move along the same path at *different* speeds – might have badly undermined the idea.

Newton thought that he was quite properly following the experimental evidence that was available. His experiments with coloured fringes seen around narrow airgaps in glass told him that longer-wavelength light found it easier to pass an obstacle, and while we might argue that this is because higher-energy, shorter wavelengths find it easier to interact with (and be blocked by) smaller features that the longer wavelengths can "wash" around, Newton concluded (wrongly) that red light was able to navigate an obstacle-course more easily because it was more powerful.

5.5: Loss of wave-particle duality

Newton's mistake was understandable: in glass, the splitting of white light into different colours by a prism suggested that the different colours moved through the glass at different rates. Local lightspeed constancy was a nice idea, but the laboratory evidence available at the time didn't seem to support it.

Without local lightspeed constancy it was difficult to visualise how the "wave" and "particle" descriptions could be interchangeable, and Newton's followers tended to avoid the issue by deciding that light was *entirely* particulate. This led to a long-running and damaging feud between proponents of the "light-corpuscle" idea in England and fans of "light-wave" theory in continental Europe, which we can see traces of in (Priestley, 1772).

RELATIVITY IN CURVED SPACETIME

5.6: Newton vs. Huyghens

Newton's "lightspeed" mistake, and the subsequent split between the "particleists" and the "waveists" led the community to decide that "wave" and "particle" explanations generated two distinct and irreconcilable predictions for the behaviour of light.

> **According to Newton**, light-particles encountering the surface of a glass object would be steered towards the glass by a *higher* speed of light inside the body: they would naturally tend to be drawn into a region where their natural speed was higher.
>
> **According to Huyghens**, a light-wave encountering the surface of a glass object would be steered towards it by a *lower* speed of light inside the glass. The signal edge that met the glass first would progress into the glass more slowly, causing the wavefront to tilt towards the region of lowest speed. In this model, light finds it easier to *leave* regions where the speed is higher, and "dawdles" in regions where it is slower.

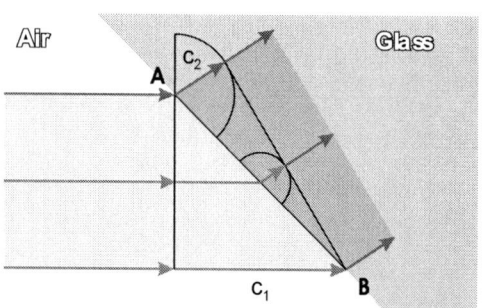

A plane wave of light moving through air (pale grey) moves towards the right of the page until it hits an angled block of glass (darker grey). The upper part of the wave hits the surface at point **A**, and causes an electromagnetic disturbance in the glass that spreads through the material. If the speed of light in air is c_1, and the speed of light in glass has a smaller value c_2, then by the time that the lower part of the wavefront hits the surface at **B**, the maximum penetration from **A** is going to be smaller, by the ratio $c_2{:}c_1$. By drawing a line from **B** to a *tangent* of this maximum penetration distance, we obtain all the intermediate penetration distances, giving us the position of the new wavefront moving through the glass at the slower speed c_2.

The wavefront is steered towards the region where lightspeed is slower: in this case, the speed of light is *on average* slower at the top of the diagram than at the bottom, and the light is therefore deflected towards the top of the page.

Figure 5-4: Huyghens' Principle: Angle-change and refractive index

We can compare Huyghens' effect to what we would expect to happen if we were driving a truck and applied the brakes to the left-side wheels only – since the right-side wheels would now be moving faster, they'd tend to overtake those on the left, steering the truck leftwards.

In the case of our gravity-well diagrams (sections 1.8 and 3.2), we said that the *local* speed of light was notionally the same at every point on our lightspeed-normalised diagram, but that light took longer to cross a stronger-gravity region because that region had more space packed into it. The deflection principle is basically the same in both cases.

5.7: The lightspeed trap

Newton's mistake was reasonable – after all, we *know* that a dropped ball is moving faster the deeper it falls into a pit ... so how can this change into a description where speeds are naturally *slower* deeper down?

one-way "velocity" versus round-trip "speed"

Part of the answer is to remind ourselves that there's a difference between "speed" and "velocity". "**Speed**" is usually frowned on in physics as a vague and muddy word, and the more precise term "**velocity**" is much preferred, referring to the rate of motion *in a specific direction*. In everyday language we might say that a falling ball increases its speed, but in physics we'll usually try to be more specific: we might say that the ball's *downward velocity increases*, but its *upward velocity decreases*. When gravitational effects are in play, we no longer have a single, natural, guaranteed *velocity* of light at any given location, but a spray of different possible velocity values, depending on the direction of the signal.

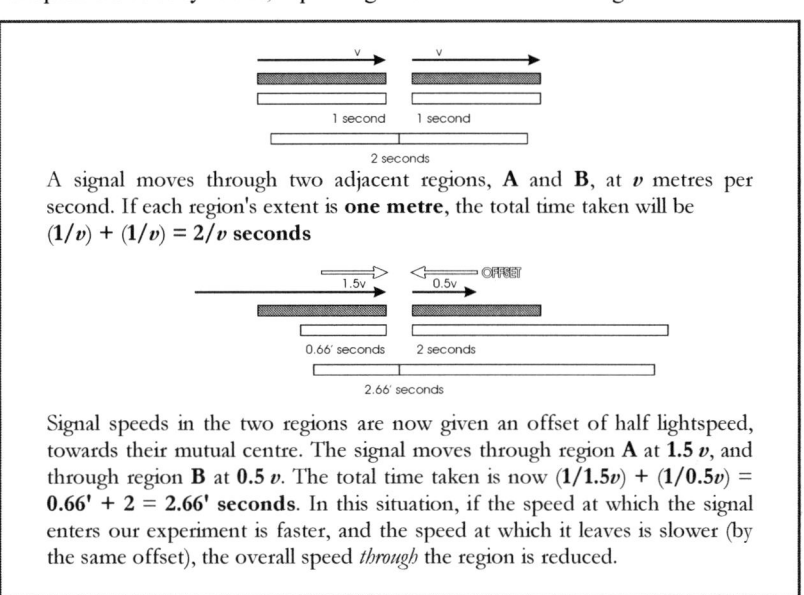

A signal moves through two adjacent regions, **A** and **B**, at v metres per second. If each region's extent is **one metre**, the total time taken will be $(1/v) + (1/v) = 2/v$ seconds

Signal speeds in the two regions are now given an offset of half lightspeed, towards their mutual centre. The signal moves through region **A** at **1.5** v, and through region **B** at **0.5** v. The total time taken is now $(1/1.5v) + (1/0.5v) = 0.66' + 2 = 2.66'$ **seconds**. In this situation, if the speed at which the signal enters our experiment is faster, and the speed at which it leaves is slower (by the same offset), the overall speed *through* the region is reduced.

Figure 5-5: Lightspeed offsets reduce round-trip speeds

If we imagine Newton's light-corpuscles being pulled toward a gravity-source, the time needed for the signal to pass *into* the gravitational field will be shorter when the field is made more intense, but the time taken for a subsequent signal to climb *out* of the field – so that we can be aware of what has happened – will be longer. As outsiders, we'll see the signal's apparent progress to be slowed by stronger gravity. We might argue that there should be a natural local *round-trip speed* of light at any given location (which a local observer should measure as having the standard value of "c"), but the observer can't be in two places at once to measure an underlying one-way *velocity* for light: only the round-trip properties are easily accessible to direct measurement. If the internal equilibrium of an observer's atoms involves their interaction with the background field, then this *interaction* suggests a two-way process, and again suggests that light's round-trip *speed* may be the relevant property.

So, although Newton's arguments seemed straightforward, it was actually very difficult to apply correctly. Even though we might *deduce* that light falls faster downhill into a denser gravitational field, we can't see or verify that increased rate of fall except by using a second signal, whose speed might then be affected differently by the field, because of its different direction. We find ourselves confronted with a split between what we *visualise* as happening, and what we actually *see* or *sense* happening.

This "split" can hurt our intuitive expectations for how a falling ball ought to "look" when very large velocities are involved. Dropping a ball into a **black hole**, we'll expect the ball to keep accelerating as it falls, until eventually it should be moving at roughly the background speed of light at it crosses the horizon. But when we stand far away from the black hole and *watch* the ball falling in, then once its recession velocity passes a certain value, we encounter the **speed paradox** – further increases in speed are associated with an apparent *reduction* in the ball's recession speed. As the ball approaches the horizon, and its recession speed approaches the background speed of light, our view of the ball appears to show it slowing towards a dead stop at the horizon coordinates, and never appearing to cross the horizon at all. Where round-trip quantities are concerned, we may *believe* "fast" but *perceive* "slow".

A great deal of physics, dealing with the way that things physically *interact* with each other, is about how the objects appear and how they're perceived by each other. In this case, things are actually even *worse*, because for a gravitational field we also have the further complication that occupants in different regions have different senses of time. For local observers deep in a gravity-well, does their slowed perception of time contribute to the apparent increase in the speed of the ball? And if so, how much? Who gets to decide what the ball's speed really is? These are horrible, horrible, *horrible* things to have to try to work out, and usually we don't even attempt it.

glass isn't simple

And, unfortunately for Newton's scheme, glass is complicated, too. Even if we were to decide (with Newton) that light's *velocity* passing *into* a block of glass should be higher, this wouldn't necessarily mean that the *speed* of light passing *through* the glass was faster. If the regions between glass molecules produced a sort of "switchback" field, with light propagating faster when traveling towards a molecule and slower when traveling away, then the overall, averaged propagation rate through the glass could be slower rather than faster (Figure 5-5). Newton assumed that the "map" of velocities passing through a glass block would be a flat plateau, but he didn't really have any first-hand experience to indicate that this was true. If the velocities of light through glass were irregular at small scales, and only produced "smooth" results over larger regions, then his arguments weren't valid.

avoiding velocity

This is all pretty mind-mangling stuff, and mainstream relativity usually finds it easier to spare students' sanity by concentrating on the simpler "round-trip" averaged definitions of speed, and avoiding nasty questions about what the underlying one-way velocities of light *might* be and the how space *might* behave at smaller scales, such as between the particles making up a piece of glass or a tank of water. Even today, we don't tend to use GR-type arguments with particulate media.

Einstein's **special theory of relativity** managed to avoided these nasty issues by rejecting the very *physicality* of a one-way velocity of light: According to Einstein *circa* 1905, if nothing is faster than light, we can't measure a one-way *velocity* for light in flat spacetime unless we can be in two places at once – which we can't. What we *can* do is to aim a lightbeam at a distant object, measure how long it takes to get there and back, and divide this by two to obtain a nominal observed distance value by *declaring* the velocity of light to be the same in both directions – this *round-trip speed* is a thing that *can* be measured directly, so Einstein's "special" theory used *measurable* round-trip light*speeds* as the basis of its coordinate systems.

Special relativity rejects the *objective physicality* of an underlying direction-dependent velocity of light – if a "proper" one-way light-velocity can't easily be isolated, special relativity tells its observers to decide, that these slippery one-way velocities are the same as the measurable round-trip *speed* of light. A signal is presumed *as an article of faith* to take the same amount of time to move from **A→B** as from **B→A**. By using this assumption to compensate for signal delays, SR's observers can assign distances and times to distant events base don the idea of c-constancy in their own frame. Observers in other frames will disagree as to what these coordinate labels ought to be, and by making these disagreements *symmetrical*, we get special relativity. This worked out well for special relativity, but it contributed to the theory's obvious incompatibility with gravitation, and meant that when we extended special relativity's coordinate-system-based approach to gravitational problems, we had to be very careful to avoid tripping ourselves up with the difference between *averaged or round-trip speeds*, which gave polite lightdistances that we could express in a tidy way with embedding diagrams, and *one-way velocities*, which weren't so conveniently expressed.

So, although it's easy to be dismissive of Newton's mistake, we might not have come much further with the problem than he did … perhaps we've spent more effort over the last couple of hundred years learning how to avoid these nasty issues than in confronting them.

5.8: Consequences for physics

It's difficult to overstate the damage done to the subsequent development of gravitational physics by the fallout from Newton's inverted relationships. We'll list some of the major ones:

Wave-particle duality – Newton's idea that waves and particles should be deflected by the same amount, and that light could be considered as having both "wavelike" and "particlelike" properties, was overtaken by arguments over whether light was "really" a wave or a particle, and **duality** didn't become a successful mainstream idea again until the arrival of quantum mechanics in the early Twentieth Century. "Opticks" was allowed to quietly go out of print until the 1930's when we noticed the correspondence between some of Newton's "wave" descriptions and QM's "pilot wave" description, prompting a new historical interest in the book's contents.

Shapiro effect – with the rejection of Newton's variable aether-density arguments, aether theorists weren't obliged to think of their medium as having a density affected by background gravitational field strength. The "gravitational signal timelag" idea (**Shapiro effect**, section 1.2) didn't seem to get into print until the mid-Twentieth Century, with Shapiro reputedly reusing Newton's old approach of imagining a gravitational field as a variation in refractive index. This effect could have been calculated directly from "Opticks", if only we'd fixed Newton's faulty lightspeed law.

Gravitational bending of light – although gravitational lightbending was a fairly obvious consequence of Newton's model, it got lost somewhere in the ensuing politics, and students somehow ended up being taught, centuries later, that Newton had said that spacetime was flat and absolute. After the early 1800's, references to gravitational light-bending seemed to shrink from the mainstream literature until Einstein reinvented the idea afresh in 1911, apparently unaware that anyone else had ever worked on the problem before him. Einstein was accused of plagiarism for not giving due credit to **Johann Georg von Soldner** for having already worked on the idea, but the subject had become rather obscure – **Henry Cavendish** had apparently also worked on similar calculations at around the same time as Soldner, but *his* calculations weren't exhumed from his personal notes until much later (Will, 1988). Meanwhile everybody seemed to forget that Newton had already explored the gravitational bending of the paths of "corpuscles" in "Principia" and "Opticks", so to complain about not crediting Soldner while forgetting to mention Newton was a little perverse. Einstein's writing in the 1920s suggests that even by this point he still didn't seem to be aware of how far Newton had got in "Opticks".

Gravitational time dilation – **John Michell**'s 1783 research piece correctly said that light reaching us from a high-gravity star should be shifted towards the weaker end of the spectrum ... but by quoting Newton's faulty relationships from "Opticks", Michell was forced to say that this was the "blue" end. If Michell had been given the correct energy/frequency relationships, he'd have been confronted with the "clockrate discrepancy" problem that we met in section 3.7, which could only be solved by saying that clocks ran at different rates in different gravitational fields. If we'd been less confrontational and had tried to use Huyghens' arguments to *correct* Newton's model, we might have had the concept of gravitational time dilation way back in 1811 rather than 1911. We'd then have been able to predict lightbending values indistinguishable from modern general relativity, and if it had been appreciated that gravity affected clockrates when **Gauss** produced his curved-space arguments, it'd have seemed natural to apply them to 4-D curved spacetime rather than just 3-D curved space. We might have ended up with a general theory of relativity in the middle of the Nineteenth Century, without ever having ever considered the existence of special relativity (*see also:* Rindler 1994).

Black hole theory – **Marquis laPlace**'s brief (withdrawn) mention of the idea of "gravitationally-cloaked" stars (published in post-revolutionary France) survived the scientific purges in England, but Michell's lengthy, high-profile piece wasn't so lucky. Michell's article disappeared from the citation chain as if it had never been written, and even in some in scientific biographies written as late as the 1960's, Michell is still primarily remembered for some obscure work on the theory of earthquakes and his relationship with **Henry Cavendish**. Michell's piece was rediscovered with much flourish in the 1970's ... but to anyone with access to those old Royal Society volumes, it had been sitting on the shelves in plain view, indexed and ignored, for all that time.

Gravitomagnetism – by using the round-trip *speed* of light as its fundamental definitional unit, Einstein's special theory ensured that **gravitomagnetic** effects – physical asymmetric variations in the velocity of light caused by the relative motion of bodies (sections 4.14, 9.1) – didn't exist in the theory. Worse, these effects were excluded from the core definitions that were then used to try to build a more complete and more general theory of relativity ... so while Einstein's general theory *has* to include gravitomagnetic effects, perhaps its support for them isn't quite as comprehensive as it could have been.

Acoustic metrics and Hawking radiation – we'll meet the subject of fluctuating metrics and trans-horizon radiation in section 10.6. For now we'll just point out that anyone working on Michell's dark star model would have quickly realised that its description makes gravitational event horizons inherently "leaky". This classical indirect-radiation effect could've made **Hawking radiation** (in its most general form) an interesting subject in the 1870's rather than the 1970's. The relevance of acoustic metrics to relativity theory might then have been apparent in the 1890s rather than the 1990s, and even if we hadn't considered the subject to be *significant*, when quantum theory appeared, our classical framework would have been in a better state to accept it.

Gravitational physics is sometimes presented to us as a great success story. But if we set aside the "educational" view of history and look at the actual historical record, the political fallout from Newton's mistake seems to have caused the subject's development to fall a century behind where it probably should have been by now. Einstein *should* have been born into a world that already had a general theory of relativity and an understanding of acoustic horizons, and he should have been able to devote his career to trying to solve the "next-generation" questions that we're only now just beginning to comprehend.

This was one of the greatest disasters in Western science, and we're still living with its consequences today.

PART II

Effects due to Relative Motion

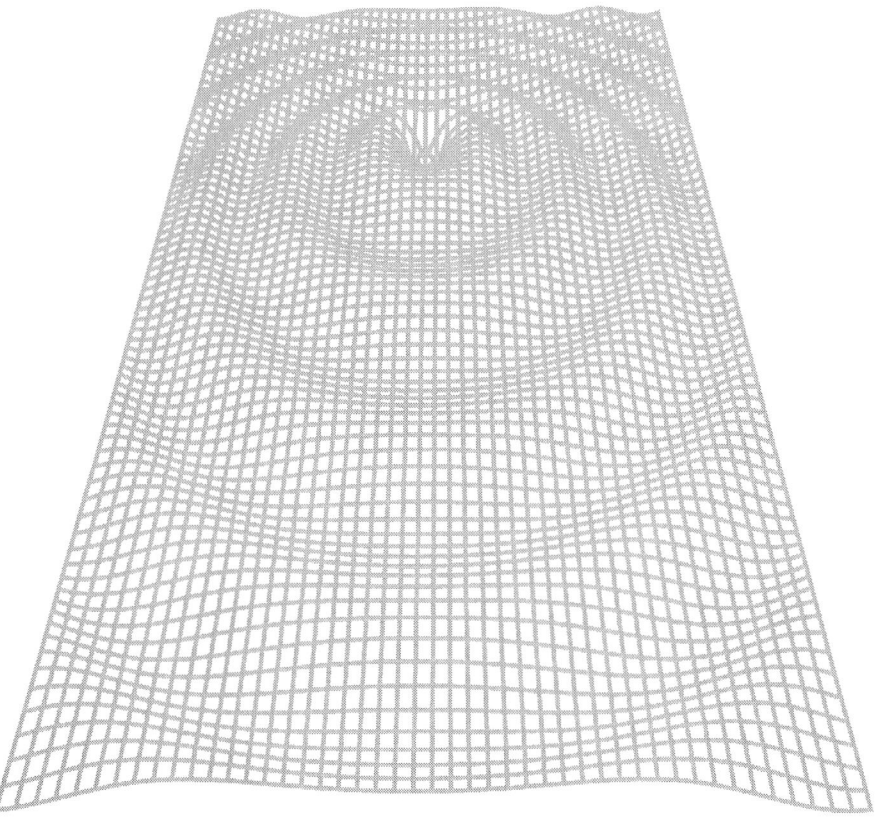

RELATIVITY IN CURVED SPACETIME

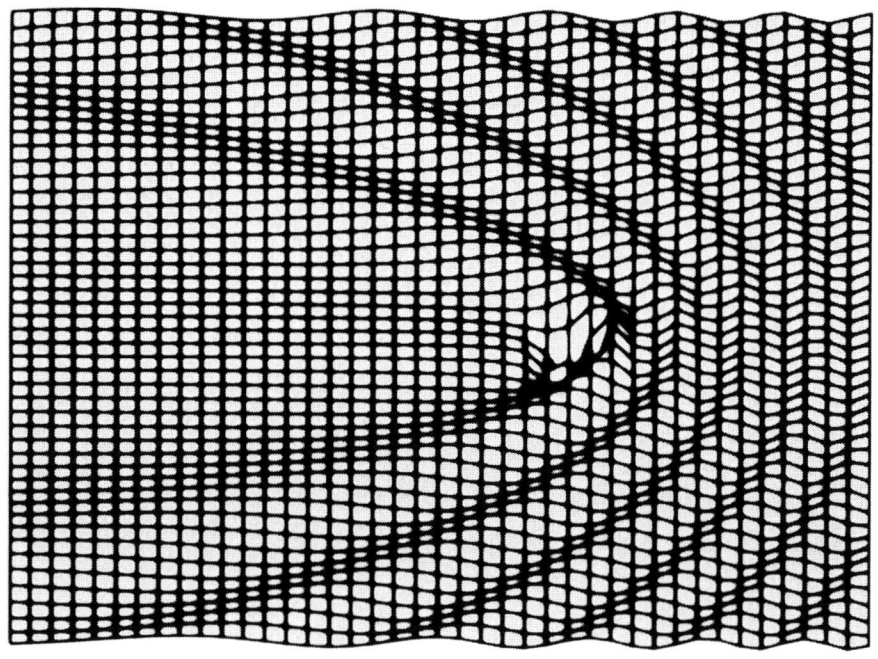

6
Doppler Shifts

> **Christian Doppler (1803-1853)** gets the credit for first calculating the way that an object's relative motion changes the apparent pitch of its signals. The effect increases the apparent pitch of signals coming from an approaching object, and reduces the apparent pitch of those from a receding object. When a racing-car speeds past our position (or past a TV camera if we're watching the race on television), we hear the engine note seeming to swoop down from a higher frequency to a lower one as the car passes by, making a characteristic "eeeeeooooowwww" sound.
>
> Similar arguments apply for light, but since the default propagation speed for light is so much higher than for sound, optical Doppler effects aren't visible to the human eye for objects moving with everyday velocities. When an object moves towards us at very high speeds, it's light should seem to us to be blueshifted to higher frequencies and higher energies, and the object should appear to be ageing faster ... when it moves away from us, we should see it to appear to be ageing more slowly, and to be giving off lower-frequency, lower-energy redshifted light.
>
> For some of the later arguments in the book it helps if we know that Doppler relationships can be different under different models, so some numbers are (reluctantly) included. This will also be our first encounter with some of the exact numerical predictions of Einstein's "special" theory.
>
> Readers with an aversion to fractions and "equals" signs might want to skip parts of this section.

RELATIVITY IN CURVED SPACETIME

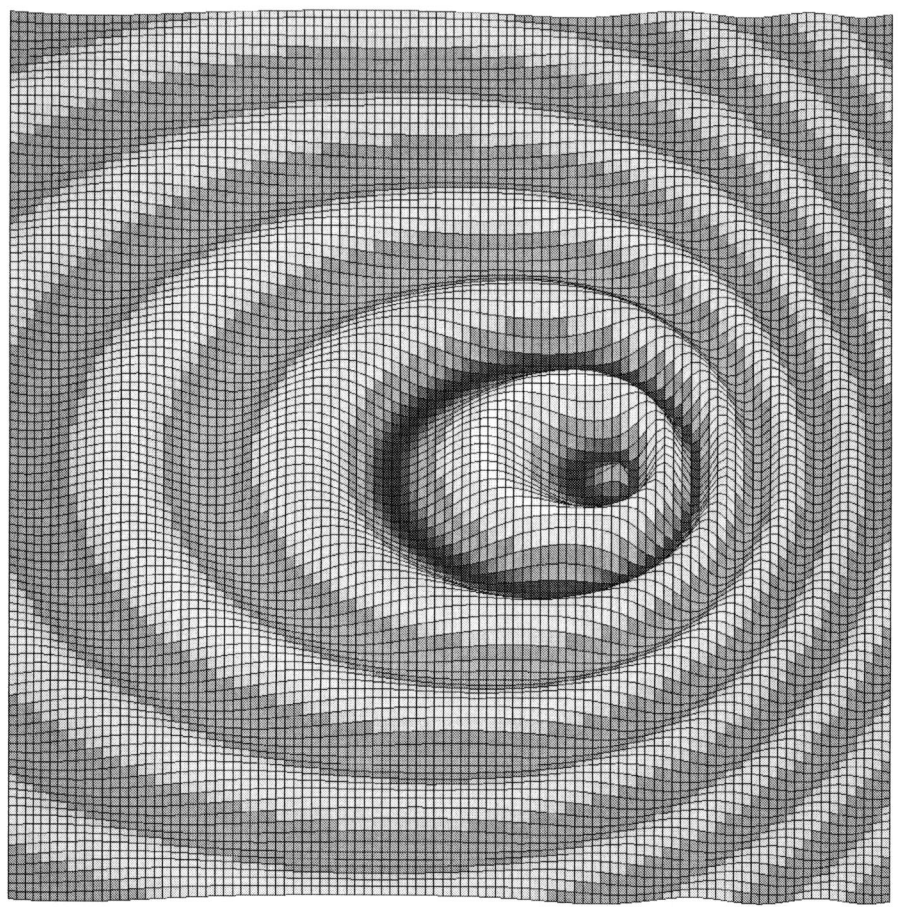

Non-relativistic Doppler effect

6: Doppler Effects – Changes in Apparent Frequency

6.1: "Stationary observer" Doppler effect

Let's suppose that we're standing in still air, and a loudspeaker is producing a constant tone while moving towards us or away from us at a constant speed. If the signal-source moves *away* from us, the rate at which we receive its signal pulses will be lower, because each pulse is being placed into the air at successively greater and greater distances from us – each incoming wavepeak arrives at our position with a slightly greater timelag delay. The rate at which the pulses arrive depends partly on the rate at which they were generated, but also on the different distances that they have to travel to reach us.

Figure 6-1 illustrates this situation, and shows a moving sound-source leaving a trail of "ripples" behind it as it moves. A listener to the right of the diagram hears the signal in a shorter amount of time because of the decreasing timelag (higher pitch), and the signal takes longer to be heard by a listener over to the left because of the timelag increase.

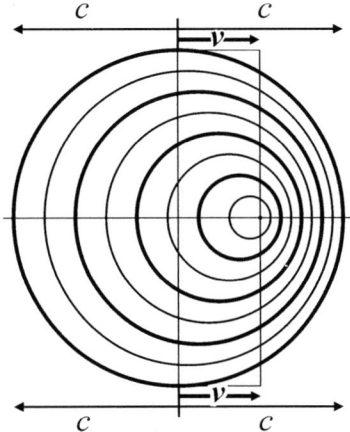

Figure 6-1: Nonrelativistic Doppler effect: Wavefronts

The amount of stretching or squashing of the signal depends on the ratio between the speed of the object and the speed of its signals. If we say that the speed of the waves is "c" and the speed of the source is "v", then our diagram shows that the wavelength-distances forward of and behind the object changing by the ratios $(c+v):c$ and $(c-v):c$, and if we turn these ratios upside down to convert them into *frequency* ratios – and we call the original frequency being generated f, and the *apparent* frequency, f', and if we define v more exactly as **recession velocity** – we get a nice simple Doppler relationship:

A ...
$$\frac{f'}{f} = \frac{c}{c+v}$$

So: if we put a ruler against our diagram of an object moving at $0.5c$, we find a rearward wavelength of $(1 + 0.5) = 1.5$, and a forward wavelength of $(1 - 0.5) = 0.5$ ... turning these two numbers upside down to get the corresponding frequencies, we find that the forward signal is heard with double the pitch, and the rearward signal is heard with only two thirds of its original frequency.

So far, so good ...

6.2: "Stationary source" Doppler effect

If our signals are instead supposed to move at a fixed speed with respect to the *emitter* ("stationary source, moving observer") the resulting frequencies will again be higher when the object approaches and lower when it recedes, but the *degree* of change will be different.

When we run headlong into or away from the signal, the change in apparent frequency is now reckoned to be *our* fault: our motion through the signal stream causes us to "read" the signal at faster or slower rates, and this gives different predictions. We'll skip the exact working (it's provided at the back of the book, in *Calculations 1*), but the Doppler effect for this alternative situation comes out as:

B ...
$$\frac{f'}{f} = \frac{c-v}{c}$$

We're again using v to mean recession velocity, with negative values of v describing an approaching object.

6.3: Comparisons

Our two results, **A** and **B**, look superficially rather similar. They'll both tell us to expect a **redshift** if an object recedes and a **blueshift** if it approaches, but the *exact amount* of shift depends on whether the signals are supposed to be moving at fixed speed with respect to *us* (version **A**), with respect to *the signal source* (version **B**), or if it does something else. So the *exact degree* of Doppler shift tells us something about how light really moves.

To make this comparison a little easier on the brain, we'll graph how our two main sets of predictions compare:

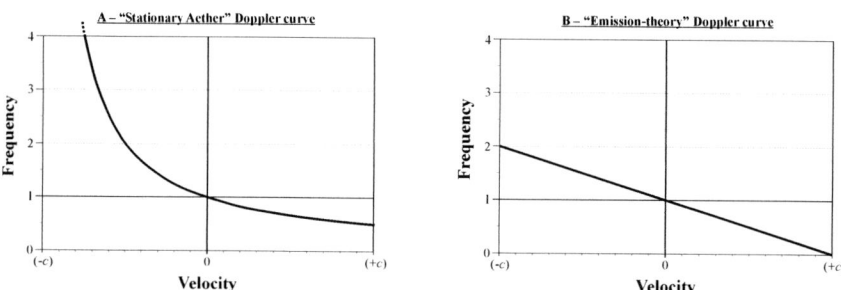

Figure 6-2: Graphs: "A" and "B" Doppler predictions

The left hand part of each graph shows the approach blueshift, and the right-hand side shows the recession redshift.

The "**A**" predictions say that light from an object approaching at lightspeed will be infinitely blushifted, and light from an object receding at lightspeed will have its frequency halved. The "**B**" predictions say that signals from the *c*-approaching object are only doubled, and that the apparent frequency of the *c*-receding object's signal drops to zero.

For the purposes of this book, we don't have to understand the exact details of *why* these two predictions disagree ... but we *do* have to appreciate that the two sets of relationships are physically different.

6.4: Transverse Doppler effects (audio)

While it isn't *too* difficult to understand why these basic Doppler effects happen, there's also a more difficult-to-imagine effect that can happen when the source is moving at *right angles* to the observer.

Suppose that we build a straight-line section of railway track running through our lab, and we use a big set-square and ruler to carefully chalk a line running directly *across* this track at 90 degrees. We then aim a highly-directional sensor to point *exactly* along this chalkline, and bolt it securely to the lab floor, in this position. Then we send a train along the track at constant speed. When part of the passing train's surface registers on our sensor, how much Doppler shift will it appear to have?

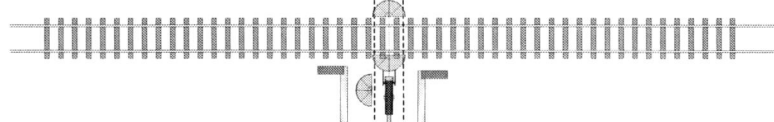

Figure 6-3: "Transverse" redshift test-rig

If our signals always move at a perfectly constant speed with respect to the lab, then we'll expect *no* Doppler shift at all, because there's no obvious reason why the signals should be stretched or squashed. Two pulses deposited in the region at the left- and right-hand limits of our narrow sensor window should cross the same distance to reach us, and we wouldn't expect the time between their arrivals to be affected by the object's motion.

$$A_{TRANSVERSE} \cdots \qquad \frac{f'}{f} = 1$$

However, if the signal's speed *is at all influenced by the speed of the emitter* (even just a little tiny bit), we'll end up seeing a redshift. This is because of **aberration effects** (section 8): if the signal is at all swept, thrown, or dragged-along by the object, then the angles at which the object *aims* light usually won't be the same as the angles at which we'll *receive* it, and the pathlengths will be different. If the lightrays are deflected forwards by the object's motion, then although the light *emitted* by the object at 90 degrees may well be unshifted when it reaches us, *this will not be the same light that enters our transverse-aimed detector*. The signal that gets deflected forwards to register on our detection apparatus will have originally been aimed somewhat rearward, and it should contain a certain amount of recession redshift.

Skipping the tedious math (which is again provided at the back of the book, *Calculations 3*), the worst-case scenario for the **aberration redshift** effect on a signal that moves along *completely* with its source, gives us a redshift of:

$$B_{TRANSVERSE} \cdots \qquad \frac{f'}{f} = 1 - \frac{v^2}{c^2}$$

So, the exact amount of redshift reported by our "transversely"-aimed equipment will depend on how the relative motion of bodies affects light – the more strongly we think a moving object drags light along with it, the stronger the "transverse" redshift effect that we'll expect to see.

RELATIVITY IN CURVED SPACETIME

Again, we can save some brain-ache and make the difference more obvious turning these mathy relationships into pretty graphs:

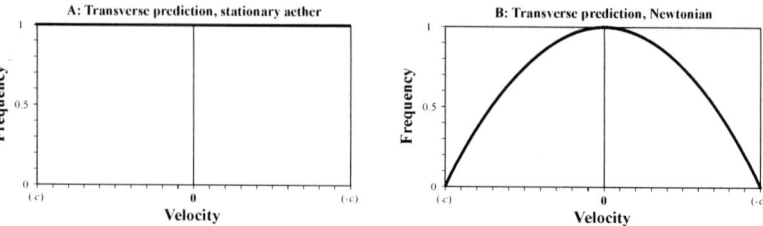

Figure 6-4: Basic transverse Doppler predictions

The graph on the left assumes a speed of light that is fixed for the observer, and is a simple horizontal "flatline", while for old Newtonian **emission theory** (or for sound in air that gets carried along by the signal source), we get a frequency curve that dives towards zero when the relative speed approaches the official speed of the signal.

We didn't really need to plot both sides of these graphs, because the effect is the same regardless of whether an object is moving left-to-right or right-to-left … but if we want to start comparing the different graphs against each other, it's handy to have them all in the same format.

why can't I find this effect in the textbooks?

The "transverse redshift" prediction for "**B**" tends to be left out of current Twentieth Century audio handbooks, because the "modern" convention is to say that an observer is only allowed to define angles if the speed of light happens to be constant for them, and this way of defining angles tends to make these "transverse" redshift effects disappear, at least on paper.

Although our "**B**" redshift calculation is correct, and is the correct *physical prediction* for the situation described, a textbook will tend to insist that this redshift result doesn't *really* count as a transverse redshift, because if lightspeed *wasn't* totally constant in the laboratory frame, our attempts to carefully aim our detector at 90 degrees weren't valid, and the equipment wasn't really aimed "transversely" after all.

For our purposes, that approach is deeply unhelpful, because …

1. … it stops us from making direct comparisons between the predictions of different theories for an agreed physical situation,
2. … the process of translating backwards and forwards between different angles for different theories is so fiddly that sometimes even "pro" physicists screw it up, and,
3. … because we'll be needing to consider theories in which lightspeed is globally *variable*, and in *those* theories, the "special" global lightspeed reference frame that we are told to use for defining all our angles, *doesn't exist*. This happens in theories in which bodies drag light locally (it also happens in normal acoustics when we take into account a moving body's "slipstreaming" effects in the surrounding air).

Since the standard approach seems almost designed to wreck attempts to model physics in curved spacetime, and destroys proper cross-theory analysis, we'll declare it "not fit for purpose" and jettison it. For the rest of the book, "transverse" (without the quotes) will usually refer to an angle that an observer or experimenter *sees* as being 90 degrees.

6.5: Optical Doppler effects

At the beginning of the Twentieth Century, physicists faced a problem: calculating Doppler effects in air was fairly straightforward because we already knew how air behaved, but when it came to modelling light, the rules weren't so clear-cut. Theory seemed to say that the speed of light ought to be constant, but who was it supposed to be constant *for*?

There were many different models kicking about, making different assumptions about the way that light behaved, and giving a correspondingly different range of different physical predictions. If we believed that lightspeed was fixed relative to the *emitter* (as it would be under old-fashioned Newtonian **emission theory**), then we got our second, "**B**" set of Doppler equations, and a strong transverse redshift effect. If we believed that light was fixed for the *observer*, then we got the first, "**A**", Doppler predictions, and no transverse effect. If we believed that there was an absolute aether moving with an intermediate speed, then we would get an intermediate prediction, and if we believed that a light-medium was physically dragged along with moving objects, then we got a prediction somewhere in the range **A↔B**, depending on the exact characteristics of the supposed dragging process.

According to modern Twentieth-Century theory, optical Doppler effects don't agree with either **A** *or* **B**, but follow a new set of relationships that are *exactly intermediate* between **A** and **B**. To get to this third set of predictions, we average the **A** and **B** predictions together by taking their *geometric mean* – we fuse the characteristics of the **A** and **B** results into a single prediction by multiplying them together, and then square rooting the result to bring the answer back into the right range.

We'll leave a proper discussion of *why* we might want to do this 'til section 14, but the result of this special averaging process gives us the newer Doppler predictions used by **Einstein's special theory of relativity** ("special relativity", or "SR").

6.6: Longitudinal Doppler effect under Special Relativity

To calculate special relativity's prediction for a receding or approaching object, we can say that

$$effect(SR) = \sqrt{effect(A) \times effect(B)}$$

, which gives us the newer predictions of SR's "**relativistic Doppler**" effect. SR's forecast runs directly between the **A** and **B** results, and deviates from both of them by exactly the same ratio.

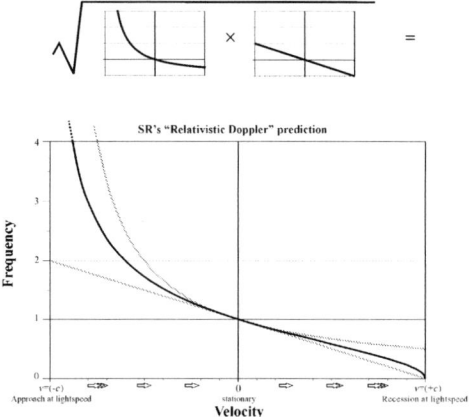

Figure 6-5: Graph: "Relativistic Doppler"

RELATIVITY IN CURVED SPACETIME

Whenever we take a "**B**" prediction and divide it by its "**A**" counterpart, we end up with a magical ratio of $1-v^2/c^2$, which gives us the degree of disagreement between the **A** and **B** predictions. If we square root this ratio, we get the difference between the **A** *or* **B** predictions and the intermediate predictions of **special relativity**. This square-rooted ratio, $1/\sqrt{1-v^2/c^2}$, is called the **Lorentz factor** (section 16.4) and it special relativity used it incessantly.

6.7: Transverse Doppler effect under Special Relativity

If we want the predictions of Einstein's special relativity's for a detector aimed directly *across* an object's path, we can do the same thing: we take the two earlier predictions, multiply them together, and then square root the result:

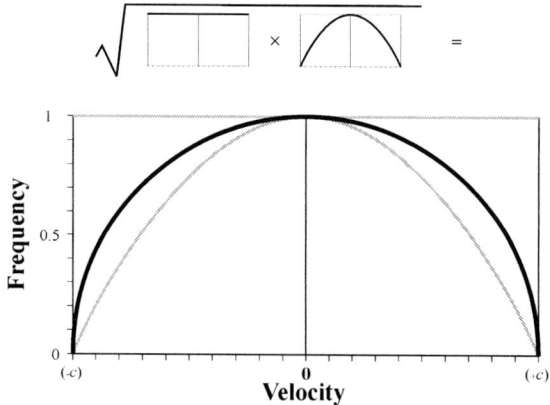

Figure 6-6: Graph: SR's "transverse Doppler" effect

Since the "**A**" transverse prediction is just "one", our graph in Figure 6-6 graph doubles as a handy graph of the Lorentz factor that we mentioned above. It's a perfect semicircle.

special relativity in brief

As far as its straightforward physical predictions are concerned, special relativity is really just the **geometric mean** of **A** and **B**. This special relationship, along with SR's use of Lorentz corrections, lets us pretend that lightspeed is globally fixed in either frame, and still end up with the same final predictions. We can claim that lightspeed is "really" fixed for the observer (giving the "**A**" predictions), and multiply in a **Lorentz factor** redshift to produce the SR result, or we can we say that lightspeed is "really" fixed for the *emitter*, giving the redder "**B**" predictions, and *divide out* that same Lorentz factor to end up with the same SR result.

Special relativity also has features that let us assume that lightspeed is globally fixed in *any other inertial frame*, and by carefully applying the right redefinitions of distances, times and velocities, the magical Lorentz factor will again help us to get exactly the same final results. In engineering terms, special relativity can be seen as a *system* that lets us generate identical predictions while assuming that lightspeed is globally locked to *any* legal inertial frame.

The system's strength is this ability to generate identical numbers under different descriptions. The system's weakness is that it assumes *global* lightspeed constancy irrespective of how objects move past each other. It assumes that the light-dragging effects that we'll meet in section 9 don't exist, and it doesn't attempt to produce a realistic model of the interaction of light and matter in a realistic way.

SR's approach simplifies things a great deal, but its descriptions don't seem to correspond to how matter and light *really* behave around each other.

7
Apparent Length-Changes in Moving Objects

> As well as appearing to change colour due to the Doppler effect, a fast-moving object can also appear to compact or extend in length in its direction of motion, as another side-effect of the finite speed of light. This has been referred to as the "spatial analogue" of the Doppler effect.
>
> Approaching (blueshifted) objects appear to be stretched out in their direction of motion to occupy a larger volume of space, and receding (redshifted) objects look as if they're compacted, and appear to occupy a smaller volume.
>
> Although this effect is straightforward, it wasn't properly described in the relativistic literature until the 1960's, after James Terrell and Roger Penrose had realised that some of the usual optical predictions associated with special relativity had been miscalculated or misunderstood. Statements about the visible appearance of relativistic objects, in books published before the 1960's therefore need to be treated with some trepidation.
>
> Once we know how to calculate the Doppler frequency-shift on a receding or approaching object for a particular propagation model, the object's apparent change in length (or depth) turns out to follow exactly the same rule.

RELATIVITY IN CURVED SPACETIME

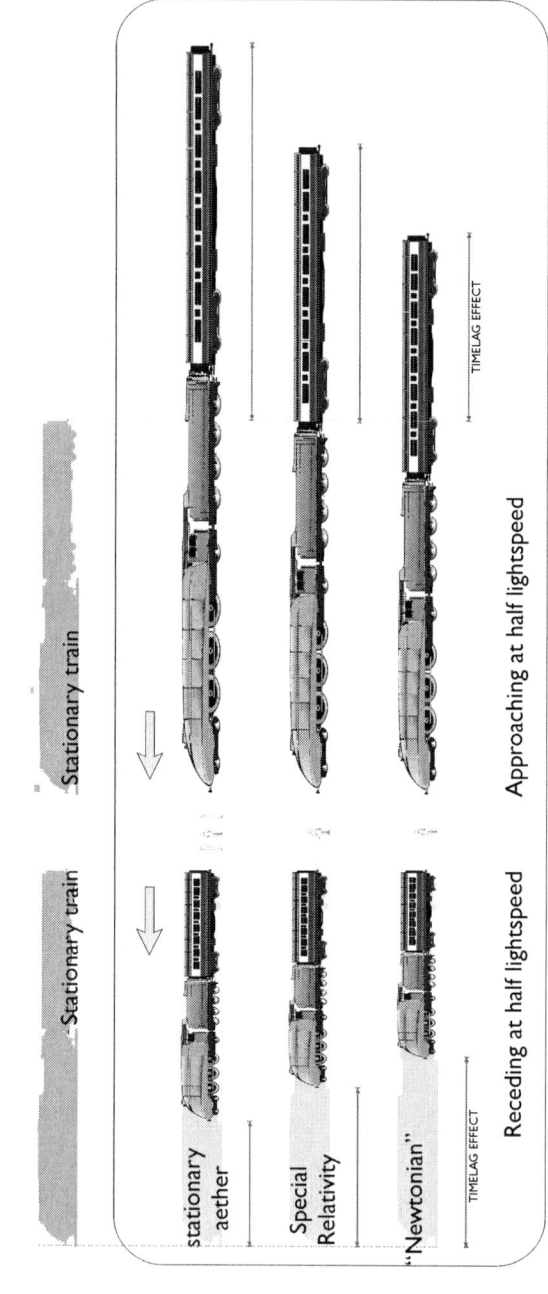

7: Changes in Apparent Length

7.1: Apparent changes in length

As well as the more famous **Doppler effect** that alters the frequency of a signal when its source recedes or approaches, there's also a companion effect that changes the apparent *depth* of a moving object. When we peer along the side of an object, the signals reaching our eye will have taken slightly different times to get to us from the nearer and further parts of the thing that we're looking at, so that these different parts appear to us as they were at slightly different times.

If the object is also seen to be moving a significant fraction of the speed of light, this timelag difference makes the different parts of the object appear at different past *positions*, so that, overall, the object appears to be squashed or stretched or otherwise distorted.

7.2: Approaching objects appear elongated

Let's suppose that we're standing by the side of a railway track, watching a very long express train approaching at a significant fraction of the speed of light. What do we see?

If we take a photograph of the approaching train at the exact moment that the train's front buffers reach us, then the photograph should show an image of the train's front (almost hitting the camera) that's pretty much up-to-date, and that represents the "true" position of the train when the shutter-button was pressed. But the situation with the *rear* of the train is rather different. If the train is one kilometre long, then images from the rear of the train, travelling at the speed of light, won't be able to reach our camera until *after* the photograph is taken. The image of the train's rear that *does* successfully register as part of our photograph will have been created at a slightly earlier time, when the train was slightly further away.

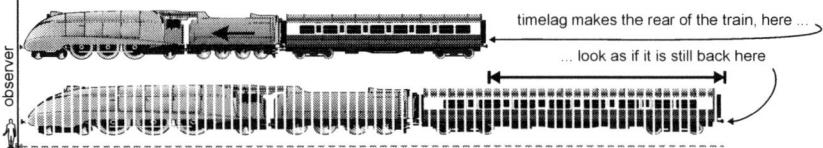

Figure 7-1: **An approaching train appears longer**

So, while the front of the train appears in the photograph to be at "position zero", the rear of a 1 km-long train appears in our picture to be *more* than 1 km away. The train looks stretched, and our photograph shows it to be occupying *more* than 1 km of track.

7.3: Receding objects appear contracted

If we take another photograph a little later, when the *rear* of the train reaches our position, then the *back* of the train will now appear in the photograph at its correct position, and it'll now be the distant, retreating *front* of the train that's seen with an observational timelag. The train's front will seem nearer than it's true position, on a section of track that it has since left behind.

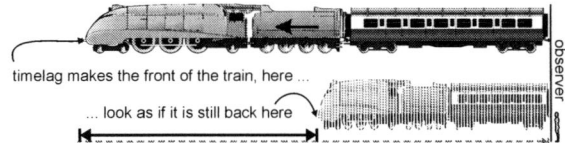

Figure 7-2: **A receding train appears shorter**

Since the front of the train looks nearer to us than its actual position, *this* photograph (of the *receding* train) shows the 1 km-long train to be occupying *less* than 1 km of track.

73

RELATIVITY IN CURVED SPACETIME

7.4: Degree of contraction or elongation

The expected *strength* of the effect – the degree of stretching and squashing that should appear in the photographs – depends on the amount of time that we think it should take light to travel along the train's length, and this in turn depends on whether we think that light moves at a fixed speed with respect to the track, or to the train, or if it's expected to show some other type of behaviour.

The diagrams below show the apparent length of a train receding or approaching at half lightspeed according to three different calculations: **(a)** assuming that lightspeed is fixed in the observer's frame, **(c)** that it's fixed with respect to the moving object, and (between them), **(b)** the "modern" prediction for the train's visible length according to special relativity.

Figure 7-3: Three different calculations of the apparent length of a receding and approaching train ($v=0.5c$)

These calculations show a familiar pattern: it turns out that, for a given set of assumptions about light, the ratio that we end up with for the object's apparent change in length is the same as the ratio that we get for the object's apparent change in frequency due to Doppler effects.

A quick example: if we had a 1 km-long train approaching at half lightspeed, and thought that light travelled simply at a fixed speed with respect to *us*, the corresponding Doppler prediction would be $f'/f = c/(c+(-v)) = 1/0.5 = 2$, and we'd expect the train to be seen to be ageing at twice its proper rate. If we now try to work out the train's apparent change in length, we find that the front of the train, 1 km away, moving at half c, would arrive after two seconds, at the same moment as signals travelling at lightspeed from the *back* of the train when it was *2 km* away. So we see the train's front at our position when its rear appears to be 2 km distant – our view shows the train as apparently 2 km long, which is the same factor-of-two as before.

Happily, this match isn't a fluke, if we sit down and work through stacks of deeply boring examples with different speeds and different light-propagation models, the two things always seem to match up. So we can usually get away with ditching all these tedious "signal-tracking" calculations, and just go back to our earlier Doppler relationships, scrubbing out the words "visible frequency", and writing in "visible depth" instead.

7.5: Special relativity and length-changes

This trick also works with special relativity. SR's predictions for the **photographable length** of the train are again the *root-product average* or *geometric mean* of the earlier "A" and "B" predictions. Since the "A" and "B" predictions for changes in *apparent frequency* are already the same as for changes in *apparent length*, the same correspondence between frequency-change and length-change carries over to special relativity: our earlier graph of the SR "**relativistic Doppler**" effect can be reused as a graph of the theory's predictions for the photographable change in depth for an approaching or receding body.

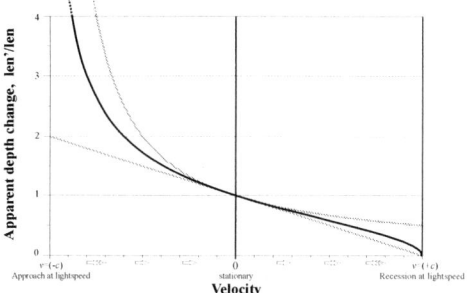

Figure 7-4: Graph: Apparent length-changes under SR follow the same "averaged" relationships as apparent frequency changes

Once again, the two earlier "A" and "B" predictions will disagree by a factor of $(1 - v^2/c^2)$, and once again, special relativity's "relativised" prediction will disagree with both of them by the square root of this, the ubiquitous **Lorentz factor**.

Once again, SR allows multiple interpretations: if a photograph showed length-changes that *agreed* with SR, we could claim that the picture shows the results of c being fixed in our frame along with a Lorentz contraction, or that it shows that c is fixed in the object's frame and the same object is Lorentz-elongated. Or we could pick some other reference frame and obtain the same final SR prediction by saying that he object is "really" contracted or elongated by some other amount. These are *optional interpretations* of the theory's physical predictions.

transverse contraction?

There's also been some controversy over what special relativity predicts for the appearance of a *passing* object. This is a difficult subject, because a straight-line path is only *truly* transverse to an observer at one point, and if we photograph a passing object that has an appreciable length, we have to decide which part of it should coincide with this point in the photograph.

If we stand at the edge of a platform watching a passing tube-train that only has two carriages, and take a photograph when one whole carriage is receding and the other approaching, then according to SR, when the lengths of the receding and approaching carriages are summed, the total visible length comes out as a Lorentz *increase*. This is a bit perplexing since we expect a visible redshift to be associated with a visible *contraction*. What went wrong?

The answer seems to be that we took the picture at the wrong moment – if we had chosen to wait a little longer, so that more of the train was receding, and pressed the shutter button at the exact moment when the unbalanced receding and approaching parts of the train now *looked* to us to be the same length, our SR photo *would* show the proper Lorentz reduction in total length to match the frequency predictions, completing the pattern.

"Transverse" length calculations often don't use very clear definitions and are *very* easy to mess up, so it's usually safest to avoid them.

7.6: Rulers and gravitation

When we were looking at the effects of relative velocity, we found that redshifts were associated with apparent contractions in depth, and blueshifts with apparent elongations. Does the same pattern apply to *gravitational* redshifts and blueshifts?

It does. We've already said that a **gravity-well** packs more distance into a given volume of space than we'd see in the surrounding region, so if an object is *inside* that gravity-well, the object's apparent depth will be shorter according to rulers *outside* the gravity-well. And, of course, it'll be (gravitationally) redshifted.

Let's once again take the extreme case of a **black hole**. If we stand far from the hole and watch an object freefall in, parts of that object's signal will find it progressively more difficult to reach us as they're being broadcast from positions deeper and deeper inside the hole's gravitational well. Once the object reaches the hole's **event horizon**, light emitted from this point onwards is supposed to be trapped by the hole's gravity, and can't reach us at all. The signal flight-time for *this* light, from object to observer, is infinite. For light emitted just above the horizon, the signal flight-time to the observer can take an extremely long time, and for light *arbitrarily close* to the horizon, the time for signals to escape can be *arbitrarily long*. The outside observer watching the object fall will only ever see signals that were generated outside the horizon, and the final, finite part of this signal stream has to be rationed out so that, for the outside observer, it lasts for an infinitely long amount of time.

The upshot of all this is that, in this idealised scenario, the distant observer never actually sees the object to fall *into* the hole, but sees it always falling *towards* the hole and appearing to slow towards a dead stop as it approaches the critical horizon radius (section 5.7).

buildings and black holes

What has all this to do with rulers? Well, let's suppose that we drop the **Empire State Building** into a smallish black hole, and come back a year or so later to see how far it got – theoretically, even though the building is already *inside* the horizon, we'll still be seeing its frozen image suspended somewhere *above* the horizon, snuggling up ever-closer to it without ever seeming to quite reach the critical surface. Let's say that when we return, we see the bottom of the building's basement appearing to be just one metre from the horizon. But it might be that on our original trip, a few minutes after dropping the Empire State Building into the hole, we'd also decided to drop the **Chrysler Building** in, directly on top of it. If we expect the two falling buildings to be seen to follow equivalent paths, then, then a few minutes after our return we should expect to see the bottom of the *Chrysler Building's* basement to be occupying the ESB's previous position, with *its* basement one metre from the horizon. But if the ESB is still seen to be *between* the Chrysler Building and the horizon, then its entire height must seem to be squashed into this one-metre region. And if its near-frozen image seems to us to have moved by less than a millimetre during these few minutes, then its entire height must seem (to us) to be compacted into *less than one millimetre* of height.

As a collection of material falls into the black hole, it should *appear* to be slowing and flattening out to try to form a zero-thickness shell at (or fractionally above) the event horizon, with the apparent frequency of its signals, and the apparent height of its image, both attempting to converge towards zero.

Similar effects might be expected for gravitational blueshifts: if we are in a strong-gravity region looking outward, we'd expect weaker-gravity objects not just to be blueshifted, but to appear to have "more depth to them" if we tried to measure them by naively extrapolating our own local ruler-distances out into the surrounding region without taking curvature into account.

The pattern that we found with velocity-based effects – redshift means apparent contraction, blueshift means apparent elongation – also applies to gravitational effects.

8
Aberration of Angles

A moving object doesn't just seem to undergo polarised distortions in colour and depth, it's angles seem to change, too. This can create disagreements between differently-moving observers over the "real" angle between pairs of lines. If the pilot of a fast-moving spaceship carefully aims a laser-beam at 90 degrees to their direction of motion, observers watching the spaceship whiz past will see this beam to be leaving the spaceship's path at a *different* angle ... **angular aberration** makes the angles between parts of an object's shape appear to narrow or widen depending on whether we're watching approaching or receding parts of the object's surface.

These effects seem to have first been discussed in print by the astronomer **James Bradley** (1693-1762), when he pointed out that the apparent direction of stars seemed to shift slightly depending on whether they were observed when the Earth was moving approximately towards or away from that section of starfield.

For the principle of relativity to work, these angles must "warp" in a very specific way, and it turns out that the principle of relativity allows a moving object's Doppler effects, apparent length-change effects and aberration effects to all lock together into a very special geometrical configuration.

8: Aberration: Distortion of Angles

8.1: Aberration of Angles

Suppose that we're standing in a room, and we throw a tennis-ball at the nearest point on the wall facing us.

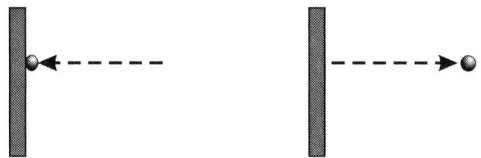

Figure 8-1: Ball thrown against a wall, and returning along the same path

The left-hand picture is the ball hitting the wall. The right hand picture is the ball bouncing back into our hand. To keep the diagrams simple we'll assume that we're doing this in "zero gee", or that we're looking directly down on the experiment, and only see the ball's horizontal motion.

The ball hits the wall full-on, and bounces straight back to our position. We wouldn't expect anything else in this situation, since we've learnt at school that the angle at which an object bounces away from a surface is the same as the angle that it originally hit it with ("**angle of incidence equals angle of reflection**"), and since the ball hit the wall at 90 degrees, there's no obvious excuse for it to bounce back at any other angle.

But now let's suppose that the room is onboard a cruise ship, travelling at a constant speed. If the sea is smooth and we're not troubled by engine noise, we'll have no real way of telling whether the ship is "really" moving or not, and so the ball should appear (to us) to retrace its path. Galileo and Newton used the "ship" example to argue that as long as our room is moving in a straight line at a constant speed, the local physics should be the same as if it was stationary.

For supporting evidence, you can look at the room around you. Do the conventional laws of physics seem to operate in that room as you'd expect if it was stationary? Does a ball bounce off the walls in the usual way? Hopefully it does, even though the room *could* be said to be hurtling around the Earth's axis at ~465 m/s, and around the Sun at ~29.78 km/s.

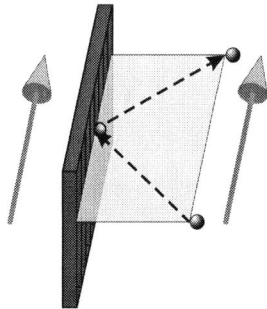

Figure 8-2: Same ball, seen by another observer to have different "throw" and "catch" positions

But for someone who doesn't share our state of motion, the ball is *not* returning to its original position – it's being thrown from one position, and caught at another. This other observer *might* agree with us that "the angle of incidence equals the angle of reflection", but they *might not* agree with us that the ball was thrown at 90 degrees to the wall, or that it returns at 90 degrees – the angle between the ball's path and the wall is different for different onlookers.

8.2: Relativistic aberration at 90 degrees

If we insist that the laws of physics must behave in the same way in an inertial system regardless of whether its objects are considered to be "stationary" or to have a uniform state of motion, then we can make some quite specific predictions about the way that light needs to leave a moving object for this to work.

The first rule that we can derive is that, when a system passing by us throws out a lightray or body at what *it* thinks is 90 degrees, then for *us* that ray or trajectory must point at a different angle – for the ray, or body, or object to be able to advance at the same rate as the rest of the system, it's path must be angled towards the system's direction of motion.

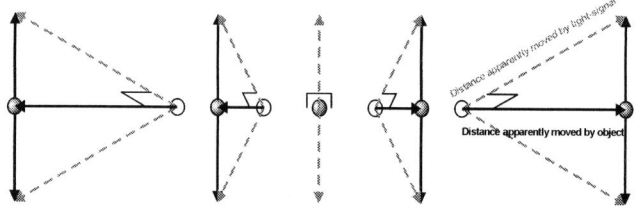

Figure 8-3: Relativistic aberration of rays emitted at 90 degrees by a "moving" object

This geometrical distortion has to affect the angles that we perceive *between objects* in the moving system, and *between features* on individual objects. The distortion is *relative* – we see the body to be "moving" and deformed, but *it* can claim that *we* are "moving", and that the disagreement about angles is due to the distortions of *our* moving measuring equipment.

- **For outgoing rays** the deflection makes rays point more in the object's direction of motion, and the faster the object is seen to move, the stronger the deflection of its light has to be. This forward concentration of a moving object's light is sometimes referred to as the **searchlight effect**.

- **For incoming rays**, the pattern is reversed. We'll see any lightray in Figure 8-3 that gets reflected back to hit the moving object as intercepting the object's path at an angle, while the object itself perceives the ray to hit it at a perfect 90 degrees. To them, *we* are receding and trying to aim light slightly rearwards, but *our* motion causes the light to be tilted forwards in our direction of motion so that it can hit them at 90 degrees.

Both observers agree on the angular discrepancy, but each blames it on the other's motion.

conservation laws and generality

We also need this to happen for **E=mc²** (section 2.5) to work consistently: if light inside a container is bouncing from side to side at *right angles* to the container's path, we need the container and its light to both be carrying **momentum**. But the only way that this light can have *sideways* momentum is if its rays are at least partly *pointing* sideways, in the system's direction of motion. So the forward tilt of rays (and the apparent visible distortion of a moving object) are also consequences of the law of **conservation of momentum**.

Since this argument affects the relative geometry of both lightrays and the trajectories of bodies, it seems to be a general "default" result, and even if we take the old **emission-theory** application of Newtonian mechanics to light, with light treated as if it was a stream of little particles, we'll calculate this same change in angles. So this would seem to be the default behaviour of special relativity, Newtonian mechanics, and (unless we can think of a good excuse why not) of any other potential theories of relativity.

8.3: The Relativistic Ellipse

Angle-changes sound horribly complicated, so it's time for some more nice diagrams.

Let's suppose that a flashbulb emits a single, omnidirectional pulse of light, which bounces off a spherical arrangement of mirrors that have been carefully set up around the bulb, and is then reflected straight back at the bulb's centre. We can say that the bulb *knows* that the reflecting surface marks out a perfect sphere centred on its position because otherwise the reflected light-rays couldn't all arrive back at their starting point, at the same time. For the round-trip distances to be exactly the same in all directions, the shape *has* to be a sphere.

But to an observer for whom the whole apparatus is moving, the light *isn't* being reflected back to its original position ... for *this* observer, the apparatus has moved forwards by a certain distance while the light was in flight. The reflection events are now described as catching light originating at one position and refocusing it *somewhere else*, and the shape that collects signals from one point and converges them at another is an **ellipse** (a circle that has been uniformly stretched or squashed in one direction). The "spherical" array of reflection-events that mark out the surface of a **sphere** for the bulb are defined for our second observer as marking out an **elongated ellipsoid** (Moreau 1994).

This ellipse (the cross-section of our three-dimensional shape) has two **focal points** representing the start and end points of our signal.

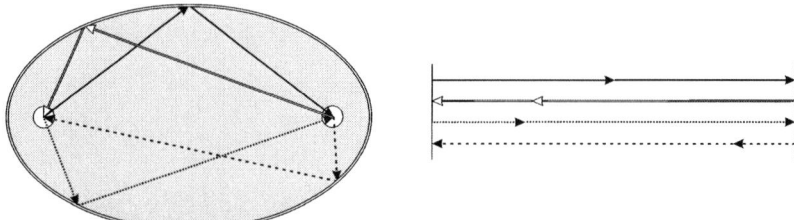

Figure 8-4: Total round-trip distances are the same in all directions

The critical property of ellipses in this situation is that when we measure from one focal point to the perimeter to the other focal point, we find the same total distance no matter which direction we measure in. This makes ellipses easy to draw if we own a couple of drawing pins and a piece of string: we stick the pins into a board and declare them to be the focal points, then tie the ends of the string to each pin, so that it's nice and slack, pull the string tight with the end of a pen, and then move the pen around the two points in a curve, keeping the string tight. Since the total string length is fixed, the total distance "**focus→pen→focus**" is always guaranteed to be the same, and we'll end up drawing a nice elliptical shape.

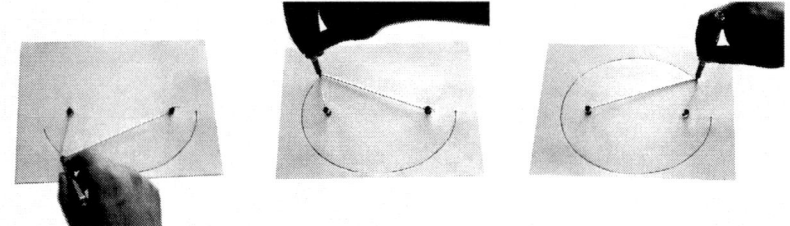

Figure 8-5: Drawing an ellipse using a piece of string

This shape then conveniently gives us the altered angles of *every* incoming and outgoing ray.

RELATIVITY IN CURVED SPACETIME

While Figure 8-3 only lets us calculate the deflection of a *transverse-aimed* ray, these "ellipse" diagrams show us the distortions for rays aimed at any angle. Figure 8-6 gives a few sample ellipse diagrams, showing aberration effects for a stationary object, an object moving at half lightspeed, and for an object moving at four-fifths lightspeed:

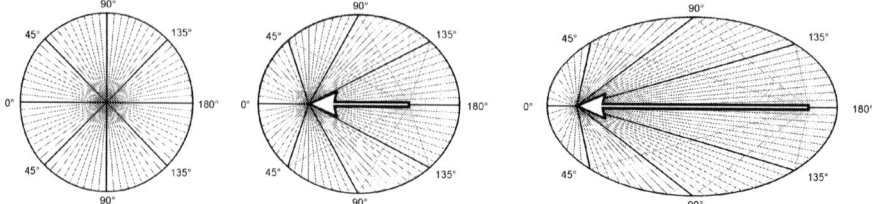

Figure 8-6: Relativistic aberration of angles, $v=0$, $0.5c$, $0.8c$

The angles that the "moving" object itself reckons are correct are written around the edge of the shape, and the angles that the lines actually make on the paper are the ones that a "stationary" background observer would see and measure for the same rays.

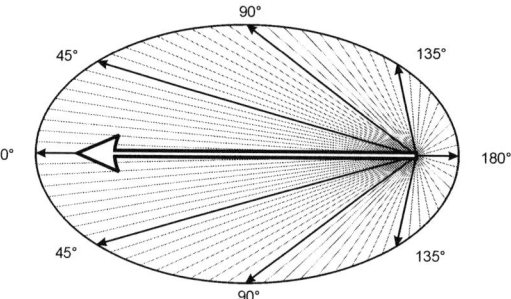

Figure 8-7: Relativistic forward aberration of light-rays

experimental verification

When we plop a finger in the middle of a cup of tea, it's easy to see that the circular waves bounce off the surrounding cup wall and reconverge back at the centre. Technically, a circle can be considered as a special case of ellipse (where the two focal points have the same position), but for a more convincing real-world demonstration of what happens with an *actual* elliptical reflector, we can build a simple **ripple tank**.

Figure 8-8: Polystyrene ripple tank

In the leftmost photo, some water has been squirted into the tank reasonably close to one focal point. In the central photo, the original part of the wavefront moving from right to left is now joined by another section of wave that has bounced off the right hand side of the tank wall. In the third picture, the leading wavefront has now bounced off the *left* hand tank wall, and we now have a circular wave converging on the second focal point. The theory works!

8.4: Putting it all together

But our clever ellipses also do a bit more.

Suppose that we know a rough model's predictions for the forward and rearward Doppler shifts on a moving object. If we want to turn it into a proper, full-blown relativistic model, we can take the forward and rearward Doppler-shifted wavelengths and use them as the maximum and minimum focal distances of an ellipse, and then construct the rest of the shape around these two distances. The resulting diagram will not only tell us the angle-changes for any ray for a given velocity, but also the *change in wavelength* for any ray, for that model.

8.5: Relativistic ellipse: Newtonian theory

Let's try this using old-fashioned Newtonian **emission-theory**. We know that the theory's Doppler relationships are ($f'/f = c \pm v/c$), and at and at three-fifths of lightspeed, **0.6c**, the forward and rearward frequency ratios will be **(1 ± 0.6) = 1.6** and **0.4**, giving us wavelength ratios of **0.625** and **2.5**.

So we start by drawing in these two distances, "0.625" and "2.5" …

… and this gives us enough information to find both focal points, work out the round-trip distance between them, and construct the rest of the ellipse.

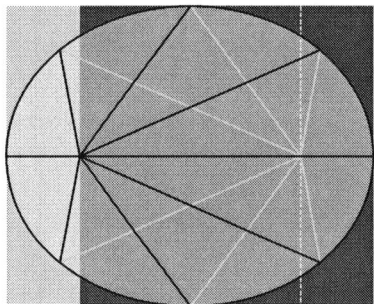

Figure 8-9: NM ellipse for $v=0.6c$

If we attack this diagram with a ruler and start measuring the ray-distances at different angles, we find that they do indeed correspond to the general shift relationships for emission theory. Measure the length of the diagonal ray at the nominal "90 degrees" position, and we find that, compared to our scale, it's elongated by a factor of $1-v^2/c^2$, which is the correct "transverse redshift" value for NM. All the other distances also agree with the numbers that we'd calculate for the theory by using nasty trigonometry.

velocities?

Our "wavelength" diagram is a bit abstract. It's tempting to try to turn this into something more literal by taking the distance between the focal points as the distance that the object moves in unit time, and the round-trip distance as the distance covered by light in the same time ("*v*" and "*c*"), but unfortunately the distance between these two points on the diagram can be different for the same speed under different models.

Why? Because different theories can interpret "moving" distances differently. We might try a one-way measurement of the distance between two points by timing how long it takes light to pass between them, but since this involves making assumptions about whether light is affected by the motion of bodies or not, and because we also can't usually manage to be in two places at once, the resulting *interpreted* distances can vary.

If we want a system to work with *multiple* theories it's sensible to avoid these "interpreted" properties and to try to work with more unambiguous data. A theory's predictions for the degree of Doppler-shift on light aren't open to reinterpretation, because to measure a frequency, we only have to be **in *one* place at *two* times**, which is a much easier thing.

8.6: Relativistic ellipse: Special relativity

As the only other basic relativistic model that we're familiar with is special relativity, we'll have a go at applying our ellipse approach to this, too.

With SR, the object moving at $0.6c$ shows forward and rearward Doppler shifts of $f'/f = 2$, and $f'/f = 0.5$, so inverting these numbers gives corresponding wavelength ratios of **0.5** and **2**, respectively.

We draw *these* in …

… then construct an ellipse around *these* two lengths,

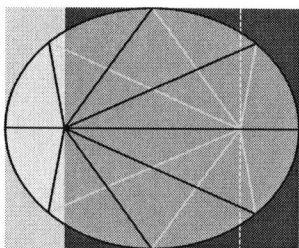

Figure 8-10: Corresponding SR ellipse for $v=0.6c$

, and when we start checking the diagram with a ruler we again find that, again, every modified wavelength at every angle agrees with the official textbook predictions of special relativity for the velocity we chose.

The method seems to be general. If we got *really* bored, we could start picking relationships at random: for instance, we *could* try constructing ellipses around the "other", "type **A**" Doppler equation, even though there's probably no known relativistic theory that uses it – the result may not describe a *credible* theory of relativity (we'll find out a bit later why this one doesn't seem to work), but the resulting predictions do show great deal of "SR-like" internal consistency.

similarities

These two ellipses for NM and SR (using the same value for *v*), look identical apart from their size. Since special relativity gives Doppler-shifted energies that are greater than their NM counterparts for a given nominal velocity by the same (Lorentz) ratio, by rescaling those two initial wavelength ratios by the same amount, we couldn't help but end up constructing a new diagram that was identical to the first one, apart from a rescaling.

This useful correspondence seems to happen for any relativistic theory that uses the same "relativistic" aberration formula: if we have a family of relativistic theories whose forward and backward "Dopplerised" frequencies diverge in an orderly way (section 13.4), once we know one theory in detail we can use it as a template for all the others.

9
Moving bodies drag light

> " I hold in fact
> (1) That small portions of space **are** in fact of a nature analogous to little hills on a surface which is on the average flat; namely, that the ordinary laws of geometry are not valid in them.
> (2) That this property of being curved or distorted is continually being passed on from one portion of space to another after the manner of a wave.
> (3) That this variation of the curvature of space is what really happens in that phenomenon which we call the **motion of matter**, whether ponderable or etherial.
> (4) That in the physical world nothing else takes place but this variation subject (possibly) to the law of continuity. "
>
> "On the Space-Theory of Matter" W.K. Clifford, 1876

> " On this point we are enlightened by a most important experiment which the brilliant physicist Fizeau performed more than half a century ago, and which has been repeated since then by some of the best experimental physicists, so that there can be no doubt about its result. The experiment is concerned with the following question.
>
> Light travels in a motionless liquid with a particular velocity w. How quickly does it travel in the direction of the arrow in the tube T ... when the liquid above mentioned is flowing through the tube with a velocity v ? "
>
> Albert Einstein, "Relativity ...",
> §13: Theorem of the Addition of Velocities. The Experiment of Fizeau

> " ... space-time is not necessarily something to which one can ascribe a separate existence, independently of the actual objects of physical reality. Physical objects are not **in space**, but these objects are **spatially extended**. In this way the concept 'empty space' loses its meaning. "
>
> Albert Einstein, "Relativity ...",
> note to the 15th Edition, 1955

RELATIVITY IN CURVED SPACETIME

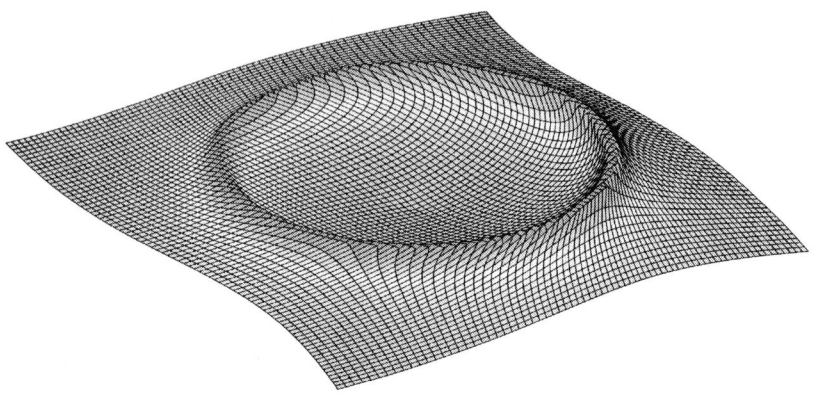

Distortions due to a rotating ring and a rotating solid

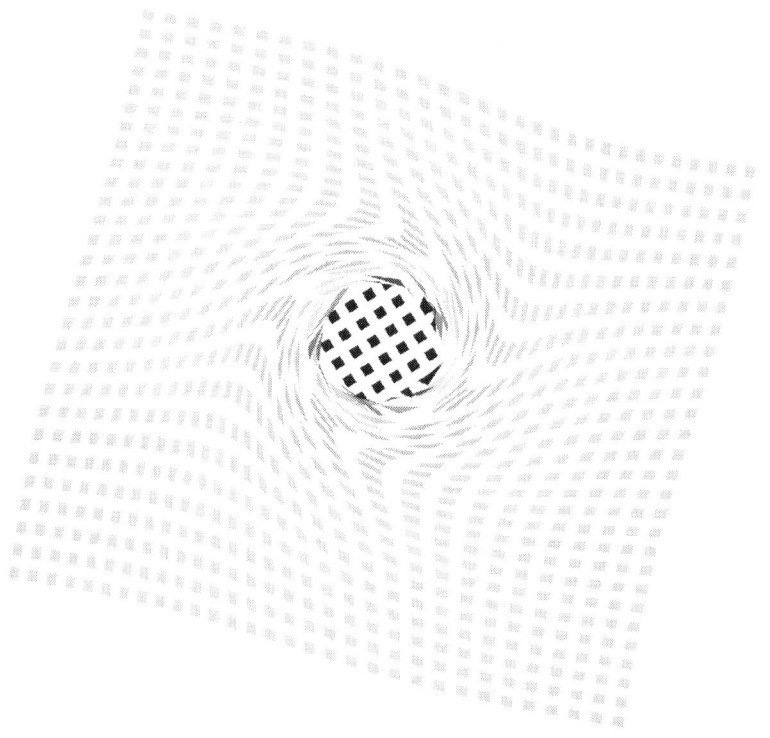

9.1: Generality of dragging effects

We've seen in section 4.14 that the **general principle of relativity** tells us that *relative acceleration* and *relative rotation* both warp spacetime in such a way as to cause nearby objects to be pulled along or around, and that spacetime warpage can be used in advanced gravitational models to describe and predict the physics of accelerating and rotating bodies.

Since those effects are fairly straightforward, in this chapter we'll be looking at the more controversial subject of how the *relative velocity* between bodies can also warp spacetime.

A range of arguments seem to confirm that an object's velocity relative to its surroundings really *ought* to warp lightbeam geometry, and the idea seems to be supported by the available experimental evidence. However, our standard derivations of special relativity depend on the idea that these velocity-dependent curvature effects either don't exist or aren't significant, so we'll be devoting a decent number of pages to this effect, to reassure ourselves that it really does seem to be real.

9.2: Naming conventions: Gravitomagnetism, frame-dragging

Moving *electric* charges produce *magnetic* side-effects, so when we're talking about field effects caused by moving *masses* (which can be thought of as *gravitational* charges), these get referred to in the mainstream literature as **gravitomagnetic** effects.

"**Gravitomagnetism**" is a very ugly name, and can give the misleading impression that we might be talking about something to do with *conventional* magnetism (which we aren't), but for now it seems to be the official textbook name for this class of effect. The broader subject of gravitation due to both static *and* moving gravitational sources has the even more staggeringly awful name, **gravitoelectromagnetism**, which at least has the saving grace that it can be abbreviated to "**GEM**". The clunky naming suggests that perhaps not so many people get involved in this aspect of gravitational theory – if it was a more popular subject, perhaps somebody out there would might have come up with a better name for it by now.

Popular discussions of GM/GEM usually focus on *rotational* dragging effects, with an emphasis on how rotating bodies should influence nearby gyroscopes (**Lense-Thirring** effect). The literature also often refers to these as "**frame-dragging**" effects. The "accelerational" frame-dragging effect is a more difficult thing to test, so it isn't discussed nearly so often – Nature provides us with many convenient examples of rotating or orbiting planets and stars that we might try to use to check these predictions (or at least to help us visualise them), but *forcibly-accelerated* planets and stars are rather more difficult to find.

9.3: Argument #1: Linear GM as a gravitational timelag effect

Because lightsignals take a certain amount of time to travel, an object *moving away* from us will seem to be closer than its "calculated" position (section 7.3). If "gravitational" signals follow similar rules to their "optical" counterparts, the object should also "feel" closer. Similarly, if the object is *approaching*, its timelagged image should make it seem to be further away, so its gravitationally-sensed location should (arguably) seem to be further away as well.

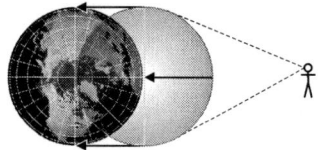

Figure 9-1: A retreating body looks nearer to us than its actual position

If we decided to take this argument seriously, we'd end up saying that an object's velocity should seem to affect the strength of its gravitational effects ... we should find that a body should appear to present a stronger effective gravitational field if it's moving away from us, and a weaker field if it's moving towards us.

This is a rather tentative argument, and it sounds slightly unconvincing – a critic could respond that perhaps we're just playing definitional games by comparing the object's apparent and deduced positions, and that perhaps it's only the *viewable* position of the object that counts, not its deduced position. Or that maybe we've missed out some important mechanism (such as **aberration**) that might conceivably cancel the effect. But it gives us a place to start.

9.4: Argument #2: "Effective gravitational potential" depends on relative velocity

Suppose that we drop a stone towards the Earth. The energy that the stone has when it impacts on the surface can be considered to be a measure of the gravitational potential between the stone and the Earth, a function of height and gravitational attraction. If the stone is initially 10 km above the Earth, it impacts with a given velocity and energy.

Now let's perform the experiment again, but this time we'll sneakily move the Earth away from the stone by ten metres during the fall. The Earth's slightly increased distance won't make a great change to the way that the stone falls, so when it reaches the point where the ground would be *expected* to be, it has almost the same velocity as in the first experiment ... but now it has an additional ten metres to fall, and as a result it hits the Earth noticeably harder. It ends up being given a greater change in velocity for the same nominal gravitational potential difference, because the Earth's gravitational field gets to act on it for a longer period of time.

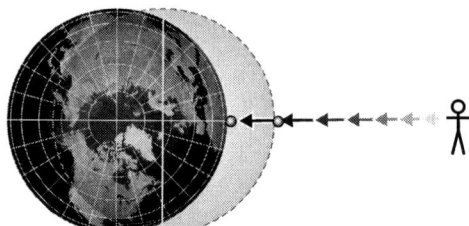

Figure 9-2: **Dropping a stone onto a moving planet**

We might also expect that if the Earth is instead moved very slowly *towards* the stone, the final impact energy may be reduced, because the stone won't get quite as long to fall, and won't undergo continuous acceleration for quite so long. The dropped stone would seem to be given a greater total gravitational acceleration when the planet moves away, and a smaller one when it approaches

How should we model this? Well, there are a number of approaches that we *could* try, but one approach would be to define the *effective* potential difference between the Earth and the stone not just on the textbook values of the Earth's and the stone's masses and their initial distances, but also according to whether they're moving towards or away from each other.

This is still a *slightly* weak-sounding argument ... it still relies on our assuming that there are no significant countereffects, and it still sounds as if it might depend on our choosing to define properties in a particular way ... but we're beginning to see some sort of pattern emerging.

9.5: Argument #3: Gravitational smudging

In advanced field theory, researchers sometimes like to blur the distinction between space and matter ... matter might be thought of as a sort of "condensation" of the medium, or a persistent geometrical feature of a metric, or space might be considered to be at least partly an extension of the matter and energy contained within it. As we move towards more advanced geometrical theories, the properties of an object can be increasingly said to belong to its *fields*, and to the distortion "footprint" that the object makes in spacetime.

If some or all of an object's mass can be considered to be *owned* by its fields, and the object's kinetic energy and momentum are imprinted on the metric in field form, the physics of how objects interact becomes the geometry of the interaction of their associated fields.

When we have a near-miss with a "moving" star, we're still impacting with a part of the star's exterior gravitational field, which should carry momentum. While a direct collision with the moving star will obviously give us a mighty thump, a collision with its "moving" exterior *field* might be described as an extended, *partial* collision, or as a sort of "collision by proxy".

If a high-density object skims our position at high speed, we should then expect there to be a sort of slipstream effect as the passing field "bump" hits us like a shockwave. Since a moving gravitational bump looks rather like a sort of **gravitational wave**, and gravitational waves carry momentum, it's not too much of a stretch to suggest that the impact with this moving gravity-well ought to pass some of the star's momentum on to us. As a result, the moving field carries momentum, the rear of a moving gravitational object should pull more strongly, its front should pull more weakly, and its field should show a tendency to move nearby objects along with it.

9.6: Argument #4: The slingshot effect

This sort of "fly-by" encounter is often used when planning the trajectories of space probes, to deliberately create close encounters between a probe and planetary bodies orbiting the Sun. The passing planet's moving gravitational field can then whip the probe into a faster trajectory without our having to spend extra money on rocket fuel.

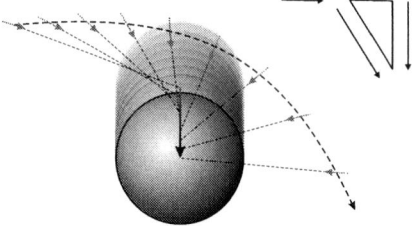

Figure 9-3: Gravitational "slingshot" effect

The art of planning slingshot trajectories is to throw your precious payload as close as you dare to the rear end of a moving gravitational mass, without actually hitting the thing.

If we throw a rock at the back end of a passing planet or star, the rock should speed up as it enters the body's gravity-well, and slow down again as it leaves. But it will also be pulled *to one side* as it enters the well, and will be pulled *to the same side* when it leaves, the end result being that, at the end of the fly-by, the rock will have increased its speed in the direction that the larger body was moving in. Since we aren't allowed "something for nothing", this increased momentum has to be donated by the larger body, which must then be moving a tiny, *tiny* bit slower after its encounter with our little rock.

The two bodies experience **momentum exchange**, *without* a direct collision, thanks to the interaction of their two gravitational fields.

9.7: Argument #5: Rotational GM and gravitational timelag

Velocity-dragging effects also fit nicely with the idea of the *rotational* dragging effects predicted by the general principle of relativity (section 4.14).

The plot below shows a circular disc, rotating clockwise, divided up into five-degree segments. An observer stands at the right-hand edge of the disc, and by calculating the timelag between this location and different parts of the disc, we can calculate the apparent geometry of these seventy-two straight radial lines, as seen by the observer.

The plot is rather idealised – it assumes that lightspeed is fixed for the observer – but it's still useful as an illustration of general trends under a range of propagation models.

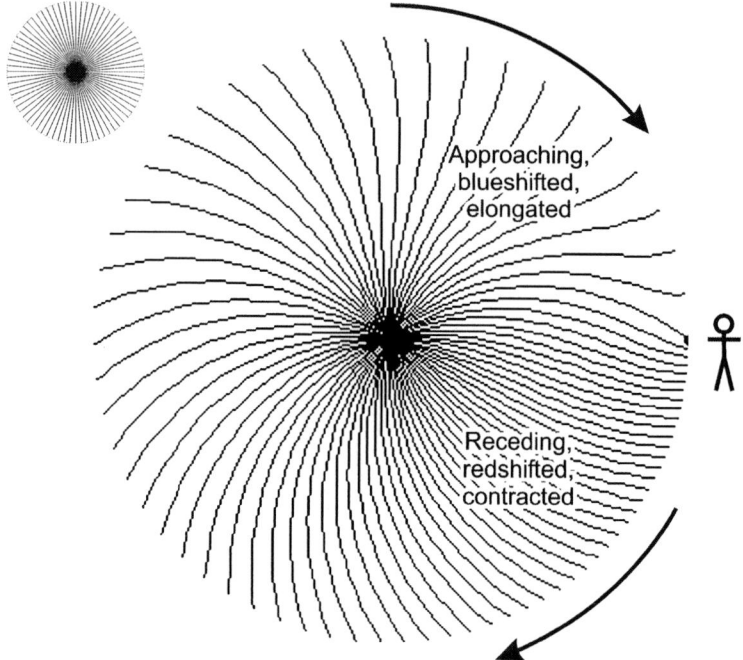

Figure 9-4: Apparent redistribution of matter in a rotating disc due to the finite speed of light

Our observer sees the widths of the receding segments of the disc to be contracted and those of the approaching segments to be stretched (section 7.1), and as a result they see the distribution of matter in the disc to be unbalanced, with the receding side of the disc appearing to own more of the disc's material than the approaching side.

Replacing the disc with a significant source of gravity (such as a star), then, if the apparent locations of the *gravitational* contributions of the different parts of the rotating star appear to coincide with the positions that they're *seen* to occupy, we'll expect the star's apparent centre of gravity to be offset towards the receding side: the star's receding horizon should pull more strongly than the approaching horizon, and the rotating star should tend to pull nearby objects around with it. If we assume that these velocity-dragging effects are real, the exterior field of a rotating star ends up with a "twist" that looks suspiciously like the result that we got from the general principle of relativity.

If we believe that these velocity-dragging effects *don't* happen, our counter-argument has to be very delicately constructed to avoid also accidentally "disproving" the GPoR.

9.8: Argument #6: QM and "probabilistic" smudging

While general relativity smudges a "gravitational" body's mass, energy and momentum into the surrounding region as aspects of its exterior gravitational field, quantum mechanics smudges these same properties into the region surrounding a tiny particle, as a **probability field**.

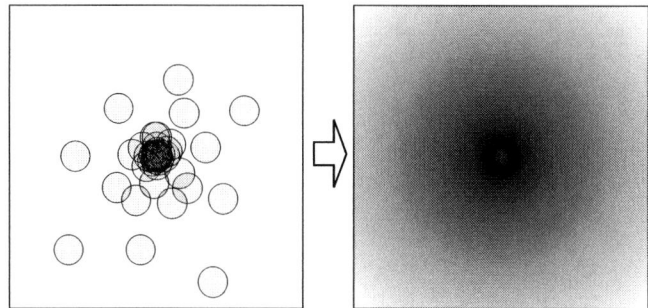

Figure 9-5: A particle's uncertain position makes it appear as a smeared-out "cloud", whose variation in density is described by a probability field

So, while a direct impact between a photon and a particle can result in an exchange of momentum (through absorption/reemission), light that is only travelling *near* a particle's official position also risks having its energy and momentum altered.

9.9: Argument #7: Experiment: The Fizeau effect

Finally, having considered the very large and the very small, we can consider the case of everyday, "middling-sized" objects. Is light dragged along by them as well?

In the early Nineteenth Century, **Augustin Fresnel** proposed that light passing between the atoms of a moving transparent material ought to be partially dragged along in the material's direction of motion, and put together a theory that predicted how strong this offset in the velocity of light ought to be.

In the mid-Nineteenth Century, when researchers were enthusiastically using new techniques to measure the speed of light in just about every situation they could think of, **Armand Fizeau** tested Fresnel's prediction with water flowing through glass tubes, and found that his results gave a workable match to Fresnel's prediction.

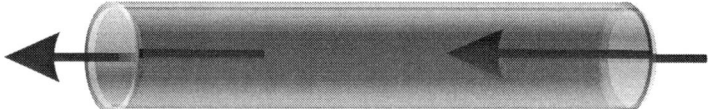

Figure 9-6: Fizeau Experiment:
Velocity of light in flowing water

Fizeau found that light moving in the same direction as the water moved faster than light sent in the opposite direction – the light seemed to be partly (but not completely) carried along by the moving water molecules. Fizeau's dragging result was then replicated by **Michelson and Morley** (1886) and by **Pieter Zeeman** (1914/5). The existence of this dragging effect doesn't seem to be disputed, even though it seems to conflict with the SR assumption that moving bodies *don't* drag light – we usually "spin" this result by saying instead that that a verification of *partial* light-dragging counts as supporting evidence for SR's **velocity addition** formula.

9.10: Inconsistencies in our approach to velocity

Our earlier arguments suggest that velocity-dependent lightdragging seems to be a pretty fundamental thing. Not only does it appear as a consequence of the GPoR, quantum mechanics, gravitomagnetism, and the idea that gravitational signals have a finite speed, but the "particulate matter" version of the effect has been accepted as a valid experimental result for the last century-and-a-half, and we already make good use of the "gravitational" version of the effect to accelerate space probes.

It seems that moving bodies warp lightbeam geometry, period. So how did Twentieth Century physics get away with deriving its fundamental relationships by assuming that there's *no dependency at all* between the speed of a signal and the speed of adjacent or nearby objects? Why do we regard SR's math as *correct*, and not just as an idealised first approximation?

The main argument that these light-dragging effects *must* be inconsequential goes something like this ...

Inconsistency #1: The "gravitational braking" problem

If we acknowledge that dragging effects are more than a tiny, *tiny* part of physics, we have to explain why they aren't a major part of our daily experience.

If a moving object pulls light *significantly*, and if a moving object with significant gravity has its effective pull increased or decreased by its motion relative to a test particle, then at first sight, Galileo's and Newton's observations that the principle of relativity applies to inertial motion would seem to have to be wrong. If an object moves with respect to the average state of motion of the background stars and galaxies, we might expect that, even if *individual* dragging effects were weak, the combined effects caused by *the entire outside universe* moving past the object should be significant, especially if we accept Machian arguments that background matter helps to define an objects inertia. If all the redshifted background stars and galaxies behind our moving object pull a little more strongly, and all the blueshifted ones in front pull a little more weakly, then we'd expect that the object should eventually slow to a halt with respect to the average state of motion of the background starfield, and that this average state of motion would then represent a sort of absolute preferred reference frame.

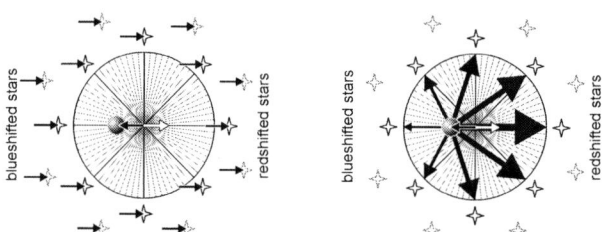

Figure 9-7: Velocity-dragging *seems* to spoil things ...

But we don't see this happening. We *don't* feel the background stars obviously wrenching us in different directions at different times of day and different times of year as the Earth rotates and also constantly shifts direction on its orbit around the Sun. Instead, simply-moving objects seem to continue on their own merry way until something interferes, regardless of what direction the background stars seem to be moving in.

This argument looks so strong that we may be tempted to throw the whole idea of velocity-dragging into the bin. Reason tells us that the effect *ought* to exist, and *must* exist, but experience tells us that we simply don't see it in our day-to-day lives. Why? Perhaps because of *another* effect associated with constant-velocity motion, that might just cancel it out ...

Inconsistency #2: The "gravitational aberration" problem

Lets say that we're in a spaceship, thundering towards the nearest neighbouring star at a significant proportion of the speed of light, and we switch off our engines. Basic Newtonian mechanics will say that we'll tend to continue coasting towards our destination until or unless something happens to change the situation. To "stationary" observers placed in front of and behind us, **aberration** calculations (section 8) predict that our lightsignals are redistributed to point more toward our direction of motion, distorting our apparent shape. A higher proportion of our spaceship's surface appears to face towards a "starfield" observer placed ahead of us that to one that is behind us.

Relativistic aberration *also* applies to the background starfield that we see outside our window: if the stars are equally distributed through the sky, then as we achieve relative speeds that are significant fractions of the background speed of light, the starfield should appear to warp and distort in such a way as to make the star positions appear to shift and concentrate in front of our spaceship. The faster our craft goes, the more strongly the background universe appears to be concentrated ahead of us (*see: e.g.* Scott and van Driel, 1970).

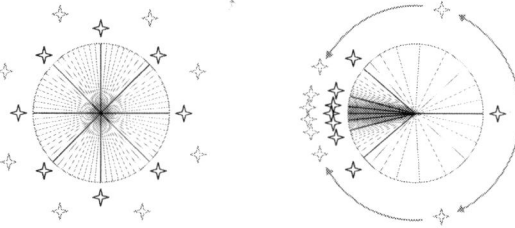

Figure 9-8: **Apparent angular redistribution of background stars ahead of a moving spaceship, at 90% of lightspeed**

This is another potentially catastrophic problem. Unless we can come up with some compensating mechanism, then if the gravity from each star seems to originate at its *optically-observed* position, our view of the "moving" universe no longer shows its mass to be evenly distributed around us. When we use photographs taken inside the spaceship to map the apparent locations of stars and galaxies, our maps should show more of the universe to be in front of us than behind us, forcing us to reason that our ship *ought* to be accelerating gravitationally towards the region of highest apparent background mass-density ... forward.

The spaceship would increase its forward velocity, stronger aberration effects would make the background distortions look *even more* severe, the ship would further increase in speed, and the aberration effect would strengthen again ... and again ... and again ...

This would be a disaster. This sort of **positive feedback** would mean that *any* "moving" object would tend to naturally continue increasing its speed relative to background starfield until it was travelling at relativistically-significant speeds. Physics in such a universe would be deeply unstable, and very different to what we actually see.

To avoid this geometrical carnage, some theoreticians suggest that perhaps the gravitational locations of moving objects *don't* coincide with their optically-observed positions, and that perhaps a gravitational body moving past us somehow "projects" its field forwards so that its gravitational effects seem to be coming from a location that it hasn't yet been seen to reach.

To outsiders, this "gravitational ventriloquism" can look suspiciously like a big ugly fudge, and creates more philosophical problems: How can we talk of there being a single metric that describes an object's apparent location, if we can assign *different* observed positions to the object depending on whether we choose to view it by using gravitational or optical instrumentation? Are there entirely separate "gravitational" and "optical" metrics?

9.11: Cancellation and unification?

But since our two "problem" effects act in opposite directions, perhaps they were never really problems after all, but *features* of a physics that we didn't quite appreciate at the time. Perhaps our "ventriloquism" workaround wasn't necessary.

Perhaps, instead of considering each of these two arguments as being at odds with experience, and rejecting them both, we should have held onto *both* of them a little while longer and then checked whether they might cancel each other out. *Together*, they might be exactly what we needed – equal and opposite mechanisms that could act in sympathy to create the circumstances that were required for the systems of Galileo and Newton to be able to function as well as they did. *Perhaps*, when an object moves at speed through a uniform homogenous environment, the increased gravitomagnetic attraction of the redshifted stars behind is *exactly compensated for* by their reduced number, and the increased number of blueshifted stars appearing ahead of the object is exactly compensated for by the gravitomagnetically-reduced attraction per star.

We might have got it wrong when we thought that it was a good idea to treat flat spacetime as fundamental and curvature effects as secondary. Maybe it's the other way round, and instead of saying that the dragging effects associated with moving bodies *must* be inconsequential because the background spacetime appears to be so flat, perhaps the spacetime background is only *able* to appear so flat because dragging effects play a key role in cancelling the aberration effects that might otherwise destroy this delicate balance.

Velocity-dragging effects might be critical to understanding how the whole thing works.

the "hollow sphere" analogy

How reasonable is it to expect full cancellation?

The closest analogy to our "starfield's field" problem seems to be the case of the cancellation of electric fields inside a hollow electrically-charged sphere. Theory says that if we charge up a hollow metal sphere, and place a test charge inside it, the test charge won't "see" any field gradients inside the sphere. We'll obviously expect this cancellation if we placed the test charge at the *exact centre* of the sphere, but if we *offset* the sphere with respect to the test change so that some parts of it are closer than others, we might expect the charge to be attracted or repelled more strongly by the nearest section of wall. But the charge stays put, because although its interaction with the nearest section of wall is *stronger*, the greater *quantity* of charged wall in the opposite direction, at a greater distance, exactly compensates.

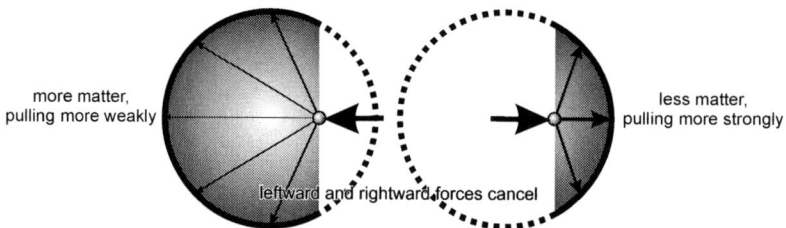

**Figure 9-9: Off-centre test particle inside a charged sphere:
Asymmetry in distances and asymmetry in angular
redistribution cancel**

For the electrically-charged hollow sphere, the "proximity" and "angular redistribution of charge" effects cancel each other out. This doesn't prove that cancellation also happens in our *gravitational* case, but it's an indication that the idea might not be completely daft.

9.12: Implementation – the tilted gravity-well

How do we implement these velocity-dragging effects as **velocity-dependent curvature**?

Well, we've already learnt that aberration effects distort the apparent geometry of a moving object, and for the object to be unaware of this supposed distortion, the apparent change in geometry should also affect *all* of its geometrical properties, including the apparent distribution of its *fields*. If the object's *gravitational* and/or *inertial* fields appear distorted, then relative motion will seem to produce a **polarised gravitational field** aligned with an object's direction of motion.

velocity-fields?

The properties of this apparent distortion-field seem to be able to produce a passable description of the moving object's appearance without actually mentioning that the object is moving. **Gravitational lensing** effects (section 1.9) can be blamed for the object's apparent change in geometry due to aberration, the field's polarised curvature can be blamed for the apparent changes in depth of the object when viewed from different directions (section 7.6), and the conventional Doppler redshifts and blueshifts seen on the object can be remodelled as *gravitational* redshift and blueshift effects.

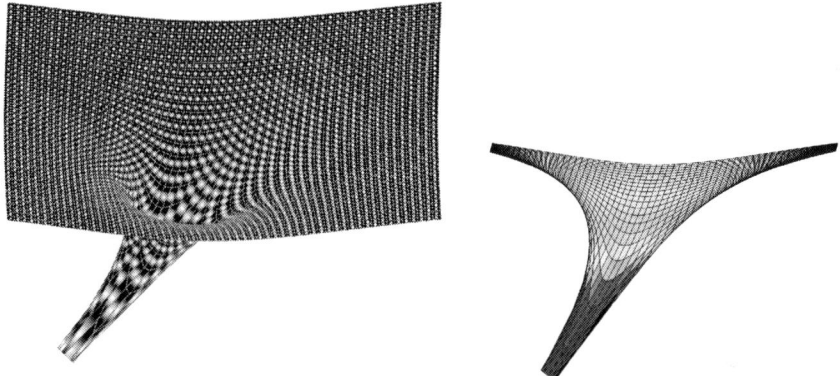

Figure 9-10: Cross-section through a tilted gravitational well (side-view)

We can try to recreate all these apparent geometrical effects in the moving particle's region by "**tilting**" the particle's associated **gravitational well**, with the degree of tilt giving its relative velocity, and with the resulting variations in curvature across the well giving our positive and negative curvature field components.

This approach seems to give us everything that we need, but it comes with a major limitation: it doesn't seem to work unless particles … *all* particles … are already associated with curvature. Increasing the object's relative speed may deepen and tilt its gravity-well, but it *must have a curvature-well to start with*. To obtain a forward gravitational blueshift, we need the object's default apparent rate of timeflow to be already built into the shape, so that we can reduce its apparent inertia by reducing the amount of forward positive curvature that the particle already has.

Since real particles all *ought* to have positive energy and a positive associated field thanks to $E=mc^2$, perhaps this isn't a such bad idea – if this system doesn't seem to work *unless* particles have positive mass and energy, and they do, we may be on the right track. But this approach isn't compatible with the flat-spacetime reasoning of special relativity.

To appreciate the *philosophical* value of this shape, we'll need to jump back a couple of thousand years, to meet up once again with **Zeno of Elea**.

9.13: Zeno revisited: the "impossibility" of motion

the "persistence" problem

We met **Zeno of Elea** and his paradox against *absolute* motion in section 4.3.

Another of Zeno's paradoxes is known as "**The Arrow**", and can be taken as an argument for the logical impossibility of *any sort of motion at all*. If we take a moving body and divide its history into smaller and smaller sections, then if this process is taken to its ultimate conclusion, we can imagine ending up with an instantaneously-short "**snapshot**" of the object frozen in an instant of time. In this snapshot, nothing can be moving. If the history of an object consists of an infinite series of such momentary snapshots, and if *at no moment in the object's history* does it have motion, then (said Zeno) motion can't be a fundamental property of the object itself, because this property doesn't exist when we isolate the object from its context. It only appears when we regard the object's properties *in time*.

Suppose that we throw a baseball across a field and record its motion with a movie camera. We then select the still image in which the ball was at the highest part of its ballistic trajectory. Our idealised "perfect camera" takes images with an infinitely short shutter speed so that the ball's image is completely unblurred, and according to Zeno's argument, if we take a still image showing the ball apparently fixed in mid-air, and then try to extrapolate forwards to find "what happens next", we should get a description in which the ball falls straight downwards towards the ground, vertically. The argument says that the ball's horizontal motion can't exist *in the snapshot*, and must be an additional property that only exists across a *series* of pictures are strung together in order.

With the physics of a couple of centuries ago, there was no obvious way to extrapolate forwards from a perfect snapshot of the ball to find what should happen next. How does the ball "remember" its state of motion from one instant to the next? *How does anything in the universe remember what it is supposed to be doing at any time*, if this information doesn't exist within any individual "$\Delta t = 0$" instant?

But as we work forwards to the Nineteenth and Twentieth Centuries, a possible solution begins to emerge. We start to appreciate that a perfect snapshot of a moving object does *not* look quite the same as a picture of a stationary object, since the photograph should preserve a record of the object's Doppler shifts and aberration distortions, and since these can be used to calculate the object's speed and direction. This "missing" information *is* encoded in the snapshot after all.

"motion without motion"

Looking at a "frozen instant" of an object moving through its environment, we still have a problem of interpretation: the photograph of the ball certainly *could* be interpreted as showing the ball moving, but it could also be interpreted as showing a ball immersed in a polarised gravitational field. Which interpretation is correct?

In a unified model, the two descriptions may be interchangeable. Is the redshift seen in a receding object due to its recession or to its gravitomagnetic dragging effect on light? As long as we can do the same calculations in each case and get the same answer, we really don't care.

Velocity-dependent curvature might finally be able to resolve Zeno's "arrow" paradox in a convincing way – the object's state of motion gets frozen into its associated spacetime distortion (captured in the snapshot), and this distortion can be used to explain the spectral shifts and other effects *without* having to involve time-variant effects. Just as audio signals can be analysed either in the **frequency domain** or in the **time domain**, we'd be able to model what we normally consider as velocity effects either in the time domain or (in a snapshot) in the **curvature domain**. The universe's "memory" of an object's state of relative motion would be carved into the shape of the "footprint" that the object makes in the background metric.

With the *electromagnetic* theory of light, **electric fields** and **magnetic fields** move hand-in-hand, continuously regenerating each other, and a "snapshot" of an electromagnetic field will contain all the information necessary to reproduce its earlier and later motion.

With our *gravitomagnetic* reinterpretation of moving-body problems we have a similar co-dependent relationship between velocity and curvature, so that if we start with a single isolated snapshot of a region, and then "re-start the movie", curvature effects can explain why the object then has to change position between images, and the motion explains why the curvature exists. The two things mutually define one another. Physical information written into the space and time axes will then not be totally distinct: there will be some redundancy between them.

9.14: Worldlines and curvature

At first sight, our "tilted gravitational well" description of motion seems to have a lot going for it. It seems to provide a geometrical description of basic physics, and as a bonus it also seems to incorporate some key aspects of quantum mechanics, because it contains a degree of observer-dependence: if we put a test particle into a region to "sniff out" the region's geometry, the distortion of the particle means that the data that it reports back might not necessarily relate to how the region would have been if the test particle wasn't there. This situation links **information theory** principles to curvature effects – if we want to measure a degree of frequency shift that can't already be seen by anyone in a region, we may have to give our physical sensor a state of motion that is different to that of existing objects in the region, which then changes the region's contours. If we can't extract or **mine** new information from a region without physically changing its shape, the transfer of information becomes linked to the idea of geometry-change.

pointing in the direction of time

A line through spacetime representing the path of an object with simple inertial motion can be said to be that object's **worldline**, and if the object is allowed to use the principle of relativity to argue that it *isn't* moving, it can claim that this line has a constant position in space, and that the only difference between different points on the line is their "time" values … the worldline gives the alignment of that object's **time axis**, and the line can be thought of as the local "**arrow of time**". Under current mainstream theory, we can take a "flat" region of spacetime and populate it with as many different criss-crossing "imagined" worldlines as we like, and use a variation on flat geometry to map how measurements taken on any one worldline will appear on others that intersect combinations of the same points, if the principle of relativity is going to work. This exercise gives us **Minkowski spacetime** (section 14.8).

But worldlines have a different status in our distortion-based model. We can no longer safely draw them in anywhere and everywhere, willy-nilly, because every *real* particle's *actual* location and state of motion is now reflected in the shape of the metric. Our **worldline forest** isn't a virtual mathematical fog describing *every possible* arrow of time that *might* exist if the numbers and states of particles in the region were unknown, but is the more limited collection of arrows of time that physically *do* exist in the region, for real particles. Instead of an infinite number of possible paths through each point in spacetime, each with their mathematical alignment of space and time axes, we have local definitions of time and space that normally only allow *one* time-arrow to point away from each point on the "space" surface – where a region's local distortions cause the **single-arrow rule** to break down, it suggests that the region either isn't occupied, or marks the location of the appearance or destruction of a **particle pair**.

If gravitomagnetic distortions are fundamental, a physical observer's perception of the apparent alignment of space and time axes is no longer just a matter of the mathematical projection of different coordinate systems onto a flat empty region: it becomes a more physical, *visceral* interaction between the observer and their environment.

9.15: Uh-oh ...

At this point, a 1960's-educated theorist might jump in and insist that all of the previous discussion is quite wrong. Certainly, our "curvature" model of velocity has some attractive features, and seems to implement **W. K. Clifford**'s vision of "**physics as curvature**", but this particular **geometrodynamic** approach to mechanics doesn't correspond to how things are supposed to be according to Einstein's 1915 general theory of relativity. GR1915 is *supposed* to reduce to special relativity and **Minkowski spacetime** (section 14.9), which models the worldlines of inertially-moving objects in a more abstract way. SR assumes that there are *no* velocity-based distortions, that the metric is distortion-free, and that although worldlines *can* be drawn through it, the metric itself contains no information as to the actual location or movement of particles, or even whether any particles actually *exist* in the region. Minkowski's spacetime is like a blank sheet of white paper, supposedly showing every poem that could possibly be written, all overlaid, and all written in white ink. With Minkowski's framework, the specific relationships of inertial physics – how the energy, momentum and so on of bodies must change with velocity – are predetermined by the shape and geometrical properties of this "empty space", not by the sort of interactively-distorted spacetime that we've just described.

The theorist can also point out that our ambition to unify velocity and curvature simply doesn't work in a physics that is built on special relativity, because SR relies on the idea that we can adequately describe the physics of inertial motion by assuming the total absence of any background gravitational fields or warpage, whereas our model doesn't work *without* them. The numbers that SR generates also don't seem to be reconcilable with our approach.

acoustic metrics

It seems that what we've accidentally done, by producing a description in which a line drawn through spacetime to represent the motion of a particle is meaningful *only if the shape of the metric agrees that the particle is physically there*, is to produce a special sort of surface that's sometimes referred to as an **acoustic metric**.

Acoustic metrics have recently come into fashion in as a way of modelling how **quantum mechanics** works in curved spacetime, and specifically, the complex behaviours predicted by QM around GR1915's **black holes**, that involve **Hawking radiation**.

These "velocity-disturbed" metrics can show deeply **nonlinear** gravitational behaviour, and give complex shapes that don't always project onto flat surfaces in a polite and well-behaved way. Projection onto a more idealised metric can result in "missing" coordinates and apparent breaks and discontinuities in information held in the "true" surface that won't always show up in an "**observer-space**" projection of it (section 11.16). The physics of an acoustic metric can show *apparent* **fluctuations** and small-scale **acausalities** due to the inadequacy of our projection surface, and these artefacts can look very much like the sort of things that happen under quantum theory. Particles should sometimes seem to appear from nowhere, for no reason, and then disappear again, and although this small-scale behaviour should look random or unpredictable at first sight, these seemingly acausal effects should build up to form a picture that still agrees with conventional statistical laws, because the underlying, "hidden" physics is still entirely classical. The physics of an acoustic metric already includes at least some of the sorts of quantum behaviour that are missing from GR1915.

So ... by taking these gravitomagnetic principles and dragging effects seriously – in the hope of jumping straight to a GR-style description of curved spacetime without paying too much attention to the conventions of special relativity – we seem to have accidentally overshot and ended up in theoretical territory that may be *beyond* GR1915, and that is currently being investigated by researchers interested in possible future theories of **quantum gravity**.

9.16: The score chart

We often come across statements in books about relativity that seem to say that the speed of light is independent of the source. This tends to be interpreted as meaning that there is *no dependency at all* between the speed of light and the speed of the object that it came from.

This isn't entirely true. Thanks to dragging effects, it's more accurate to say that there seems to be a *short-range* dependency between the speed of a lightsignal, the speed of the source, the speed of the observer, the speed of any material in the signal path, and the speed of anybody else who may have just happened to be passing through the neighbourhood at the time. The speed of the original source doesn't have any *special* power to dictate the speed of its signals once they've left the emitter's locality, but while lightsignals have an *intimate proximity* to objects, there should be at some sort of correlation between the velocity of an object and the velocity of any (nearby) light. This should be especially true inside a moving **particulate medium** (such as a block of glass), where the signal is always reasonably close to at least *some* of the particles making up the medium.

Light-dragging effects complicate the lightbeam-geometry of spacetime and may be very annoying and inconvenient for anyone hoping to come up with a theory of physics that can rely on a simple lightmetric and Euclidean geometry, but if we hope to produce an *accurate* model, they do seem to be a part of the way that Things Really Happen out in the real world.

There doesn't seem to be any situation in which moving matter *doesn't* drag light. Very large bodies drag light by their gravitational fields, very small particles may accelerate other nearby particles through the magic of quantum field theory, and medium-sized collections of atoms (such as the water in **Fizeau's** tubes) have been known to drag light for over one hundred and fifty years. A star will drag light, a water molecule will drag light, and so will an astronaut's visor, a spaceship's window, a camera's silicon sensor, or a potato.

Is it at all *important* that special relativity doesn't include dragging effects? We might feel that it's a shame that these effects aren't used as a starting point for the theory of moving bodies, but reckon that they can probably be retrofitted with additional layers of theory afterwards. We can try to append quantum theory to SR to reproduce at least some of the effects of the "smudging" of velocity effects at atomic scales, and GR1915 can be built on top of SR to try to put in some "missing" gravitomagnetic effects. At medium scales, we can set aside fundamental theory and apply a set of more pragmatic rules and laws for dealing with altered lightspeeds moving in and around moving particulate media. The only problem with these incremental layers of theory built on a common SR foundation is: they do not seem to be compatible with each other.

The clearest example of this seems to be the way that **horizons** are dealt with under these different theories: For a full implementation of velocity-dependent curvature, the *effective position* of a gravitational horizon will seem to behave like that of an **acoustic horizon** – it'll be a slightly artificial, mathematical surface that will appear to fluctuate and fizz and leak information in response to events occurring nearby. This seems to at least approximately coincide with the statistical description given by quantum mechanics. But this description is incompatible with GR1915's worldview, which needs gravitational horizons to be smooth, perfect, non-radiating surfaces that permanently seal off their contents from the outside universe.

In order to see how, where and why this disagreement arises, we need to learn a little bit about the theory of **dark stars** and **black holes**, and why they're different. And before we do *that*, we need to take very quick peek at **quantum mechanics** and find out why it's usually considered to be such a freaky subject.

9.17: "Relativistic" implementations of lightspeed constancy

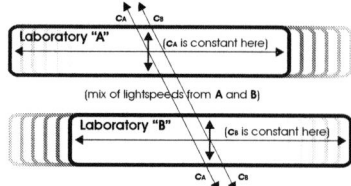

(1): In simple **ballistic emitter theory**, light is emitted at c with respect to its source, and having been thrown off at this speed, continues with the same speed until it hits an object or encounters a gravitational field.

If we have two sealed laboratories, measurements made inside each laboratory may show the locally-generated speed of light to be constant, but light leaking in from other laboratories or from otherly-moving objects would be measured as having different speeds.

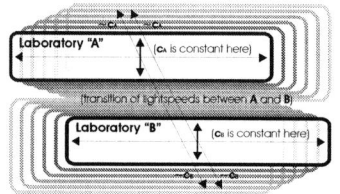

(2): In **dragged-light theories** (e.g. Fresnel), light is dragged along with a moving object, and if it's *completely* dragged, light moving between two objects can ride a dragging gradient, leaving the source with a speed of c_{SOURCE} and arriving at a detector with a speed of $c_{DETECTOR}$. *Relativistic* dragging might act as a mechanism for regulating local lightspeed constancy in a region containing bodies with relative motion. But these theories are complicated and tend to look slightly arbitrary.

(3): Under **special relativity**, we assume that light is *not* dragged: instead we insist that lightspeed is globally constant for all observers, and show that this result can be produced by requiring that all observers define distances and times in a particular way. Relative velocity causes observers' perceptions of space and time to be "skewed". One lab may then say that a lightbeam covers a shorter distance is a shorter time, another might say that it crosses a longer distance in a greater time, but both are forced to agree that the final calculated speed of the signal is the same according to each observer's calculations.

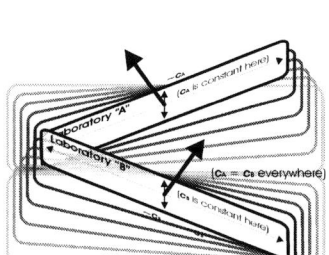

(4): With a **general theory of relativity**, *local* c-constancy is required, but *global* c-constancy isn't necessary. Variations in lightspeed between locations are associated with matching gravitational field gradients that keep everything nicely regulated.

We might also like to use GR's **gravitomagnetic effects** to implement the sort of relativistic light-dragging effects that we had in **(2)**, using spacetime curvature and gravitomagnetic fields as our missing light-dragging mechanism. But GR1915's design feature of reducing to special relativity seems to prevent us from doing this.

This makes is difficult to judge the true status of gravitomagnetic velocity-effects under GR1915.

PART IV

UPDATING STANDARD THEORY

RELATIVITY IN CURVED SPACETIME

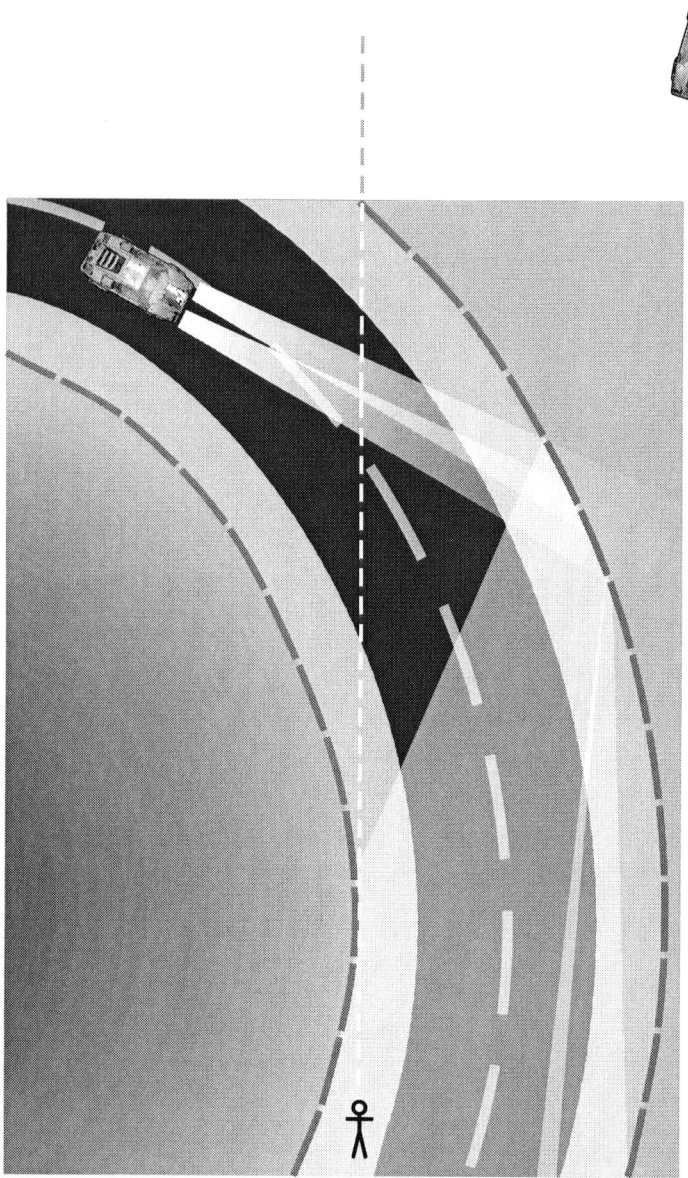

A car rounds a curved section of road at night. There is no direct line of sight from the car to our observer. The buildings lining the road hide it behind an *observational horizon*. However, the car's headlights reflect off the glass windows of the shops running alongside the road and illuminate the observer, who sees the light to be coming from fragmented, fluctuating images of the car *outside* the horizon.

10
Quantum Mechanics and Observability

> *" The human understanding is like a false mirror, which, receiving rays irregularly, distorts and discolors the nature of things by mingling its own nature with it. "*
> **Francis Bacon, "The New Organon" 1620**

> *" ... Possibly I did use this kind of reasoning, but it is nonsense all the same. Perhaps I could put it more diplomatically by saying that it may be heuristically useful to keep in mind what has actually been observed. But on principle, it is quite wrong to try founding a theory on observable magnitudes alone. In reality the very opposite happens, it is the theory which decides what we can observe. "*
> **Einstein, to Werner Heisenberg (1926), in the conversation that Heisenberg credited as the inspiration for his uncertainty principle**

> *" The question of the particular field law is secondary in the preceding general considerations. At the present time, the main question is whether a field theory of the kind here contemplated can lead to the goal at all. By this is meant a theory which describes exhaustively physical reality, including four-dimensional space, by a field. The present-day generation of physicists is inclined to answer this question in the negative. In conformity with the present form of the quantum theory, it believes that the state of a system cannot be specified directly, but only in an indirect way by a statement of the statistics of the results of measurement attainable on the system. The conviction prevails that the experimentally assured duality of nature (corpuscular and wave structure) can be realised only by such a weakening of the concept of reality. I think that such a far-reaching theoretical renunciation is not for the present justified by our actual knowledge, and that one should not desist from pursuing to the end the path of the relativistic field theory. "*
> **Einstein, Relativity ..." Appendix 5**

> *" I think I can safely say that no one understands quantum mechanics. "*
> **Richard Feynman**

> *" There are two opposing armies drawn up on the field. But there is a heavy fog, they can't see each other. Oh, they **want** to, of course, very much. You are in the gap between them. You can just see us, you can just see them. ... That's where you are, Quiller. In the gap. "*
> **"The Quiller Memorandum", Michael Anderson (1966)**

> *" That is **preposterous**! Your Honor, the Fermi-Dirac function is, for any system of identical fermions in equilibrium, the probability that a quantum state of energy, E, is occupied! My WORD man, **don't you know your quantum statistics?** "*
> **Pinky and the Brain ("Of Mouse and Man"), 1995**

Tree reflected in water

10.1: The origin of quantum mechanics

At the start of the Twentieth Century theoretical physics faced a crisis known as **the ultraviolet catastrophe**. Up until this point, conventional thermodynamics had seemed to suggest that a "hot" atom could emit radiation however it liked, and since the higher wavelengths carried away energy more efficiently, the atom *ought* to find it easier to lose its unwanted energy by making use of the higher-frequency end of the electromagnetic spectrum – we'd expect the distribution of frequencies in thermal radiation to be strongly biased towards the higher end. But the emission spectrum that we *measured* for hot bodies didn't show the expected statistical curve: we didn't understand why the atoms in a hot cup of tea weren't spewing out their thermal radiation right up into up into the x-ray and gamma-ray range.

Statistical analysis showed that the measured spectrum was oddly consistent with the idea that atoms could only absorb or emit energy in discrete quantities, or **quanta**. Since the energy in a light-signal is proportional to its frequency (section 3.5), an atom might throw out all of its spare energy as a single **photon** of light at a particular frequency, or as several successive photons at lower frequencies, but it *didn't* have the option of emitting *half* a photon at a *higher* frequency: there was an **amplitude** threshold that had to be crossed, and an atom didn't have the option of giving out a "blip" of gamma-ray radiation unless it had enough energy to cross the emission threshold and give off a ***full quantum*** of energy in that frequency-range. It seemed that energy levels inside an atom only had the ability to jump between certain well-defined states, and as an atom snapped between these states, it could only absorb or emit corresponding fixed amounts of electromagnetic energy. When we tried to unlock the secrets of atomic physics, it seemed that all access to the atom's interior was **quantised**.

As this idea began to be taken seriously, we realised that it had some serious implications. Quantum mechanics seemed to say that the best tools that we had for probing small regions of spacetime were crude and clunky, and the quantised nature of all *interactions* between matter and light meant that we couldn't study the finer detail of systems in a passive, non-destructive way: in order to probe an atom at smaller scales, we had to use smaller wavelengths of light, which carried higher energies: but since we couldn't compensate for this increased energy by reducing the "loudness" of our probe signal below the quantum threshold, we had to accept that below a certain size, the energy that we'd have to pump into a region to sense and interact with fine detail would be so drastic that it could destroy or dramatically change the very systems that we were trying to measure. The energy concentrations that we needed to *see* tiny particles were in the same range as the energies required to create or destroy them, and we found ourselves relying on **probabilities** as it became increasingly difficult to distinguish between pre-existing effects and those caused by our own attempted measurement processes.

Sensitivity to measurement technique when studying delicate systems isn't something that's unique to quantum theory. In the field of electron microscopy we sometimes have to ask ourselves whether certain details really do have an independent existence or whether they might be side-effects of our complicated sample-preparation process, and in other more "mundane" businesses such as market research, the act of putting someone with a clipboard outside a supermarket to ask passers-by their attitudes to a product may end up causing those people to have thoughts that they mightn't have had if we hadn't asked. Awareness changes behaviour – this is part of why large advertising campaigns are successful – but what made quantum mechanics different was that it seemed to set *fundamental limits* to what we could measure *on principle*, making the measurement process an integral part of the physical model.

It's easy to become frustrated at the idea that the interiors of atoms keep trying to hide themselves from us, but perhaps this behaviour is a good thing: if information *could* seep gently out of atomic nucleii, then so could energy – this *inability* of atoms to allow energy to trickle out may be part of what gives them their stability, and perhaps without these quantised properties, regular, dependable atoms as we know them might not be able to exist.

10.2: Is quantum mechanics a theory?

This is a difficult question to answer. When we theorise that light can only get into and out of an atom in quantised amounts, and successfully use this to predict and explain the spectral properties of atoms, then this *is* a nice example of a "theory" making predictions based on the consequences of a suggested (theorised) physical behaviour.

The idea that light is only absorbed and emitted as "quanta" certainly counts as a theory, but the associated probabilistic tools that we had to develop in order to model light emission and absorption and atomic theory then ended up being pressed into use to model other effects that were difficult to analyse or model by conventional means, and in some of these areas, we no longer bother to imagine a "conventional" mechanism for how physics might be operating.

meta-theories

Quantum mechanics has grown into a larger model or system for doing physics (incorporating **statistical mechanics**) that sometimes has the ability to crank out answers for what *should* happen, without supplying conventional explanations for *why* it should happen. This doesn't make everybody happy: while some theorists may argue that physics theory is simply about making "hard" physical predictions about how the universe works, other theorists get a bit upset that they got into this line of work in order to *understand* something about how our universe works, and QM sometimes seems to be depriving them of this. Perhaps an intermediate position is to regard QM as a **meta-theory**: a set of rules and laws that sets limits and values on the predictions that "conventional" theories are allowed to make, without necessarily getting into specifics about the precise mechanisms involved.

predictive confidence

Quantum theory is often criticised for being **counterintuitive**. If we don't have a deep intuitive understanding of the character of a theory, then when that theory is used in unfamiliar circumstances it runs the risk of being misapplied, and used to churn out results that could be horrendously wrong. To be able to apply a theory confidently in new areas without making these mistakes, it helps if the theorist appreciates the difference between "sensible" answers and "junk" calculations, and with quantum mechanics it can sometimes even be difficult for experts to agree whether or not a "new" QM prediction is valid, until it's been experimentally verified or found to have some precedent in another field of research.

This has been a long-standing problem with QM: while it makes a quite excellent framework for taking in known experimental data and remodelling it, its unfamiliarity can make it difficult to tell, when a "new" prediction is made, whether or not the calculation is legitimate.

toy models

When experimental support is unavailable we often find theorists resorting to analogies with other systems that show similar or equivalent statistical behaviour in a more familiar context. If a well-known set of classical equations for the behaviour of a fluid happens to produce the same statistical behaviour as some aspect of QM, then we can use the "classical" prediction as a **toy model** of the QM problem. The "toy model" acts as a more familiar visualisation tool that helps us to avoid making silly mistakes, without our needing to believe that the underlying classical mechanism in the toy model has a literal QM counterpart.

Quantum mechanics poses difficult questions about how knowledge and science interrelate. If a quantum approach defies attempts to use "normal" language to describe what might really be going on in a situation, and is capable of generating the right answers without our knowing or understanding (or perhaps even caring) what the underlying mechanisms might be, then how far are we justified in claiming that those deeper mechanisms really do exist?

10.3: The "Copenhagen" and "Hidden Variable" interpretations

There are many opinions and camps in the QM research community, but researchers generally tend to agree on two things: firstly, that that quantum mechanics is unreasonably effective, and secondly, that the people in the other camps are quite, quite mad.

The problem of how to interpret quantum theory, and indeed *whether it was even valid* to interpret quantum theory, became the subject of serious debate in the 1920's and 1930's, with **Niels Bohr** championing the idea that quantum mechanics could be considered as an entirely self-sufficient system whose probabilistic methods needed no further explanation, and with **Einstein** (and **Podolski** and **Rosen**) arguing the opposite case, that it was simpler to assume that "classical" physical behaviour was still operating below a quantisation threshold that prevented us from being able to see the "real" physics.

According to Einstein, quantum theory's *probabilistic* descriptions were **incomplete**, and represented a statistical model of some more "sensible" underlying reality whose exact parameters weren't *directly* measurable, thanks to the fact that every tool available to use showed irritatingly quantised behaviour. This "interpretationalist" view became known as the **Hidden Variable Interpretation (HVI)** of quantum mechanics.

Bohr argued that since we couldn't *directly sense* any underlying properties, and since our probabilistic models seemed to be able to produce predictions that perhaps couldn't be bettered *on principle*, the probability-based approach correctly described reality, and didn't need to be "dressed up" with unnecessary mechanisms that wouldn't obviously improve our ability to make predictions. Bohr lived and worked in Copenhagen, Denmark and his position, that QM's statistical descriptions should be taken literally as "observerspace" theory, became known as the **Copenhagen Interpretation (CI)** of quantum mechanics.

Both camps were able to argue that their own viewpoint was the most efficient:

- **The Copenhagen Interpretation** (Bohr) argues that what we sense, see and measure is what physically exists. If every method at our disposal for *measuring* low-level light always reports the measurement of a number of **photons**, and every method we have for *generating* light can only produce it in photonsworths, then the measurable *reality* of light is that it *exists* as photons. If this sort of approach leads to odd-sounding consequences, then so be it. If the independent existence of a "classical" low-amplitude wave can't be verified (because all our measuring equipment is made of atoms and "quantises" its results), then the existence of this wave is merely *interpretational* while the existence of photons is *physical*. The classical wave description is therefore superfluous.

- **The Hidden Variable Interpretation** (Einstein) applies Occam's Razor differently. According to the HVI, quantum mechanics is describing conventional-looking physics, garbled by a quantised measurement process that is responsible for QM's more "spooky" aspects. Since QM-style effects can also appear in conventional physics as *statistical artefacts* (for instance, in **digital audio**) it's simpler to assume that the existing classical mechanisms operate down to and below the QM quantisation threshold than to introduce another set of rules. Instead of saying that QM has to be explained by recourse to "spooky" new physics that operates according to different rulebook, it's simpler to suggest that only one set of laws are in operation. The more "radical" CI interpretation of QM is therefore unnecessary.

After several rounds of arguments, with Einstein insisting that the Copenhagen Interpretation *must* be incomplete and *shouldn't* work properly, and Bohr showing that it really *did* seem to be consistent in each of the counter-examples that Einstein had produced, the general consensus seemed to be that the "Copenhagen" group had won the debate.

10.4: The two-slit experiment

The debate between the two camps provoked some deep philosophical soul-searching to do with the nature of reality and observation, and with the correct application of **Occam's Razor**.

If low-amplitude light can only be *generated* in quantum amounts (as multiple photonsworths of energy), and can only be *seen* or *measured* in quantum amounts (again, as quantised collections of photons), then how legitimate is it to claim that low-level light travels *between* these locations as a conventional classical wave, when nobody is actually watching? How far are we really entitled to claim that the light *moves* from **A** to **B** as a classical wave, given that every time we try to measure the properties of this wave, our (quantising) measurement equipment obstinately returns a "photon-based" answer?

Having found a method that could model the *statistical probability* of sensing a photon in any given situation, it seemed a shame not to use these equations as a description of how reality "really" was, even when this required some extreme reinterpretations of older effects that we might previously have considered to be pretty straightforward. How these reinterpretations work in practice is nicely illustrated by **the two slit problem**.

classical wave-interference effects

If we take a two-dimensional representation of a classical waveform with a single frequency spreading out into space, like this:

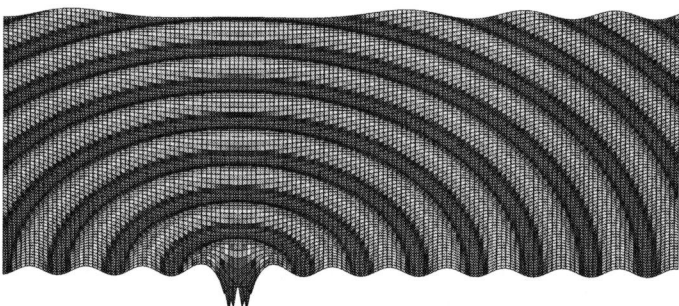

, and then add a second identical wavesource that is exactly synchronised (or at least, synchronised with a constant time offset), then we get this:

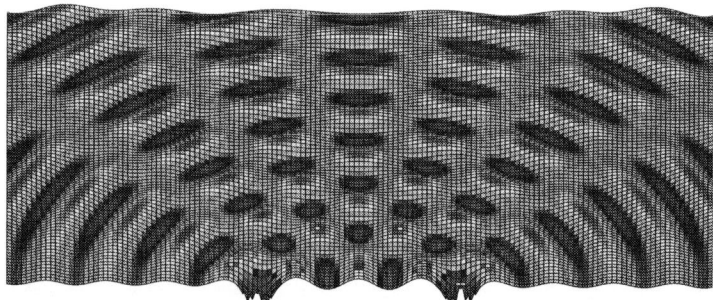

There will be some locations where the two arriving waveforms will always be synchronised and will reinforce each other (**"in phase"**), and there will be other places where the two signals will be exactly out of step (**"out of phase"**), and will cancel each other out. If the resulting light illuminates a screen, a single lightsource will provide reasonably uniform illumination, while the pair of synchronised sources will project an **interference pattern** on the screen, producing a pattern of alternating "bright" and "dark" bands.

10: Quantum Mechanics and Observability

We find it easier to create the *effect* of having two identical synchronised signal sources by taking the light from a *single* source and splitting it, by putting a barrier around the light-source that only lets light through at two narrow holes or slits. At the other side of the barrier we then see something that looks like a matched pair of phase-locked lightsources.

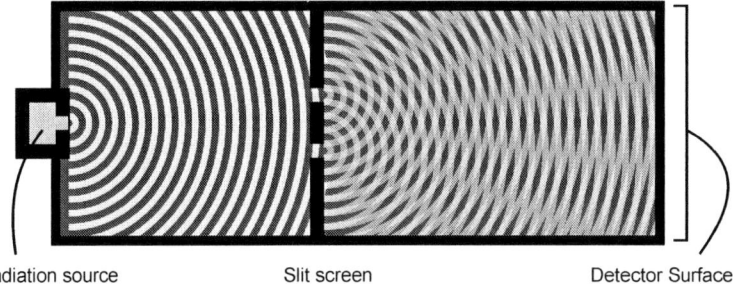

Radiation source   Slit screen   Detector Surface

Figure 10-1: The "two-slit" apparatus

Next, to try to find out whether light "really" travels as waves or particles, we can reduce the intensity of the lightsource until it is so weak that it's only capable of sputtering out individual single-photon bursts of light, and record what happens:

(a): "photon-wave-photon" interpretation of the two-slit experiment

With the Hidden-Variable Interpretation, the weak lightsource eventually accumulates enough energy to be able to send out a single quantised pulse of light (a "photon"), which produces a wavelike electromagnetic disturbance that spreads out through the box, passes through the two slits, and forms an interference pattern with itself against the screen. Since our detector screen can only register the existence of this light by *absorbing* light in quantised amounts, much of this light won't register. However, given a certain amount of random background "noise", from time to time we'll expect this background noise to lift the level of the interference signal above the detection threshold of our electronic camera or photographic film. When this happens we see a flash, or a chemical reaction is triggered on part of the photographic film. We can't predict exactly *where* and *when* these individual flashes will happen (because they're partially triggered by noise), but we know that if we wait long enough, the flashes will eventually build up into a reasonable image of the light and dark bands that we know ought to be in the region. The resulting pattern may look "noisy" if we use a shortish exposure time, but if we run the experiment for longer periods, or overlay a sequence of images taken with the same equipment, the noise in the different images will partly cancel out, while the interference bands will always reinforce each other. The longer the exposure time, or the more images we superimpose, the smoother the resulting image will be. Any unabsorbed light bounces around the box interior and contributes to the levels of background noise.

(b): "photon-only" interpretation of the two-slit experiment

With the Copenhagen Interpretation, we produce a single photon at a time, which then travels through the apparatus and hits the detector surface. This photon is directly responsible for the flash seen on the camera sensor, or the speck seen on the photographic film. Most of the photons won't make it to the other side and won't register, but as before, we can fiddle with the lightsource intensity until individual "flash" events register at a useful rate, and can take a long exposure so that these individual flashes build up to form some sort of image.

We *might* then expect that since we are only generating a single photon at a time, and since this can only pass through *one* of the two slits, there can't be an interference pattern (because there is no second photon moving through the second slit for it to interact with). But quantum mechanics has a few tricks up its sleeve that it can use to recreate the earlier result. We can say that the **probability** of sensing a photon in any region of space inside the box can be mapped as a **probability function**, and the probability **wavefunction** goes through *both* slots and spreads out, overlapping with itself. The likelihood of a photon appearing at the screen at any position is governed by this probability waveform, and if the waveform interferes with itself, we'll get the same final pattern of light and dark bands as a statistical effect.

lies, damned lies, and statistical mechanics

This is where someone new to QM will usually start frowning. In our everyday experience we don't usually see probabilities showing these sorts of cancellations and interference effects. But our QM lecturer can shrug and say that, well, *this* is how things work under quantum mechanics, and the pattern of light and dark bands shows that the description is valid whether we find it intuitive or not. If we insist that the interference pattern can't be created by the existence of a slit through which no light is actually passing, the lecturer can say, well, when the flash happens on the screen we have no *absolute* way of knowing which slit the photon went through, so in a purely *probabilistic* sense, a photon hitting the centre of the screen could have a 50:50 chance of having come through *each* slot, so in some sense we might say that ghostly "half-particles" actually passed through *both* slots, causing the interference.

Again, we might not *like* this explanation, but it produces the correct result.

The lecturer can then take things one stage further. In the **Many-Worlds Interpretation** of quantum mechanics suggested by **Hugh Everett**, the mysterious probability field represents the sum of all possible outcomes for the photon, and these different outcomes *really do all happen*, but in different "alternative versions" of our universe. In the universes where the photon travels through slot **#1**, these photons are able to interfere with their counterparts in the universes where the photon travels through slot **#2**, and the interactions between all these parallel universes then produce the same final interference pattern, in each universe.

The "many-worlds" interpretation is too much even for some quantum theorists, but it does illustrate that when it comes to QM, no suggestion is "too mad", as long as it's consistently worked through and converges on the correct final answer.

Physically, the explanations **(a)** and **(b)** seem to be indistinguishable. Every time we try to find some little wrinkle in **(a)**'s hybrid description that we think can't *possibly* be reproduced in a "photon-only" description, some bright quantum theorist seizes on it and announces yet another crazy-sounding quantum effect that exactly reproduces the missing effect. In description **(a)** we might expect background noise to sometimes result in two flashes being triggered at different locations on the screen from a single source photon, or (if we are less lucky) in no flash at all. We might think that if we see these inconsistencies we'll know that the "photon" explanation is wrong and that we're seeing the results of classical wave behaviour ... but in **(b)** we can invent descriptions in which the fluctuating background gives a finite probability of a photon being absorbed in flight, or of the *idea* of the photon being in two places at once gets "physicalised" thanks to energy donated by the fluctuating background field. Or a photon that "goes missing" in one universe is passing through a passageway in the quantum foam and emerging as a "spare" photon in another universe.

In science we usually try to come up with ideas that don't sound too crazy. With QM this isn't so important – "crazy" is just fine, as long as the thing is consistent and memorable. Instead of a physical model being used as an *explanation* or a *justification* for why a set of mathematics emerges in a particular form, with QM the mathematics often comes first, and the associated description becomes a convenient story that we invent afterwards in order to try to make sense of what the mathematics is telling us. The challenge for the QM theorist trying to do original research is often instead to come up with an idea that is *so* crazy-sounding that it won't already be in print somewhere, with someone else's name attached to it. This isn't always easy.

10.5: Quantum mechanics and everyday experience

QM's "spooky" effects often turn out to have "mundane" everyday counterparts:

Take QM's **principle of complementarity**: QM arguments say that we can *in principle* only measure the **speed** and the **location** of a particle with a particular combined accuracy: the two properties are **complementary** – the more accurately we measure the speed, the more vague its location becomes, and vice versa. But complementarity is also a feature of marketing statistics in business. If we want to study the *bulk* sales of confectionery in North America we'll tend to apply classical rules, but when it comes to predicting the time and location of the sale of an *individual* chocolate bar (these sales are, or course quantised), we can predict *either* the location of a "sale-event" *or* its timing with good accuracy, but not both together: we can set strict windows on timing and say with great confidence that someone, somewhere, will buy a candy bar in North America between 9 am and 9:01, or we can predict with great confidence that someone will buy a bar at a particular counter within a four-hour time window, but we can't use our statistics to predict both time and location together: the principle of complementarity applies. Similar correspondences between the rules and principles of quantum theory and day-to-day life show up in life insurance (people are quantised), car insurance (accidents are quantised) finance (transactions are quantised), politics, military intelligence, economic management, visual recognition systems and almost any branch of human endeavour that deals with information and prediction and categorisation in highly complex networks.

We live our lives surrounded by deeply complex and chaotic systems, and sometimes we see "weird"-looking relationships emerging, with strange patterns of events emerging that appear to show "spooky", quantum, **nonlocal**, "**synchronistic**" behaviours, when these things are actually resonant aspects of a broader and more complex (but more mundane) underlying system. For instance, if you turn up at your regular bus stop in the hope of hopping onto a bus, then if you already know the rough frequency of the service and how popular it is, you'll probably guess how long you'll have to wait for the next bus, based on the length of the queue. At first sight this calculation seems nonclassical (or even nonsensical): how can a "random" group of people, *none of whom* might any special knowledge of the position of a bus that hasn't arrived yet, calculate where that bus is simply by counting each other? None of our individual travellers might have had any idea where the bus would be when they left their homes, and yet the sum of all these trivial and apparently unconnected life-events mysteriously translates into a usable prediction about the location of a bus that hasn't yet been seen by anyone involved ... It's a **system-level** effect: what the length of the queue *really* tells us is the number of people that *aren't there* ... when the previous bus left, it removed the previous queue, and this "lack of people" in the queue then decayed statistically with time as more passengers arrived – so the queue length actually tells us how long ago the *last* bus left, and in a well-behaved transport system, this suggests how far away the next bus is likely to be – we're already unconsciously using "particle physics"-style logic in our daily lives without being particularly aware of it.

If we visit a library and flick through a book on advanced QM, and then do the same with a book on "cost-benefit analysis" or some other branch of advanced accountancy, many of the equations and graphs will look the same, and for some years, companies that dealt on the stock market were hiring QM-trained physics graduates, because these people were already familiar with the sort of statistical analysis used in corporate-level financial planning. What we think of as "atomic" quantum theory seems to have broader applications and implications for the chaotic human networks and systems that we depend on, and perhaps it's only a matter of time before someone wins a Nobel Prize for a full quantum theory of economics.

It's often said that quantum theory only applies to the very small, but this isn't strictly true – lots of things, even very big things, can be unpredictable and sensitive to measurement, and pollsters who attempt to predict the outcomes of major elections know from bitter experience that just because a thing is "big" doesn't always make it predictable.

10.6: Illusion and reality

buildings and lakes

Suppose that we're in Central Park, watching a tall building's reflection in the lake. As the wind blows against the surface of the lake, it creates ripples that cause the building's image to wobble and break apart.

Our (reflected) view of the building shows some strange behaviour:

As the water ripples, parts of the building seem to warp and bend, and although we might *hope* to see the whole building staying connected to itself, if we look closer we may see pieces of wall appearing to bulge, disconnect and leave the main body of the building and disappear, and other sections of building seeming to do the same trick in reverse, appearing in mid-air, then linking up with the main chunk of building and merging into it. As the reflected images shift and fluctuate, some pieces of the building seem to briefly disappear altogether, and other pieces seem to be briefly duplicated.

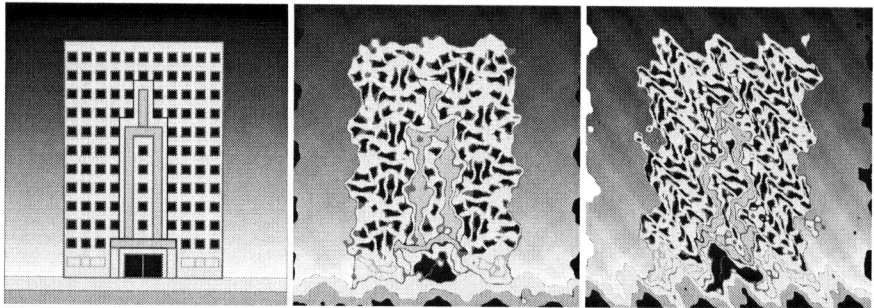

Figure 10-2: Fluctuating views of a solid building

How badly the image seems to fluctuate partly depends on the scale of our observations in space and time. If we were to observe our building for a longer period of time (perhaps by taking a photograph with a very long exposure, or by combining a large number of separate differently-distorted images), we'd get a more accurate picture of the object's real shape. Our composite picture or long-exposure photograph will be a little fuzzy and indistinct, but it would tend to become sharper and less fuzzy as we allowed ourselves longer observation periods. Our reflected building's image appears to show quite a few of the characteristics that quantum field theory describes at small scales, the difference being that with our reflected building we know enough about the *real* behaviour of skyscrapers to be able to confidently dismiss these "odd" effects as **optical illusions**, effects that are merely interfering with the signals in the region connecting us and the thing that we're trying to look at.

But if we had no personal experience of the sort of object that we were trying to view, we might not be so confident in explaining away this behaviour as an illusion – if we'd never *seen* a building and the only views that anyone could possibly have (or could ever have had) of our building were "wobbly", it'd be difficult to argue that the object wasn't *genuinely* wobbly, and it'd be difficult to justify saying that there was definitely a more conventional solid structure hiding somewhere beneath our fluctuating observational curtain.

A mathematician may prefer to explain the apparent behaviour of the building by analysing its patterns for a different sort of underlying structure. "Aha!", says the mathematician, "The explanation is *obvious*! The different blobs of building that we see are clearly connected together outside of normal space! We are viewing part of a higher-dimensional object! These

10: Quantum Mechanics and Observability

"blobs" are actually cross-sections of tendrils intersecting our space that belong to a single building that exists in five or more dimensions!"

The mathematician's analysis might well be technically capable of describing everything that we see, even though *we'd* tend to think that it wasn't a particularly helpful way to explain things. We could complain that really, all that's happening is that wind is blowing across the surface of a lake, but this knowledge might not let us build a more accurate model than the mathematician, and if we are only able to see the view via a fixed camera, we my not be able to prove that there *really is* a lake there, or that a wind *really is* blowing. These may be interpretations on our part, based on our experience of other similar-looking situations. And no matter how much we learn about the behaviour of water and air, we still can't predict the exact position and shape of any ripple hours in advance, because the situation is too **chaotic**, and too sensitive to tiny disturbances.

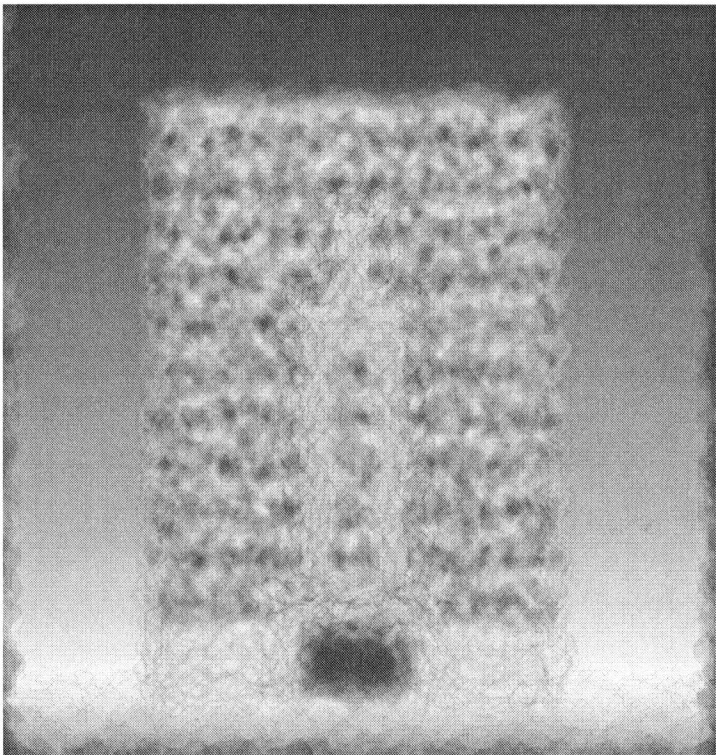

Figure 10-3: Composite of many snapshot images

The mathematician with their crazy statistical or multidimensional approaches may be just as good as us (or better than us) at describing how the building may look in a few hours time, and our belief in what is "really" happening might not give us any obvious predictive advantage.

Where a deeper understanding *might* help us, though, is if this well-behaved situation were to become *dynamic*, and its parameters were to be suddenly altered. We'll expect, if we see a rock thrown into the lake, that there will be a big "sploosh", and the building's image will do even more crazy things for a while. The quantum theorist or mathematician working with simple extrapolations of a gently-rippling image might not have any way to extrapolate from an image of a falling rock in the foreground to sudden results in the background image, and even if they do, they might not be confident that their calculations are meaningful.

10.7: Pair Production

Some of the more intriguing aspects of QM involve **particle pair-production**. If we pump enough energy into a region and distort spacetime strongly enough, this extreme energy-density can start to condense into physical particles. Since we aren't supposed to be able to create *overall* **electric charge** or **momentum**, mirror-image *pairs* of particles are created that then hurtle away from the creation-point in opposite directions.

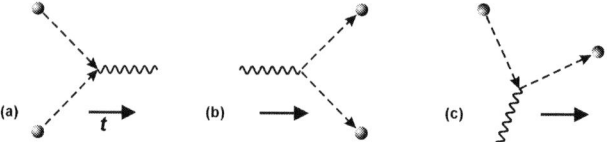

Figure 10-4: (a) Particle-pair annihilates, (b) gamma ray creates particle pair, (c) particle deflected by the absorption of a gamma ray

We can represent these matter-antimatter creations and annihilations with doodles known as **Feynman diagrams**, and one of the cool things about these sketches is their reusability: we can usually turn their "time" and "space" coordinates upside down or back to front, or tilt them, and still get rational-sounding physical descriptions of an interaction when it's been projected onto spacetime at all sorts of crazy angles.

10.8: Virtual particles

Since QM insists that there should be a certain degree of uncertainty about how much energy exists in a given region, it can be used to predict that empty space should, at some level, be continuously producing ghostly pairs of particles whose energy is borrowed from the background vacuum, and which mutually annihilate again (paying back the borrowed energy) in a vanishingly-short amount of time, before anyone or anything has a chance to notice them. In QM's statistical description of pair-production, when we pump energy into a region we are energising some of these particle-pairs strongly enough for them to escape each other's grip.

Figure 10-5: Short-lived virtual particle-pair

Virtual particles (and the toolbox of neat geometrical tricks that come with them) can be very handy when we need to explain types of behaviour that seem impossible under a particular theory, but are nevertheless required to happen and need to be describable in a consistent way. As long as we make sure that quantities of energy, momentum and information all have their proper values at the end of the "accounting period", we can invent descriptions that have "wild and crazy" things happening in the short term. Just like our example with the rippling image of the building, we can have information apparently appearing and disappearing into nothingness, amoeba-like multidimensional surfaces passing through our "3+1"-dimensional' world, particles bouncing backward and forward in time, and all sorts of other fun exotica. The idea to cling to when QM starts getting "scary " like this is that **None Of It Matters**. As long as the final results are correct, and the description corresponds to some valid statistics, it doesn't matter whether the accompanying description is "real" in a conventional sense. If we're feeling especially perverse, we can even *deliberately model standard physics wrongly* (acoustics, section 19.9), and then invoke these "quantum" effects to bring the predictions back on track.

10.9: Pseudo- pair production

There are a number of games that we can play with these particle-pair descriptions:

If we reckoned (rather naively) that signals were entirely undisturbed by the motion of objects in their path, then, if an object was approaching us at *more* than the background speed of light, we'd expect the object to reach us before its earlier signals, and only afterwards would we see those earlier signals starting to arrive at our location, in reverse order.

Watching this old sequence of images of the object, we could be forgiven for thinking that we'd actually seen a *pair* of objects appearing at our location, apparently from nowhere: there'd be the "real" object that actually hit us, and there'd also be what would *look* like the object's time-reversed twin, moving directly *away* from us. Since "reversing" a particle makes it look like its **antimatter** equivalent, if we translate this very idealised description into modern coordinate-system language, we could claim that what we'd just witnessed looked just like the creation of a particle and its matching **antiparticle**, as a quantum effect.

This thought-experiment becomes less frivolous if we consider the problem of radiation emitted by a body with a strong gravitational field: if an "ultrafast" particle was thrown off and then decelerated by gravity to *less* then the background speed of light as it approached us, then this (over-idealised) exercise would say that we should see a *pair* of images, one of a particle moving towards us and escaping the object's gravitational field, and another of its apparent "antiparticle" twin moving away from us and falling back towards the gravity-source. We could then argue that the particle didn't "really" originate at the strong-gravity body, and didn't "really" start out moving at more than background lightspeed, but sprang into existence some distance away from it, as one half of a particle-pair that was then torn apart by the region's intense gravitational tidal forces.

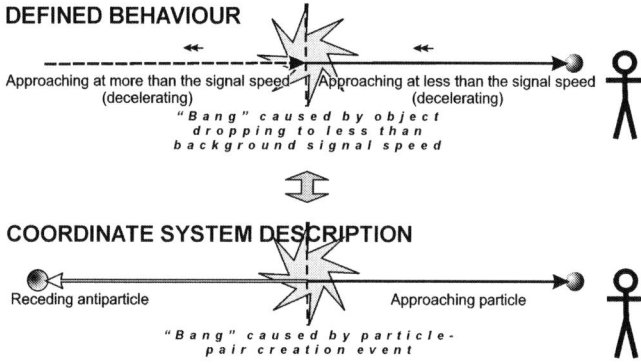

Figure 10-6: Apparent time-reversal and "false" particle pair-production

This is a good example of the power of quantum mechanics: When a classical model doesn't allow certain things to happen, and we *want* those things to happen in order to comply with more general laws, then since QM is derived to conform to very general laws and principles, we can often ask quantum mechanics to step in and patch things up, replicating the missing effects without our having to go back and invent further classical mechanisms or workarounds. When general rules require something to happen, but that thing can't be predicted from our current descriptions, we can usually retrofit it as a separate QM effect.

This useful ability of quantum theory to fix up "bad" classical models seems to take on a new significance when we come to look at the subject of **black holes** under GR1915, and **Hawking radiation**.

Figure 10-7: Chaotic rippled reflection

11
Dark Stars and Black Holes

If a man will begin with certainties, he shall end in doubts; but if he will be content to begin with doubts he shall end in certainties.

Francis Bacon, "The Advancement of Learning", 1605

" Hence ...if the semi-diameter of a sphaere of the same density with the sun were to exceed that of the sun in the proportion 500 to 1, a body falling from an infinite height towards it, would have acquired at its surface a greater velocity than that of light, and consequently, supposing light to be attracted by the same force in proportion to its vis inertiae, with other bodies, all light emitted from such a body would be made to return towards it, by its own proper gravity. "

John Michell, "Proceedings of the Royal Society", 1784

[detective]: *'Is there any other point to which you would wish to draw my attention?'*
[Holmes]: *'To the curious incident of the dog in the night-time.'*
[detective]: *'The dog did nothing in the night-time.'*
[Holmes:] *'That was the curious incident.'*

Arthur Conan Doyle

" The Lord is subtle but not malicious. "

Albert Einstein

" At the present time the opinion prevails that a field theory must first, by 'quantization', be transformed into a statistical theory of field probabilities .... I see in this method only an attempt to describe relationships of an essentially nonlinear character by linear methods. "

Albert Einstein, "Relativistic Theory of the Non-symmetric Field", 1954

" Smelly cat, smelly cat, what are they feeding you? Smelly cat, smelly cat, it's not your fault "

Phoebe Buffay, "Smelly Cat song", Friends

" Let's see how deep this rabbit hole goes "

The Matrix, Wachowski Bros. (1999)

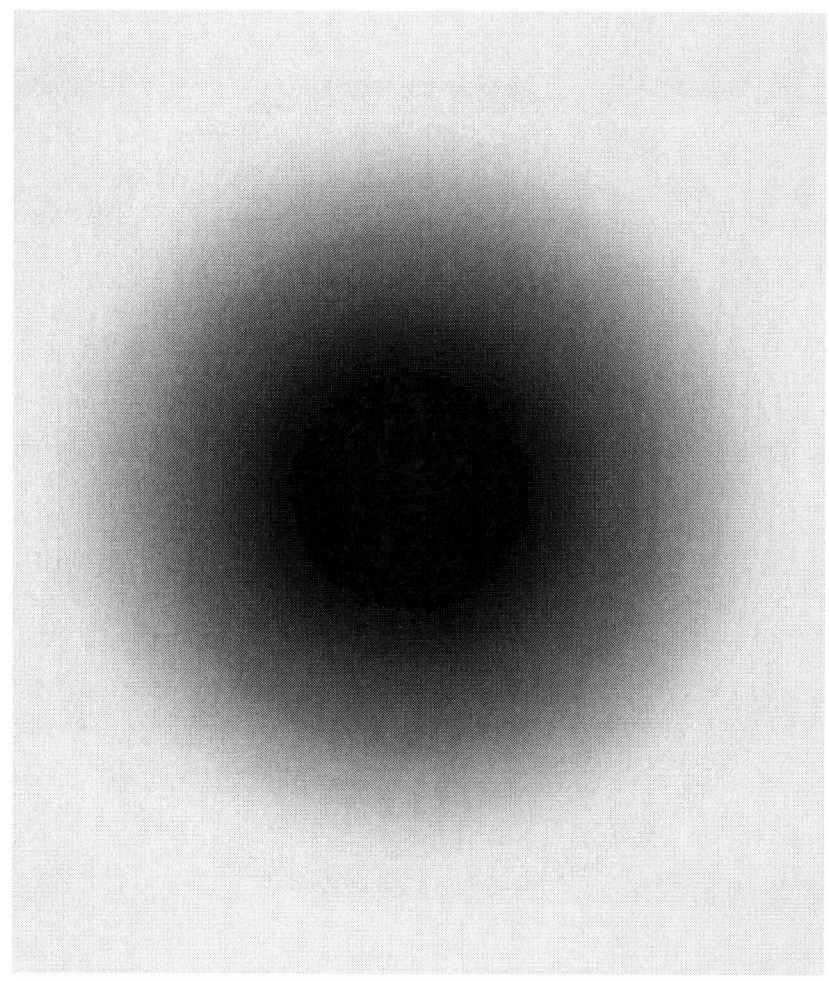

11.1: John Michell's dark stars

A couple of hundred years ago we realised that, as a consequence of Newtonian theory, stars that were massive enough or dense enough to have a surface **escape velocity** greater than lightspeed shouldn't be directly visible to a distant observer, because light leaving their surface would be gravitationally trapped in the region. They'd still be able to radiate *indirectly*, so to distinguish them from the newer (and *utterly black*) objects now predicted by GR1915, we now tend to refer to the Newtonian ancestor of the black hole as a **dark star**.

The largest known contemporary study of dark stars was **John Michell**'s 1783 letter to his friend and colleague **Henry Cavendish**, which was published in the 1784 volume of the **Royal Society's** prestigious journal "Philosophical Transactions".

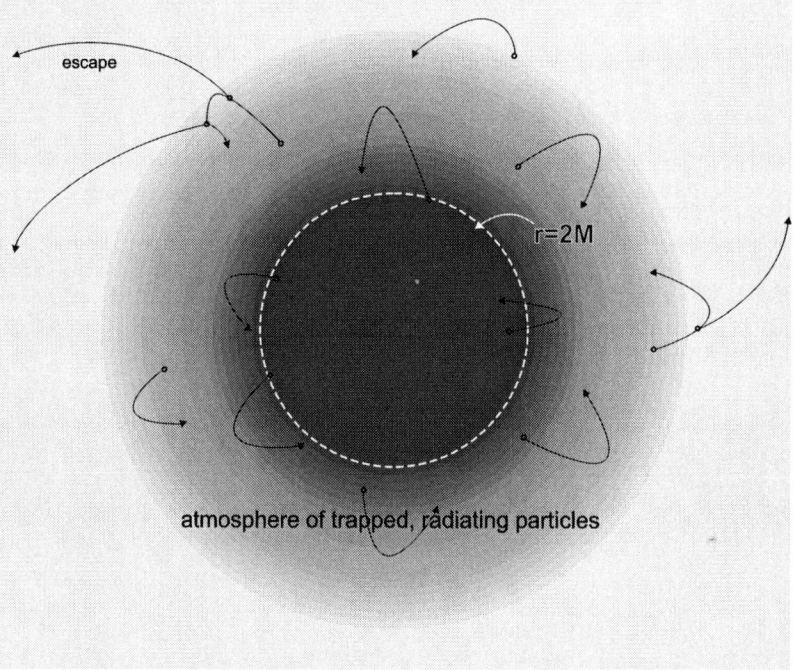

Figure 11-1: A pre-GR "dark star", and indirect radiation

Although **gravitational shifts** are sometimes presented as a Twentieth-Century idea, Michell was already stating back in the Eighteenth that light from high-gravity stars ought to be shifted to the weaker end of the spectrum. Michell also pointed out that we should be able to detect these stars in binary systems by the motion of the visible star around its unseen companion, and argued for a project to compile a directory of double stars in order to find out how many of them had "dark" partners, giving a statistical baseline that we could use to estimate how much other matter in the universe might be composed of "dark" material.

Despite the size and high profile of Michell's piece, the political fallout from Newton's mistake (section 5) caused it to disappear from most historical accounts until its belated rediscovery in the 1970's. Michell acknowledged that some of the techniques he was discussing were beyond the technology of the time, and that his piece was mostly for the benefit of future generations of astronomers, but by the time his paper was unearthed and resurrected in the latter Twentieth Century, everything he had described in it had already been rediscovered from scratch.

In books published before the 1970's, the idea of the dark star is usually credited to **Marquis LaPlace**, who briefly mentioned the idea in his book, "Systeme du Monde" a few years after Michell's article. LaPlace is supposed to have deleted the topic from later editions of the book.

11.2: Properties of a compact gravitational object

Michell's exercise assumed that "light corpuscles" gained and lost energy and momentum when moving into and out of a gravitational field, just like conventional objects. If a body's surface escape velocity was then equal to or greater than lightspeed, light leaving the surface would eventually be pulled back towards the star by gravity.

The escape velocities of bodies such the Sun and the Earth are much smaller than lightspeed (about ~617.5 km/s and ~11.2 km/s respectively). Their mass is distributed over a significant volume of space, limiting the *proximity* that we can achieve with the body's total mass. Once we reach the Earth's surface we can't get any closer … or at least, we could get closer to the planet's *centre of mass* by burrowing down into the planet, but this isn't where the mass *is* – even when we are at its very centre, most of the Earth's material will still be thousands of kilometres away, far up above us.

What makes "compact" objects so dangerous is that we *can* continue getting closer to their total mass before this limit starts to apply. If we crushed the Earth down to a fraction of its current size, we'd still expect to feel the same Earth gravity while we hovered in a spaceship at about six-and-a-third thousand kilometres above the centre, where the surface *used* to be … but each time we descended and halved our distance from the newly-compacted Earth, the gravitational pull would go up fourfold. If we squashed the planet to one tenth of its original radius, the surface gravity would be a hundred times stronger – compact it to a hundredth of its original radius, and the pull of gravity at its surface goes up ten thousand times.

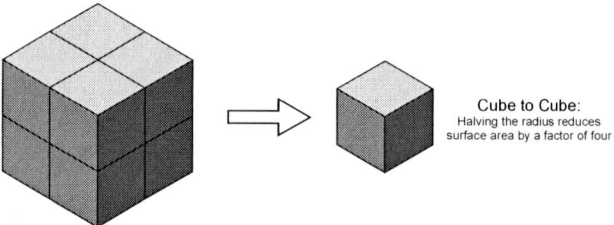

Cube to Cube:
Halving the radius reduces surface area by a factor of four

Figure 11-2: Radius versus surface area: Dividing the radius by two shrinks the surface area (through which gravitational field lines exit) by a factor of four

If the Sun was suddenly compacted to a few kilometres across, we wouldn't expect its conventional gravitational effect on distant objects to change. The Earth and other planets would continue in their current orbits around the Sun's centre of gravity as if nothing unusual had happened. The Sun's crude gravitational influence on the Earth wouldn't change, since its mass, distance and position would be the same as before we made our change.

Our calculated strength for the shrunken Sun's gravitational pull at its *original* surface would be the same, and we'd expect light *originating at that surface to* show the same gravity-shift when it reaches us as it does now. But the light from the new surface would be originating much deeper than this, and as we approached the crushed Sun in a spaceship, we'd be able to pass where the old surface used to be and continue approaching, with the escape velocity at our deepening location doubling, and doubling, and doubling again as we got closer and closer, until our engines could no longer resist its pull. If the Sun was compacted far enough (to less than around ~18.5 km) the escape velocity would exceed the speed of light, and the Sun would have become a dark star or a black hole.

11.3: Escape velocity calculations and the gravitational horizon

For a gravitational body with mass **M** and radius **R**, a quick Newtonian calculation tells us that the body's **terminal velocity** – the speed that an object achieves by falling onto it from arbitrarily far away – is $v = \sqrt{2GM/R}$, where **G** is the **Gravitational Constant**, the amount of gravity that we get per unit of mass. This terminal velocity is also usually reckoned to be the **escape velocity**, the initial speed that we'd need to throw an object with if we were standing at the surface and wanted to the object to be able to completely escape the body's gravitation.

If we want to know how compact a body has to be for this escape velocity to equal the speed of light, we can write in *c* instead of *v*, square both sides of the equation and then rearrange things to get the new, "critical" radius, **r=2GM/c²**.

Since **G** and *c* are natural constants, it's convenient to set them to one side (or equal to "one") and write our special radius at which escape velocity equals lightspeed as just **r=2M**.

Under general relativity, we often refer to this critical **r=2M** distance as the "**Schwarzchild radius**" (after **Karl Schwarzchild**), and the surface that it marks out around a supercompact object is often referred to as the object's **event horizon**.

11.4: Tidal forces

Another dirty trick that superdense objects can play on us is their generation of severe **tidal forces**. Tidal effects are the wrenching forces felt by an object in a gravitational field due to different part of it being pulled by different amounts, or in different directions. Let's suppose that we drop a baseball onto our collapsed Earth or Sun, and watch it fall. Since the gravitational attraction at or just outside the **r=2M** radius is absurdly high, the rate at which the field strength *changes with distance* close to the hole is going to be far higher than anything we're used to.

As the falling baseball nears the gravity-source, the "lower" part of the ball is always nearer to the gravity-source than the rest of it, and because it feels a stronger gravitational pull, it'll always be trying to fall at a faster natural rate than the rest of the ball. As the baseball gets closer to the source, eventually the internal stresses are going to become stronger than the ball's material can bear, and the ball will be ripped apart, ending up as an infalling trail of pieces that are progressively torn into smaller and smaller fragments by the steadily increasing tidal forces. This effect is popularly known as **spaghettification**, for the gruesome reason that for an astronaut falling towards a smallish black hole or dark star, tidal forces would be expected to pulverise and stretch the unfortunate freefaller into something like a strand of human spaghetti by the time they reach the **r=2M** distance.

Figure 11-3: Dripping tap: Tidal forces help to
tear apart a falling cylinder of water

Spaghettification at **r=2M** isn't entirely compulsory. The tidal forces across a 2-metre-high astronaut can be reduced by a factor of eight simply by telling the astronaut to go for a lie down. Falling into the hole horizontally rather than vertically could reduce the "effective" height of the astronaut to about 25 cm, and if we replaced the human astronaut with a smaller creature, this height (and the resulting differential stresses) could be reduced even further.

If we replaced the two-metre astronaut with a computer program running on a solid wafer of semiconducting diamond less than a *millimetre* thick (equipped with optical sensors and so on, all etched into the diamond substrate), then the tidal forces across our new, super-thin cybernetic astronaut will be more than *two-thousand* times weaker, and the falling (diamond-based) observer would be much stronger, and better equipped to survive them for longer.

The region intersected by a gravitational horizon isn't always especially particularly hostile even to squishy human life, as long as the black hole is sufficiently large (Thorne 1994, prologue). For a black hole of *galactic* mass, the **r=2M** surface is so far from the centre of mass that the *rate of change* of gravitational force is comparatively weak, and one might drift right through the horizon without being particularly aware that anything especially "odd" was going on in the region.

Another way of alleviating spaghettification might be to have your spaceship *power-dive* directly at the dark star or black hole. With extreme physical acceleration towards the hole, the geeforces trying to turn you into a flat pancake against the rear wall of the spaceship, and the tidal forces trying to pull you out into a thin vertical strand, should produce at least *partial* cancellation between the two effects. But power-diving also has the disadvantage of giving you less time to sight-see on the way down.

11.5: "Visiting" particles around a dark star

In a Newtonian model, the collapsed Sun would be *extremely dark* as seen from Earth orbit, but not quite *totally* black.

In NM-based models, although no light can reach a distant observer *directly* from **r=2M** or lower, the region around the high-gravity star will be populated by a thin atmosphere of trapped particles and light that can be thrown out of the star in the usual way, but which only gets a chance to "visit" the region outside for a limited amount of time before being pulled back into the star. This thin surrounding halo of localised particles and light is able to generate secondary radiation *outside* the **r=2M** surface, so although simplistic arguments *seem* to prove that a dark star can't emit light, a more detailed study shows that collisions in the surrounding atmosphere result in trapped particles or other radiation being thrown free, causing the dark region around the star to "twinkle" slightly.

Light and matter would still be emitted (in vastly reduced quantities) from the stellar surface, and as we drew closer (say, to the orbit of Mercury) we'd find ourselves encountering a greater quantity of emitted light and miscellaneous ejecta coming from the star's approximate location. Standing on the surface of Mercury and looking up at the cold shrunken Sun, we should theoretically be able to see *some* light and other particles passing through our spacesuit visor and registering on our detectors, but the quantity would be very, very small, and we'd only be seeing a faint, fuzzy atmospheric blur instead of a sharp image of the object.

We'd find it easier to see more of this atmosphere of **visiting particles** if we flew a spaceship deeper through it. While we were travelling through the region, our distant colleagues might see us mysteriously illuminated by a glow of local light – the physical acceleration of the ship as we used our rocket engines to avoid falling into the star would help accelerate some of the adjacent light and matter free from the star's pull, and our ship could also accelerate particles out of the region simply by colliding with them and batting them away. We can refer to these effects cumulatively as **indirect radiation** effects. When we emerged, the front of our ship would be covered with a smur of "visiting-particle" dirt, and if a distant observer wanted to verify that this mysterious atmosphere really existed close to the star, they'd only have to lower a bucket into the region on the end of a long rope and haul it out again – when the bucket was yanked out of the region it would scoop out some of the "visiting" material and also accelerate some previously-trapped light towards the observer.

11.6: Dark stars and "acoustic" metrics

Our "dark star" description isn't easy to model using simple geometrical definitions, and doesn't fit some of the basic definitions that we apply in SR and GR1915.

With SR's approach, we describe spacetime by laying out a **metric** that describes how distances and times appear for a given observer or set of observers, and the theory then says how a set of events seen by one observer will map to the same events as seen by another. We think of this set of agreed events as being **complete**, and if a hypothetical event doesn't exist in one observer's metric, the event *doesn't exist* for that observer, and also doesn't exist for any other observer. Every event on one observer's metric has to have a corresponding point on another's, and when this idea breaks down (in GR1915, at an event horizon), we reason that events that exist outside an observer's metric are effectively outside their universe, and can't have any further influence of what they see.

But in a dark star model, this kind of reasoning doesn't work. A particle that originates behind the **r=2M** surface, and which doesn't appear on in a distant outsider's conventional coordinate system, can still pass outward into the "visible" region to interact with "on-metric" particles and perhaps even be knocked free from the star's gravitational pull. It can suddenly appear within an outside observer's coordinate system (without prior warning) in response to events that don't exist within that observer's observer-coordinates, and "off-metric" events can produce disturbances and consequences that *can* be appreciated by the observer, indirectly.

here be dragons

In a dark star model, an observer's usual GR1915-based "sense of metric" isn't a complete description of reality: it's frayed and tattered around the edges near a gravitational horizon, and the visible limits of the metric will fluctuate and fizz in response to events outside our directly-visible range. Where GR1915 likes to assume that all the parameters that could possibly affect an observer's future are always visible and in full view, a dark-star-compatible metric supports variables that can duck behind horizons, and **"lurk"** there to emerge at some later time.

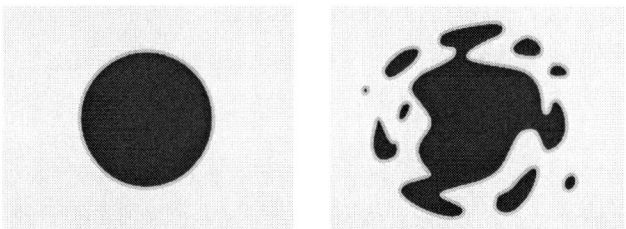

Figure 11-4: **Boundary to the observable coordinates for a very small black hole (a) with GR1915, and (b) with an acoustic metric**

We can make some crude comparisons between dark star horizons and conventional *planetary* horizons, in that a sharp distinction drawn between what *we* can and can't see directly doesn't dictate what actually happens in the transitional region – the locals don't really care. Our "cutoff" line will be somewhat mathematical, and will be drawn differently by different observers. When we take the expanse of planetary surface that's visible to us and examine its boundary in fine detail, we find that the *effective* horizon is no longer a simple continuous line or surface – it has islands and voids where small-scale curvature makes areas disappear from our view behind mountains or at the bottom of valleys (section 11.16), and patches of mountain may be visible even through they're further away than the "idealised" limit. Things that aren't currently visible may heave into view at a later date, and their presence can be deduced indirectly by the consequences of their actions on things that we *can* see.

11.7: Acoustic metrics and nonlinearity

This sort of "complicated" metric already appears in conventional physics when we study acoustics, so it's referred to as an **acoustic metric**. Acoustic metrics tend to be associated with **nonlinear** signal behaviour, where the existence of a signal modifies the shape of the metric, so that definitions of distances and times made in the absence of the signal aren't automatically valid when the signal is present. Analytically, acoustic metrics can be a bit of a nightmare, and we usually prefer to avoid them, but they do have one compelling property – they combine classical (nonlinear) field theory with QM-style behaviour, which is something that a more conventional GR1915 metric can't do. The projection of the visible parts of an acoustic metric *onto* a more conventional metric produces apparent fluctuations, discontinuities and uncertainties that do still correspond to a deeper single underlying reality. Although the surface relationships in view, watched over short periods of time, can *seem* acausal and discontinuous (and perhaps even irrational), the underlying statistical pattern is still causal, and the underlying surface is still entirely classical. Particles in an acoustic metric can pop in and out of sight, and acoustic horizons allow information to leak and bleed across into our region from regions that aren't directly accessible to us.

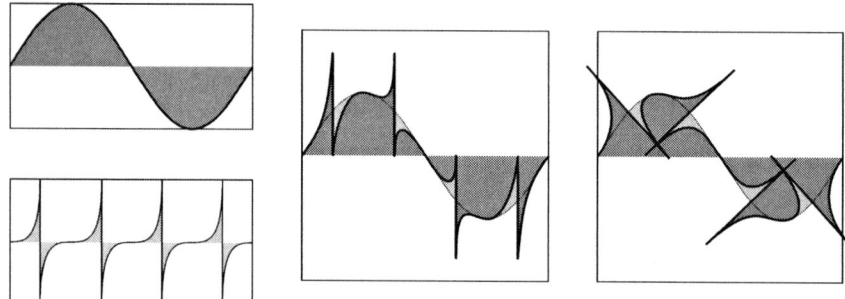

Figure 11-5: Simple, linear, mathematical addition of waves, and a more complex, nonlinear, "warped" superposition of wave values

This indirect radiation effect across **acoustic horizons** sounds rather like the effects predicted by QM for **Hawking radiation** across gravitational event horizons, and acoustic metrics are increasingly being used as **toy models** for experimenting with ideas in **quantum gravity** with the help of older concepts that we're more familiar with, and are more confident of. **Matt Visser** has recently categorised the "acoustic" counterpart of Hawking radiation as a legitimate example of the effect (albeit in a context that isn't compatible with current gravitational models), and the subject of "classical" or "**analog/analogue**" Hawking radiation has recently become a popular area for research.

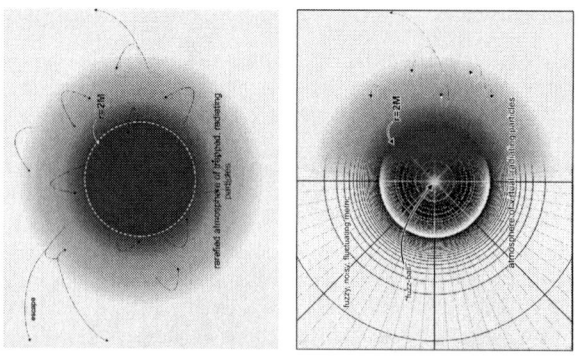

11.8: Black holes under GR1915

zero radiation

Quite a bit of the previous discussion works equally well with "old" Newtonian physics or with current mainstream general relativity, but we have to make one critical distinction between dark stars and the newer GR1915 **black holes**: the *total absence of any persistent atmosphere or any indirect radiation effects* in the GR1915 model.

The GR1915 coordinate-system approach is somewhat different to the way that we might have dealt with these things under earlier models. Under Einstein's special and general theories, what you *see* is what is deemed to *exist*. Special relativity doesn't support the concept of a **velocity horizon**, and when gravitational horizons arise under GR1915, since events past the horizon don't appear on a distant observer's metric, they're said *not to exist* for that observer. If they don't exist, they can't have consequences, and they must be unable to affect anything else that the observer could see. Under GR1915, signals associated with events at or behind an event horizon can never influence events outside – the region is totally and permanently off-limits.

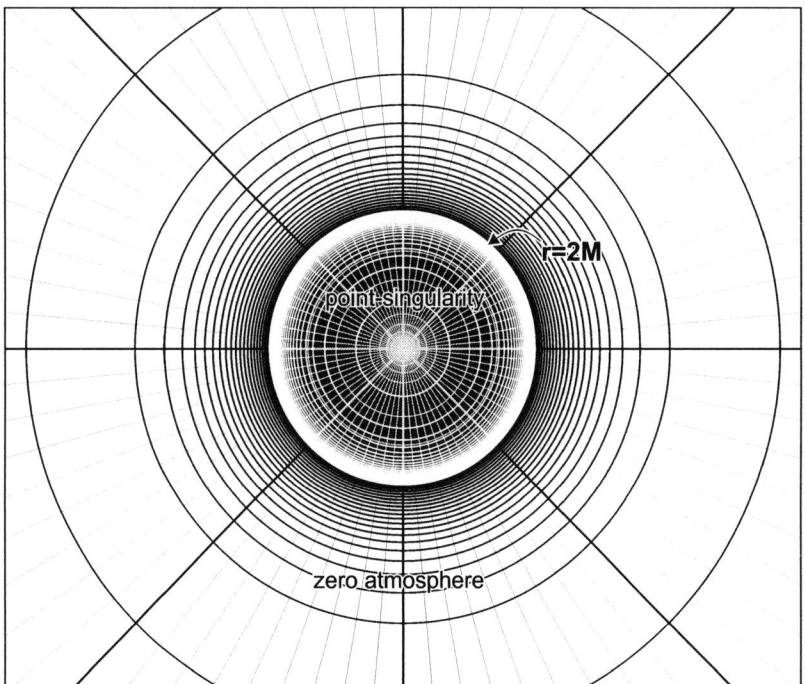

Figure 11-6: A GR1915 "black hole": pure, smooth geometry

This is what earns the GR1915 "**black hole**" its name. With no surrounding atmosphere of "visiting" particles there can't be "indirect emission", and no matter how close we get to it, it always insists on being completely, totally, utterly devoid of any radiational effects.

It is truly, perfectly, ultimately black. And if light and all other forms of signalling find it impossible to escape, and if matter and signals can't move faster than light, then the object can no longer be sensed as an object, but appears instead as an apparently featureless "hole" created from pure spacetime curvature, whose only external properties are mass, rotation (optional), charge (optional) and the size of its horizon (which is defined by mass and rotation).

RELATIVITY IN CURVED SPACETIME

Material and light can't migrate outward though the surface of GR1915's gravitational horizon, and as a result, the interior of a black hole is permanently cut off from the outside universe – nothing at all can get out, no signals, no information, nothing at all. As far as the rest of the universe is concerned a black hole is just that: a black and apparently bottomless hole that swallows anything that gets too close to it.

The only signals that we expect to hear coming from the vicinity of the GR1915 horizon are the last remnants of the ancient screams of long-infallen matter that originated just outside the critical surface, and which have been trying to swim uphill against gravity for all this time. If we wait long enough, even these ancient "ghost" signals will be progressively redshifted so far into the far-longwave radio band that the surface region effectively becomes silent and dead.

total collapse

The second new feature of a GR1915 black hole is something called "**total collapse**". If we imagine a star that is slowly compressing under its own weight towards the critical **r=2M** radius, then once it reaches that radius, GR1915 says nothing can stop it collapsing towards infinite density. Normally we'll try to model a star's structure by imagining it as a series of concentric shells of different densities, and then calculating how much each shell pushes against its neighbours until the system reaches equilibrium.

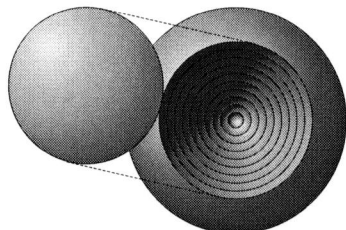

Figure 11-7: Star, divided into concentric layers

But under GR1915, once an effective horizon forms inside the star, matter within that horizon has no way of communicating with layers further out, and this means that material inside *can't exert any outward pressure at all* on the material above it. With GR1915's "perfect" horizons, there are no tricks left in our toolbox to prevent the rest of the star's material undergoing a complete unresisted freefall collapse *through* the horizon, and since there's also no outward communication allowed *within* the horizon, we have nothing left to prevent total collapse all the way down to a dimensionless point (or, for a rotating hole, to a zero-thickness ring).

singularities

As the collapsing material compresses further and further, and its surface gravity increases more and more, ultimately (in the GR1915 account), there's no force or effect that can make this continuing collapse come to a halt. As the volume of the material tends towards zero, the surface gravity tends towards infinity until we expect to end up with a single dimensionless point (or ring, if it rotates) at the core of our black hole to represent the black hole's entire mass. This point of infinite density is referred to as a **singularity**.

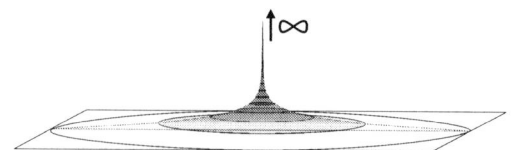

Figure 11-8: Singularity spoiling a smoothly-curved surface

Classical field theory tends to rely on curves and gradients varying *smoothly* from place to place, but if we draw a line across a region with a central singularity, the line is sharply folded back on itself to form an infinitely-sharp point as it crosses the singularity, so on the face of it, the prediction of a physical singularity would seem like evidence that a classical field theory has broken down.

cosmic censorship

One way of making these singularities seem slightly less outrageous is to point out that GR1915's gravitational singularities do at least *usually* have the decency to be hidden from observers by **event horizons**. The **cosmic censorship hypothesis** says that any singularities that *do* occur should be decently hidden from a disapproving outside universe by an event horizon (rather than being left **"naked"**), the hope being that if we can't ever *see* a singularity, and can't *see* our laws of physics being broken, then perhaps things aren't quite so bad. Perhaps, if we're doing *observerspace* physics, and if singularities can only form in regions that don't appear on our usual observerspace coordinate maps, we might be able to get away with saying that if they only happen in regions we can't *see*, it might be okay to ignore them.

There are a couple of difficulties with this "out of sight, out of mind" argument:

We might argue that no matter how close we get to the centre of a black hole, a horizon always ought to stand between us and the point of infinite density: however, as we fall into the hole, we should *hit* that point in a finite amount of observer-time. At some point in time, the distance between us and the singularity will be zero, and an intervening horizon can't then exist. Admittedly, we and everything we are made of ought to be destroyed *before* this moment, but from a geometrical point of view, we have to accept that our worldline may point straight at a singularity.

The second problem is that if we fall into a Kerr black hole (section 11.9) by dropping in through one of its two poles, we may end up hitting the flat surface *bounded* by the ring, and at the moment that we hit it and are *surrounded* by the ring-singularity, the geometry goes rather weird, suggesting that perhaps the singularity's naughty bits might be exposed to us.

causal disconnect

Taking GR1915's predictions seriously, we'd expect the interior of a black hole to be *completely causally disconnected* from the outside universe (in one direction, at least), with no events located inside this surface having any physical consequences at all for the universe outside. We'd have a strictly "one-way" relationship between the regions inside and outside a horizon surface. Occurrences *outside* **r=2M** could influence subsequent events *inside*, but this causal relationship would not be reciprocated: the GR1915 black hole is an essentially "parasitic" region. What *happens* inside the event horizon *stays* inside the event horizon.

To Einstein, this behaviour clashed with the basic design principles that he had imagined that his general theory should follow, and he argued that since a region *couldn't* be causally disconnected from its surroundings, and since a GR1915 event horizon meant that a region *had* to be causally disconnected for its surroundings, that this showed that black holes couldn't form. Perhaps GR1915 wasn't valid in such extreme situations, or perhaps some other overriding principle or mechanism would cut in. Something Else must happen at this point to Stop Black Holes Happening.

To modern theorists, Einstein's counterargument seems to be as much about design aesthetics as anything else. It's not taken very seriously, and the idea that horizon formation should be prevented by "extreme" effects associated with its creation is undermined by our realisation that for a sufficiently *large* hole, the horizon physics can be quite mild. If we want to argue against complete causal disconnect, we need to use a different approach to Einstein's.

RELATIVITY IN CURVED SPACETIME

inescapability and immortality: when holes collide

GR1915's black holes seemed inescapable by conventional means, but this still left open one possibility, that we might be able to use "gravity-assist" to pull an object out from just underneath a GR1915 event horizon by teasing it out with the help of a *second* black hole. But GR1915 doesn't seem to allow this either.

Since the radius of a black hole is proportional to its mass, when we add two black holes together, the resulting black hole's mass will be the sum of the two smaller masses, and therefore its radius will be the sum of the two smaller radii.

A sphere's width (it's diameter) is twice its radius, so if we place two idealised spherical black hole horizons next to each other, with their spherical surfaces just touching, the final black hole that will be produced when they merge will have a diameter exactly equal to the two smaller diameters, and the resulting sphere can exactly enclose the two smaller spheres.

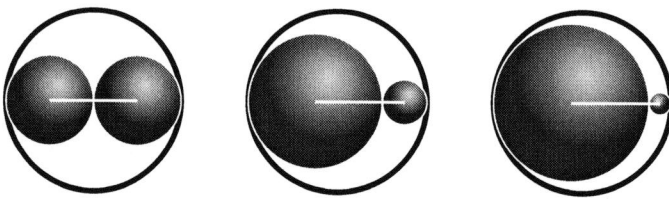

Figure 11-9: Black hole addition

The final horizon will include the volumes of space included by the two smaller horizons, and a big chunk besides, so if any information (hypothetically) *was* able to cross between the volumes of our two black holes, we'd still expect that that transfer event would be censored from the outside world.

These diagrams are over-idealised: in a more realistic depiction we'd have to acknowledge that horizons are only reckoned to be spherical when they're wrapped around a *spherically-symmetrical* distribution of matter, which isn't the case for our colliding black hole pairs in the earlier stages of their collision, above. Calculating the exact shape of a horizon surrounding a black hole collision site is difficult, but we'll tend to expect the enclosing horizon to wrap around the two masses and hug their gravitational contours more closely than Figure 11-9 suggests, producing a shape more like an old-fashioned weight-lifter's dumbbell ... but this (slightly more realistic) enclosing shape seems to make the escape of information from the collisional region even *less* likely.

Figure 11-10: The sort of horizon that might be expected to surround a pair of black holes

Stephen Hawking has proposed a more complex variation on this argument, involving changes in surface area – since the area of a sphere is proportional to radius *squared*, the final black hole will have an area that is *larger* than the sum of the two smaller areas, and Hawking's **area-increase theorem** is taken as saying something very fundamental about GR1915 black holes: that their area should always go up with time and never down, which also seems to tie in nicely with some aspects of **entropy** and **thermodynamics**. Under GR1915, we reckon that black holes can't shrink, and are effectively immortal.

11.9: The Kerr black hole

We often treat gravitational masses as if they are idealised point-sources. When we want to calculate the Earth's orbit around the Sun, we say that the Sun has a certain position and a certain mass, and then we assign that mass to a point at the location and use it as an idealised stand-in for the properties of the real Sun. This idealisation usually works well, and under GR1915, we say that when a star undergoes gravitational collapse, it really *does* become a dimensionless point with a mass and momentum equivalent to those of the original body.

But this idealisation fails when an object rotates relative to its environment. A rotating body pulls things around with it and has an equatorial bulge, and these effects show up as a corresponding distortion in its exterior gravitational field. The circling motion of points on a surface (or on a horizon) justifies our description of fieldlines being "twisted up" around a rotating body, but a perfect point-source isn't able to exert this sort of sideways pull, because it has no moving perimeter, and a true point has no components that can be tracked over time. If we can't distinguish between "fixed" and "rotating" points, then the property of rotation, applied to a *point*, wouldn't seem to have a physical meaning. Geometrically, a "simple" point just isn't complex enough as a shape to support the property of physical rotation. A rotating GR1915 black hole (known as a **Kerr black hole**) has to have something else at its core.

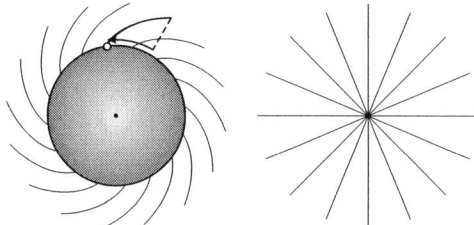

Figure 11-11: A "rotating" point can't drag fieldlines

For a *rotating* star that undergoes total collapse while retaining at least some of its angular momentum, GR1915 says that the end-result won't be a point, but the minimal shape that maintains the original angular momentum and exterior field shape – an empty circle whose radius expresses how fast the black hole is rotating. The rotating mass can then be idealised as a *central ring*, and the core of a Kerr black hole is represented as a **ring singularity**.

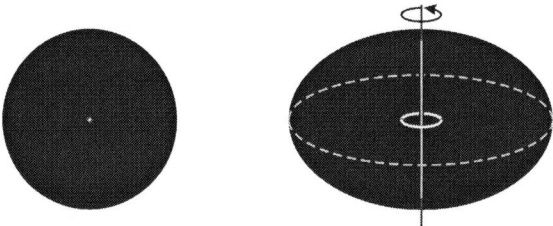

Figure 11-12: Point- and ring-singularities

We can also suggest that when an object rotates, outside observers in the equatorial plane should see the object's fieldlines pointing to positions that are offset towards the receding side of the object instead of to its official centre. Equatorial observers distributed around a rotating body should all see the object's apparent centre of gravity to be offset towards the receding side, and will disagree as to its actual location. These different positions can't easily be said to correspond to the same point in space, but *can* be said to all lie on a circle or ring. The diameter of the ring is then a measure of how badly these lines miss each other.

11.10: The expansion problem

GR1915's black holes give us a few definitional headaches. Suppose that we take a smallish "standard" GR black hole and add some matter to it. The event horizon of the hole should then move outwards to take into account the new, increased horizon radius. But to actually *implement* this behaviour in a GR black hole model is a little complicated: at the start of the experiment, the hole has "frozen" light at its event horizon, and light fractionally above the critical surface is moving outward incredibly slowly. At the end of the experiment, the hole's radius is increased, the horizon is now larger, and the hole might be expected to have grabbed and "frozen" some of the light that was originally slightly outside its horizon. Since the hole now has stronger gravitational pull, it should now be capable of trapping light further from its centre, corresponding with its new horizon radius. But if this is to happen, the horizon seems to need to expand outward *faster than the outward speed of light in the region that it is expanding through*, zero metres per second. If the gravitational "expansion signal" can move faster than the local speed of light, then our tidy GR1915 definitions start to go just a little bit mental.

If we feed matter into one side of the black hole and watch the expansion from the other side, isn't the visible change in the hole's size telling us something about events inside the hole? Isn't the "mass-increase" information being transmitted to us from one side of the hole to the other, *though* the original horizon? Doesn't this violate the GR1915 rule that no information can pass outward through **r=2M**?

There's at least one way around this problem: we can suggest that perhaps the additional gravitational effect doesn't reach us *through* the hole, but washes *around* the hole when more mass is added, just above the original surface, like a small "horizon tsunami". This seems to solve our problem ... but it's a little uncomfortable to realise that having just invented this new class of object with its simple, rigorously-provable properties, one of the first things we have to do is to invent clever workarounds to let us *sidestep* that math and those theorems, so that we don't accidentally end up "disproving" things that really ought to be legal.

11.11: The acceleration problem

A similar problem shows up when we consider the behaviour of a small stationary black hole that changes speed. Suppose that material is fired at high speed into the far side of a small ultra-dense body – as the hole absorbs more fast-moving material (and also absorbs the associated momentum of that material), the hole *ought* to start moving towards us.

But once again, this means that the horizon surface facing us ends up moving outwards through a region where the corresponding velocity of light was originally zero: the horizon moves faster than the earlier nominal speed of light, and the *change* in the speed of light has to propagate faster than signals are supposed to be able to travel through the region. Again, we can construct slightly elaborate explanations where the momentum of the additional mass crawls *around* the surface of the hole to assemble a new horizon, moving through the region *above* **r=2M**, where the outward velocity of light is *more* than zero, ... but by relying on these explanations we're losing the simplicity that was supposed to be the new "GR1915 black hole"'s unique selling point. As the rear of the horizon starts to recede, shouldn't this appear to the rearward observer as an illegal horizon *contraction*?

We can use common sense to argue that GR1915's simpler results should be applied *intelligently*, and that if we get a "nonsense" result, we've made a mistake and should go back and adjust our approach until it produces a more sensible outcome. But if GR1915 *really was* a complete, rigorously predictive system with no room for "fudge factors", then perhaps we wouldn't need to make quite so many of these interventions.

This workaround method of treating the "super-horizon" region as a surface for dynamic information exchange becomes more important when we look at the **Black Hole Membrane Paradigm**, and the **Holographic Principle** (sections 11.21 & 11.22).

11.12: Black holes according to Quantum Mechanics

Although GR1915 makes some unambiguously "new" predictions about the behaviour of high-density objects, quantum mechanics needs a few things to happen differently.

no central singularity

According to QM's worldview, there shouldn't be any clean, sharp, well-defined field-singularities: singularities represent perfectly-locatable, perfectly-sharp objects with perfectly-defined properties and these don't fit well with quantum mechanics' "fuzzy", uncertain, probabilistic worldview.

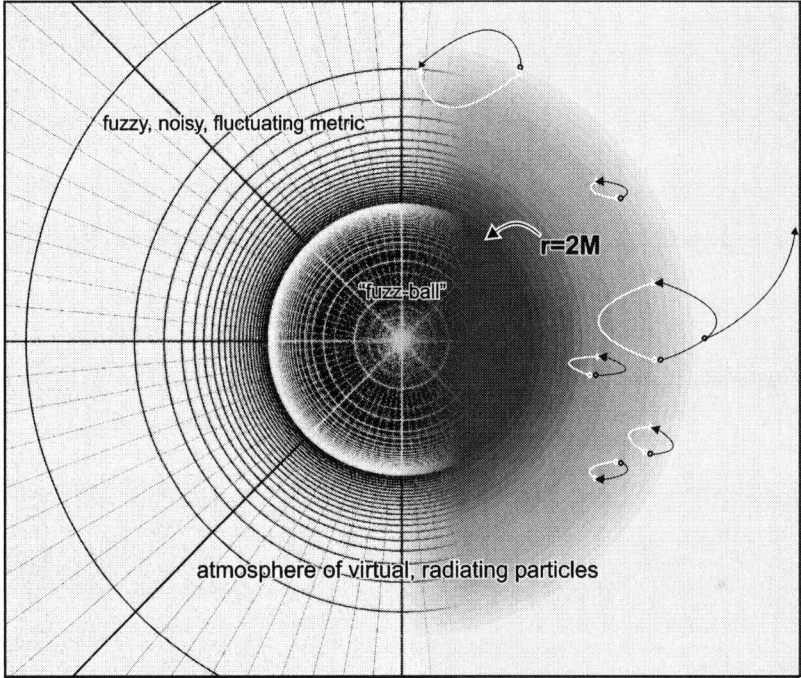

Figure 11-13: A black hole, according to QM

Since QM is usually reckoned to become significant at very small scales, and since a point-singularity is *infinitely* small, it's been argued that QM probably ought to prevent GR1915's "total collapse" process going all the way to completion. The black hole's mass could collapse to form a **quantum fuzzball**, but it couldn't collapse all the way down to a point.

no perfect horizons

There's still a sense in which a GR1915 horizon is "odd" – in special relativity we can say that the limit at which the an object is travelling at the speed of light is unreachable, so the notion of a lightspeed limit, considered as a sort of "brick wall" limit, doesn't have an obvious physical existence. With GR1915 (which inherits some of special relativity's core conventions), SR's hypothetical, unreachable lightspeed limit suddenly has a physical set of *spatial coordinates* in the form of the event horizon, and is perfectly sharply defined.

Quantum mechanics doesn't like perfect barriers or any other sort of perfectly-defined surface, and when confronted with one, tends to insist that particles and information should jump or "quantum tunnel" across it.

11.13: Hawking radiation

When a dark star *rotates*, the position of an observer's "effective" horizon changes. It's easier for a dark star's emissions to hit a nearby observer from the approaching side of the star than from the receding side, and the dark star's *effective* horizon coordinates extend outwards at the receding redshifted side of the star and contract inwards at the approaching blueshifted side, much as we'd expect if its rotational Doppler shifts were "gravitational" in origin. To an equatorial observer, the receding side of a rotating, critical-radius dark star could be hidden behind a (slightly leaky) effective horizon while its approaching side is exposed: the existence of the horizon between the observer and a given location is **route-dependent**, and the approaching side of the star can spray dirt at the observer, like a muddy bicycle wheel.

Dark star horizons essentially obey the rules for **acoustic horizons**, they leak, fluctuate and can be incomplete and observer-dependent. As "mathematical" descriptions of more complex underlying geometry, acoustic horizons don't have to be in the same place of all observers, and their "fuzzy" nature allows them to be incomplete surfaces, with fluctuating frayed edges. They involve **relative geometry** rather than absolute geometry.

An analysis of QM's description of the effect of a rotating mass (based on an analogy with an obscure QED effect called **superradiance**) turned out to produce some disturbing similarities to the "dark star" predictions. A counterpart to our "spraying" effect appeared around rotating bodies under quantum theory, explained as a **particle-pair production** effect: to give empty space an appropriate degree of quantum uncertainty we could imagine the vacuum as populated by a seething mass of short-lived particle-pairs that briefly popped into existence and then mutually annihilated again so quickly that we weren't normally aware of them (section 10.8) ... but around a *rotating* black hole, the way that rotational dragging forces around the hole extend out into the surrounding region could cause tidal "shearing" effects that'd sometimes be able to wrench apart some of the more enthusiastic particle pairs before they had a chance to cancel each other out, and throw one of them towards the observer ... the approaching side of the rotating QM-modelled black hole could emit a spray of "real" particles at a nearby observer, (and lose mass) even though the region above it was supposed to be classically empty.

Although this was a disturbing result, theorists at the time could still console themselves with the hope that perhaps *rotating* radiating black holes were just a weird special case. But in 1974, QM's radiation effects "hit the fan", as **Stephen Hawking** announced that another version of this "tidal shearing" effect should happen above a *non*rotating black hole, due to the more basic tidal forces pulling apart particle-pairs (section 11.4). This meant that even the region around a *simple* black hole should be throwing off particles, and if we didn't feed the hole enough fresh material or energy to compensate for the loss, it ought to shrink and eventually disappear altogether. The intensity of this **Hawking radiation** effect depends on the strength of the tidal forces around the hole (section 11.4), and since tidal variations are more extreme for smaller holes with more tightly-curved surfaces, smaller black holes should radiate more strongly. Although we'd expect "solar-mass" holes to be extremely cold, the exterior of a *shrinking* hole should gradually increase in temperature as its radius contracts, until, as a hot, *microscopic* black hole, it throws off its remaining massenergy in one final blast of radiation, and vanishes.

Hawking published his result as a letter to the journal "Nature" rather than as a peer-reviewed paper, and his announcement wasn't universally well received. The announcement caused a flurry of activity amongst theorists eager to find where Hawking's argument had gone wrong, but this eventually resulted in a grudging realisation that Hawking's QM-based prediction seemed to be unavoidable, and also seemed to coincide with general arguments from **thermodynamics** and **information theory**: It seemed that a black hole horizon *should* radiate, and *should* present the outside world with a conventional temperature and a set of conventional-looking thermodynamical properties, regardless of anything that GR1915 might have to say on the matter.

11.14: Pair-production and pseudo-pair-production

We can use QM's pair-production arguments (sections 10.7-10.9) to "fake" the effects of classical indirect radiation (section 11.1), and present them as "quantum" corrections to GR1915's black hole predictions. The slightly convoluted logic goes something like this:

We'll suppose that a dark star is radiating normally, and that some of the radiation gets knocked out of the region around the star by a chance encounter with a passing object, or by collisions between "visiting" particles in which one particle gets bumped free and accelerated out of the star's gravity-well. The accelerated particle escapes and reaches the detector of a distant observer, who then has to try to explain how the heck it got there. If this observer believes that the star is actually a *GR1915 black hole*, their reasoning might go something like this:

> " *In the coordinate system that I apply to the exterior of the hole, there can be no "visiting" particles (contra **figure a**) and therefore no indirect radiation either, and with nothing to accelerate the particle towards me, it must have travelled along a simple, unaccelerated, "ballistic" trajectory (**figure b**) ... But if I extrapolate back along the escaped particle's path towards the horizon, I find that in order to have escaped along this ballistic trajectory, the particle must initially have been travelling at more than lightspeed, which GR1915 says is impossible. However, in my coordinate system's description, a particle that approaches at more than lightspeed can be said to be "tachyonic", and time-reversed – so if I mark a point in my coordinate system where this "ballistic" explanation breaks down, I can say that this point marks the creation event for a **particle-antiparticle pair** (**figure c**) ... one particle then travels towards me to be captured, and its twin travels along a mirror-trajectory and is captured by the black hole. The appearance of the escaping particle outside the horizon is therefore a **non-classical** effect correctly described by **quantum mechanics**.*
>
> *Since the "breakdown point" will be different for different observers, the particle-pair's creation point will also be different for observers at different locations (**figure d**). But the final trajectory of the escaping particle outside the hole will be exactly the same in each case.* "

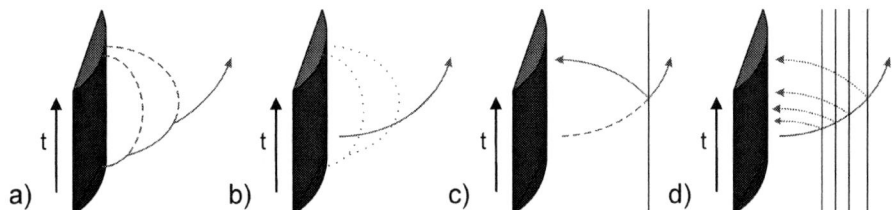

Figure 11-14: Pseudo-pair-production

If we started out with a description of a particle being physically accelerated out of the clutches of a dark star, and tried to inappropriately apply the conventions of general relativity to the situation, then what we'd end up with would be an artificial description of how the particle escapes that seems to correspond to the modern "textbook" description of Hawking radiation. Quantum mechanics seems to force the observable physics of black holes to look just as we'd expect if the "dark star" or "acoustic metric"-style arguments were right and GR1915 was wrong.

We may not feel happy making this comparison, but it's how QM makes things *seem*.

11.15: Attempts to eliminate the "dark star" explanation

Our initial reaction to this explanation is to reject it. If Hawking radiation really *is* a purely non-classical effect, then there should be some critical part of the "advanced" QM description that won't be compatible with the more "naive" dark-star "indirect radiation" explanation.

Our *first* *objection* would be that in a dark star model, there are particles that are visible to nearby observers but that don't appear on a distant observer's coordinate system. If we were working with a conventional coordinate-system based theory of physics we'd then have a situation in which particles could be said to exist in a region for some observers but not for others. They'd be "real" for observers that could see them, but only "virtual" for other observers who could only infer their existence indirectly.

We might think that surely current physics *wouldn't* predict anything quite this weird. But this behaviour *does* now seem to be part of the QM description – QM describes particles outside the horizon that aren't directly detectable by distant observers, and that don't exist on their coordinate system maps, but whose presence does have indirect consequences for them. QM lets a nearby observer see and report particles around a gravity-source that supposedly aren't really there for a more distant GR1915 observer.

Our *second* *objection* is that if the dark star interpretation was *really* correct, our artificial extrapolated trajectories in Figure 11-14 should break down at different gravitational depths for different observers, so that different observers would have to disagree about the height at which a particle-pair is "really" created. Distant observers might claim that a particle is "really" created an appreciable distance from the hole, whereas a nearby observer might claim that it really originates at the vicinity of the horizon.

This *sounds* like an unacceptable result that *can't* be part of current physics, but again it does seems to be part of the QM black hole descriptions.

A *third objection* might be that in the dark star description we should be able to *prove* that the local atmosphere really exists, by lowering a bucket into the region and scooping out some of the local dirt. Again, modern quantum theory has a way of reproducing the same final result: In QM-speak, we say that the acceleration of the bucket physically converts some of the "virtual" particles around the black hole into "real" particles. If we lower a video camera into the region to obtain images of the particles that supposedly aren't there, QM says that the camera is able to see and report particles that were previously only "virtual" for us, because its physical acceleration converted "virtual" particles into "real" ones. If we try to peer into the region by lowering one end of a fibre-optical cable into it, QM explains that the view *through* the cable reveals more particles because accelerative forces in the suspended cable warp the local geometry (GPoR) and allow "virtual " photons in the region be accelerated up the cable and be seen – the physical acceleration converts "virtual" photons into "real" photons.

And so it goes on. Every time we think that we have found an effect that happens in an "acoustic metric" description and that can't *possibly* be reproduced in a "GR plus QM" context, QM proves us wrong by providing a convenient effect that just happens to make the final physics look more like Eighteenth Century physics than the 1960's GR predictions.

A dark star model says that observers suspended just above the horizon should see radiation apparently coming out *through* the horizon, and even on this point, quantum theory doesn't quite disagree: according to the **Membrane Paradigm** (section 11.21), the combination of quantum effects above should conspire to make these hovering observers see radiation coming *away* from **r=2M** (or at least, away from a surface a Planck distance above it). Since the basic phenomenology of the QM description seems to agree so well with that of a dark star model (if we ignore the *explanations* and just look at the *resulting physics*), it's becoming more difficult to say that we know that QM is "right" but that dark-star-type approaches are "wrong".

11.16: Acoustic metrics, once again

In acoustic metrics, "**virtual**" particles can be entirely *real* particles whose current location just happens to be temporarily out of view, hidden behind a temporary observational horizon. When we project our acoustic metric onto certain sorts of **observerspace** coordinates (that describe what we can see *directly*), there'll be some parts of the original, complete surface that won't appear on the "observerspace" map, because the coordinate-mapping between the acoustic metric and the observerspace metric isn't complete. Some coordinate regions and their associated information will be missing.

We can compare this mapping failure to what we see when we look at a photograph of a piece of mountainous land (or of very rough seas). Our photographic image has no gaps or voids, and as an *image*, it is complete and continuous. The surface that we try to photograph is *also* continuous. But the projection of information held in the "real" surface onto the photographic film inevitably results in missing regions and discontinuities. Some parts of the surface won't appear in our photograph because they face away from us or at right angles to us, and others that *do* face us won't be shown because they're hidden behind other surfaces. We can't easily map back from a single photograph to the original shape, because the information that belongs to these regions is missing from our photographic record. Some regions of the **underlying** metric are missing from the **observerspace** metric.

Figure 11-15: Mountain view: Although our photograph has no gaps, it doesn't reveal the full surface of the landscape

However, we can still deduce that life goes on as normal, unseen, on these hidden sections of surface, and although we can't see the hidden events *directly*, we can deduce that they must be happening because they'll be affecting things that we *can* see.

As we watch a mountain range, a hiker may sometimes seem to appear without warning at the edge of a mountain peak, and then walk down the front of it: although we see the hiker appearing "acausally" in the centre of our picture as if they sprang into existence from nowhere, we can reason that nothing especially "scary" has happened: the hiker was previously walking up the other side of the mountain, or was hidden by some other obscuring piece of landscape before they strolled into our field of view. The *apparent physics* may seem acausal and discontinuous, but the *deduced physics* is still entirely classical. If our hiker calls out to their friends over the hill for supplies, and those people respond by throwing food, we can consider the other hikers to be *virtual*: they can't be seen directly, but their behaviour has visible consequences. The "**virtual**" hikers can become "**real**" to us by striding over the hill to a place where we can see them, and although they can't (yet) register on our photographic equipment, they're visible to our "real" hiker, who can describe them to us, or can photograph them and then wave the resulting pictures at us. If the hiker holds up a conveniently-angled mirror for us, we may even be able to see their friends' images reflected in the mirror, even through we can't see those people directly.

In "GR plus QM" descriptions, the idea of a virtual particle is less straightforward. We can assign a ghostlike "probabilistic" status to *some* sorts of particles whose existence is too fleeting for them to register as full-blown particles, or to regions with "critical" energy-concentrations right at the edge of a particle-creation threshold (so that attempts to measure whether particles exist might affect whether those particles actually *are* created or destroyed) ... but GR doesn't like the acoustic "horizon-shielding" explanation of persistent particles that are directly visible to some observers but not others, because GR1915's horizons are supposed to be perfect, closed surfaces ("*a boundary has no boundary*"), and anything behind a horizon is supposed to be permanently lost to the outside world, forever.

The "acoustic" idea of **route-dependent horizons** doesn't really seem to fit GR1915's definitions, and if we want to use this explanation of Hawking radiation, we seem to have to come up with a different way of implementing the general principle of relativity.

11.17: "Acceleration radiation"

A further consequence of the "Hawking radiation" result was that if we take the **equivalence principle** at face value – if we say that the local physics of an observer feeling geeforces should be the same irrespective of whether those forces are due to being suspended in a gravitational field or to acceleration (section 4.10) – then since the "accelerated" observers hovering closer to a black hole get to see more "real" particles, perhaps observers undergoing similar accelerations through a supposedly empty region should see more radiation, too?

This is known as **acceleration radiation**, or **Unruh radiation** (after **William Unruh**), and in the context of classical SR-based physics it seems like nonsense: How can particles exist for one observer but not another in a *conventional* region of space?

"acoustic" acceleration radiation

Well, there are two possible answers: we could use QM to suggest that the physical acceleration of the observer distorts the surrounding region of spacetime in such a way as to help to *create* new particles that otherwise wouldn't have been there ... or we might suggest that the motion of matter *within an acoustic metric* alters the gravitational landscape in such a way as to allow the observer to start to register the existence of particles that would previously have been "cloaked" from their view, lurking behind recession horizons.

The acoustic metric's "scenery" is more complex than GR1915's background geometry. It allows objects to be accelerated away from us at *more* than the background speed of light, and we might technically categorise these particles as "virtual" – their recessional **velocity horizon** prevents us from seeing their current states *directly*, but we can sense their existence *indirectly* by watching the trail of wreckage that they leave behind. But since these *acoustic* horizons are purely *relative*, if we then accelerate after one of these receding particles and start to catch up with it, the intervening horizon may dissolve and we may start to see some signals ahead of us that wouldn't previously have registered on our detectors: the "virtual" particle becomes "real". As we accelerate, some features of the gravitational background geometry will seem to freeze and fold closed and other closed pockets of spacetime will seem to dilate and unlock and open out to reveal their contents to us. We'll be able to see a little deeper into some gravity-wells ahead of us, and not so far into others behind us.

But it's difficult to reconcile these "acoustic-metric" descriptions of acceleration radiation with SR-based theory, because according to special relativity's worldview, velocity-horizons *don't exist*, and *nothing* in the universe travels at more than background lightspeed. By accelerating through an SR metric, we'd not be seeing any new information, just the same "global" selection of events with different spacetime alignments. GR1915 obviously *does* support some forms of motion-dependent curvature, but isn't supposed to make any predictions that might undermine the basis of SR. So we've essentially the same situation here with acceleration radiation as we had with Hawking radiation – a classical explanation seems possible, but only outside GR1915.

11.18: The Black Hole Information Paradox

history

The **Black Hole Information Paradox** sprang from a growing realisation in the 1970's that **Stephen Hawking**'s result suggested that something might to be seriously amiss with our understanding of high-gravity objects.

For some years, scientists had been wrestling with the unpleasant notion that black holes swallowed information and permanently removed it from view. With time, the idea began to seem less outrageous, after all, no information was actually being *destroyed* ... although it was disappearing into black holes where it could no longer be *used*, at least we knew where it had gone. We could point to the hole and say: *that's* where the information is hiding.

But after Hawking's 1974 letter on "black hole explosions", there was a problem. Hawking's "new" radiation effect meant that black holes gave off energy and lost mass, and a hole that loses mass has to shrink. While the Hawking radiation from a decent solar-mass black hole would still be very cold indeed (colder than the current background cosmological temperature, ~2.7K), smaller holes would have a higher temperature and shed massenergy more quickly, which in turn would make them smaller and hotter, so that they would shed mass at an even higher rate, making them even smaller ... until eventually the process would run away with itself and the hot, shrinking hole, fizzing like a lump of sherbet in water, would eventually explode away the last of its massenergy and disappear altogether. Even with decent-sized holes, *eventually* (in an expanding universe) one would expect the cosmological background temperature to eventually cool to less than the hole's **Hawking temperature**, at which point (if the hole was no longer being "fed" with fresh local material), the inevitable shrinking process would begin. Given enough time, in an expanding and cooling universe, it seemed that all black holes were mortal after all.

Black hole evaporation gave us two immediate "accounting" problems. Firstly, if all the information that ever fell into a hole was forever hidden behind an event horizon and could never escape, then once the hole disappeared, where the heck did all that information go? Once the hole had evaporated, it would seem as if the information contained within it was no longer merely *hidden*, but actually *removed* from the universe.

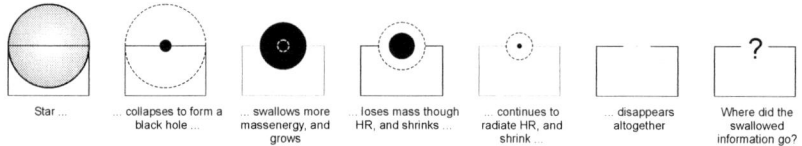

Figure 11-16: The "GR+QM" black hole vanishing trick

Hawking's initial answer to this was the idea of the **baby universe** – the hole's perimeter could be considered as a one-way **wormhole** (section 18.13) leading to the interior of the black hole that seals itself off when the hole disappears, with the missing information then persisting *outside* our universe in its own little independent bubble of spacetime.

This was quite a nice idea (in that it allowed universes to spawn other universes), but it still left us facing the second problem: the quantity of Hawking radiation emitted by a black hole during its finite lifetime was considerable (seemingly equal to the quantity of information originally "lost" inside the hole). If this radiated information was really *not* the same information that had fallen into the hole at an earlier time (because this would seem to make it difficult to argue that a hole was a GR black hole and not a dark star), then *where the heck was it coming from?*

11.19: The BHIP and Microcausality

One of the more subtle aspects of quantum theory is that although individual small-scale quantum events may individually act in an unpredictable way, with no apparent regard to the laws of cause and effect, *cumulatively* these events conspire to reproduce the more conventional-looking larger-scale physics that we see around us and that we expect from classical arguments, where cause and effect relationships appear to be honoured.

Even if we consider *individual* quantum events to be **"randomlike"**, their *aggregated* properties have to follow more conventional laws: they *must* fuse together to reproduce the more traditional, large-scale, "everyday" physics that we're familiar with. For instance ... under the **Copenhagen Interpretation**, a conventional electromagnetic wave with significant amplitude can be reconsidered as a cascade of individually "non-classical" photons whose *individual* actions seem "spooky", but their bulk behaviour conspires to recreate the effects of more conventional-looking classical wave behaviour. This allows standard QM to recreate the "wavelike" outcome of the two-slit experiment (section 10.4), and helps to make it possible for us to think of the "Copenhagen" and "Hidden Variable" interpretations as at least *potentially*-equivalent descriptions of the same physics.

We say that although standard QM doesn't *appear* at first sight to obey the conventional rules of causality, these tiny fluctuations somehow conspire together in such a way as to ensure that larger-scale causality is preserved. There is a method to their apparent "madness" that becomes apparent when we step back and look at the larger picture, and we call this small-scale compliance with the larger-scale rules of causality, **microcausality**.

The first solution that Hawking suggested to the problem of where the information encoded in outgoing black hole radiation was coming from was drastic. The obvious solution to the Information Paradox was to suggest that since the quantity of information going into and coming out of the hole seemed to be the same, *it was actually the same information*. But this made GR1915's assertion that the event horizon was inescapable look a bit silly. Hawking's alternative suggestion was that if QM was in conflict with GR1915 about the identity of the escaping information, then since we knew more about GR1915 than we did about this aspect of QM, we might *rewrite* quantum theory to fit GR1915 and say that in this *revised* form of QM, microcausality needn't be preserved. The information that was radiated from a black hole didn't relate to the swallowed material, and didn't originate *anywhere*. It was *truly* random.

Even apart from the idea that "random information" sounded like a contradiction in terms (the **"no random numbers"** principle, *"there's no such thing as a random number"*, see also section 17.14), this suggestion made some physicists very unhappy indeed for another reason. Losing microcausality would suggest that some things in our universe would then have to happen *for no reason at all*. Since we live in a highly chaotic world, and since chaotic systems allow the consequences of "quantum-scale" information to "snowball" into significant differences at macroscopic scales (the classic thought-experiment on this subject being **Erwin Schrödinger**'s infamous **"Schrödinger's Cat"** paradox), to lose *microcausality* under QM might be to lose at least some *conventional causality* on larger scales. If quantum mechanics broke the chain of evidence, there could be larger-scale events whose origins were entirely arbitrary, and everyday physics wouldn't be forced to respect the conventional laws of cause and effect. While many physicists could cope with the idea that quantum theory predicted *apparent* breakdowns in causality, *to lose causality itself* had implications for physics that went far beyond black hole theory. What was the point of theorists spending their lives trying to find a set of "rational" rules underlying physics, if Nature was then going to ignore them?

In 2004 Hawking changed tack, ceding a bet on the subject that he'd had with **Kip Thorne** and **John Preskill**. Instead of rewriting QM to remove microcausality, perhaps we could rework or reimagine *general relativity* to allow fluctuating horizons and indirect-radiation effects. Doing this classically was difficult, but perhaps a larger, more complete theory of

quantum gravity might provide a new way for information to be absorbed and reradiated without explicitly contradicting GR1915.

Hawking's new suggestion was difficult to understand, and was likened to a **Many-Worlds** interpretation (sections 10.4 & 17.12). Perhaps the information leaving a horizon originated in a different universe, in which it hadn't yet fallen into any black hole? For this to happen without just smudging and sharing the resulting information-loss across multiple universes seemed dubious – quantum theory seems to forbid making "perfect copies" of information, and the new approach seemed to break this rule at the "multiverse" level ... a bit like a student forgetting to do their homework assignment but claiming a top mark on the basis that somewhere out there in a parallel universe, they *would* have done it, and would have gotten an "A" ... but discarding all the alternative universes in which they'd have flunked the course.

Could information cross over between universes *at* the horizon, so that information falling into one hole emerges at the horizon of its parallel-universe twin? What if all the information disappears at the horizon and is reradiated without ever entering? The problem here is once again QM's insistence of the impossibility of **quantum photocopiers** (Preskill 1992). If we dive into a huge black hole and survive the passage through the horizon, then *we* will be pretty damned sure that all our information came with us. If an outside observer insists that we were somehow vaporised just outside the horizon surface for later re-radiation, we'll have to disagree. As far as we're concerned, we are still very much alive and in one piece. The two sets of observers then see *irreconcilably different* outcomes for the same experiment, and although it might be difficult or impossible for them to compare notes, the idea of loosening our concept of reality to allow one set of initial conditions to lead to two incompatible outcomes – reality splitting into two – is still considered to be an unpalatable way of fixing the problem.

As of 2006, we're still wrestling with the Information Paradox, and still looking for a way out that might be considered acceptable.

11.20: "Observerspace" arguments

As we found in section 7.6, although we *believe* that matter falling into a black hole really does fall through the horizon, what a distant outside observer expects to *see* is different: for them, matter appears to be converging *towards* the horizon but never to actually getting there, and long-infallen matter seems to be constantly compacting toward a strange two-dimensional state. The view that the GR1915 event horizon presents to the outside world is of a strange, vanishingly-cold "skin" suspended in space arbitrarily close to the $r=2M$ surface.

If the "visible" physics of a black hole says that nothing is ever *seen* to fall in, and appearances do (to some extent, at least) seem to define the quantities that are *directly measurable*, then what would happen if we treated what *appears* to be happening as if it was real?

"theory without theory"

Treating direct observation as literal reality doesn't always give us a complete description of a situation, but useful insights can still often be gained by asking how physics would have to behave for our observerspace *view* of it to be internally coherent (if physics was compelled not just to *be* consistent but to *be seen* to be consistent).

We sometimes fall back on this **observer principle** when conventional theory deserts us, and it's often a useful "emergency" option when we really can't work out what's going on behind the scenes ... we can abandon theoretical interpretations altogether and "play dumb", in the hope that although the resulting description might not be *complete*, at least the artefacts that might appear in the description will be *natural* artefacts rather than the results of things that we've invented ourselves. It's a useful approach to try when we're "stuck", and this approach, applied to black holes, gives us the black hole **Membrane Paradigm**.

11.21: The Membrane Paradigm

Thorne (1994) relates that this approach to modelling black holes was prompted by the realisation by Hanni, Ruffini, Wald and Cohen in the early 1970's that an electrically-charged pellet dropped into a black hole should still *appear* to a distant outsider to be just outside the critical $r=2M$ radius, complete with its original charge – if its "image" persists, and still *shows* it as carrying a charge, then the pellet's electrical fieldlines should *also* continue to exist and should still point towards the location of the "frozen" image (Thorne 1994, pp.406). If the black hole rotates and the image of the pellet is pulled around with it, then the associated electrical fieldlines should be pulled around too, so that the hole behaves like an electrical dynamo.

the physics of a mirage

Further work suggested that the horizon surface region shouldn't just appear to contain everything that had ever fallen into the hole, it should also show also an apparent electrical resistance (Thorne 1994, pp.408). Since these effects seemed to be exhibited down *to* the event horizon (or at least, to a **Planck distance** above it), and since general relativity insisted that no dynamic exterior interactions could extend *through* the horizon, it was convenient to invent a surface arbitrarily close to the horizon (the "**membrane**" in the title) and to assign ownership of the hole's newly-discovered electrical properties to it. Kip S. Thorne, R. H. Price and D. H. Macdonald then published an anthology of papers by various researchers ("**Black Holes: The Membrane Paradigm**", 1986) exploring how far the idea could be taken.

the Membrane Paradigm and Hawking Radiation

After being introduced to model the theoretical electrical characteristics of the horizon region, the "membrane" approach found itself being pressed into service to model the **Hawking radiation** effect predicted by quantum mechanics.

In the coordinate system of a distant stationary observer, Hawking radiation tends to be described as a quantum-mechanical "particle-pair production" effect involving "virtual" particles ... but for stationary observers hovering at fixed heights nearer to the hole, QM seems to say says that the effect should look like purely conventional radiation from an atmosphere of "real" particles. The Membrane Paradigm describes the black hole as it'd be described by an array of these stationary, suspended, accelerated observers, and since their shared coordinate system ends at $r=2M$ (because an observer can't legally hover at or below the event horizon under GR1915), this radiation ends up being described (within "**observerspace**") as being due to a hot radiating surface at or just above the critical radius, where the coordinate system ends.

Again, although these radiation effects should appear to extend all the way down *to* the event horizon, they aren't allowed by GR1915 to be coming *through* the horizon – so blaming them on a hypothetical thin radiating membrane *at* the horizon lets us model Hawking radiation as if it was coming from a conventional "hot" surface, without *explicitly* contradicting general relativity's prediction that $r=2M$ should be inescapable. As long as we don't ask what happens in a region that *spans* the event horizon, the "quantum" approach and the "classical" membrane approach seem to be physically equivalent (Thorne 1994, pp.408-411).

"Do not look at the man behind the curtain!"

Thanks to QM (and, more explicitly, thanks to the Membrane Paradigm), we no longer have an obvious way to distinguish between the properties of a "modern" black hole and the general, "classical", statistical properties of an idealised dark star (section 11.1). Perhaps this shouldn't be entirely unexpected, since both "QM black holes" and "dark stars" are compelled to radiate according to the same statistical laws of entropy and thermodynamics ... and perhaps, statistically, there might not even *be* a difference. But a *complete* classical description of these indirect-radiation effects would put the original radiation source *below* $r=2M$, violating GR1915 – the Membrane Paradigm avoids this issue by only modelling as far as the horizon.

11.22: Holographic arguments

Leonard Susskind has since taken this "membrane" idea further, and turned it into a more general set of arguments.

If we define a volume of space and draw a boundary surface around it, the interaction of the entire "interior" physics of the region with its surrounding environment can be described in terms of how this surface interacts with the outside world. Everything that enters the region (objects, radiation, forces, fields and so on) has to pass inwards through this surface, and everything that the region does that affects the universe outside has to pass outwards through the same surface. Geometrically, this suggests that the information contained within a volume ought to be capable of being compacted into two dimensions – if we emptied a volume of space but "faked" identical boundary behaviour, outsiders shouldn't be able to tell what we'd done. If we imagine a perfect hologram printed on a spherical surface that appears to show an object inside that surface from every conceivable angle – it might be difficult to tell whether the object was really there or not.

Suppose that you visit a museum in order to see a priceless diamond that is touring the world. For security reasons the diamond is sealed inside a cabinet made of thick bulletproof glass, and the exhibition only allows visitors to walk past the diamond in near-darkness, and to view the diamond by artificial, monochromatic light. As you walk around the glass case, you briefly see the diamond appearing to be in multiple locations at once, but you blame this on the refractive properties of the very thick glass plates arranged around it. There is no way of looking at the diamond without looking through these glass plates. How do you know that the diamond is really there? How do you know that each thick glass plate doesn't simply have a piece of transparent holographic film stuck to its inner surfaces showing a *hologram* of the diamond? In fact, since holograms can be created by computer, how do you know that the diamond that you *think* you are looking at *has ever actually existed*?

In black hole physics, the **holographic principle** seems to be intimately related to questions concerning information density: if all the interactions between a volume and its surroundings can be expressed by the behaviour of its boundary surface, then it would seem that we might be able to compact any volume of *information* into a bounding surface without losing anything, and this is quite suggestive of some results from black hole theory that seem to say that **the surface area of a black hole event horizon** is a measure of the total amount of information that the black hole contains. According to these newer arguments, since the horizon represents the edge of "observerspace", and exists as a bounding surface that appears to hold all the information that ever fell into its associated hole, then perhaps the horizon surface represents a state of maximum holographic information density, and perhaps the expansion of the horizon when it absorbs massenergy, and its contraction when it gives off Hawking radiation, might allow us to relate the quantity of information inside the hole with the minimum horizon surface area required to store that information.

The holographic principle may seem a little bit abstract (because it is), but it generated some interest when Leonard Susskind suggested that it might allow a solution to the dreaded **Information Paradox**. If a bounding surface can "stand in" for the volume that it contains, and if we can legally describe information passing into and out of the surface without that information actually entering the volume within, then couldn't we invoke the holographic principle to argue that perhaps the event horizon of a black hole absorbs and then remits massenergy and data without anything *really* passing out of the volume behind it?

Could we solve the paradox by agreeing that information flows in and out of a black hole's *surface* – without explicitly contradicting GR1915 – by suggesting that the outgoing information originated *within* the event horizon surface rather than from somewhere behind it?

11.23: The Holographic Principle in action

Unfortunately, the Holographic Principle's apparent reconciliation of GR1915 with quantum theory isn't really a proper solution: if we accept that a black hole's boundary *can* show information moving in and out of the region, it'd seem that the surface is then *mimicking* an interior physics in which the laws of nature seem to be following a set of rules other than those of GR1915. But why should the bounding surface choose to do this instead of "mimicking" the physics of GR1915? Why shouldn't it instead be replicating the results of an interior physics in which information falls into the region never to return? If the Holographic Principle is just a way of letting us mimic *any* interior physics that we like, then it doesn't seem to tell us very much. On the other hand, if it's telling us that it *prefers* to describe a physics that allows information to leave a black hole, and that it prefers to mimic *non-GR1915* physics, then perhaps, since any other region of space can also be emulated with an arbitrary bounding surface, perhaps *every* other part of the universe, mimicked with the Holographic Principle, should show non-GR1915 behaviour, too. In that case, we'd have an *entire universe* appearing to operate according to a different set of laws to GR1915, and we'd have to accept that GR1915 didn't really describe how that universe worked.

black hole escape using the Holographic Principle

For the sake of argument, suppose that a spaceship somehow violates GR rules by descending through a gravitational event horizon and then escaping (perhaps by jettisoning most of its mass, or activating a hypothetical warp drive). This might be possible under a pre-GR "dark star"-type model, but wouldn't be allowed under GR1915.

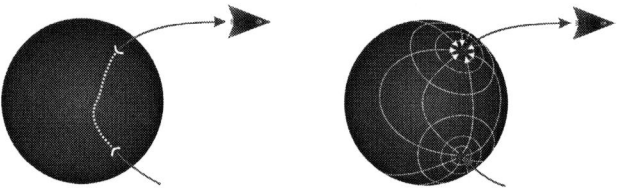

Figure 11-17: Surface absorbing and reemitting a spaceship

The holographic principle can take this "GR-illegal" behaviour and make it seem reasonable: it can say that the spaceship's information is absorbed as it passes inward through the horizon, and that the horizon then mechanistically convolves and combines the spaceship's information with all the other information embedded in the same surface, describing everything else that has ever fallen inside the hole. The spaceship's entry causes a ripple of information through the surface that converges at a new position, where the ship's information combines with other interior information, and a suspiciously similar-looking spaceship is then spat out as a focussed "Hawking radiation" event. Once this information has been removed from the horizon it can't be emitted a second time, so only one copy of the ship is ejected.

This "new" spaceship has the same basic dimensions as the original, it has the same number of crew, its ship chronometers show a suitably advanced time, and its databanks hold what appear to be convincing video recordings of the ship entering and successfully escaping the black hole horizon. The ship perhaps also shows some charring and some damage that appears to correspond to events shown in the video recordings, and the ship's crew claim to have a memory of approaching the horizon, passing through, and exiting again.

Since the holographic principle seems to let us describe physics at a black hole appearing to operate *exactly as we would expect if GR1915 was wrong*, without forcing us to say that the theory *really is* wrong, the price of "saving" GR1915 in this way might be that we'd no longer be able to class GR1915 as a proper, falsifiable, "scientific" theory.

11.24: The "no-signal" problem

The holographic principle brings us to one of the oddest potential paradoxes to do with black holes under GR1915: considered as a problem in **information theory**, it *could* be argued that for GR1915 to have a chance of being correct about gravitational horizons, the exterior physics of a black hole might be forced to behave exactly as if GR1915's description is wrong.

The argument goes like this: if the horizon of a black hole prevents any electromagnetic information about the interior from escaping, then we can't tell whether the interior description provided by GR is correct. We also can't use the hole's external gravitational field to tell if total collapse really *has* happened, because the hole's gravitational fieldlines are obliged to leave the hole's horizon at exactly 90 degrees to the surface for any spherically-symmetric arrangement of matter, regardless of whether it's a solid sphere or a hollow shell about to collapse, or a spherical membrane, or a central singularity – GR1915 doesn't let us "read" the exterior field to assess what's really happening inside the hole.

Here's where things get paradoxical: if we *could* tell (from outside) that the interior of a hole *did* obey GR1915 and really *did* collapse, then since this information would have to leave the hole interior to reach us, any verifiable evidence available to an outside observer that the GR1915 description was *correct* in this regard might actually disprove the theory.

We can't immediately tell whether the object is a black hole or a dark star by watching an object fall in, because we never *see* it fall in, it always seems to hover at the horizon. But our knowledge of acoustic horizons throws up a physical distinction between GR black holes and dark stars: in a dark star model, the observerspace description is incomplete. Although a radio transmitter dropped into a dark star would never be directly seen to cross the horizon, part of its signal stream emitted inside the hole should still be able to escape indirectly (in a somewhat garbled form). If this signal (and similar signals coming from other infallen matter) *don't* exist outside the hole, then this *silence* carries information about the hole's interior – and the identifiable *lack* of a such a signal, apparently *supporting* GR1915, would tell us that the interior physics of the hole wasn't obeying the dark star rules. But by carrying information about what the hole's interior was *not* doing, the *lack* of a signal (the silence) would itself count as a signal!

This is an odd idea, and perhaps the easiest response to this sort of argument is to ignore or reject it ... but if we try to persist with this line of reasoning through to the bitter end, we do seem to arrive at something oddly similar to the situation according to QM: Perhaps the only way that a horizon could conform to the GR1915 rules and prevent us from knowing what is really happening inside **r=2M** would be if it somehow managed to emit a signal that was *indistinguishable* from the signal emitted by a dark star (i.e. something that looks like Hawking radiation). If the degree of garbling in the dark star model is so severe that we can't make out any details, then perhaps all the black hole has to do to cloak its real interior physics is to emit "thermal" noise that can emulate the basic statistical behaviour of a dark star: perhaps something like Hawking radiation is an inevitable outcome of starting with GR1915 and applying information theory principles.

Unfortunately, for a hole to emit truly "random" noise wouldn't seem to be a complete answer, because in some situations (such as our "spaceship" example), a dark star's radiation might manage to carry *identifiable* information, and if a black hole could reproduce *that*, then there wouldn't seem to be a justification for saying that it really *was* a GR1915 black hole.

A lot of researchers would probably consider this line of reasoning to be specious ... but if we *accepted* this argument (and the logic behind is does seem to be alarmingly "wild"), GR1915's "new" predictions about black hole physics would have to be considered to be automatically self-invalidating.

11.25: The verdict

Einstein reputedly never accepted his general theory's prediction of "pure" event horizons and totally inescapable black holes (Einstein 1939, Bernstein 1996). Although some take this to mean that Einstein may not have quite appreciated the beauty of his own theory, perhaps Einstein's deeper familiarity with the structure gave him a sense of which parts were less solid, and didn't quite "smell" right. The further we get into quantum mechanics and information theory and thermodynamics, the more likely it seems that GR1915's single most radical prediction – the perfect black hole – may simply have been wrong. Einstein's distrust of his general theory's "black hole" predictions on intuitive and aesthetic grounds – as an objectionable situation that a theory *shouldn't* predict – might yet be vindicated.

In many respects, it seems that at least *some* of GR1915's predictions about the behaviour of gravitational horizons seem to be worse than the predictions that we'd have gotten using bad old Newtonian theory, even without the benefit of modern updates (such as the 1911 gravitational time dilation effect). The distinction between "modern" black hole theory and older theory used to be clear: older models predicted radiation, and GR1915 didn't. But when we apply QM corrections to GR1915, the "impossible" radiation effects reappear, and according to the Membrane Paradigm, these should look almost indistinguishable (and perhaps even *totally* indistinguishable) from the more conventional indirect-radiation effects from earlier centuries that GR was supposed to have made obsolete. It was almost as if QM was deliberately taking all the effects that had been thrown out with the advent of the GR black hole, and was determinedly putting them all back, one by one.

At this point we had to try a new interpretation. We'd been forced to backtrack a little, and to admit that QM predicted radiation effects with the same basic *statistical characteristics* that we'd have expected from pre-GR theory, but we could still try to argue that even if the *quantities* and *distributions* of escaping information were the same, this didn't immediately mean that the *content* of that information matched. Our last way of distinguishing between pre-GR1915 theory and QM was that in dark star models, the information in Hawking radiation would be inherited from information previously carried into the hole by infalling material and light, whereas if we truly believed in GR1915, we'd want to believe that the *identity* of the information encoded in outgoing Hawking radiation wasn't related to the information previously lost inside the hole. This was our way of demonstrating that GR1915 really did make some new, qualitatively-correct predictions ... the idea that the ingoing and outgoing information was unrelated let us cling to the idea that information didn't *really* migrate outward through the event horizon.

But we're now increasingly getting used to the idea that perhaps the pattern of outgoing Hawking radiation *does* carry information that originated with previously-infallen material after all, and recent work such as Susskind's seems to be increasingly assuming that the infalling and outgoing dataflows are transmitting the same information back and forth, just as we'd expect if the newer insights that GR1915 had given us about these situations had been quite wrong. Since quantum theory also doesn't like the idea of central singularities, this means that – if we believe the QM arguments – GR1915's list of unique predictions for high-gravity-objects seems perilously close to having a 100% failure rate.

And for a major theory of gravitation, that's not good.

In the next section, we'll look at *why* GR1915 refuses to predict Hawking radiation, and we'll try to identify why GR currently gets this subject so wrong. While we're at it, we'll also list some possible structural defects in GR1915 and some other aspects of the GR1915 that are less than inspiring, and we'll also will try to identify how many of these potential problems might be blamed on design decisions taken when GR1915 was originally devised.

PART IV

UPDATING STANDARD THEORY

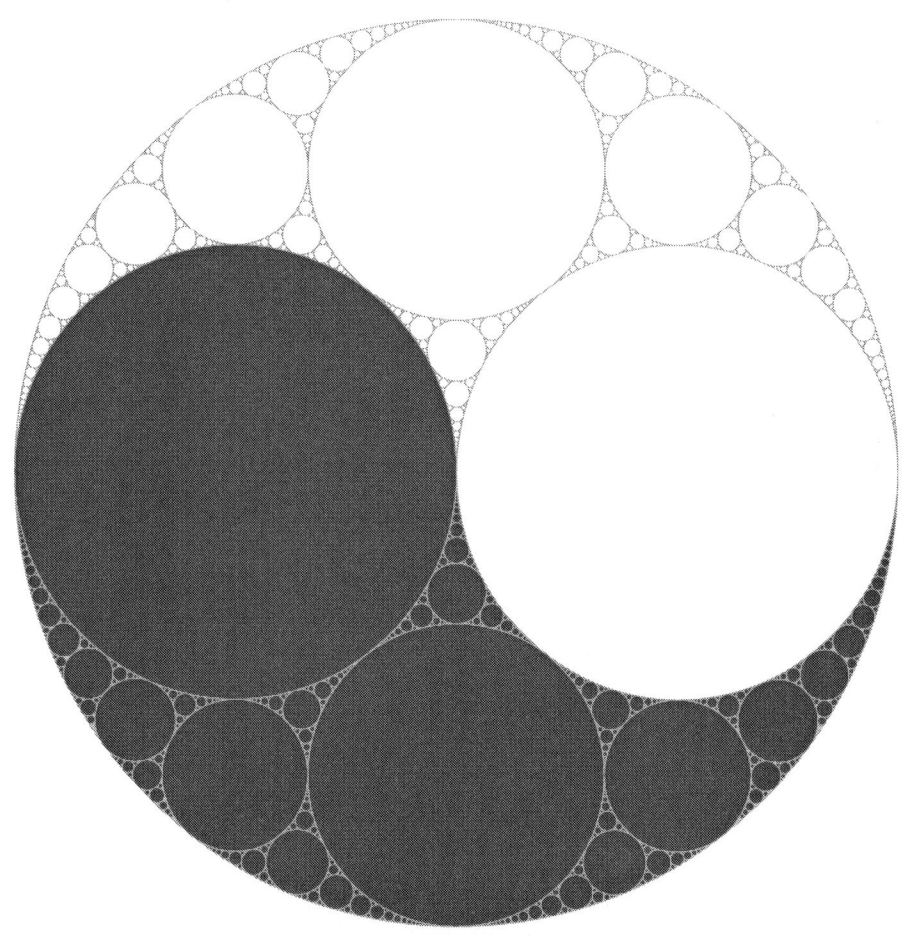

RELATIVITY IN CURVED SPACETIME

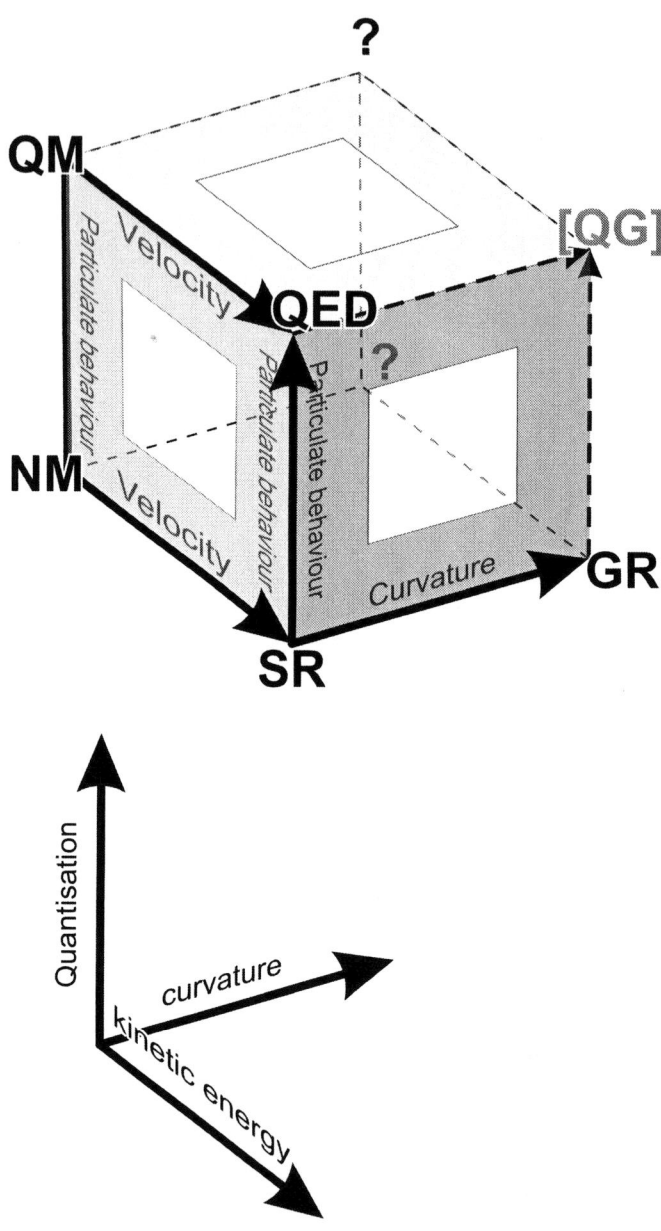

"Missing links" in C20th Physics

12
What's wrong with General Relativity?

> 'When **I** use a word,' Humpty Dumpty said in a rather scornful tone, ' it means just what I choose it to mean – neither more nor less.'
>
> 'The question is,' said Alice, 'whether you **can** make words mean so many things.'
>
> 'The question is,' said Humpty Dumpty, 'which is to be the master – that's all.'
>
> "Through the Looking-Glass", Lewis Carroll
> (A.K.A. mathematician Charles Dodgson)

> " The most dangerous strategy is to jump a chasm in two leaps. "
>
> Benjamin Disraeli

> " Paul Gerber alone ... (1898), from the perihelial motion of Mercury, forty-one seconds in a century, finds the velocity of the propagation of gravitation to be the same as that of light. This would speak in favour of the ether as the medium of gravitation. "
>
> Ernst Mach, in "The Science of Mechanics" (2nd ed), 1902

> " But why should anyone be interested in getting exact solutions of such an ephemeral set of equations? "
>
> Albert Einstein, on GR1915's gravitational equations,
> recounted by Cornelius Lanczos

> " ... the vision of Riemann, Clifford and Einstein, of a purely geometric basis for physics, today has come to a higher state of development, and offers richer prospects – and presents deeper problems – than ever before."
>
> John Wheeler (1962), quoted in "MTW"

> " It is suggested how Bernhard Riemann might have discovered General Relativity soon after 1854 and how today's undergraduate students can be given a glimpse of this before, or independently of, their study of Special Relativity "
>
> Wolfgang Rindler
> (imagining an alternative history), 1994

> " General relativity is incompatible with quantum field theory. ... There remains the conceptual need for a new theoretical scheme where these two pillars of modern physics, quantum field theory and general relativity, merge consistently. "
>
> Michele Maggiore, "A Modern Introduction to Quantum Field Theory" (2004)

> " First you guess. Don't laugh, this is the most important step. Then you compute the consequences. Compare the consequences to experience. If it disagrees with experience, the guess is wrong. In that simple statement is the key to science. It doesn't matter how beautiful your guess is or how smart you are or what your name is. If it disagrees with experience, it's wrong. That's all there is to it. "
>
> Richard Feynman

RELATIVITY IN CURVED SPACETIME

Constructing an "Alternative GR" is not quite as easy as this

12.1: "Core" experimental tests of general relativity

General relativity isn't required for most solar-system scale physics, and it's difficult to find predictions that are specific to the theory and testable. We can test some *basic assumptions* and *principles* that are *used* or *implemented* in Einstein's general theory, but showing that these individual features hold up well to experimentation doesn't necessarily mean that we're also verifying the way that GR1915 chooses to *weave these components together* to create a larger structure, or that the theory's subsequent predictions are then also known to be right.

the three (or four) basic tests

Einstein ("*Relativity ...*" book, Appendix 3, 1920 and Princeton lectures, 1921) said that the correctness of GR depended on three fundamental tests:

(a) **Motion of the perihelion of Mercury** – the planet Mercury has a more elliptical orbit than most of the other planets, and along with its proximity to the Sun, this creates an appreciable difference in gravitational field strength between the nearest and furthest points of its orbit (the **perihelion** and **aphelion**). The elliptical path that Mercury follows is drawn on a background that isn't quite "flat", and because of this (and also because of the complicating influences of the other planets) this ellipse-like shape doesn't quite join up with itself after one complete orbit. Mercury's path isn't a perfect static ellipse but a shape whose alignment shifts and drifts very slightly each time it swings around the Sun. Using more technical language, we say that *Mercury's perihelion precesses*.

(b) **Gravitational redshift (and gravitational time dilation)** – As we found in section 3.7, by building on Newton's arguments we find that a gravitational difference between two regions ought to alter the energy of light passing between them (Michell, 1783), and should also change the relative rate of timeflow between regions: a stronger gravitational field makes time go slower.

(c) **Deflection of light by a gravitational field** – Although Newton had already proposed the bending of light by a gravitational field, and Einstein's 1911 calculations predicted essentially the same amount of bending due to gravitational time dilation, in Einstein's 1915 theory the effects due to the *warping of space* and the *warping of time* aren't equivalent descriptions of the same effect: they *combine* to give GR1915 a light-bending prediction that is about twice as strong as either of the two earlier predictions.

To these three tests, we now tend to add a fourth:

(d) **Shapiro time delay** – Signals reaching us from objects at the opposite side of the solar system, which have to skim the Sun's position to reach us, spend longer in flight then we'd expect from just their spatial route and the assumption of flat spacetime (section 1.2).

testing "predictions" vs. "assumptions"

The weakness of these tests is that they're not so much tests of the *specific predictions* of Einstein's general theory as tests of the general assumptions *made by* the theory – when we get a positive result, we usually find that it can be calculated directly from these more general principles and assumptions without having to learn the specific mathematical machinery that GR1915 uses. These successes might support Einstein's more general "gravitational time dilation" argument, but they don't test whether Einstein's particular approach to spacetime curvature in GR1915 was right ... since more specific predictions for testing are hard to come by, we test what we can.

RELATIVITY IN CURVED SPACETIME

... gravitational redshifts and time dilation (b)

Although we're often told that gravity-shift tests (example "b" on Einstein's list) count as verifications of general relativity, when we come to look at the scientific significance of these results, things aren't quite so impressive.

GR and NM will *both* predict this effect (although Einstein seems to have been the first person to publish correct "Newtonian" calculations, in 1911), so GR's main claim to ownership of the result seems to be that the general theory's *need* for gravitational redshifts to be real is less ambiguous than NM's. We can argue (with hindsight) that gravitational shifts should be an inevitable consequence of NM, but this doesn't seem to have been obvious to physicists in the Eighteenth and Nineteenth Centuries. On the other hand, GR's requirement for the effect to be real is written into its structure, and doesn't seem to be open to discussion. GR deserves some extra credit here because it makes a more *obviously definite* prediction.

Does GR then make a *more accurate* prediction here than NM? So far this doesn't seem to have been demonstrated: the most reliable data for this sort of comparison still seem to be the Pound and Snider's experiment in the 1960's, which used the **Mössbauer effect** to boost the sensitivity of their apparatus to the point that it could measure the gravitational redshift across a 22-metre-high university building. We're sometimes told that this test verified GR's predictions to umpteen decimal places, but what isn't mentioned so often is that because of the comparatively small gravitational drop involved, the experiment would probably have needed to be at least an order of magnitude more sensitive in order to be able to tell the two theories apart, even based on the most optimistic assessment of how far the two sets of predictions might diverge. If we calculate our shift from an agreed gravitational potential, it's actually quite difficult to find a testable shift difference between NM and GR.

There are also more recent tests using rockets or satellites that seem to indicate that gravity-shift effects are real and in the correct range for GR. The **Global Positioning System (GPS)** has a network of atomic clocks installed in satellites orbiting the Earth, and because they're inhabiting in a weaker gravitational field, we should see these clocks running at a fractionally faster rate. When we're building atomic clocks for the GPS network that are will be launched into space, we compensate for this effect in advance by deliberately reengineering the clocks to have a slower natural "tick-rate" than their Earthbound counterparts (so that they deliberately "tick slow" by roughly one tick in every two billion). Once these "pre-adjusted" clocks are safely launched into orbit, they find themselves keeping decent time with our groundbased reference clocks, as expected.

But it's still not obvious that using GR calculations is necessarily any more accurate than a "modernised NM"-based approach would have been. The people building these systems just want them to work, GR is good enough for that and it's a standard theory, so that's what they use.

... light-deflection (c)

What about test "c" on the list? GR1915 gives twice the "standard" Newtonian prediction for the amount of gravitational deflection in starlight skimming the Sun, and we can choose to present this difference by saying that the "historic" Newtonian calculation only assumed spatial curvature (or its "ballistic" equivalent), whereas the GR1915 calculation presumes space*time* curvature, with the difference between these two predictions being due to the presence of that additional time-curvature component. Since Einstein's rough 1911 calculation for how much bending should be caused by time dilation came out the same as the "old" Newtonian calculations, when we put the two effects together we then end up predicting twice as much lightbending regardless of whether we are using GR1915 or "updated" NM.

Finding this "doubled" amount of light-bending doesn't specifically support GR, but *does* support **(1)** the idea of gravitational time dilation and **(2)** the idea that its associated light-bending effects act *in addition* to spatial-curvature effects.

How do we know that the two effects ought to add together like this? The short answer is that they have to if we want the orbit of Mercury to come out correctly.

... perihelion of Mercury (a)

Now let's go back and look at the first test on Einstein's list, **(a)**. It should be fairly obvious that if we can't distinguish between "updated NM" and GR1915 for the approximate shape of spacetime around the Sun, we probably shouldn't be able to distinguish between the two sets of predictions for the orbit of Mercury through the region, either. Mercury's extra precession is essentially another test of gravitational time dilation.

Since the amount of theoretical light-bending due to *spatial* curvature was already pretty much nailed down, and Einstein's rough 1911 calculation for light-bending due to *time* curvature gave essentially the same prediction, the only major decision that still needed to be taken was how to implement these two effects. Was this a *single* effect (as Einstein's 1911 paper seemed to imply), or were space-curvature and time-curvature two different components that each contributed to the final answer?

We could give GR1915 some credit here for saying that the space- and time-distortion predictions *combine* to produce the correct Mercury-orbit and light-bending predictions, but in this respect Einstein had the advantage of already knowing in advance exactly how much "missing" Mercury-precession the theory had to predict. We tend to think of testable predictions as being things that are predicted before the data is available, and if we manage to get the answer right, "blind", then this gives additional credibility to our calculations. But in the case of Mercury's orbit, the mismatch was already well documented and Einstein seems to have consciously used the Mercury "gap" as a guide when developing the theory: once the theory predicted the right precession value, it was judged nearly ready: when the loose ends were tidied up, and the "good" Mercury prediction was shown to still be in there, the theory was finished.

If we decide that the "Mercury precession" data is correct and important, and also decide that the mismatch between theory and observation should be caused by a fundamental problem with older theory (rather than, say, the existence of an undiscovered planet spoiling the calculations), then the way to predict the *right* answer for Mercury is to have our two different light-bending calculations acting together. Starting with the Mercury data and working backwards, we should probably expect to get the "doubled" lightbending prediction pretty much by default, regardless of whether we choose to use GR1915, or "updated NM" or some other model.

... Shapiro effect (d)

For the Shapiro effect, **(d)**, the situation is similar to the "light-bending" case in **(c)** – we can predict the appearance of a *general* gravity-dependent timelag effect from very basic principles without going near general relativity. A version of the Shapiro effect appears naturally in a corrected version of Newton's old aether model as a function of variations in density of the underlying medium, and Einstein's general 1911 arguments relating gravitational time dilation to the slowing of light moving through a region give us a second "pre-GR" way of arguing that this sort of effect ought to exist.

As with **(c)**, once we know the basic shape of spacetime around the Sun (either by using GR1915 or some other method), the end-results would seem to be unavoidable.

12.2: Experimental significance

Looked at this way, Einstein's "three tests" seem to be more important as tests of his chain of reasoning (and his inspired guesses) than the exact mathematical details of his general theory.

Einstein had already argued that gravitational redshifts imply gravitational time dilation and that gravity should bend light *before* GR1915, and his instinct that the "Mercury precession" anomaly should be fixed with deep theory then meant that the two lightbending effects – due to curved space and curved time – needed to reinforce each other. This seems to give us enough information to predict the outcomes of Einstein's "three tests" of GR1915, without actually requiring us to know anything specific about how GR1915 is constructed. What we consider to be tests *of* the theory, don't seem to need to *involve* the theory, and this leads to some natural scepticism over whether the theory itself is being tested here.

testing is difficult

But we have to admit that none of these experimental issues and difficulties can be blamed on GR1915 itself. Given that *any* decent theory competing with GR will probably predict identical or near-identical values in these situations, a "better" theory would still have suffered from the same sorts of testing difficulties that have plagued GR1915. The theorists' enthusiasm for GR coupled with frustration over the difficulty of getting the damned thing tested might have sometimes led to a degree of **"cheerleading"** by some of its keener proponents, along with a level of exaggeration over the significance of the available evidence that isn't really supposed to happen in science ... but these lapses don't mean that *the theory itself* is wrong, and may have been just as likely to happen if we'd been presented with a different theory instead of GR1915. In fact, if we'd been given an even *better* theory, it's possible that with even higher levels of frustration over our inability to *prove* how wonderful it was, the temptation to exaggerate might have been even worse.

successes

The desire to validate GR1915 has helped bring about a useful body of experimental data supporting the idea that gravity slows time and warps spacetime coordinates, and it has also helped to show that this "warpage" happens in the range that we'd expect from general arguments. What it hasn't done is to show us that GR1915 is significantly better at making predictions that the more general arguments that it was based on. GR1915 is a useful "umbrella theory" that *implements* those arguments, and it is a useful proof of concept that shows that turning those arguments into a theory is *possible*, but experiment has not (yet) shown us that GR1915 has an especially significant claim to being "correct" in a deeper sense. In fact, the general rule-of-thumb with Einstein's general theory seems to be that the further we move away from the "general" arguments *implemented* in GR1915, to arguments that are really *specific* to GR1915 (like the theory's "black hole" predictions in section 11.8), the more often we find ourselves making predictions that seem to be wrong.

Amongst theorists, GR1915 sometimes gets credit for being the *most efficient* implementation of the general principle of relativity that we can construct: it's considered to be the default version of a general theory: deterministic, wholly defined by its essential starting assumptions, with no internal inconsistencies and no leeway for "fudging" its results. Credible competing general theories are considered to be "GR1915 plus": GR1915 with additional bells and whistles. If we can't justify the existence of those additional bits and pieces, we are supposed to revert by default back to "pure" GR1915.

As our reference "base theory", GR isn't supposed to have any structural problems or internal inconsistencies, but when we try to bypass the hype and take a cold, hard, critical look at the theory, we find that things may not be quite as rosy as we have been led to believe.

12.3: Incompatibility with quantum mechanics

One problem with GR1915 that doesn't seem to be disputed is the theory's incompatibility with quantum mechanics. Einstein compared the properties of spacetime under the two models by saying that according to GR1915, when we zoomed in on the surface under general relativity it became smooth and featureless like fine marble, whereas with quantum theory the surface became more rough and chaotic, like a coarsely-cut piece of cheap wood.

Theorists used to be more philosophical about this mismatch. We said, well, general relativity was a theory of the very big, and quantum theory a theory of the very small, and although it wasn't ideal to have different laws operating at different scales, we'd have to be satisfied with this combination until we eventually found some larger theory of **quantum gravity** that incorporated both sets of laws. This GR-QM mismatch was thrown into sharp relief with the prediction of Hawking radiation (section 11.12), since now we had *large-scale* objects whose gross behaviour was affected by QM – according to QM a black hole will eventually fade away and radiate, under GR it is impassive and immortal.

We usually try to explain this difference by saying that it's natural for GR and QM to be different – that GR is a "classical" theory and QM is a "quantum" theory, and this difference automatically puts the two theories at odds. We say that *of course* GR doesn't allow Hawking radiation, but that this doesn't represent a shortcoming in the model, because Hawking radiation is inherently a "quantum" effect. But recent work on **acoustic metrics** shows that an incompatibility with quantum predictions and information theory is not an *inevitable* feature of classical physics, because acoustic metrics seem to be quite capable of replicating these effects from mundane, "non-spooky" principles. It might instead be that the incompatibility with QM is telling us that GR1915 is the *wrong* classical theory

Perhaps the reason why Hawking radiation doesn't appear under GR1915 isn't because the theory is built along classical lines, but because it's built along classical lines *wrongly*. We chose to *categorise* Hawking radiation as a nonclassical, quantum-specific effect because it didn't appear under our standard classical model, GR1915 ... but this doesn't necessarily mean that the effect can't be predicted classically by other models.

Since "acoustic metric" issues have become better understood, we've run into some difficulties over what we should call things. It's easy to say that QM is an inherently "quantum" description in which discontinuous "nonclassical" things seem to happen and have to be modelled statistically, and it's also easy to see that GR1915 is an obviously "classical" theory in which continuous parameters evolve in continuous and strictly causal ways. But how do we classify *acoustic* metrics, which assume classical mechanics and have a classically-smooth metric, but which generate *apparent* causalities and *apparent* spooky behaviour as signals and objects slide out from behind horizons to surprise us? Acoustic metrics seem to destroy the clean distinction between classical and quantum. QM-style effects arising within acoustic metrics are sometimes described as **"semiclassical"**, which is a pragmatic way of acknowledging that they seem to have a foot in both camps.

But we shouldn't assume that just because we've invented a special new class for things that didn't fit into the old classification system, that we've actually resolved anything. If our original classification scheme was *inappropriate*, then to invent new **exception categories** for things that don't quite fit doesn't mean that we have fixed the underlying problem: it may just mean that we've scooped up all the evidence of a possible structural incompatibility, put it all into a tidy box and given it a nametag. That's fine as a short-term measure – it means that anyone working on a major overhaul of the system knows where to find the parts that currently don't fit – but it shouldn't be taken to mean that, by successfully categorising these elements (as "uncategorisable"), we've resolved these issues and made a major overhaul unnecessary. Just because we have named something doesn't mean that we understand it.

12.4: Fudge factor?: The Cosmological Constant

Although we now consider that the most "natural" form of Einstein's general theory describes a universe that is spatially closed and expanding, the theory was originally supposed to describe a universe that was infinite in extent and **"quasi-Euclidean"** (having no cumulative long-distance curvature). If a universe is sprinkled with objects that have *positive* gravitation, we'd normally expect the cumulative gravitational effects of these masses to build up as we looked at larger and larger regions of space, producing an overall curvature effect that became more significant as we looked at progressively larger chunks of universe. This model would also tend to collapse under its own gravity unless something was stopping it, so in order to predict a nice, stable, "tidy" universe that looked "flat" over large regions, and that was neither expanding or collapsing, Einstein added a compensating long-range antigravitational effect that he called the "**Cosmological Constant**", Λ.

By 1929 **Edwin Hubble** had analysed the apparent recession velocities of a range of galaxies, and shown that, in fact, there *did* seem to be an average redshift effect that increased with the distance of the object ("**Hubble shift**"). This suggested that at large scales, our universe *wasn't* static and flat, but expanding and curved.

If Einstein had managed to use GR to predict the existence of Hubble's redshifts *before* they'd been discovered, it would have been a triumph ... but instead he had to make do with changing his theory's predictions retrospectively to match the new data. Faced with Hubble's evidence, Einstein accepted the "expanding-universe" solutions to GR produced by **Alexander Friedmann** and **Georges Lemaître**, and agreed that a general theory without a cosmological constant was aesthetically preferable, famously referring to the Cosmological Constant as the biggest blunder of his career.

The Cosmological Constant's lesson is that sometimes a theory's predictions can be wrecked by an innocuous hidden assumption that only *seems* mathematically unavoidable. Before Hubble's result we could have argued that non-static solutions were obvious failures and that these unstable solutions could be dismissed out of hand without further justification. We could have seen the Constant as an essential, legitimately-derived feature based on the unavoidable condition of energy conservation. But nowadays we can look back on the Cosmological Constant and see it as an artificial **fudge factor** designed to force the theory to produce an artificial result that its designer had just *assumed* to be correct.

We now tend to consider the early use of a cosmological constant not as a problem with the theory itself, but as a case of "**user error**". We distinguish between the predictions that were originally made *for* the theory (or *with* the theory), and the predictions that we now think of as being made *by* the theory, without our interference. We say that in its most natural form, GR1915 *would* have predicted Hubble shifts, and we now don't see the Cosmological Constant as being an essential part of the model (although it has a habit of resurfacing when cosmologists need to tweak things).

This raises some difficult issues to do with how we test theories, and how we're supposed to be able to tell when they've failed: Was GR1915 without the cosmological constant technically a *different version* of GR? Should it have been assigned a distinct name? How far are we allowed to change the predictions of a theory to fit new experimental evidence? Perhaps this depends on what we want a "theory" to do: if we're happy for it to only act as a way of repackaging existing, validated, experimentally-confirmed relationships in a convenient form, then adding and removing parts of GR to make it correspond better to reality as a *model* may be acceptable as a progressive "best-fit" approach. But if we're trying to predict features of the universe *before* we have discovered them by accident, based on *theoretical* arguments, we'd prefer theories whose predictions aren't being made up as we go along to fit the evidence.

12.5: Possible breaking of conservation laws

Einstein's Cosmological Constant allowed the **conservation of energy** under GR1915. The cumulative effect of gravity over large regions would tend to give light crossing the universe a degree of redshift (and a corresponding loss in energy) that increased with distance travelled, and the "Constant" (a hypothetical long-range repulsive effect invented by Einstein) generated a compensating long-range blueshift that cancelled this effect, balancing the books.

But since our *actual* universe seems to show cosmological redshifts that *haven't* been cancelled by anything, this route to energy-conservation no longer works, forcing us to avoid the issue or invent arbitrary mathematical terms to absorb it. Our universe is flooded with radiation that is constantly losing energy due to expansion. Where does all that energy go?

A second possible "conservation" issue is that in an expanding universe, the "quantity of space" is continually increasing with time. Since space (under general relativity) is a tangible, bendable, measurable thing, the idea that the quantity of something physical could be increasing without incurring a debt – something for nothing – isn't ideal.

We *might* like to try to put together some sort of larger conservation rule, where perhaps the energy lost and the associated expansion have a more intimate relationship. The usual linear relationship between frequency and energy would then only be a first approximation (analogous to the force/extension behaviour of a spring used within its linear range), and this leads to an interesting "bubble" cosmology mentioned in section 17.8. However, since GR1915 wasn't developed to deal with variable-size universes, it doesn't have much to say on these sorts of issues.

12.6: Possible incompatibility with Mach's principle

The subject of Mach's Principle and general relativity is quite controversial. Some physicists say that GR *doesn't* fully implement MP, but that this doesn't matter too much, while others say that GR *does* fully implement MP, as long as we apply the theory properly.

Finding a consensus position is difficult. Some say that MP was a useful *route* to GR1915 but that Einstein's general theory makes it redundant, and that MP IS now completely contained within a more fundamental set of *geometrical* principles. Others say that MP gave a useful *route* to GR, but now that we have a fully-geometrical theory, the theory's properties are now more fundamental than the obsolete arguments that led us to them, and that although MP *isn't* fully supported, we shouldn't worry about it. Some experts produce solutions that seem to show non-Machian behaviour (such as "rotating universe" solutions), while others respond that these solutions represent a misuse of "short-cut" methods or **boundary conditions**, and that the theory itself, applied in a "purist" way, shouldn't have these problems.

These technical arguments have been grumbling on for decades. Although it's probably best to stay out of these disputes, we'll just make one point: if GR1915 is as well-understood as it's *supposed* to be, then we'd hope that expert opinion would at least be able to *agree* as to whether the structure of the theory corresponds to its founding principles, or whether it doesn't. If we don't know (or if we insist that we *do* know, but then produce conflicting answers), then this suggests that perhaps the theory *as it is understood and applied*, may not yet be entirely deterministic. If the theory *is* definite in this regard, different users have still somehow retained the ability to study it and come to different conclusions about it.

Either the theory doesn't make hard predictions in this area (leaving users to fill in the blanks), or it *does* make solid predictions, but end-users have trouble coming to a consensus over exactly what they should be. In either case, it's difficult to give a theory credit for making clear and unambiguous predictions if experts can't decide amongst themselves what those predictions *are*.

12.7: Fudge factor?: Galactic curves and Dark Matter

When we look at the way that galaxies seem to rotate, we hit a problem: according to conventional orbital mechanics, the outer edges of galaxies ought to be taking far longer to complete a circuit than those further in, but the shape of spiral galaxies indicates that the peripheral material seems to be happily scooting around the edge much faster than it has a right to. If this material really *was* moving this fast, we might expect it to be flying off into intergalactic space. So how do these galaxies hold themselves together so well?

There are a couple of ways that we can attack this problem:

> **Decrease gravitational field strength *outside* the galaxy.** Since general relativity isn't supposed to assume any particular **prior geometry**, we might expect it to be in an ideal position to deal with the idea that the gravitational field strength might be able to drop to arbitrarily low levels away from regions of concentrated mass. The idea that a galaxy's fields sit on top of a background **gravitational floor** may be a leftover from older thinking that says that we start with a flat background and add fields to it ... but if the substance of spacetime *is* the gravitational field, then we have no reason to assume that there is a default "lowest field density" out there, or even a fixed number of spatial dimensions when we are looking at highly distorted or highly attenuated regions. The "no prior geometry" feature would seem to give a general geometrical theory of gravitation an ideal excuse to suggest (or even tentatively predict) that that field strength may well drop away outside galactic regions faster than we'd calculate from the normal inverse square law, with the drop in field strength resulting in a corresponding dilution of the geometry, so that the regions between galaxies have a weaker geometrical and inertial connection, and galactic regions become gravitationally more self-contained. With a more rarefied connectivity between galaxies, our more familiar "flat-background" laws rules might not apply.
>
> While Mach's principle and the general principle of relativity can suggest this sort of **"no-floor"** behaviour, it doesn't currently seem to be part of GR1915. While GR1915 *nominally* has no prior geometry, in practice the reduction to special relativity seems to bring with it an assumed reduction to "flat" Euclidean spacetime with a default background field strength and dimensionality.
>
> **Increase gravitational field strength *inside* the galaxy.** Under current theory we instead try to explain the extra cohesion of galaxies by the idea that we're surrounded by a vast amount of invisible ... something ... of unknown composition, whose distribution "shadows" that of normal matter and whose additional gravitational influence helps to hold the rims of rotating galaxies together.

It's difficult to say that this second hypothesis (that the universe is filled with **"dark matter"**) is *wrong*, since the idea doesn't yet seem to tell us anything other than what we already know – that there's a discrepancy between the predictions of GR1915 and what we see with our telescopes. Since the community hasn't yet decided what this dark matter *is*, we don't have an obvious way to test whether it really exists, or whether it just represents a convenient piece of creative accounting.

There have been occasional claimed verifications of the existence of dark matter – for instance, by showing that gravitational lensing effects disagree with the GR calculations by an amount consistent with there being a corresponding concentration of additional "dark" matter shadowing the visible matter in the lensing regions – but this *consistent* discrepancy between GR's predictions based on the amount of visible matter and the apparent strength of gravitational effects might just mean that GR1915's predictions are *consistently* wrong.

12.8: Arbitrary suspension of the Equivalence Principle

the 1960 Harwell centrifuge test

Another interesting study of time dilation effects was performed by the Harwell group in 1960. The group published a pair of experimental results in quick succession, testing the appearance of gravitational redshifts, due to **(1)** the Earth's gravity, and **(2)** the apparent gravitational fields generated inside a centrifuge (Hay, Schiffer, Cranshaw *et.al.*, 1960&1960). As the team's second paper pointed out, the same final "centrifuge redshift" effect could be explained *either* by treating the **centrifugal field** as a genuine gravitational field (principle of equivalence, "**EP**") and assuming that *gravitational* time dilation effects apply at the centrifuge's rim as a consequence of acceleration, *or* by applying special relativity and assuming a Lorentz time dilation effect caused by its *speed*.

For a circling clock, its speed is proportional to the **(a)** the radius of the circle, times **(b)** the revolution rate ... but the gee-forces felt by the circling clock scale in the same way, and this double statistical match makes it difficult in simple centrifuge experiments to establish whether it is the speed or the acceleration that is *really* responsible for the effect.

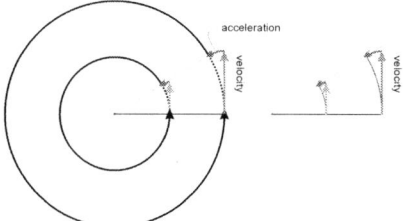

Figure 12-1: **Doubling the radius of a circling object with constant angular velocity doubles its speed, but also doubles its acceleration**

The second Harwell paper gave us two effects that could claim responsibility for the reduced rate of the centrifuged clocks, and both of them seemed to make an equally good match to the data. Since both calculations gave good agreement with the results, it probably seemed reasonable to assume that the two approaches were **dual** – that the effect could be calculated as a speed effect (using SR) for inertial observers, and as an acceleration effect (using the EP) for observers rotating with the centrifuge.

But after the second Harwell paper got into print, a nasty suspicion began to emerge: perhaps the two explanations weren't so much *dual*, as *geometrically incompatible*. According to our general arguments regarding light (section 3.4), if timeflow and lightspeed are slowed in a region, there should be corresponding gravitational gradients pointing "downhill" toward that region, and these can then be used to recalculate the effect within the gravitational domain (section 4.16). If the centrifuge's curvature turned out to be **intrinsic**, then (like the water's curved surface in Newton's bucket, sections 4.4-4.5) it would exist for *everyone*, even for observers who were reckoned to be "inertial". This suggested that perhaps the more fundamental relationship was between the *geeforces* and the time dilation.

But special relativity generated the result by assuming flat spacetime, so if spacetime was significantly curved, the calculations would seem to be invalidated ... and since the Equivalence Principle predicted curvature strong enough to explain *the entire experimental result*, it's effects couldn't really be said to be "insignificant". We could try to argue that GR had the more complete calculations and that SR was just a "quick" method of getting to some of the same answers, but GR1915 assumed that SR's derivations were *exact*. If the "gravitational" exercise was valid, and it invalidated special relativity's calculations, then had GR1915 invalidated *itself*?

the Schild rebuttal

According to **Alfred Schild**'s prompt rebuttal paper (1960), the answer was straightforward. General relativity *had* to reduce to special relativity, and special relativity had predicted the effect first. If SR's prediction was physically wrong then GR1915 would be in doubt, and we'd have invalidated both theories. The "least worst" solution was therefore to assign "ownership" of the result to special relativity, and to say that where the equivalence principle seemed to be in conflict with SR, it should be suspended.

This SR-based assumption that physical accelerations have *no* effect on clockrates is now referred to as the **clock hypothesis**, or **clock postulate**. Some theorists will say that because we *know* that acceleration doesn't affect clocks, there's no explanation for the centrifuge redshift other than SR (and never was) … but if we look at the underlying math, and the historical record, we find that they tell a different story.

interpretational bias

The SR-based explanation involves a **closed logical loop** – we say that since we "know" that SR must be right, and since SR can predict the full redshift, and since this doesn't leave anything for the acceleration arguments to explain … we therefore "know" that there are no significant acceleration effects. Then we come full circle and say that since we've *demonstrated* that acceleration effects aren't a factor in this experiment, there's no remaining explanation for the redshift result except special relativity's speed-based time dilation effect … and the experiment therefore shows that SR's time dilation effect is physically real.

But this argument is **pathological**, and (from the point of view of statistical analysis) illegal. This chain of logic only showed that the experiment supported special relativity because that was the initial assumption that we made at the start. It might be an *internally-consistent interpretation* of what happens according to SR but it's not exactly an impartial analysis, since we could have reversed the argument and come to the opposite conclusion … we could have instead argued that since we know that the *Equivalence Principle* is correct, the fact that the EP predicts the full redshift so accurately means that we know that there *can't* be any significant additional effect due to speed …and the experiment therefore shows that SR's time dilation effect *isn't* physically real.

The community *seemed* to have taken an experiment that could have been interpreted as either validating or *invalidating* an effect, picked the interpretation that they liked, and then claimed that experimental evidence showed that their preferred interpretation was correct and that the other one was wrong. But it's not obvious that the suspension of a general principle within a general theory of relativity is something that we ought to be doing. If we can't get a theory to work within the principles from which it was derived, then it's difficult to say that the theory is known to be consistent: where "consistency" is only maintained by our deciding to suspend key principles whenever the structure is otherwise about to crash, and is only maintained by intervention, it's not so impressive.

We might suggest (in defence of GR1915) that perhaps the final shape of the math has become so powerful that it has evolved past or *transcended* its original design specifications and founding principles, which might be seen as a good thing: in this case, the original design specification might be seen as just a rough guide, to be discarded once the full theory has taken shape. But it might instead be that the theory doesn't match its original design criteria because it failed, because something in the design specification was wrong, or because it tried to combine components that were inherently incompatible.

There may be some other way to resolve this problem … but our willingness to suspend the EP undermines the idea that GR1915 operates according to strict principles. It's natural for a limited rule to be overridden by a deeper principle that supersedes it, but we don't usually suspend a *general* principle to accommodate idealised special-case solutions.

12.9: Invoking reduction to flat spacetime

GR1915 is meant to be a **superset** of special relativity, with the equations and relationships of SR built into the theory as a limiting case. Although special relativity made a few simplifying assumptions that weren't really appropriate for a general theory of relativity (such as the existence of inertial mass without gravitational mass, and the allowance of arbitrarily high concentrations of kinetic energy without curvature), GR1915 nevertheless assumes that the basic relationships that SR obtained by doing this were the correct ones, and that they have to carry over into our shiny new "curved spacetime" theory.

When Einstein was designing his general theory, he was trying something very ambitious. As someone who wasn't a professional mathematician, he was trying to produce a curvature-based solution where the mathematical "greats" before him had failed. He had the advantage over Nineteenth-Century researchers of having realised that gravity had to bend both space *and time*, but he'd published that result in 1911, and now he was racing to produce a general theory before someone else beat him to it. He was familiar with special relativity, he trusted it, and by declaring that general relativity had to reduce to it, he managed to narrow the options for how his new theory should function without having to start over completely from scratch.

Einstein doesn't seem to have demonstrated that general theories *must* reduce to special relativity, and doesn't seem to have been able to rederive SR's equations of motion in the context of a curved-spacetime model. But by deciding that *his* general theory *would* reduce to SR as a matter of *design*, he made his job slightly easier.

"flatness" in classical field theory at small scales

The geometrical argument for GR's reduction to SR goes something like this: general relativity describes the shape of the metric as curved, but as we zoom in to examine smaller regions of spacetime, larger-scale curvatures become more difficult to notice. Eventually, if we zoom in *arbitrarily* far, we find ourselves looking at a section of metric that is *arbitrarily flat*, and we can then argue that, in this small region that is *effectively curvature-free*, relativistic geometry must conform to the flat-spacetime relationships of special relativity.

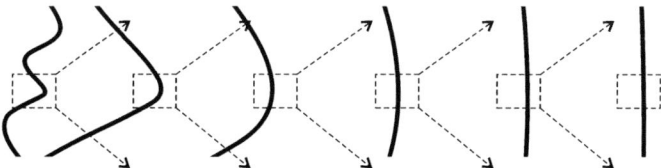

Figure 12-2: Zooming in on a classical curved surface produces an arbitrarily flat region

Fair enough. But it doesn't follow from this that a general theory of relativity is compelled to reduce to the *physics* of SR, unless we can also show that curvature isn't an **intrinsic feature** of particles and their interactions ... it ignores the possibility that perhaps physics *is* curvature. If we zoom in so far that curvature effects no longer exist in our selected region, then this *might* just mean that we've zoomed in too far, and are now studying a region in which no meaningful physics is taking place.

The idea of "intrinsic curvature" would spoil the case for SR being inevitable as physical law. If we were to believe that the massenergy of particles warps spacetime, and that relative motion is expressed as *further* warpage of spacetime, then by zooming in so far that there is no significant curvature in our field of view, our "flat" region would seem to be defined by an *absence* of any interesting physics. The region wouldn't contain any particles with significant relative motion, or any associated gravitomagnetic effects, or particles with *any* relative motion at all. In fact, it wouldn't contain any particles, period. The limiting case where SR would be geometrically valid would then be the case where there is actually no physics to describe, and as soon as we started including real particles with real energies and motions, our "proof" of a reduction to special relativity would fail.

We'd have nothing happening, and nobody to watch it not happening. We'd also not have a compelling reason to insist that the principle of relativity had any reason to apply to the "empty" region, because there wouldn't be anything concrete to apply the PoR *to*: if the principle hypothetically *didn't* hold in the empty region for observers that *didn't exist*, we'd probably be none the wiser ... there'd be nothing unusual to see and nobody to notice, and no-one to file a complaint. This would seem to be **null physics**.

So, to say that we know that it's a geometrical requirement for general theories to reduce to the *physics* of SR, for *real objects*, is equivalent to saying that we know that gravitomagnetism and particulate effects and energy don't warp spacetime in a significant way, and don't play an essential role in the physical interactions between bodies. It's a rejection of **geometrodynamic** principles. The special theory doesn't seem to emerge from curved-spacetime geometry *as physics* unless we've already convinced ourselves that the SR concept of "the physics of flat spacetime" is valid. The background arena in which inertial physics is *enacted* may well be flat, but the *actions themselves* need not be – the flatness of the background "stage scenery" doesn't have to apply to the actors on the stage.

It may well be that all physics is geometry. **It does not follow that all geometry is physics.**

12.10: Use of tailor-made definitions

It's understandable that when one theory becomes dominant, and develops into a standard reference model, that the understood meanings of words can shift to reflect how they appear in the context of that theory. But this can sometimes go too far, and can make a popular theory seem unassailable not because it *really* makes a good fit to more fundamental arguments and principles, but because those arguments and principles have been retrospectively redefined to agree with the theory, where they might otherwise have conflicted with it.

Theories based on principles tend to be more impressive, and the more general the principles are, the better the theory looks. But if a theory initially *conflicts* with some of these principles, and an accommodation is only reached *by redefining the principles to fit the theory*, then any eventual agreement becomes less impressive. If a theory is "good", then changing words to fit it may be reasonable thing to do. If it's "bad", then these redefinitions may become **pathological** – they can turn a theory that is merely *wrong* into something that corrupts scientific language and becomes harmful to further research.

definition of the general principle

While GR1915 is supposed to support the **general principle of relativity**, it's not usually pointed out that the theory's definition of "the general principle" isn't *truly* general – it's instead what the computer software industry might refer to as a "platform-specific implementation".

Instead of saying that all motion between objects should be relative (which leaves open the possibility of non-SR solutions based on acoustic metrics), textbook statements of the GPoR typically define the principle according to the way that it's *implemented* under GR1915 (section 4.8) – the justification for including frames in a definition of the GPoR seems to be that this is that way that GR1915 tackles the problem. Using the *frame* as the basis of physics rather than the *body* introduces an additional assumption that these particular frame-based coordinate system arguments are legal and appropriate for tackling fundamental physics problems. Adopting a frame-based approach to simple inertial physics tens to set us on the path to special relativity, so by also putting frames into a definition of the *general* principle of relativity, we find ourselves thinking that a general theory *must* reduce to special relativity. But since a relativistic implementation of an *acoustic* metric doesn't use SR's concept of "flat" inertial frames, this definition to cuts us off from other potential solutions.

It was reasonable for Einstein to describe the general principle of relativity in this way when laying out the chain of reasoning that led to GR1915 ... but if we want to ask *whether* SR and GR1915 are correct (instead of just arguing the case that they are), or if we want to do legitimate wide-ranging cross-theory research, then we should try to define "the general principle" as genuinely *general* principle, not as a theory-specific implementation of it.

redefinition of a metric theory

This departure from the literal meaning of words isn't an isolated example: let's suppose that quantum theory and the **Black Hole Information Paradox** have whetted our appetite for investigating "acoustic-metric" solutions to the general principle of relativity in its *broadest* form: presumably we can look up a body of research on theories of gravity based on these "alternative" metrics, to see if they're any good or not? Unfortunately we can't, because according to the way that GR textbooks define the meaning of words, there can be no such thing. In MTW's **"Gravitation"**, section 39.2, "metric theories of gravity", we're told that in order to earn its name, a metric theory of gravity has to obey two rules: **(1)** it has to agree that spacetime possesses a metric (fair enough), and **(2)** it has to agree *that the relationships of SR are valid in local Lorentz frames*.

This is not a good way to do science. According to MTW, a metric theory of gravity that uses a non-SR metric is, *by definition*, now *not* a metric theory of gravity ... which begs the question of what the heck we *are* supposed to call these things if we want to study them. If these things really *can't* exist, the second part of the MTW definition is redundant and shouldn't be there, while if such metric theories *are* conceivable, then the second part of the MTW definition would seem to be a **pathological** statement that makes alternative research programmes impossible by defining them out of existence.

It may well be that some specialists find these selective definitions to be handy for eliminating what they feel to be pointless questions and useless research (and questions that they can't answer), but this strategy carries risk. It might seem like a suitable use of enlightened academic influence to use these "targeted" definitions as a way of benignly steering students away from pointless subjects and towards what is thought to be the "true path", but if it later turns out that what we've actually been doing is leveraging control over technical language to prevent researchers from being able to study alternatives to mainstream theory (because those alternatives have been eliminated by a mainstream classification scheme designed to support current models), then this "benign" influence can start to look more malignant.

If special relativity is *really* an such a great foundation for more advanced models (giving us GR1915 as our "default" reference theory of gravitation), then this superiority should emerge naturally from fair comparisons between SR/GR and *any* potential competitors – we shouldn't have to defend GR1915 by artificially narrowing the field to only allow theories that are built on special relativity.

defining credibility in a theory

We also find statements in textbooks warning students that a theory has to reduce to the equations of special relativity in order to count as "credible".

"Credibility" is usually considered to be a subjective thing, dependent on who it is that's making the judgement, and when. We know that in the past, some "bad" theories have been well-respected and some "good" ideas have been ridiculed – fashions come and go. To say that non-SR solutions aren't considered "credible" may reflect *a current political reality*, but it doesn't necessarily indicate the result of an exhaustive scientific process. It's a transient thing.

Based this on current teaching, yes, a theory that doesn't reduce to SR will currently tend to be dismissed as having "credibility problems" regardless of whatever else it might have going for it, and perhaps it's responsible to warn students of this. If your aim is to have a safe career in physics, and you want to avoid controversy and potentially career-damaging associations then this is probably good career advice. If you want the respect of your peers, and a "safe" career path, and you want to avoid being called names, then you probably *should* only ever consider solutions based on special relativity. But if we consider ourselves to be *scientists*, and if we consider physics to be a search for *objective truths*, and if we don't especially need the approval of the peer group, then popularity should be less of a concern. We shouldn't care if something is currently considered "credible", we should only care whether or not it's *right*. The rest is politics.

Advice regarding "credibility" is something that we might expect to hear from a record company executive, who might tell a new band that if they want to make good music and become successful, their strategy should be to pick an existing band that makes good music and is successful, and try to copy them. It's a "low-risk" strategy for the record company and it allows them to produce large quantities of predictable product that doesn't stray too far from the format of their previous hits.

redefinition of "experimental compatibility"

One last textbook instruction that we need to beware of is the one that says that all reasonable theories of gravitation need to agree with the experimental evidence, but that because the experimental evidence agrees so overwhelmingly with SR, *in order to be compatible with the experimental evidence, a theory must agree with SR.*

This is a slightly bizarre condition: if we want to evaluate whether or not a theory agrees well with experiment, the sensible thing to do is to *compare* it with experiment. Sometimes we may take a few short-cuts, and we might sometimes use a well-established theory as a decent *first approximation* of the experimental data, but this can only be regarded as a crude "rule-of-thumb" check: it's like checking your answers in a school test by comparing them to what the kid at the next desk has written. If both answers agree then your chances of being right are higher, but there's also the chance that you've both made the same mistake.

Certainly, a gravitational theory that *isn't* based on SR will still have to make predictions that are *sufficiently similar* to those of SR in a range of different situations for us to be able to understand how SR could have managed to be so successful for so long *without* being a valid component of a final theory ... but unless we're sufficiently familiar with experimental procedures, techniques and assumptions, we might not have a realistic idea of which experimental results can be taken at face value in the context of new theories, and which ones can't. "Copying" earlier models is fast, but it isn't a substitute for proper research, and if you really want to find out how closely a new theory seems to corresponds to the experimental evidence, there still doesn't seem to be any substitute for actually going and checking.

It may be that some theorists can't *conceive* of the idea that an experiment could be compatible with the overwhelming majority of experiments and still disagree with SR, but this doesn't make the thing impossible, and by this point in the book we should have found enough examples of the supposedly "impossible" being true to take specialists' ideas about "inconceivability" with a large chunk of salt – they may sometimes tell us more about the sociology of science than they do about the actual mathematical possibilities.

12.11: Do cosmological horizons count as "acoustic"?

In a Hubble-expanding universe, the effective recession velocity between two points depends on their separation. If we look far enough away, there'll be a critical distance at which this recession speed seems to equal lightspeed. What do we see beyond this distance? If it marks the position of an observational horizon, and if that horizon has existed since the beginning of the universe, we might expect it to be a surface at which time appears (to us) to stand still. This would make it look rather like a *gravitational* horizon (except that this surface would be shielding us from the Big Bang event instead of a GR1915 gravitational singularity).

Even if the "velocity horizon" argument is naive, the idea that the universe has a finite age suggests that there really ought be a cosmological horizon out there somewhere. If looking out into space corresponds to looking back in time (because of the time it takes light to reach us from distant objects), if we look far enough away we should be looking right back to the Big Bang itself, and we could then argue that nothing further away than this should be visible, because that would involve seeing back to *before* the Big Bang, and those events (and places, and times) aren't supposed to exist (Figure 17-3). This apparently "frozen" surface would conveniently delineate the outer limit at which our viewable spaces and times politely ... stop.

Cosmological horizons are awkward for GR1915 because although they obey some very general rules, they don't follow GR1915's conventions for how horizons should behave.

Lets draw a surface that marks the position of our own cosmological horizon. Can signals pass through it towards us? We might argue that this is impossible by definition, because we drew this horizon to represent the extreme limit at which locally-generated signals can reach us.

But now let's examine the case of two adjacent stars or galaxies, one that is nominally inside our boundary, and one nominally outside. Can the two communicate? Yes. Light from **System A**, beyond our horizon is capable of reaching **System B**, which we can see ... **B** can therefore appear to us to be illuminated by light that originates beyond the horizon. The extreme cosmological redshift on the nearer **System B** may make it seem to us to have hardly aged since the Big Bang, while **A**'s stars appears not to have yet formed. And yet, ancient post-BB light from those stars, and from objects that may not appear on our map of the visible universe is somehow leaking through. Strict **observerspace causality** appears to be breaking down.

And it's not just light, matter is also capable of crossing our arbitrarily-drawn surface. If a baseball is thrown from **A** to **B**, then although it may take many millions of years to reach **B**, and *billions* of years for us to see it to arriving, the baseball will eventually arrive, and must eventually be *seen* to arrive, despite its origin behind the surface. Where does the baseball come from? In a coordinate system that describes the visible universe as "all that there is", the baseball arrives from "elsewhere". It originates at spatial coordinates that might not exist in our observerspace-defined universe, and from a region of spacetime that for us, doesn't seem to have yet happened. The baseball's appearance looks like a case of **Hawking radiation**.

In effect, the baseball and a surrounding pocket of spacetime "bubbles" though the effective horizon, which spits and swallows objects and energy, and seethes and fluctuates. The motion of the baseball alters the local geometry causing the "effective" horizon to jump, and we have ourselves an **acoustic horizon** within an **acoustic metric**. And since our own location will itself lie on the cosmological horizon of some other very distant future observer, when *we* throw a baseball here-and-now, similar surrounding spacetime distortions should result. So the behaviour of cosmological horizons suggests that **gravitomagnetic** effects may have to be considered as an *intrinsic feature* of relative motion between objects, and our slightly abstract exercise in cosmological events at the far side of the universe seems to be telling us something valuable about the physics of little baseballs here on Earth – the need for cosmological horizons to fluctuate "there" seems to require gravitomagnetic effects to exist "here".

Our fluctuating, leaky, indirectly-radiating cosmological horizon seems to behave very like gravitational horizons under quantum mechanics or archaic "dark star" models, with the **2.7K cosmic background radiation** playing the part of Hawking blackbody radiation.

Theoretical physicists used to be able to reject this comparison by saying that these were clearly two incompatible situations. We "knew" that cosmological and gravitational horizons shouldn't be compared, because we "knew" that one sort radiated and the other didn't. A cosmological horizon had a temperature, but GR1915 had *proved* to us (we thought) that gravitational horizons were at **zero Kelvin**. But this distinction is being undermined by our growing realisation that QM arguments suggest that gravitational horizons should radiate, too.

12.12: Doppler effects and the Black Hole Information Paradox

Nobody knows how to persuade GR1915 to start producing the right answers in this situation. We need to persuade GR's horizons to fluctuate and leak so that they can behave more like **dark star** horizons, (and more like **cosmological horizons**). But it seems to be impossible *on principle* to retrofit classical fluctuation to a GR black hole, and the reason for this seems to be ... special relativity.

Lets suppose that we drop off a spaceship at the event horizon of a dark star. The pilot fires up their ship's engines, and (this being a dark star) has a decent chance of escape if their ship is powerful enough. The outward physical acceleration of the ship creates a distortion in spacetime (section 4.10) and an increase in the outward speed of light, causing a contraction of the effective horizon surface to somewhere behind the ship. The craft can then continue trying to escape from the dark star, from its position outside the modified horizon surface. With the horizon treated as a *static* surface, we might conclude (wrongly) that the escaping ship has jumped discontinuously from one side of the surface to the other in a classically-impossible way ... but in this example the escape is wholly classical, and it's instead the effective horizon surface that does the jumping.

Now let's repeat the exercise using an SR-based gravitational model such as GR1915. At the moment that the ship is dropped off at the horizon, we can calculate the gravitational blueshift of light falling into the black hole at the ship's position, and by applying special relativity's Doppler relationships, calculate that this blueshift is infinite. The situation seems hopeless. Not only would it be impossible for the ship to survive long enough to fire its engines before being melted by the infinite heat, not only would the infinite inward radiation pressure be impossible to resist, but even if the ship did somehow manage to survive this impossible onslaught long enough for the pilot to fire up their engines, the outside universe might no longer exist: the rate of timeflow of the outside universe would be infinitely fast compared to ship time, so even if the ship did somehow manage to escape, there might not be anything left of the outside universe left to escape to.

Given these things, it isn't surprising that even to *place* the ship at the event horizon doesn't seem to be legal under GR. The ship can't legally exist in a stationary state at the horizon's coordinates, even for a moment. The situation is analogous to moving at the speed of light under special relativity, the appearance of all these infinities is telling us that what we've just tried to do isn't legal.

In the "dark star" version, we can look at our "Newtonian" graphs of blueshift versus approach velocity and see that the inward gravitational blueshift at the horizon merely doubles the apparent frequency of infalling signals, and doubles the rate at which the outside universe is seen to age (before we take into account any acceleration effects). The ship's pilot doesn't have to worry about the universe outside disappearing in a flash of infinite energy.

These arguments suggest that a resolution to the BHIP might not be possible until we're prepared to consider modifying the basic Doppler shift relationships of special relativity.

12.13: Grand unification?

When we talk about **unified field theories**, we usually mean the unification of obviously *different* forces (nuclear, electromagnetic, gravitational). But our standard theory doesn't yet seem to have achieved a fully "unified" description of how *gravity, inertia and motion* tie together. Before we try to unite, say, gravitation and electromagnetism, it'd be useful to have a more convincing model of gravitation to unify electromagnetism *with*.

(1) gravity ⇔ velocity

Conventional Doppler shifts and gravitational shifts obey similar rules: both associate a redshift with a matching contraction in apparent depth, and while relative velocity is associated with geometrical distortions due to aberration, gravity is associated with similar-looking distortions due to gravitational lensing.

Real differences in timeflow between regions should be associated with *real* gravitational gradients between those regions (section 3.4), but if moving objects *appear* to age at different rates, shouldn't "observerspace" arguments suggest at least *apparent* associated gravitational effects? A difference in velocity between two points on a single object's worldline is associated with gravitational effects (for example, gee-forces due to acceleration), so shouldn't a velocity-difference between objects *in space* be associated with similar effects? We might say that we know that Doppler effects are only illusions because they change with viewing-angle, but we see the same patterns with gravitomagnetic effects, which are considered real.

If we see a planet with a redshift on one side and a blueshift on the other, we'll tend to interpret this as a simple Doppler effect due to rotation, and not as something gravitationally interesting ... but if we accidentally interpreted these motion effects as *gravitational* effects, our "mistaken" expectations – of an increased gravitational pull towards the redder region, and an offset centre of gravity – would be right. The "confusion" gives good physical predictions.

Current theory already lets us calculate gravitational shifts by treating **gravity-as-velocity** – calculating the change in velocity that the gravitational field would produce in an object during the signal's flight-time, and then using that **terminal velocity** value to calculate a matching velocity-shift in the signal (section 3.7, pp.31). Ideally, we'd like to also be able to run the argument backwards and consider **velocity-as-gravity** – to make the effects associated with relative motion inseparable from the idea of gravitomagnetism. If we were free to choose between "motion-shift" and "gravity-shift" approaches (as **"time domain"** and **"curvature domain"** descriptions of the same problem), we'd obtain a deeper model for gravitomagnetic effects and a stronger concept of velocity-dependent curvature. Curvature effects would become an integral part of moving-body problems, rather than an afterthought, we'd have a deeper geometrical basis for the existence of kinetic energy and momentum, and we'd be closer to a "purist", "curvature-based" description of physics.

(2) velocity ⇔ cosmology

The next obvious candidate for unification is the Hubble expansion characteristic. If a distant Hubble-shifted galaxy is receding and redshifted, it would be very nice to be able to calculate the redshift *either* as the effect of curvature, *or* as the effect of recession velocity, and use the same equations for both cases.

It's strange to have to divide the redshift of a distant receding galaxy into two different "velocity" components, generating apparently indistinguishable effects, but obeying different relationships. We currently distinguish between Hubble shifts (which involve curvature) and more conventional velocity-based shifts (which supposedly don't): but if we were to rewrite the rules governing conventional motion shifts to incorporate gravitomagnetism as something fundamental, then this distinction between "velocity-with-curvature" (in cosmology) and "velocity-without-curvature" (under special relativity) might not have to exist.

(3) gravity ⇔ cosmology

Combining **(1)** with **(2)** gives us a third pair of effects that seem to be begging for unification:

Mach's Principle suggests stronger inertial effects and slower timeflow in a more compact universe. It's not obvious how we're supposed to be able to detect this effect locally, but when we look back across an *expanding* universe and see very distant objects as they were billions of years ago, we're seeing an "old" region of spacetime that looks more compact and more densely populated than ours, and appears to have a background massenergy-density substantially greater than that of our own region: we'll therefore tend to expect those objects to look as if they're inhabiting a region with a significantly more intense gravitational field.

If this field intensity reduces over time (as the universe expands), and looking across space is equivalent to looking back in time, we'll expect to see what appears to be a distance-dependent gravitational redshift.

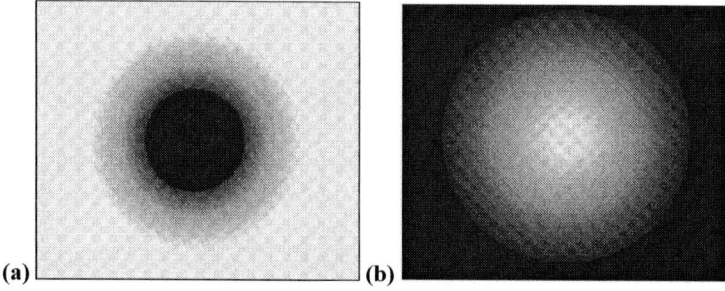

Figure 12-3: Variation of redshift with distance:
(a) black hole, (b) Hubble shift

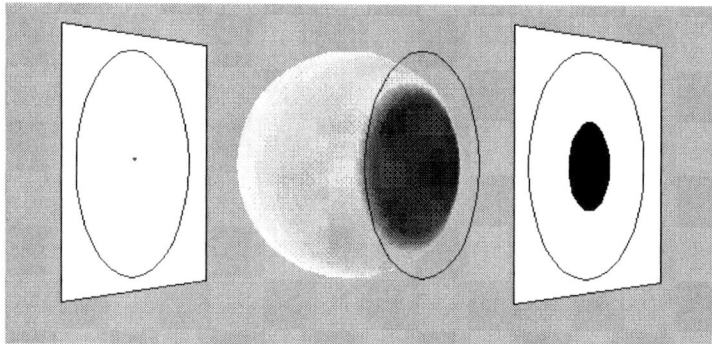

Figure 12-4: The universe considered as a (w)hole:
Gravitational and cosmological horizons

"Observerspace" arguments predict the appearance of an effective gravitational differential along our line-of-sight between "there-and-then" and "here-and-now", with very distant objects seeming to age more slowly than us as a function of a **time-aligned** gravitational differential between our time-period and theirs. Since we can predict effects that look like *cosmological* redshifts from *gravitational* principles, we might therefore reasonably expect cosmological effects and gravitational effects to be equivalent and interchangeable.

But under current theory, they're separated: cosmological curvature supports "acoustic metric" behaviour (section 12.11), and gravitational curvature under GR1915 doesn't – cosmological horizons leak and GR1915's gravitational horizons don't.

why these things don't work

To unite two different effects in physics is a cause for celebration, so to have *three* effects crying out to be merged is frustrating. The obstacle in each case seems to be special relativity.

With **(1)**, SR wasn't designed to support the idea of velocity-dependent curvature, and its relationships break down if we assume that gravitomagnetism is fundamental. So we can't fully unite gravitational effects with velocity-based effects unless we're prepared to lose special relativity, and this is considered to be too high a price to pay.

With **(2)**, we can't fully unite velocity effects with the Hubble relationships because they currently seem to be assigned different shift laws: velocity shifts assume flat spacetime and are supposed to follow the SR relationships, while Hubble shift is accepted as being a curved spacetime problem (whose exact characteristics seem to vary depending on who you ask). We might be able to achieve convergence by deriving new velocity-shift relationships around the idea that gravitomagnetic effects are fundamental, but while we continue using SR as a foundation theory, this has to remain an unsolved (and probably unsolvable) problem.

Lastly, with **(3)**, we have all the previous problems rolled together – the SR legacy means that GR1915 has to make predictions for gravitational horizons that can't be reconciled with the behaviour that we need from cosmological horizons. Gravitational horizons *could* behave more like cosmological horizons if they obeyed the rules for acoustic metrics, but to allow acoustic metrics would invalidate special relativity.

12.14: Gravitomagnetic incompatibility?

Gravitomagnetism – the tendency of moving objects to warp lightspeeds – seems to be a fundamental aspect of physics. We've known since the 1850's that moving material drags light (**Fizeau**), we know that high-gravity objects *ought* to drag light, we've argued that a form of velocity-dragging seems to be essential for the *general* principle of relativity to work, and we've already found that the gravitational fields produced by moving medium-gravity objects are useful for slingshottting spacecraft across our solar system (all covered in section 9).

This isn't just abstract theory, its engineering ... so it's disappointing to find that that there seems to be a fundamental incompatibility between gravitomagnetic principles and the equations of special relativity, since GR1915 tried to incorporate both.

"SR plus gravitomagnetism" would give non-SR equations of motion

Let's think about the case of a "moving" high-gravity star. We might expect to see the energy of its light altered by a few different effects, including **(1)** the standard gravitational redshift of the star (which should be the same in all directions) and **(2)** the star's conventional Doppler shift due to simple relative motion, but also, perhaps, **(3)** the effect that the moving star's gravitomagnetic field has on nearby light.

Conventional physics would seem to require effect **(2)** to obey the equations of special relativity, with any further effects due to **(3)** acting on top. But there's a snag: if the star's motion generates the SR Doppler relationships for its own emitted light, then any *additional* gravitomagnetic influences on the star's light as it leaves the region should change the energy and momentum of those signals, so that the star's *overall* Doppler characteristics would no longer agree with the predictions of special relativity. If the energy and momentum of the star's gravitomagnetically-affected light *don't* follow the SR predictions, then the $E=mc^2$ arguments that we met earlier (section 2.5) suggest that the energy and momentum relationships for the moving star's *mass* should disagree with special relativity, too. The basic equations of motion for moving *strong-gravity* objects would have to be different to the relationships given by Einstein's special theory. If we then want to keep using SR for *weak-gravity* objects, the resulting mixture of different laws isn't pretty.

"integrated" equations of motion would be non-SR

We can try to restore order by suggesting that the previous argument was based on a misunderstanding, and that perhaps a distinction between *motion* shifts and *gravitomagnetic* shifts isn't justified: If gravitomagnetic dragging effects cause nearby light to be more blueshifted when moving with the body and more redshifted when moving the other way, then perhaps as lightrays pass nearer and nearer to the object, and their strength reaches a maximum, they might *turn into* the conventional motion-shift of the object.

In this version of the argument, light emitted by a moving star isn't influenced *twice* – once by the "basic" motion-shift and again by "advanced" gravitomagnetic effects in the surrounding region – it's only influenced *once*. The gravitomagnetic effect is effectively *part* of the star's conventional Doppler effect, smeared out into the surrounding region of spacetime by the star's surrounding field. For a distant observer, the signal isn't considered to have completely left the star until it's also escaped the star's surrounding gravitomagnetic field. Although the energy of lightsignals moving *into, out of* or *through* this region would all be gravitomagnetically shifted, we wouldn't need to impose any additional gravitomagnetic terms when calculating the motion-shift of light emitted *by* the star. It would have already been dealt with.

This would seem to be the natural solution to our problem – we'd be back to a single set of equations of motion for all moving bodies, regardless of their gravitational field strength, and we'd be able to forget about any additional complicating gravitomagnetic effects, because these effects would already be built into the motion-shift equations.

But there's a catch. A single, *universal* set of gravitomagnetic-compatible equations of motion would then need to be compatible with the strong-gravity case, which involves explicitly-gravitomagnetic geometries and velocity-dependent curvature. Since the relationships of special relativity don't seem to be able to fit this geometry, any hypothetical "universal" equations would then have to disagree with those of special relativity (sections 13.8, 15.3).

12.15: Complexity

Einstein's general theory is part of a web of different theories that make up current physics.

To start with we have Newton's mechanics, which works very well for objects moving with smallish velocities (and which "does" gravity), but which becomes incompatible with flat spacetime at higher energies: Then we have special relativity, which changes the NM relationships to *force* compliance with flat spacetime, but "doesn't do" gravity or particle-behaviour. Then, because curvature and gravity seem to be important things, we have Einstein's general theory, a curvature-based model that tries to deal with these properties relativistically while still reducing to SR (but which still can't handle the "particulate" properties of matter).

Quantum mechanics deals with "particulate" effects. Quantum mechanics, combined with special relativity's optical characteristics, gives us **quantum electrodynamics ("QED")**. But QED doesn't fit GR1915. The last step – bridging the gap between general relativity and quantum theory – should give us a theory of **quantum gravity** ... but since our current models were derived from mutually-incompatible starting assumptions, the two branches of theory are structurally incompatible. We have two or three separate theories branching off from the same NM starting point, using different assumptions and approaches, and they don't work together.

Ideally we'd like a theory founded on (or compatible with) aspects from all these different disciplines, and which might have a fighting chance of uniting them, but our current implementation of general relativity – founded on SR, which isn't compatible with curvature effects or particulate-matter effects – might be carrying too much historical "baggage" to be the basis of a workable theory of QG. The *most* general principle of relativity might live on in QG, but SR/GR1915 may not necessarily be a part of it.

12.16: Is GR1915 scientifically falsifiable?

GR1915 is quite tricky to test, and it seems that the main tests of GR actually boil down to the idea that gravitational redshifts and gravitational time dilation effects exist (in the approximate Einstein 1911 "Newtonian" range), and that these effects then increase the strength of lightbending effects and the precession of elliptical planetary orbits. But we don't need the scary "squiggly math" of GR1915 to make these predictions.

What else do we have that's testable? Well, there's the decay of orbiting binary stars due to the emission of gravitational radiation, and there are frame-dragging effects, and these seem to be reasonable evidence that some of the effects that we'd expect according to the GPoR may be real. But these still don't tell us that Einstein's general theory is the right *implementation* of the general principle. The basic idea could be good, but the theory itself could still be bad. If we can test the apparent validity of basic principles, but can't test the more specific predictions of a theory that we based on them, then are we really entitled to say that the *theory* is testable?

In the absence of significant experimental evidence, we could instead try to test GR1915 by cross-referencing its predictions with those of another reference-theory that we have some faith in, the obvious candidate being quantum mechanics. But GR1915 famously conflicts with quantum theory, so instead of taking this conflict at face value, we prefer to say that we require *another* layer of theory – quantum gravity – to include both QM and Einstein's general theory. But QG doesn't exist yet, as much more than a general ongoing research project, a catchy name and perhaps a logo or two. There's no single "theory of quantum gravity", as such, and we're still at the stage of thrashing out the basic design rules that a future theory of QG would have to include.

Does GR1915 have *any* unique predictions? Yes, it does, in the shape of the "perfect" non-radiating black hole, and in the prediction of gravitational collapse to a point-singularity. But neither of these things is exactly easy to verify (they perhaps can't be verified on principle), and in any case, it seems increasingly likely from QM arguments that these new behaviours simply don't happen in the real world. But we don't (yet) say the theory is invalidated by this apparent misprediction, we just say (again) that GR needs to be supplemented by QM effects.

How about taking our new GR1915 knowledge and applying it to unusual situations? We might try to apply the model to atomic theory, but GR1915's arguments don't seem to work at small scales in their present state, thanks to an incompatibility with the QM effects that dominate in this range. We say that GR simply doesn't "do" small objects, and describes the physics of the very large. So, does GR cope well with large-scale physics? Well, since it doesn't predict or explain galaxy rotation curves, it doesn't seem to be appreciably better at this than (post-1911) Newtonian theory. It doesn't seem to be especially good at the very big *or* the very small, and at medium scales, "updated NM" is usually sufficient. GR1915 also isn't especially good at *extreme* large-scale effects, and since it wasn't designed around modern expanding-universe cosmologies, it's used in conjunction with arbitrary terms (such as a **Cosmological Constant, dark matter** or **dark energy**) to *force* compatibility with astronomical data. Its original cosmological achievement was to explain why cosmological redshifts didn't happen, just a few years before experiment showed that they did.

The case for GR1915 seems to have been somewhat exaggerated, and we could be forgiven for asking whether this theory is actually of any real use to us. But perhaps GR1915 will eventually be appreciated *not* for its specific physical predictions, but for its demonstration that a theory of this type could be constructed at all. Einstein's general theory may end up being remembered as a "proof of concept" that established and popularised some of the ground rules that subsequent, more general theories would have to comply with, a prototype "**mk1**" theory that we used to familiarise ourselves with the subject so that we could later go back and try to build a "**mk2**" version that would have a better chance of being the real thing.

12.17: Blaming special relativity

Almost all of the problems and potential problems that we've identified here with Einstein's general theory seem to be consequences of the theory's incorporation of special relativity, and its assumption that the relationships of SR have to apply as a limiting case of the theory.

The special theory isn't compatible with general relativistic principles, it's not compatible with gravity, it prevents us from building gravitomagnetism into the model, and stops us using acoustic metrics. It seems to be the reason why GR conflicts with quantum theory, why GR predicts "perfect" black holes instead of "fuzzy" QM-compatible ones, and it stops us from integrating cosmological effects, gravitational effects and velocity effects together into a single block.

Almost every time we find a potential problem with general relativity, the reason why that problem exists turns out to be the same: it's that things *have* to be that way in order to avoid creating conflicts with special relativity, and special relativity – despite its apparent conceptual clashes with the general principle of relativity – is generally reckoned to be an *unavoidable* foundation-stone for GR. The idea that a general theory *must* reduce to SR is considered so self-evident that to question it is risk your fellow physicists wondering of you've gone mad, or whether you ever really understood the basics of relativity theory to begin with. Reduction to special relativity, according to the articles in section 12.9 isn't just a matter of *faith*: it's considered to be rigorously-demonstrated *geometrical truth*.

It's not something that we're supposed to question. Although there's been some mainstream work on "classical" alternatives to GR1915 (**Brans-Dicke theory** being probably the best known), these theories are usually *supersets* of GR1915, with extra parameters and tweaks and variations on a theme. In order to conform to the established view of what makes a theory "credible", these alternative theories are supposed to be able to demonstrate that they reduce to special relativity, too.

But mathematics or geometry, applied without consideration for physical realities or physics principles, doesn't always give us real physics: If can sometimes give us **math fiction**: reasonably consistent-looking descriptions of worlds that aren't real, or at least, aren't ours.

To find whether SR is *really* the root of all our problems, we need to step away from the subject of general relativity as it *actually* developed, and look at how it *might* have developed without this reliance on a Minkowski-metric underpinning. What does general relativity look like when SR is deleted? What does a freestanding, *truly* general theory look like, that *isn't* built on an SR base, and which applies the spacetime-curvature paradigm "all the way down"? Are there any other sets of basic relationships that might have worked other than those of the special theory? What are the *mathematical* consequences of General Relativity's engineered reduction to special relativity, and what might the implications be of *not* making this assumption?

Although Einstein's 1950 piece in Scientific American seemed to indicate that he'd come to distrust SR as a foundation for more general theory, and that he now considered the usual two-stage approach to be a "historical" decision that couldn't really be justified with hindsight, the "follow-up" information that we're looking for – telling us what happens when we *don't* base GR on SR – doesn't seem to be generally available. The standard textbooks and the usual research papers don't seem to tackle this subject, so if we're to proceed, we're going to have to try deriving a few things from scratch.

In the next section we'll go back to first principles and try to develop a *general* set of equations of motion for an entire *family* of potential theories of relativity, and then we'll attempt to narrow down the possibilities again with some reasonably non-technical arguments.

The results may be a bit surprising.

13
Horrible Nasty Mathematics

" The danger already exists that mathematicians have made a covenant with the devil to darken the spirit and confine man in the bonds of Hell. "
Augustine

" All truths are easy to understand once they are discovered; the point is to discover them. "
Galileo Galilei

" The idea is there locked inside. All you have to do is remove the excess stone. "
Michelangelo (attrib.)

" The mathematician can do a lot of things, but never what you happen to want him to do just at the moment. "
Albert Einstein, 1914

" Many of the things you can count, don't count. Many of the things you can't count, really count. "
Einstein (attrib.)

" I'd rather be right than rigorous "
Stephen Hawking (attrib.)

" The theoretical possibilities in a given case are relatively few and relatively simple, and among them the choice can often be made by quite general arguments. Considering these tells us what is possible but does not tell us what reality is."
Einstein, recounted by R.S. Shankland

" When all other contingencies fail, whatever remains, however improbable, must be the truth. "
Sherlock Holmes, in "The Adventure of the Bruce-Partington Plans", Arthur Conan Doyle

" Moses supposes his TOEses are roses
But Moses supposes erroneously
For nobody's TOEses are posies of roses
As Moses supposes his TOEses to be "
Edward Lear (capitalisation changed)

RELATIVITY IN CURVED SPACETIME

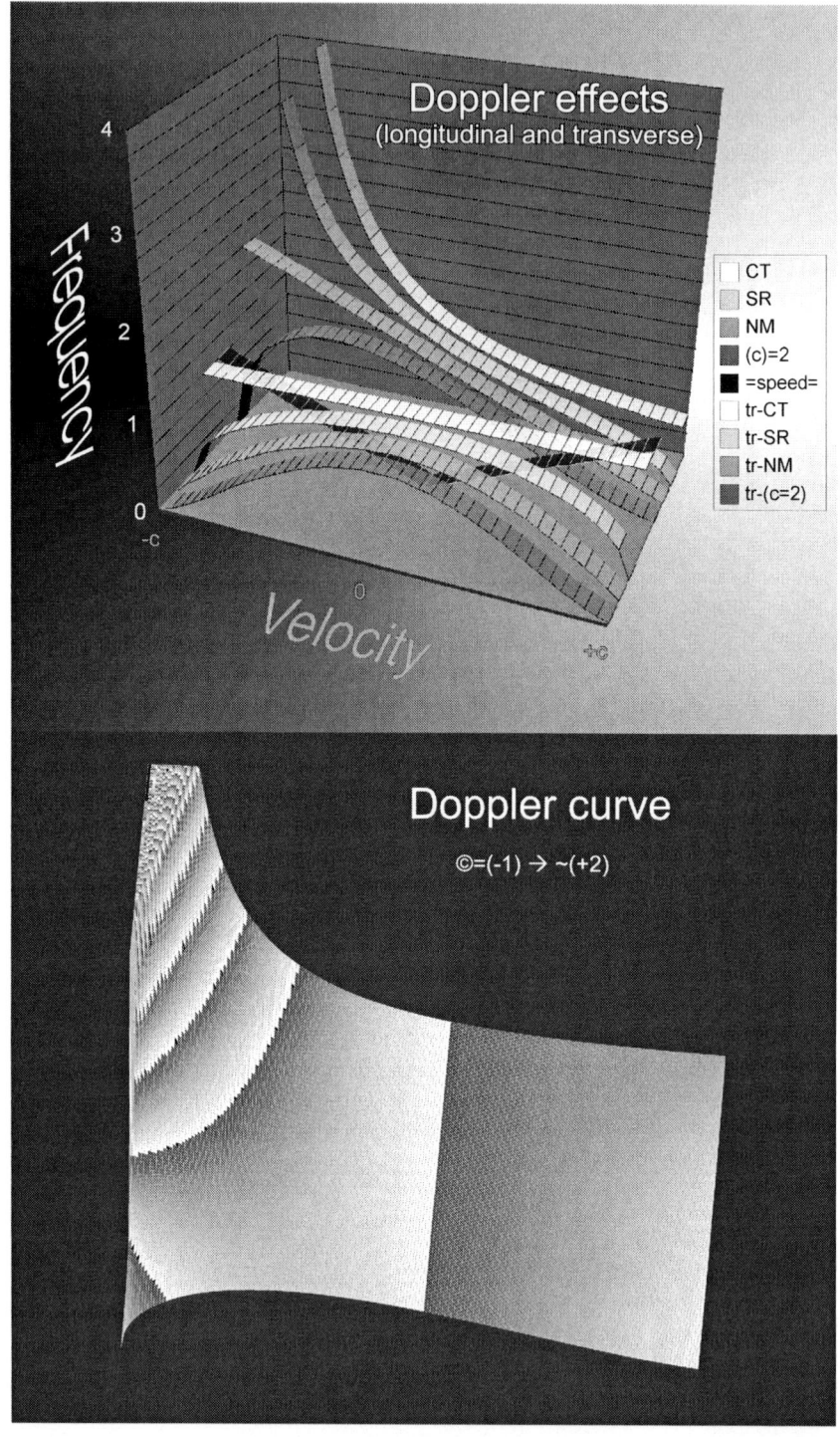

13: Horrible Nasty Mathematics

13.1: A family of relativistic theories

It's all very well to talk about the existence of possible alternative relativistic equations of motion, but at some point a scientist will want to know what these mysterious hypothetical relationships actually are (if they do, in fact, exist). How do we turn our weak, "hand-wavy" arguments into more concrete sets of testable mathematics?

In this section we'll derive a continuum of possible theories, find where special relativity fits into the larger scheme of things, and then try to isolate the simplest possible solution for curved spacetime. We'll expand our range of possibilities outward from SR, and then try to contract it again on a single solution based on some of the previous arguments.

The identity of this solution turns out not to be special relativity.

starting assumptions

Before we try to create a new relativistic model we should first check what we already have in our toolbox. We already know *three* sets of fundamental Doppler equations: these form a sequence, each separated from its immediate neighbours by a **Lorentz** relationship:

...	stationary aether → $c/(c+v)$	...	$c/(c+v) \times$ (Lorentz) → SR	...	SR×(Lorentz) → $(c-v)/c$	...

The two right-hand sets of relationships belong to the only two reasonably consistent and fundamental theories of relativity that we know about, special relativity and NM.

other "Lorentzlike" factors

We know that the "nice" parts of SR (such as $E=mc^2$ and the conservation of **momenergy**) stay unchanged when we multiply or divide out that magical Lorentz factor, as long as the factor is applied *consistently* throughout the theory. Mathematicians tell us that these Lorentz relationships are fundamental and form a very special mathematical group, and that if we want different observers' quantities to translate cleanly between frames, we need to use them. But *if we don't need spacetime to be flat*, the standard Lorentz ratio isn't quite so unique – if applying the Lorentz factor *once* leaves all these special properties unaffected, then applying it *twice successively* ("Lorentz-squared", "Lorentz^2"), or any number of times (Lorentz^3, Lorentz^4, etc.) should *still* preserve these critical relationships. We'll refer to the more usual Lorentz factor raised to an unspecified power, as a **"Lorentzlike" factor**.

$$\text{"Lorentzlike" factor} = (\textbf{Lorentz})^{(\text{Insert Number Here})}$$

So our sequence can be extended. We can also apply the argument in reverse ... if a "Lorentzlike" ratio multiplied by itself always gives us *another* Lorentzlike ratio, then perhaps if we take one of these ratios and *square root* it, since the root *also* gives a Lorentzlike factor when multiplied by itself, perhaps *that* should be "Lorentzlike" too. Perhaps *fractional* powers – square roots and so on – should also work. This gives us additional, intermediate, solutions.

how to write it down

We can specify different positions on our chart by assigning different values to the **exponent** (the missing "top-right" number), and mathematicians already have a nice system for expressing these things: if we wanted to invent a new model that lay exactly *between* SR and NM, then since those two models disagree by a Lorentz factor, the *square root* of the Lorentz factor (which is how far the properties of this new model will disagree with the two other theories), will just be (Lorentz)$^{0.5}$. If we have a wild notion to apply the Lorentz factor *three* times, we can write (Lorentz)^3, a cube root of the usual Lorentz ratio becomes (Lorentz)$^{1/3}$, and so on. Any number that we write in at the top right of the "Lorentzlike" term should give us *some* sort of half-credible-looking relativistic model, at least on paper.

What about *dividing out* Lorentz factors? This seems to preserve our key relationships too, as long as we are careful and consistent. The system for writing down exponents can handle this by making the exponent number negative, so to *divide out* a Lorentz or Lorentzlike ratio we can write "×**Lorentz**$^{-1}$" or "×**Lorentz**$^{-(x)}$".

In fact, we can write in *any number that we like* as our exponent, positive or negative, whole number or fraction, and the magic still works – there is an *infinite number* of potential relativistic theories that will preserve our key relationships. All we have to do is take *one* well-understood theory of relativity as an agreed reference, multiply it by a "Lorentzlike" factor, and by using different numbers for the factor's exponent we can identify any other potential relativistic theory in the system.

13.2: Selecting a reference theory

We now need to define a reference theory. We'll expect any credible theory to reduce to NM as a limiting case, so we might want to use this as our starting point ... but out of deference to the SR community – and because special relativity's optical predictions are arguably better known and better defined than NM's, and because SR tends to be considered the "default" relativistic theory – we'll use SR as our reference model.

13.3: Defining the range

The core predictions of our *continuum* of potential theories of relativity now becomes:

$$prediction[\text{ "Theory X" }] = prediction[\text{ SR }] \times \text{Lorentz}^{(?)}$$

The missing value "?" identifies individual solutions, and this is the parameter that we'll be experimenting with. If we set it to zero, the additional Lorentzlike part disappears and we just have "standard" special relativity, if we set it to "**plus one**", we have the (re-evaluated) relationships of NM, and if we set it to "**minus one**", we have a new relativistic theory based on the relationships usually associated with a fixed stationary aether. We could also try setting it to two, three, two thirds, or anything else that takes our fancy. This gives us our range.

However, our "unidentified" number still doesn't yet have a name. Since special relativity is a unique solution for flat spacetime, and since bigger positive numbers seem to relate to larger amounts of velocity-dependent curvature, we'll call it a theory's "**curvature factor**", and since the letters of the Greek alphabet have already been used for other things in physics, we'll borrow the symbol "©" to represent it.

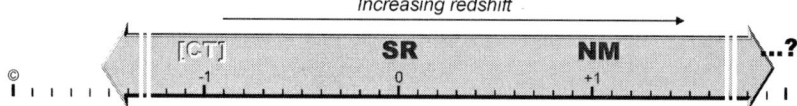

Figure 13-1: An imagined range of potential theories of relativity

Special relativity is the special solution for no velocity dependent curvature, so we'll say that it has a curvature factor © of **zero**. NM uses the special set of equations that we get by assuming that light is fully dragged along (or thrown) at the surface of the emitter, so we'll say that it represents a curvature factor © of **plus one**. The leftmost Doppler equation in section 13.1 would correspond to a curvature factor © of **minus one**, and we can also look at regions outside this range, and between these solutions, just by selecting other © values.

Now, lets leave the math behind and get back to the more interesting stuff ...

13.4: Ellipses

What else do all these hypothetical theories have in common?

Well, if we're limiting ourselves to "relativistic" theories, we'll probably expect them to use the **relativistic aberration formula**, making the "ellipse" exercise in section 8.3 seem appropriate. Altering © changes a theory's forward and rearward Doppler predictions, and the ellipse method then gives us the resulting predictions for all other angles. By smoothly varying the strength of a "Lorentzlike" component we'd seem to be able to fade smoothly from NM to SR without upsetting any really critical properties like momentum-conservation.

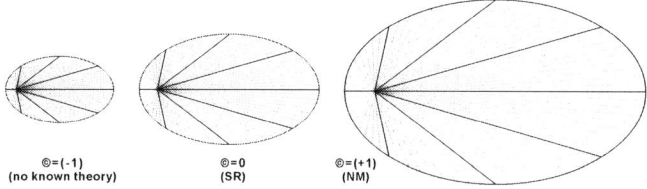

Figure 13-2. Relativistic ellipses, $v=0.8c$

If we take our three "standard" Doppler relationships from section 13.1 and draw ellipses for each of them (for a single nominal velocity) we find that the parts of all three ellipses keep the same ratios. When we change the value of © from "**minus one**" to "**zero**" to "**plus one**", the *magnification* of the ellipse changes, but its *proportions* stay the same. This is also true for intermediate solutions, and for other solutions outside the range.

So ... since *all* our hypothetical theories differ from each other by "Lorentzlike" factors, these factors also tell us the scaling between ellipses belonging to different members of the family. Knowing *one* of the theories reasonably well (say, special relativity), and the degree of another theory's "Lorentzlike" deviation from it, we know how to map between them.

13.5: Special relativity as a special solution

Although SR's (**©=0**) version of the ellipse grows as we apply larger velocity values, its **minor radius** (its width) is constant, and its outline can be fitted back into the original "$v=0$" circle with a simple **Lorentz contraction** of distances in the direction of motion. A compaction of the ellipse's *x*-coordinate distances (which we could get by tilting the ellipse off the page by an angle that depends on velocity) will turn the ellipse's outline back into the circular outline that would be expected by an observer moving along with the signal source.

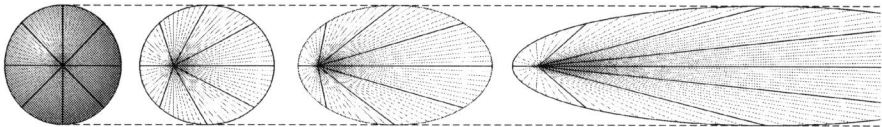

Figure 13-3: Sample SR wavelength ellipses for $v=0$, $0.5c$, $0.8c$, and $0.9c$

This is an significant property, and it plays a key part in **Hermann Minkowski**'s geometrical interpretation of SR – in **Minkowski spacetime** (section 14.9), different observers' perceptions of space and time are "tilted" by an angle that depends on their relative velocity, and our SR ellipses can be imagined as angled slices through an observer's **lightcone**. To someone brought up on Minkowski's ideas, this is a compelling argument for the inescapability of special relativity's relationships. *There is no other value of* © *that can generate ellipses with this property*, so the assumption of flat spacetime, coupled with projective geometry, would seem to tell us that **©=0** and the equations of special relativity are *the only possible solution* for relativity considered as a flat spacetime problem.

13.6: Positive values of © and positive curvature

Now lets look at the NM (©=1) version of the same ellipse.

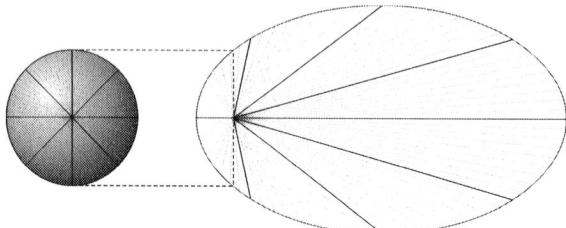

Figure 13-4: NM ellipse: too big to fit

It again has Lorentz proportions for "length-versus-width", but the extra Lorentz magnification (because Doppler-shifted wavelengths are "longer" under NM than SR) means that we can't now cram it back into its original-sized circular outline just by tilting or contracting it, no matter how we try. The only way to pack all these extended ray-distances into the original sphere is to distort and curve and *extrude* the surface into something that looks suspiciously like the **tilted gravitational well** diagrams back in section 9.12. This applies to *any* ellipse with a value of for © greater than zero – **positive-© solutions** seem to represent theories in which an increase in the relative velocity of an object is associated with a corresponding increase in curvature, an increase in the total distances packed into the region, and an increase in the overall depth of an object's gravitational well due to the gravitational effect of its recoverable kinetic energy.

the most general implementation of NM is not a flat-spacetime solution

This shape also demonstrates graphically that if light is taken to be a wave, **the Newtonian relationships are not compatible with flat spacetime** except as first approximations in situations where the velocities involved are very much less than lightspeed. For any "significant" velocities, the wavelength distances simply don't fit into the available space. If we start with textbook NM and try to apply it to situations where the velocity is a significant fraction of the speed of light, we have to choose between *modifying* NM's equations to fit into flat spacetime (which gives us special relativity), or allowing spacetime to warp and distort in order to accommodate the theory's wavelength relationships. Newtonian mechanics has traditionally been considered as a "flat-spacetime" theory, but its relationships, applied to EM waves, imply the existence of some form of velocity-dependent curvature.

13.7: Rejecting negative solutions for ©

If we construct a relativistic ellipse for values of © *less* than zero, the volume within the ellipsoid *shrinks* as velocity increases, implying a negative **kinetic energy** for moving masses. This doesn't seem likely, so we'll eliminate the range of negative values to the left of the scale.

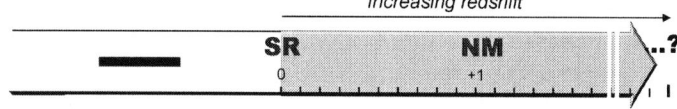

Rejecting the negative solutions also seems to be a reasonably safe thing to do according to the available experimental results. Experiments validating SR's Doppler predictions seem to be almost exclusively concerned with showing just how bad ©=-1 is when compared to ©=0, and demonstrating that **©=0** seems to be a sort of minimum.

13.8: Gravitomagnetism suggests positive ©

We can now take this argument slightly further and say that, for our purposes, the solution at *exactly* ©=0 (special relativity) is *also* not very interesting to us, since we *want* moving bodies to warp spacetime. While our scale illustrates that special relativity (©=0) predicts frequencies and distances "redder and shorter" than they are under **"Classical Theory"**, the **gravitomagnetic** effects that result from the general principle of relativity, appended to SR, suggest relationships that are in turn "redder and shorter" than those of SR.

This means that we're now only looking at positive values of ©. But how positive should © be? Are we looking at a tiny deviation from SR or something much bigger?

13.9: Graphed Doppler responses

How does modifying © change the character of physics? One way to try to get some sort of gut feeling for the effect of changing © is to plot the Doppler response curves for a range of different values for ©, and to stack them together to produce a nice-looking 3-dimensional graph-ey surface.

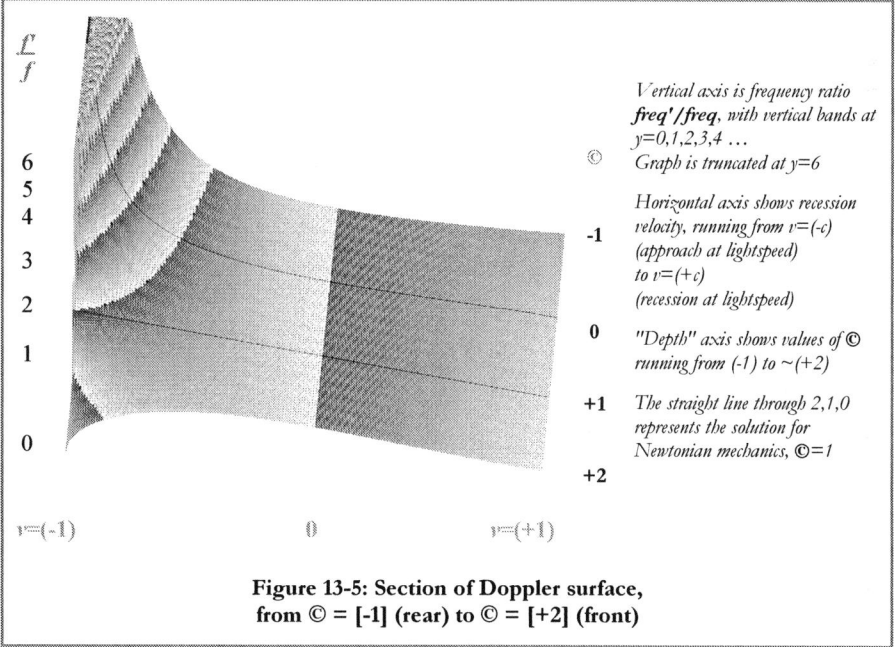

Figure 13-5: Section of Doppler surface, from © = [-1] (rear) to © = [+2] (front)

This particular plot only shows values of © from ©=-1 (the line at the very back of the graph, going through **infinity, 1, 0.5**), through ©=0 for special relativity (curved line a third of the way from the back, going through a **"weaker" infinity, 1, 0**), through ©=1 (straight line through **2, 1, 0**), and ending up at the front edge of the graph, which is just short of ©=2, with a shape that looks like a cross-section through an aircraft wing. Since we can't fit an infinite linear scale onto our page, the surface has been truncated, at $f'/f = 6$.

The left hand edge of the graph deserves special attention, because as soon as we move back *any distance at all* from ©=1, the leftmost edge of the graph immediately hits infinity, and as soon as we move *any distance at all* forwards from ©=1, it hits zero.

13.10: Setting "one" as a higher limit for ©

This sudden "snap" from infinity to zero at the left side of the graph is slightly weird. It means that if we apply any values of © greater than "one", there'll be a certain approach velocity that gives a maximum finite blueshift, and as we further increase velocity the blueshift reduces again and eventually turn into a redshift. Beyond a certain point, the momentum of the object's forward-aimed light appears to *reduce* as its relative speed increases.

This is unfamiliar and downright freaky behaviour, so we're going to provisionally rule this range out, too (on the grounds of general weirdness), with the caveat that we haven't actually *disproved* these higher values of ©, we just don't currently understand what they might mean.

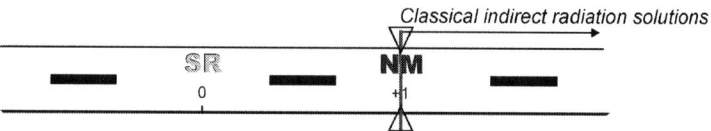

If any future researcher wants to try to resurrect this higher-value range and make sense of it, they are quite welcome to it. This leaves us with the range of values for © "greater than zero", but "not greater than one".

13.11: Using the BHIP to set a minimum of "one" for ©

We'll now try to make use of what we learnt about the **Black Hole Information Paradox** (section 11). Suppose that we want to find a classical solution to this problem, can this be converted into a range of values for ©? Apparently it can.

In a "dark star" model, the ability of a particle to escape the **r=2M** surface relates to the finite inward blueshift seen by an observer placed at the horizon (section 12.12). The frequency-sift of signals hitting the horizon surface can be calculated form terminal velocity arguments (section 3.7) – since the default terminal velocity of an object falling into a black hole reaches the speed of light when they reach the horizon, we'll expect the blueshift to agree with the "conventional" Doppler blueshift for an object approaching at lightspeed – with NM this gives a doubling in frequency (graph, section 6.2), and for special relativity the blueshift is infinite (graph, section 6.6).

To solve the paradox *classically*, we'd seem to require the blueshift on infalling light hitting the horizon to be finite … but *there is no value for © less than one* that can generate a finite gravitational blueshift at the horizon … if © is even *microscopically* smaller than one, the blueshift prediction still shoots vertically up to infinity, implying inescapability.

This means that in order to solve the BHIP classically for our range of possible theories, **we have to rule out all solutions *between* NM and SR**, and "©=1", the Newtonian relationships, represent the *minimal modification* to SR that could be used solve the BHIP in this way.

The blueshift calculations also get freaky *beyond* ©=1. if we set aside values *larger* than "one" on the grounds of general-purpose weirdness, our exercise seems to have taken a continuous range of possible relativistic theories of curved spacetime and (for black hole physics, at least) rejected everything but one discrete solution, which just happens to correspond to the core relationships of Newtonian theory.

13.12: Oops?

This is where we start we wonder if the whole exercise has gone horribly wrong.

How can we possibly have *started out* with special relativity and *ended up* with Newtonian mechanics? This seems to turn conventional physics on its head. We're *supposed* to begin with NM as a "starter" theory, build SR on top of that, and then gradually work up with more layers of theory to get QM, GR, **Quantum Electrodynamics** ("**QED**"), and hopefully, eventually, a final theory of **Quantum Gravity** that might remove the disagreements between the incompatible GR and QM layers. We're expected to keep *adding* stuff – we're *not* supposed to start with special relativity and end up with a final "Theory of Everything" that tells us to drop SR-based physics and revert to the earlier, simpler NM relationships. Simplicity is an admirable thing, but we'd prefer not to be told that we've just spent a century chasing the wrong equations.

is © "fixed" or "variable"?

One possibility that needs to be considered is that *perhaps* we only ended up with ©=1 as a unique solution because we involved black holes. The rotating event horizon of a Kerr black hole is reckoned to drag spacetime completely, so perhaps "©=1" is appropriate for black holes and other extreme gravitational objects, and perhaps lower, more "SR-friendly" values of © might apply to everyday "weak-gravity" physics. If © was a measure of the surface gravity or the escape velocity of an object, we could define a "low-gravity" realm where © was close to zero and in which special relativity would still be a very good theory, and © could be used to represent the strength and significance of an object's explicit gravitational field. We'd have a conventional-looking distinction between "low-gravity physics" (© close to zero) and "high-gravity physics" (© significantly greater than zero).

But a distinction between low-gravity and high-gravity physics creates new problems – since © describes the "Lorentzlike" relationships of a theory, modifying its value doesn't just change an object's velocity characteristics and its Doppler response –to modify © is to alter the way that the object's frequencies *appear to change with relative velocity*. If different objects followed different Doppler relationships, then if we were looking at a distant starfield and decided to change the speed of our telescope, light from different types of star in the starfield would have to frequency-shift by different ratios. When the telescope moved, different light-signal components arriving at our detector with the same position, energy, amplitude, frequency and polarisation would have to "know" to react differently, based on the properties of the bodies that originally emitted them, which would mean that light would have to be a significantly more complicated thing.

This idea doesn't seem to be compatible with the idea of a light-metric, and it seems to resurrect the sorts of inconsistencies that plagued historical **emission theory** (section 16.12).

And, thanks to the "**E=mc²**" arguments of section 2.5, modifying the relationship between an object's relative velocity and the energy and momentum of its *light* also suggests a matching modification in the relationship between velocity and momentum of the *body* that emits that light, so objects with different surface gravities would then have to obey different basic laws of motion. This turns our optical relationships into an ugly road crash, and the slew of paradoxes and problems that come with this description seem to strongly suggest that, really, *all* objects ought to be obeying a single, common universal set of Doppler relationships. But if that's true, then if *black hole physics* isn't using the SR Doppler relationships, it would seem that perhaps nothing else can be using them, either.

If light-dragging effects around rotating black holes involve full local source-dependency, *suggesting* ©=1, and if classically-radiating black holes *need* to use ©=1 to resolve the BHIP, then perhaps *everything else in the universe* would need to use ©=1, too.

13.13: Preliminary conclusions

Our innocent little exercise was shorter and more emphatic than we wanted! We might have hoped for *new* and *exotic* relativistic solutions that were previously beyond our ability to imagine, and which might in turn have led us to new and wonderful sorts of physics, but instead it dumped a single "old" solution into our lap that we'd already known about and rejected about a hundred years ago. Our wonderful hypothetical theory, that might have transcended SR, united general relativity with quantum theory and given us a new set of relationships founded on curved spacetime turned out to be ... a modernised, curved-spacetime implementation of grotty, old-fashioned, "unsexy" Newtonian mechanics.

But these arguments aren't definitive, and perhaps they aren't complete. Future research may throw up arguments in favour of different solutions that might turn out to be more powerful than the ones proposed here, or we might eventually find a different solution to the **Black Hole Information Paradox** (although this seems less likely with each passing year).

Did our method screw up somewhere? Maybe. But with a healthy dose of hindsight, the idea of a curved-spacetime implementation of NM may make some sort of sense: if we want a curved-spacetime solution that reduces to NM, and is "redder" than SR, and is as simple as possible, then perhaps the simplest solution that reduces cleanly to NM is just ... NM.

And at this point we begin to perceive the outline of a possible unified scheme of inertial and gravitational physics. In this system, a cosmological horizon is the same thing as a gravitational horizon, which is the same thing as a velocity horizon, and all three descriptions are interchangeable and follow the same basic rules. The same **acoustic metric** that describes a cosmological horizon then applies to gravitational horizons (*contra* GR1915) allowing black holes to radiate according to purely classical mechanisms, and this twofold description then also applies to *velocity* horizons, which don't have full legal status in SR-based physics.

Along with this three-way compaction, our new acoustic metric also creates duality between quantum mechanics and gravitational theory. Taking *two* different physics descriptions and compacting them into one would be a neat trick, so managing to do this with *four* would be a pretty impressive. The penalty for this unification (and the reason it wasn't done like this before) is that in order to merge any two of these components we seem to have to abandon special relativity, and that option hasn't been considered as "acceptable".

Is it *at all conceivable* that SR's Doppler equations could have been less accurate than NM's for all these years without our noticing? Could something this big have slipped by us? It's just conceivable ... researchers trying curved-space implementations of NM in the latter part of the Nineteenth Century were held back by the assumption that this curvature only applied in *three* dimensions, and by the time that Einstein had spotted the link between timeflow and gravity (in 1911), special relativity and the new Lorentz-Einstein relationships had already been presented as the natural successor to NM. Einstein presented GR1915 a few short years later to keep theorists busy until the 1950's, black hole theory kept them occupied until the 1970's, and then they had string theory, quantum gravity and Hawking radiation to worry about. Researchers may have been kept so busy with all these new subjects that they may not have had time to step back and re-evaluate everything again from scratch. Resurrecting the NM core set isn't a "sexy" thing to do, even suggesting it is liable to provoke ridicule, and fixing up someone else's "old" theory isn't likely to result in fame or fortune. There's no clear reason why a classically-trained relativist would want to do it.

Our next step is to search the databases and journals for information on non-SR-based solutions to the general principle of relativity. If this sort of model *doesn't* work, we'd hope to find a full study of the different characteristics of various non-SR schemes, showing why none of them work. Instead we find ... nothing. Or almost nothing. We're told that non-SR solutions aren't credible and *don't even have to be considered*, but to appreciate how and why the SR community came to believe this, we have to do some more digging into SR's background ...

PART V

SPECIAL RELATIVITY AND FLAT SPACETIME

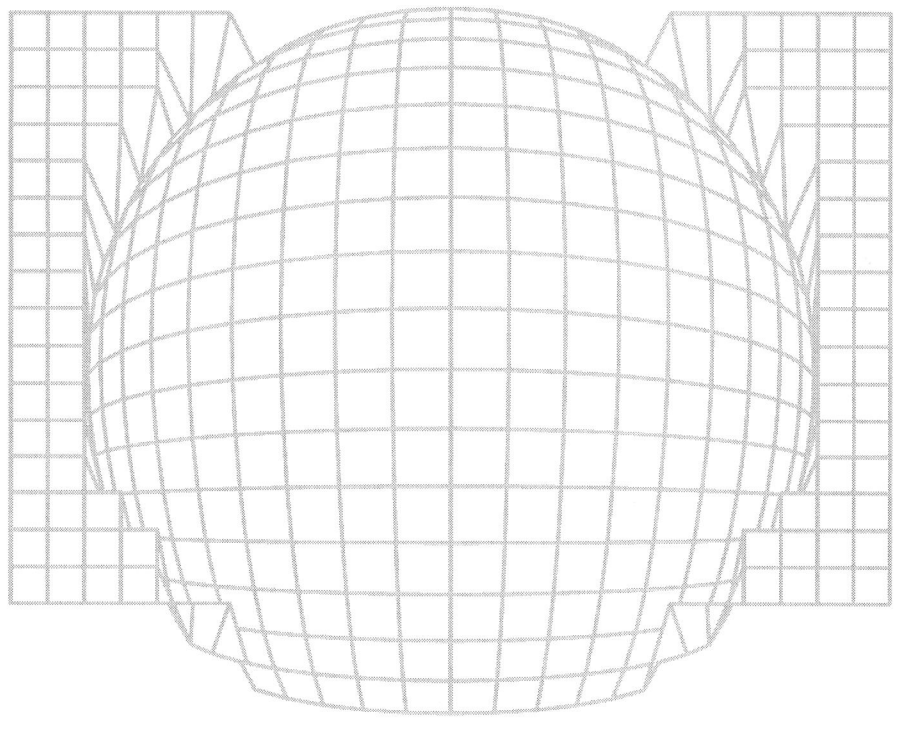

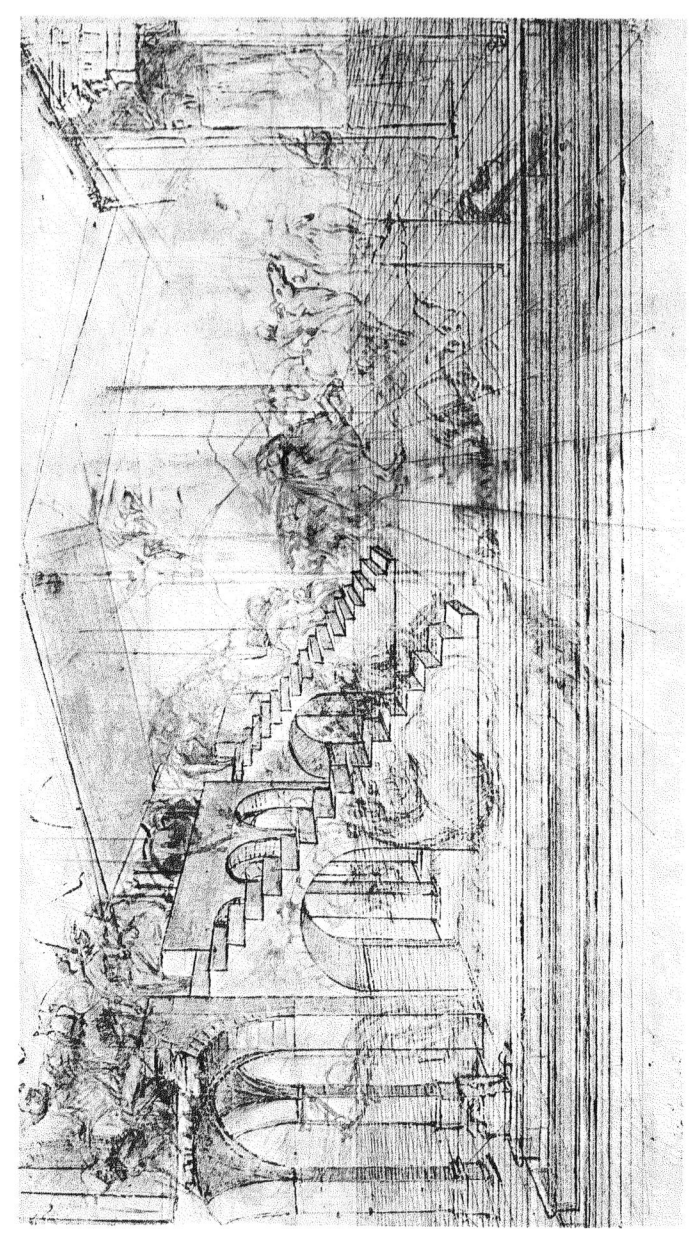

"Adoration of the Kings",
perspective sketch, Leonardo da Vinci

14
Einstein's "special" theory of relativity

Special *adj.* ... specific; not general. ...

> *" There is hardly a simpler law in physics than that according to which light is propagated in empty space. Every child at school knows, or believes he knows, that this propagation takes place in straight lines with a velocity c= 300,000 km./sec.*
>
> *... let us assume that the simple law of the constancy of the velocity of light c (in vacuum) is justifiably believed by the child at school. Who would imagine that this simple law has plunged the conscientiously thoughtful physicist into the greatest intellectual difficulties? Let us consider how these difficulties arise. "*
>
> ... Einstein, "Relativity", chapter 7

> *" ' That light requires the same time to traverse the path A→M as for the path B→M is in reality neither a **supposition nor a hypothesis** about the physical nature of light, but a **stipulation** which I can make of my own free will in order to arrive at a definition of simultaneity. ' "*
>
> Albert Einstein, "Relativity ..." chapter 8

> *" ... the theory of relativity resembles a building consisting of two separate stories, the special theory and the general theory. "*
>
> Albert Einstein, The Times, Nov.28 1919

> *"... every month, hundreds of claims come to this desk. Some of them are phoneys and I know which ones. How do I know? Because my little man tells me. "*
>
> Double Indemnity, Billy Wilder (1944)

> *" Cardboard boxes. Nobody clever be's cardboard boxes "*
>
> Lords of Chaos, the "Sandman" series, Neil Gaiman

RELATIVITY IN CURVED SPACETIME

Simple light-cone

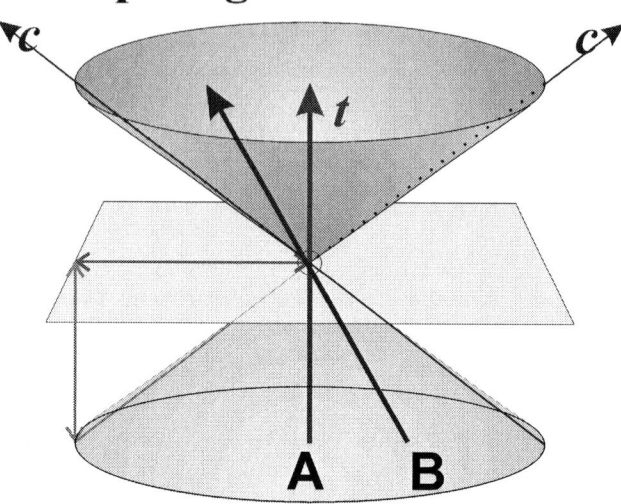

Light-cone for B

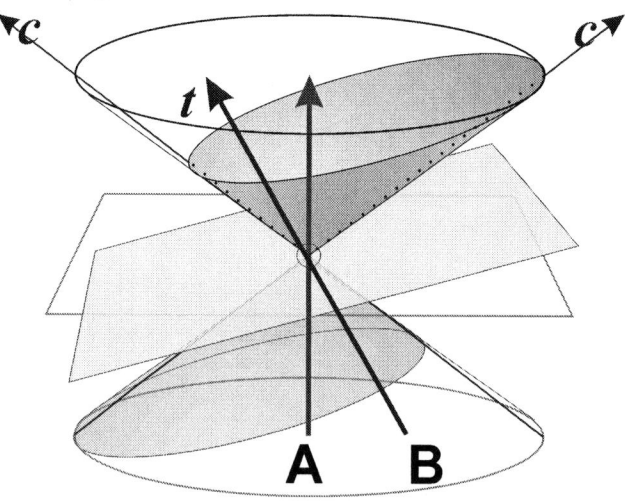

For observer "A", any lightcone intersecting their worldline appears centred, representing a speed of light fixed in their own frame. Observer B agrees about the coincidence of events on this cone, but from their perspective it appears centred around *their* worldline

14: Einstein's "Special" Theory of Relativity

14.1: The birth of special relativity

Having raised the heretical possibility that special relativity might not just be incomplete but actually *physically wrong*, we're obliged to explain how it is that we originally arrived at special relativity as our standard model, and why we ended up embracing it quite so strongly. How is it that, if SR might *not* be the correct theory of relativity, it still managed to be so successful for so long? This requires a little bit of forensic archaeology.

14.2: Failure of earlier theories ...

Between about 1800 and 1900, fundamental physics theory found itself in an odd situation regarding light. Newton had aimed for a combined wave/particle description, with gravity explained as the result of a variation in the speed of light between different regions, but his "light-corpuscle"-based approach had gone badly wrong. Researchers on the European continent were increasingly coming to think of light as purely a "wave" phenomena ... but a wave is usually considered to be a wave *in* something, and although Newton had hedged his bets over whether an underlying medium ought to be "particulate" or not (*"for I do not know what this Aether is"*, "Opticks"), at least *some* Nineteenth Century aether-theorists seemed to be inclined towards the idea of a medium that behaved like a conventional fluid or gas. If **James Clerk Maxwell**'s equations (mid-to-late Nineteenth Century) said that electromagnetic fluctuations should propagate at the speed of light, and electromagnetic theory suggested that light *was* in fact an electromagnetic wave, then to some, it seemed reasonable that the speed of these waves should be referenced to the speed of the medium, and that if a body was passing through the medium, it should encounter different lightspeeds in different directions. But from our "flasks" arguments in section 2.3, a lightspeed offset would tend to cause bodies containing trapped electromagnetic energy to be "braked" by internal radiation pressures until they were stationary in the medium's rest frame. This would clash with Galileo's and Newton's observations.

14.3: ... "Draggable" aethers

The obvious way to get a medium to follow more conventional "relativistic" rules was to suggest that the light-medium (whatever it was) was *locally dragged along* by the motion of bodies through it. **Fresnel** had suggested light-dragging early in the Nineteenth Century, and **Fizeau** set out to isolate Fresnel's predicted dragging effect by measuring the difference in lightspeeds for signals sent upstream and downstream through water flowing in glass tubes. Fizeau's experiment indicated that the effect existed, in the range that Fresnel had predicted.

"Dragged aether" models seemed sophisticated, and "modern", and intuitive, and seemed to fit nicely with our existing ideas about how Nature seemed to operate. But they were also more complicated and didn't seem to supply many guiding principles to explain *how* moving bodies dragged light, how *strongly*, or exactly why they should choose to do it. We could take our lead from the theory of fluids and include various dragging factors and coefficients to describe how an electromagnetic medium *might* behave, but there seemed not to be a universal way of predicting the strength of these effects, and theories sometimes seemed to be merely **parameterising** the problem, inventing a range of general-purpose variables whose values could be set "by hand" to bring the predictions into line with the experimental data.

Particulate fluids can behave in quite complicated ways, and can show a multitude of subtle behaviours depending on the characteristics and interactions of the particles that make them up. Since we didn't know which founding principles and laws a luminiferous aether might follow, or what sort of particles (if any) it might be composed of, we didn't know if it should behave exactly like a conventional fluid or not, and if it didn't, which aspects of its behaviour *wouldn't* agree with conventional fluid dynamics. We were slightly lost.

14.4: ... Absolute aether

The one type of aether model that didn't seem to have any arbitrary parameters was an **absolute aether**. This could provide a fixed, undragged, perfect reference system for anything embedded in it, and while the idea of an absolute aether may well have seemed wrong to some researchers at the time, it could be used as a standardised reference model with unambiguous, testable predictions, and at least *some* people seem to have liked the concept. Where more complex aether theories introduced a multitude of possible terms with unknown values and interactions, an absolute aether at least had the saving grace that we knew *exactly* what it should predict, and we could start doing calculations with it straight away.

One of the immediate predictions of a fixed aether model was the idea that there should be an **aether wind**. As the Earth rotates, a sensor placed near its equator will sweep out a circular path, and if an absolute medium had any motion through the Earth's rotation plane we'd expect the detector to be sweeping through this medium it at different speeds at different times of day. On top of this, the Earth also swoops around the Sun approximately once every ~365 days, which might also be expected to sweep the detector "upstream" and "downstream" through the passing medium at different times of year. Putting the two effects together, even if an aether was *perfectly stationary* with respect to our solar system, or was flowing through it at a *perfect* 90 degree angle to its plane, the combination of the Earth's rotational and orbital motions would mean that a detector fixed to the Earth's equator would be repeatedly swept through the medium faster and slower, as the motion of its section of the Earth's surface was alternatively rotated with the direction of the planet's orbit, and against it.

Albert Michelson set out to measure the motion of this aether wind, and in 1881 reported that he couldn't find it. His early experiment had some "sensitivity" issues (his analysis was criticised by Lorentz), but after teaming up with **Edward Morley** to do a more accurate and hopefully more reliable version of the experiment, the new results were published in 1887. The **Michelson-Morley experiment** again obstinately refused to show an aether wind in the expected range, and we now refer to the M&M experiment as having produced a **null result**.

A "null" result doesn't mean that an experiment produced a "good" result of "zero": it means that the experiment's outcome was inexplicable according to the assumptions that were used when the experiment was designed. It means that that we need to go back and re-examine our core assumptions from scratch, because something in our procedures obviously hasn't worked the way that we thought it would. Michelson's test was designed to look for the significant wind effect that should have been there, and to report back how strong it was and in what direction it acted, and instead he found that any effects that *might* have been there were too weak to be compatible with the assumption of an absolute aether. All bets were off.

Einstein was later fairly non-committal about whether he'd known about the Michelson-Morley result or been influenced by it before 1905, but we now consider the M&M test to be one of the defining experimental results that paved the way for the special theory of relativity: if one believed M&M's outcome, then the idea of an absolute aether, like Newton's faulty corpuscular theory before it, was a dead duck.

The idea of an absolute aether had also been troublesome for other reasons. Generally speaking, it's considered "bad form" to invent a substance that dictates how physics ought to behave, but which shows no **backreaction** – that can influence without itself being influenced ("*active*" but not "*interactive*"). Philosophically-speaking, action without *re*action is usually considered to be Very Bad Design. With a *truly absolute* aether we would have had a supreme, perfect, inscrutable, implacable, all-powerful "thing" dictating the laws for physical behaviour, without memory or conscience, acting as a form of "**prior cause**" and making all attempts at further scientific analysis futile.

This isn't usually considered to be a good thing.

14.5: Aether, either, neither neither

The theorists of the late Nineteenth Century found themselves inheriting a bit of a mess. Having largely abandoned Newton's idea of calculating light trajectories as if light was composed of little particles, and embraced the idea of light as waves in an aether, they found themselves pretty much at a dead end. The speed of light was supposed to be constant, but nobody knew who or what it was supposed to be constant *for*. The "medium" might be particulate or non-particulate, in the first case we had almost no idea what the properties of the particles might be, and in the second case we had no clear precedents for how a nonparticulate medium might behave. New aether theories were popping up so quickly that just cataloguing and classifying them and listing their different predictions was turning into a major effort. Every time that someone came up with yet another bright idea for how an aether *might* behave, there was yet another aether theory to add to the list.

In some ways the situation was similar to the impasse that string theorists found themselves in a century later: **String theory** appears to give us a *reasonably* intuitive, mathematically modellable set of rules for how physics might behave, but doesn't obviously tell us which particular variation on those rules is the one that actually applies. String theory sometimes looks like a complex way of retrospectively *re-modelling known* physical laws, which we still have to derive ourselves using some other means. As the joke goes:

> *A physicist and a mathematician sit down to lunch, and the physicist asks how the research on string theory is going. The mathematician replies, Oh, its wonderful, with our new tools and techniques we can predict absolutely anything. Really, says the physicist, in that case, what about predicting next week's lottery numbers? If you can predict that, then I'll happily pay for lunch. No problem, says the mathematician, just tell me in advance which six numbers you want the theory to predict and I'll have the working for you by tomorrow.*

"**Aether theory**" (like "string theory") *was not a single theory*, but was a broad field of research that provided tools for constructing different individual theories according to basic rules, much as the phrase "atomic theory", tends to be used as a collective term rather than as referring to one specific model. We still say that atomic theory is reckoned to be "good" (matter appears to be built from atoms), even though the majority of *historical* atomic theories are known to have been badly deficient in one respect or another.

The overwhelming range of possibilities in string theory (the string theory "**landscape**") will *not* refer to the physics of the universe that we actually inhabit, and so, even if *individual* string theories are fully deterministic, the subject is actually very bad at making specific predictions. If we can hand-tool the various factors and variables that can affect string theory's output, to perfect accuracy, then the resulting theory should be good. But if we somehow *already knew how to predict* what these parameters should be, we'd probably already understand exactly how things worked, and wouldn't *need* string theory.

So it was with "aether theory": if we *already knew* exactly how our hypothetical electromagnetic medium behaved, we could parameterise its properties and produce a single aether model that reproduced that behaviour in the language of fluid dynamics supplemented with whatever additional quirks and departures from normal fluid mechanics were considered necessary. But given our lack of knowledge as to what these required behaviours and characteristics *were*, aether theory couldn't tell us very much. By being able to predict almost *anything*, aether theory found itself in a Borgian state of predicting almost *nothing*. Given the proliferation of different aether models that researchers had to keep track of in order to stay on top of the subject, the promise offered by wave-only models was rapidly turning into a nightmare, and it isn't surprising that by the turn of the century some disillusioned aether-theory researchers seemed to be getting heartily sick of the whole idea.

RELATIVITY IN CURVED SPACETIME

14.6: Lorentz Ether Theory (LET), → 1904

Around the end of the Eighteenth Century, a number of researchers, notably **George Francis Fitzgerald**, **Joseph Larmor**, **Henri Poincaré** and **Hendrik Antoon Lorentz** spotted a unique mathematical relationship that seemed to be able to "relativise" absolute aether theory.

The logic behind their Big Idea went something like this:

> *If there was an absolute reference frame for the transmission of light*, then the appearance of an object moving at a fixed speed with respect to us should depend on how we were moving with regards to that frame. If c was "really" fixed in our frame we'd see the "A" Doppler relationships and length-changes of section 6, whereas if it was "really" fixed in the emitter's frame, we'd be seeing the "B" set. All we'd have to do to identify the absolute frame would be to measure the shift on a few objects moving in different directions and we'd be able to compare their appearances to work out the true, *absolute* speed of light.

> *But if the principle of relativity is also to hold for light*, then this sort of measurement shouldn't be possible, because objects should be obeying the same laws of physics no matter what direction or speed their laboratory is moving in. For a given *relative* velocity, optical effects should look the same for everybody.

> *So* ... if we could come up with a new set of predictions that looked *equally wrong* for all inertial observers, we'd satisfy the principle of relativity and make it impossible for observers to compare notes and isolate an absolute preferred frame. How do we find this new solution? Well, it just so happens that if we divide the "B" predictions by their "A" counterparts to find how much they disagree, we always get the same discrepancy, the factor 1- v^2/c^2. If our "new" predictions were to be the "**geometric mean**" average of "A" and "B", then they'd disagree with *both* earlier predictions by the same amount, which would be the *square root* of this ratio. The difference between our new forecast and the earlier **A** and **B** predictions is called the **Fitzgerald-Lorentz factor**, the **Lorentz-Fitzgerald factor**, or, just the **Lorentz factor** for short.

$$\text{Lorentz Factor} = \sqrt{1 - v^2/c^2} \text{ , or } \frac{1}{\sqrt{1 - v^2/c^2}}$$

These three Doppler curves are graphed in sections 6.6 and 7.5. If the Doppler relationships turned out to agree with the new *intermediate* prediction, we could explain this by saying:

> ... that the speed of light was "really" fixed in our frame (graph "A") and the moving object was time-dilated and length-contracted to bring things into line with the central plot, or

> ... that the speed of light was "really" fixed in the emitter's frame (graph "B"), making things even redder, but that our own motion through the aether was slowing and squashing our reference clocks and rulers, so that we saw everything with a Lorentz *blue*shift, again bringing us back to the same final, intermediate answer.

$$\text{prediction[LET]} = \sqrt{\text{prediction[A]} \times \text{prediction[B]}}$$
$$= \text{prediction[A]} \times \text{Lorentz factor}$$
$$= \text{prediction[B]} \div \text{Lorentz factor}$$

14: Einstein's "Special" Theory of Relativity

So if we had an absolute aether, but objects moving through that aether just happened to contract in length and slow down by the Lorentz factor, we wouldn't be able to compare notes and work out who was "really" moving and who wasn't.

Lorentz's best-known paper on this idea appeared in the 1904 volume of "Annalen der Physik", and this sort of model, now referred to as **Lorentzian Electrodynamics** or **Lorentz Ether Theory ("LET")**, seemed to be a fast way of combining the idea that light propagated as if there was an absolute aether with the idea that all observable physics should seem to operate as if no such absolute reference existed.

The **Lorentz contraction hypothesis** was a fairly *ad hoc* solution ... Lorentz didn't explain exactly *why* objects should contract in this particular way, except that it made things come out rather nicely if they did ... but it had the important advantage of being a mathematically unique solution, and of making definite predictions.

14.7: Special relativity, 1905

Einstein published *his* famous paper "**On the Electrodynamics of Moving Bodies**" in the 1905 volume of "Annalen ...". This presented the same basic relationships that had appeared in Lorentz's paper but put them on a different philosophical basis.

According to Einstein's way of looking at the problem, Lorentz's corrections to Newtonian theory were A Good Thing, but we shouldn't have to worry ourselves about *why* objects moving through an aether should choose to contract and slow in this way. We didn't need to try to construct, say, an "aerodynamic" theory where a moving particle might represent a pinched point of aether flow resulting in Lorentz relationships thanks to the Bernoulli effect or some other fluid-medium principles. We didn't need any specific aetheric mechanism.

In fact, we didn't need to worry about an aether at all.

What *was* important (said Einstein) was only that any collection of inertial observers' measurements should be seen to be obeying the principle of relativity. If we agreed that the speed of light ought to be (globally) constant, and also agreed that that the principle of relativity required that this law of the constancy of light should work equally well in any frame, than *these two conditions were sufficient* to let us derive the Lorentz relationships as the only possible mathematical solution, without worrying our little heads about how this might relate to the properties of some underlying aether that might (or might not) exist. If we could derive a single, final, *non-negotiable* set of relationships as representing the only possible answer, then further analysis and explanations were unnecessary. If we already had everything we needed to predict real-world physics, *without even mentioning the idea* of an aetheric medium (whose state of motion couldn't be determined under LET anyway, as a point of principle), then perhaps the aether concept was redundant and could be ignored.

Special relativity wasn't adopted immediately and didn't have an entirely easy ride. Not everyone appreciated its more abstract philosophical approach, or the idea that Einstein seemed to have found a way of generating "physical" predictions without explanations that some physicists would have considered to be necessary for a "real" physical theory. Special relativity in some ways seemed more like a **meta-theory**, one stage removed from more conventional approaches: it predicted what the outcome of a more "physical" theory *would have to be* in order to satisfy certain conditions, without providing an actual physical model, or a conventional mechanistic description of what was really going on.

The essence of Einstein's special theory was that *if* lightspeed was globally constant, *and* obeyed the PoR, then these would be the only relationships that could work. This Was How Things Had To Be.

Period.

14.8: Additional interpretational overhead

We've already seen that in order to create the *physical predictions* of special relativity, we only need to average two earlier predictions together. We don't need much else to find special relativity's predictions for how differently-moving inertial observers see each other. That "averaged" calculation gave us special relativity's frequencies and wavelengths, energies, momenta and so on, and the relativistic aberration formula gave us the angles. Once we had all these physical properties nailed down, there wasn't much left to do.

But Einstein wasn't able to get away with saying, "... just take these two conflicting predictions and average them, that's the whole theory" ... if he *had*, his 1905 electrodynamics paper would have been much shorter. To produce a theory acceptable to theorists trained in strict *procedural methodology*, and to make it look more scientific and less like Lorentz's "*ad hoc*" arguments, Einstein had to make the thing much more complicated and much more technical. Einstein developed a set of strict procedures that would allow observers to define distances and times according to the assumption that lightspeed was globally constant in their own frames, then showed that different observers applying this approach to the same events could disagree as to what the "real" distances and times were, and that the PoR then produced "Lorentz factor" discrepancies between different observers' interpretations of the same effects.

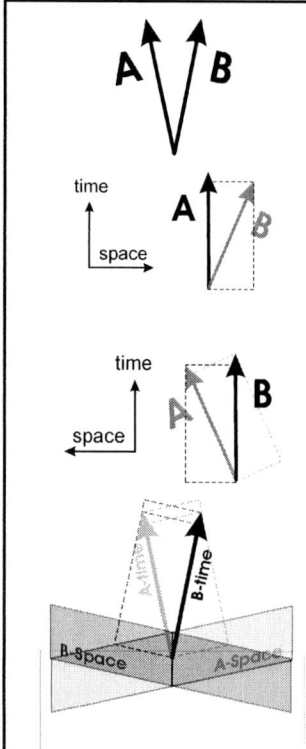

The 4-D "flat spacetime" approach

Two inertial observers move away from each other at constant speed. They both have to agree on the angle that their two worldlines make in spacetime.

Observer **A** believes that they are not moving and that their worldline therefore points only in the time direction, and can be thought of as representing their "**arrow of time**". They believe that observer **B** is moving, and that **B**'s worldline therefore cuts across both time *and space* coordinates at an angle.

Observer **B** believes that *they* are not moving, and that *their* worldline points purely in the time direction. According to *them*, it is **A**'s arrow through spacetime that crosses space and time coordinates at an angle.

If both observers insist that lightspeed is "really" fixed in their own frames, they may agree on the *separation in spacetime* between two agreed events, but they won't necessarily agree on the differences in distance or the separation in time between the two points.

A "straight-line" path through spacetime can then be treated as being made up of different amounts of space and time, depending on the observer's reference system.

If different observers could assume different speeds of light, they could also disagree on the times and distances of separated events, on the rates of clocks moving between those events, and on the lengths of rulers linking the events together. Special relativity started out using Einstein's procedures and predictions for how these hypothetical networks of **rods and clocks** should behave, evolved into a set of more mathematical arguments about meshing **coordinate systems**, and ended up as a four-dimensional exercise in **Minkowski spacetime**.

14: Einstein's "Special" Theory of Relativity

We won't be covering the conventional "textbook" view of special relativity in any real depth here, partly because it's already covered in a lot of other books, partly because SR seems to be incompatible with *curved*-spacetime physics, and partly because most of the literature about special relativity isn't so much about the *actual physical predictions* of the theory, but about more abstract, *interpreted* quantities and concepts that are at least one stage removed from what our hardware can actually register. A few of the common SR "buzzwords" are:

Inertial frame – applying the principle of relativity to moving *objects* can be complicated, because the exact physical side-effects of relative motion usually depend on whether an object is moving towards the observer, away from it, or at some other angle. Einstein chose to simplify the problem by grouping all observers with the same state of motion into a **frame** in which everyone could agree that everyone else was "stationary". SR then set out to explore how the PoR might define the relationships between these **inertial frames** rather than between the individual inertially-moving objects within them. The advantage of the "frames" approach was that it eliminated **proximity** information and proximity-related effects (such as dragging) from the analysis, and when these things were abstracted away, the resulting system allowed just one possible solution. The disadvantage of the approach was that it presupposed that "local" effects (such as **gravitomagnetism** and **gravity**) weren't relevant.

Clock-synchronisation – by exchanging signals with other observers in the same frame and assuming that the speed of light is the same in both directions, observers can measure the round-trip time-delays between objects, divide those by two to get the *assumed* one-way signal timelag, and then compensate for these timelag values when setting their clocks to an agreed reference clock. The round-trip signal time also provides interpreted "distance" labels for distant objects. The observers then believe that an array of clocks set in this way are "really" running in perfect synchronisation.

Coordinate systems – this method produces a reference-network of interpreted distances and times for distant objects and events. This network of values then provides a reference **coordinate system** for distances and times within that frame.

Relativity of simultaneity – When Einstein's "clock-synch" method is applied by two arrays of observers with *different* states of motion, their different interpretations of the one-way flight-times of a signal linking two agreed events will result in the two arrays of clocks being synchronised differently, and separated events that are described as "**simultaneous**" according to clocks "synch'ed" in one frame won't be described as "simultaneous" for those in another. The two sets of observers will assume different **planes of simultaneity**.

Velocity addition – in older theories, if the Doppler shift ratio for a given velocity was $shift[v]$, then the result of passing the signal through two *successive* velocity-shift stages, $shift[v_1] \times shift[v_2]$ was often *not* the same as the result $shift[v_1+v_2]$. To get around this inconvenience we could invent an artificial "equivalent" velocity for $[v_1+v_2]$ that wasn't the same as just adding the two values together. This was calculated using a **velocity addition formula**. The need for these formulas tended to suggest that the presence of intermediate moving objects altered the behaviour of light. Under SR this isn't allowed, so the velocity-addition characteristic has to be considered as a structural property of spacetime itself, and the resulting velocity value must be a "real" velocity.

Much of the *subject* of special relativity is taken up with describing and explaining these concepts and how they fit together, and disputes about special relativity tend to use this sort of language. But as *physicists*, we should remind ourselves that *none of this overhead is actually terribly important*. Coordinate systems and relative simultaneity are part of a layer of *interpretational infrastructure* that was *supposed* to make the theory easier to accept. But they aren't obviously *physical*, and as Terrell pointed out (1959), experts who delve too deeply into this interpretational jungle can sometimes lose sight of the theory's real predictions.

14.9: Minkowski Spacetime

The mathematician **Hermann Minkowski** took Einstein's "abstract" approach to the Lorentz relationships one stage further, reputedly causing a perplexed Einstein to grumble at one point that the special theory had now been taken so far that he no longer fully understood it.

Einstein had argued that an inertial observer could claim that lightspeed was "really" fixed in their own frame, and could use this belief to assign nominal "distance" and "time" labels to distant events. Another observer with a different state of motion could claim that light was fixed in *their* frame, and by making different assumptions about how long light had taken to reach them, they could extrapolate *different* distance and time labels for the same events ... but the underlying pattern of events, and the deeper web of relationships between them would still be the same for all observers. Einstein then went on to argue that for these remappings to be able to work in *exactly* the same way for everybody, without a "favoured" frame, the Lorentz factor had to be in there. It was a bit confusing, and SR's constant redefinitions made it difficult to judge whether the underlying relationships were really as consistent as claimed.

Minkowski's contribution was to put aside Einstein's arguments about rods and clocks and to re-set special relativity in the context of a **"block spacetime"** designed around SR's postulates.

In Minkowski's "reimagining" of SR, we could consider spacetime as a four-dimensional volume containing a "web" or "mesh" of crosslinked events, where reality consisted of these events and their interconnections. Differently-moving observers could *experience* this web differently, because their own **worldlines** would be passing through the block at different angles. Their perception of the spatial and temporal distances between two events would depend on the angle that their worldline made with the rest of the block.

Spacetime was the thing, said Minkowski, and space and time as individual entities were just ghostly shadows of a deeper underlying mathematical truth. Project a 4-D line connecting two events onto arbitrary "space" and "time" axes, like the shadow of a stick onto a flat surface, and you'd end up with contracted shadows whose lengths depended on the angle of the stick. To make these repeated remappings work *relativistically* then required Lorentz effects.

do you want a cone with that?

Minkowski's spacetime depends on the idea that the motion of physical masses doesn't distort the physical lightbeam geometry – potential worldlines can be drawn through any region at any angle corresponding to a relative velocity less than *c*, and all these different alternative worldlines can have their "space" and "time" values mapped onto each other using the **Lorentz transforms**, which describe the result of a special sort of 4-D geometrical projection.

With Minkowski spacetime it can also be useful to think of each point in spacetime as the origin of two cones: if the point-event produces a lightsignal, then the expanding sphere of this signal can be represented as a **future lightcone** that marks out all the locations in spacetime from which that event can be *seen* ... similarly we can draw in a second, **past lightcone** converging on the point that marks out the complete set of events that will be *visible* at that location, at that same moment. The angle of the sides of the cone indicate how far light travels in unit time. We then have an infinite spray of possible worldlines passing through each and every point in spacetime, with an infinite spray of equally-legitimate time-arrows pointing into and out of the event, all trapped within the limits of that event's two lightcones. When we want to see how the block appears for a *specific* observer, we can tilt it so that *that particular* observer's worldline becomes vertical, and although the alignment of events may change, and our history-map can then end up assigning different nominal "date" and "location" labels to the same distant events, the causal relationships between those events won't change.

14: Einstein's "Special" Theory of Relativity

What made Minkowski spacetime "new" was an extra property – when we tilt SR's event-map to match the perceptions of a different observer, it also *distorts* in a very particular way, so that although the events on the surface of a lightcone will mathematically shift and redistribute themselves around the conical surface, this **morphing** between coordinate systems leaves *the angle of the cone itself* magically unchanged.

If we take a lightray travelling through the block at an angle corresponding to c for an SR observer, the events along the lightray's path will still lie at the same angle after the block has been rotated and distorted. No matter how many times we apply these **"tilt and skew"** remappings, the cone keeps its original alignment, and the cluster of potential worldlines trapped within the lightcone will always be trapped inside it.

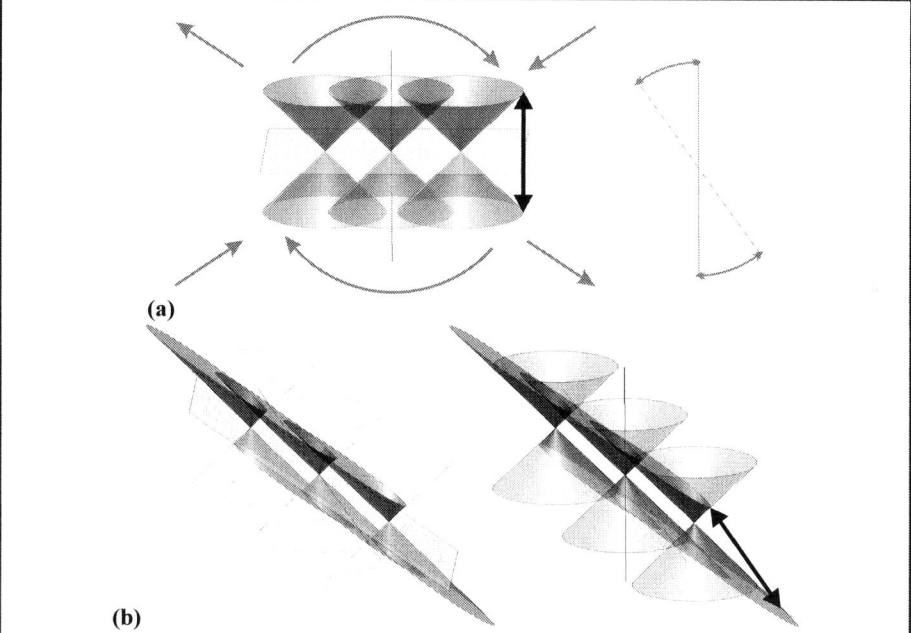

Three cones have "simultaneous" origins for observers in frame **(a)**, and intersect at particular agreed events. For an observer in frame **(b)**, the three origins are *not* simultaneous, and the "official" distances and times of events "tilt and skew" when we compare **(a)** with **(b)**. But the underlying web of events is the same in each case, and the angle of the cones doesn't change

Figure 14-1: Minkowski spacetime: Fun if you are a mathematician, not so much fun if you aren't

Minkowski showed that this "special" type of geometry was both conceivable and internally consistent. Although lines drawn at any angle *less* then the speed of light will change their orientation when we use a different worldline as our time-reference, lines drawn at *exactly* the speed of light within Minkowski spacetime don't tilt. Minkowski's model expresses the idea that no matter how many times you increase your velocity by less than lightspeed, you can never catch up *with* the speed of light. No matter how many times you try to tilt the block by, say, ten degrees, those lightcones will never flip over ... the map becomes progressively more distorted compared to your original view of it, but the cones always stay obstinately symmetrical, with their sides presenting the same angle. In Minkowski's mathematical universe, no matter how many times you increase your velocity by less then c, you can never touch the surface of your own lightcone – the surface is *still* moving away from you at c.

14.10: Implications of Minkowski spacetime

As an exercise in mathematics, or geometry, or in visualising **Poincaré**'s ideas about group theory and SR, Minkowski's concept was undeniably pretty. Although it didn't appear to give SR any new physics, it demonstrated that special relativity's description could be expressed geometrically, and could be shown in a mathematically-provable way to be self-consistent and complete within a particular sort of universe.

Minkowski spacetime as mathematics

To mathematicians, Minkowski spacetime is a fascinating theoretical object in its own right and is worth studying irrespective of whether or not it's thought to have any direct relationship with real-world physics. To mathematically-inclined physicists, special relativity has increasingly become seen not as the result of Einstein's thought-experiments involving rods and clocks, but as "how physics operates inside a Minkowski-spacetime universe".

Minkowski's approach lifted the burden off physicists of having to prove that every little kink of special relativity's core math was consistent with the rest of the theory, and showed that the SR system of physics could be set free from the earlier "aether-based" concepts and language used by Fitzgerald and Lorentz. Even if we'd suspected that some of the early derivations of SR might not have been entirely rigorous, the end result had a definite internal self-consistency that transcended any shortcomings that might have existed in the original derivations that we'd used to get there.

linkage with special relativity

We could now prove (or at least, friendly mathematicians could prove for us) that if we wanted to combine **(1)** lightspeed constancy, **(2)** flat spacetime, and **(3)** the principle of relativity, what we inevitably ended up with was Minkowski's spacetime as a unique solution. If we took one worldline's view of a region and applied the appropriate Minkowski "tilt-and-skew" operations to convert those coordinates into the language of an observer with a worldline running at a different angle, cross-referencing the two sets of coordinate-system distances and times produced the specific "relativistic Doppler" relationships of special relativity.

These three conditions (or the two postulates "*global* lightspeed constancy" and "relativity") made the "relativistic Doppler" relationships of section 6.6 compulsory, with the structure of special relativity reduced to a matter of how projective geometry allowed an agreed set of points to mark out different-looking "shadows" within a Minkowski metric.

To many physicists, it seemed that the book was now closed on the subject of whether SR was right or not. If the theory was now based on mathematical theorems that could be *proved*, and which said that SR's relationships were the *only possible solution*, then how could the theory possibly be wrong?

The answer is: quite easily, as long as we remember that physics and mathematics are still distinct branches of research with different remits. Mathematics is a powerful language. It can describe a great many things, and some of them may even be true. But languages can also describe things that are *not* true, and with Minkowski spacetime we have a description of how the principle of relativity could operate in a universe in which physical principles don't all correspond to the known rules of our own world. Minkowski's spacetime seems to give us the (provable) geometrical properties of a universe whose laws don't quite correspond to ours.

Although the flawlessness minimalism of Minkowski's geometry is often taken to mean that the relationships of special relativity *must* be correct, in the next section we'll consider the slightly radical idea that the sheer *perfection* of this system might mean that it has to be "bad".

15
So, what's wrong with the special theory?

Mathematics is not the language of Nature. It is the language of Mathematicians.

> " It may easily become a disturbing element in unprejudiced scientific theorising when a conception which is adapted to a particular and strictly limited purpose is promoted in advance to be the foundation of all investigation. "
>
> **Ernst Mach**

> " If it was so, it might be, and if it were so, it would be, but as it isn't, it ain't. That's logic! "
>
> **Tweedledee, "Through the Looking Glass, and What Alice Found There"**
> **Lewis Carroll**

> " Things should be made as simple as possible, but no simpler. "
>
> **Albert Einstein**

> " As far as the laws of mathematics refer to reality, they are not certain; and as far as they are certain, they do not refer to reality. "
>
> **Albert Einstein, "Geometry and Experience" 1921**

> " In classical mechanics, and no less in the special theory of relativity, there is an inherent epistemological defect which was, perhaps for the first time, clearly pointed out by Ernst Mach. ... "
>
> **Albert Einstein, "The Foundation of the General Theory of Relativity", 1916**

> "... it was better to permit such inconsistency – with the obligation, however, of eliminating it from a later stage of the theory."
>
> **Einstein, on the "problem of consistency" due to the unphysical nature of SR's clocks and rods "Autobiographical Notes" (1949) pp.59-60**

> " It aint necessarily so ... "
>
> **Gershwin**

> "You are playing all the wrong notes!"
> "I am playing all the RIGHT notes. Although not necessarily ... in the right order."
>
> **Eric Morecambe**

RELATIVITY IN CURVED SPACETIME

**Illustration by John Tenniel, from Lewis Carroll's
"Through the Looking Glass…"**

15.1: SR and Observerspace

Explanations of special relativity often emphasise the importance of "observers" and "observations", and this can give the impression that Einstein's special theory is a *literal* **observerspace** theory – that is, that it deals *directly* with what observers at particular locations should *experience*. This seems to make a strong case for the idea that the special theory's physical predictions *have* to apply if the principle of relativity is correct.

But this impression wouldn't be entirely accurate. By convention, a **perfect observer** is usually supposed to be someone or something that records the experiences that they're presented with, literally, without extrapolation or bias – they try to report their experiences *objectively*, without imposing their own personal belief systems or interpretations onto the data that they collect. An **"observationalist"** theory will tend to say that what we *see* to be happening is, for us, what *is* happening, and if we define our "observers" broadly enough to include solid inanimate bodies and atoms, then our resulting theory of how these objects "see and feel" each other should then tell us something useful about the actual physics of how these bodies interact with each other.

But, as **James Terrell** eventually pointed out (in 1959) this doesn't correspond to the behaviour of "observers" under special relativity. Einstein's special theory insists that inertial observers can extrapolate their own locally-constant speed of light outwards throughout the surrounding region, and can also treat the *velocity* of light as a global constant, and their "observations" are predicted, interpreted and reported in the context of these new beliefs. This reinvention of the act of observation, incorporating the assumption of **flat spacetime**, restricts our relativistic options to the equations of special relativity. The theory *does* then go on to make specific physical predictions for the effects that should be directly visible according to the theory ... but we have to remember that what an SR observer is supposed to *observe* isn't necessarily what the theory predicts they should actually be *seeing*.

"non-SR" observerspace approaches

Could we try to build a more literal observerspace model than SR? If we try, we immediately run into some odd behaviour. For instance, if we said that the rate of timeflow of an object (for a given observer) was the rate that that observer would *see* the object to have, then a circle of "stationary" observers surrounding a "moving" object would report different "observed" values for the object's rate of timeflow depending on their viewing angle – an observer in front of the object would see it ageing more quickly, and an observer behind it would see it ageing more slowly. We wouldn't be able to use a single value for the object's "observed rate of timeflow" according to observers in the same **inertial** *frame*, and our more abstract SR logic to do with collections of "observers" and "frames" wouldn't work.

If we took this "seen" behaviour literally, we'd have to allow an object's apparent rate of timeflow to be *route-dependent*, and by assuming local c-constancy, these "apparent" timeflow differences would then turn into "apparent" gravitational gradients between the object and the surrounding observers – the object's motion would *appear* to have an associated gravitational field. The moving body would seem to be producing something like the "tilted gravitational well" description of section 9.12, we'd not be able to say that the metric stays flat when confronted with masses with significant relative velocity, we'd have dragging and **gravitomagnetism** as fundamental effects, and we'd be led naturally towards the equivalence principle and the "most general" version of the general principle of relativity.

So, although SR tends to be considered as an observerspace theory, it doesn't fully embrace observerspace principles ... if it *had*, it would have ended up as a different class of theory, with a different lightspeed propagation model and different equations of motion. Its application of the PoR to "frames" is necessarily one stage removed from direct observation.

15.2: Is the special theory "robust"?

Special relativity takes a "blank page" approach to modelling spacetime. We start with a perfectly flat page and populate it with abstract mathematical observers, rulers and clocks drawn onto the page ... we assume that the rules work across a *stack* of pages ... we draw a coordinate-system grid onto the pages ... and then we work out what the laws of physics would have to be for us to be able to create *new* sets of pages at will by slicing through the stack at different angles, with the same coordinate-system rules and relationships operating for all of them (provided that we don't "cut" too steeply). It assumes that the *contents* of a "page" have no influence on its geometry, and that the shape of spacetime is pre-set and non-negotiable.

This gives us a theory derived from clean geometrical principles that doesn't assume, borrow or hypothesise anything to do with the more "messy" behaviour of *real* clocks, rods, and observers. It's been described as a *perfect* set of mathematics. But this method's weakness (known as the "**consistency problem**") is that it doesn't describe the interplay between *real* light and *real* moving bodies – its idealised building-blocks aren't consistent with real objects.

If we take a perfectly deterministic process (such as a computer program or mathematical proof) and feed it "bad" initial data, then shouldn't its output be similarly bad? In computing this principle is known as the "**Garbage In Garbage Out**" principle, or **GIGO** for short – GIGO says that a rigorous, rigid, logical system, given "faulty" logical relationships or properties, should correctly derive solutions from them that are (appropriately) wrong. The principle doesn't always apply to more **fault-tolerant** systems: sometimes we can get away with a crude approximated framework as our starting point and refine its structure later with further levels of detail without overthrowing our initial, fundamental assumptions, and this is often a good plan of attack when we're confronted with an intimidating mass of uncertain detail to model – to begin with a "skeleton" theory that has a minimal set of assumptions, derive preliminary results, and then to *hope* that those answers might survive as we go back and start feeding progressively more realistic starting assumptions into the model.

When we take a very idealised theory and "shuggle" or "perturb" its initial data or its simplifying assumptions, and find that its basic properties *don't* change in the face of these perturbations – if the "perturbed" calculations still **converge** on the same solution – then we can say that the solution is **robust**. But if the shape of the theory's predictions turn out to *depend* on those idealisations in a **critical** way – if it's **sensitive** to the exact idealisations used, and seems to stop working when we stray away from them – then we may suspect that the structure of our theory is to some extent an **artefact** of the choice of idealisations that we used to build it, and that its final shape may have more to do with the idealisations that we made than with the underlying physical situation that we were hoping to model.

Special relativity isn't "robust" with respect to deviations from flat spacetime. It might seem that by setting aside issues to do with light-dragging and the notional properties of a light-medium, the special theory is nobly refusing to make any unwarranted assumptions about how matter interacts with light, and that it's therefore deriving some deeper and more fundamental relationships that don't depend on any particular assumed interplay between light-beam geometry and moving matter ... but special relativity *does* make a key assumption about the interaction between matter and the propagation of light – it assumes ***that no such interaction happens*** – and that assumption then generates new relationships such as a revised law for momentum. This is a perfectly pragmatic approach as long as we remember that after we've built and played with this sort of "skeleton" model, we're *supposed* to go back at the end and test how far the shape of our solution depended on the exact nature of its idealisations, as a **sanity-check**. Does the theory survive when we include small departures from these idealisations, or does it break down or change? Does the theory survive "contamination" by more realistic conditions? The special theory's *critical dependency* on the assumption of flat spacetime (section 13.5) means that it doesn't seem to pass this test.

15.3: Minkowski spacetime as an argument *against* SR

The principle of relativity, combined with the idea of global lightspeed constancy, seems to lead us *inevitably* to Minkowski spacetime and special relativity (section 14.10). But if we have a Crichtonesque interest in the ways that supposedly-*perfect* logical systems can break down or crash, a proof that these two postulates and the Minkowski/SR structure are *inseparable* and *uniquely define one another* should start to make us feel dimly uneasy – it raises the stakes and suggests a more dangerous possibility: that instead of these geometrical proofs signifying that SR has to be *true*, they may instead mean that the special theory and the structures that we have built on it may have to be *wrong*.

How does this work? It might seem that all Minkowski did was to rework the SR relationships as a slightly mind-bending exercise in four-dimensional geometry without adding anything obvious to the physical predictions ... but Minkowski's work did two major things:

1. It converted SR's predictions into a geometrical form that mathematicians could use as the basis of work on more complicated curved-spacetime theory (e.g. GR1915), and,
2. It proved *geometrically* that SR's relationships really were the only possible solution to its two stated postulates.
   SR gives Minkowski spacetime, Minkowski spacetime gives SR

This is where things can start to get ugly:

To a mathematician, or to a mathematically trained physicist, the skeletal beauty of Minkowski's geometry and its rigid, non-negotiable structure suggests that the thing *must* be true. It allows no room for error or modification. It has a crystalline perfection to it that seems nonnegotiable.

But this *perfect fit* between Minkowski's flat geometry and special relativity means that if we have an object that shows *any* **gravitomagnetic** *effects at all*, then since its motion would alter lightspeeds in the region, the lightbeam geometry would warp, and the metric that it would inhabit *could not* then be a Minkowski metric. If Minkowski's is the *only* metric that contains the modified Doppler relationships of special relativity, then if our "dragging" object inhabits a different metric, presumably its associated Doppler relationships *can't* be those of SR.

This is sacrilegious stuff: It suggests that if moving bodies drag light, *no matter how weakly*, then their Doppler relationships should show a corresponding deviation from the relationships of SR, and if we're to insist that the behaviour of these objects must still conform to the principle of relativity, *there must be other relativistic solutions* (or perhaps even a range of non-SR solutions, section 13.3) to the problem of inertial physics that we haven't considered.

The Minkowski case demonstrates the danger of taking an overly-mathematical approach to theoretical physics: we may think that by identifying SR as a special case, we've proved that it represents the only possible answer ... but what we've actually done is to show that it's the only possible answer to an "unphysical" question. If we accept the existence of gravitomagnetic effects and apply a little logical judo, we find that of a *range* of possible answers, special relativity *is* a special case ... whose "flat" characteristics allow it to be rejected *immediately*.

If we suppose that SR is *almost* right, and that weak dragging effects don't disturb the SR predictions *very much*, then we end up with a continuum of relativistic relationships that diverge from SR by different amounts depending on dragging strengths (section 13.8). We'd not only have accidentally overlooked *one* possible alternative relativistic solution, we'd have missed an *infinite number* of them, which would be mildly embarrassing. On the other hand, if we tried to make ourselves feel better by saying that perhaps there are only a finite number of *discrete* solutions out there (that we missed), then these should represent *significant* departures from SR, and special relativity would then have to be *significantly* wrong.

15.4: The "stratification" problem

The current approach to building a final theory of physics is *layered*. Special relativity provides a template for how relativistic physics should operate in the absence of particulate matter or gravitational fields, quantum mechanics retrofits missing particulate-matter effects (giving us quantum electrodynamics), and general relativity deals (separately) with curvature effects. This has given theoretical physics a somewhat haphazard, *ad hoc* appearance with different types of laws operating in different regions, and without a single unifying set of rules.

But perhaps the construction of a general theory of gravitation might turn out to be more like the construction of a classic gothic cathedral. The cross-section of one of these buildings is a branching complex of organic-looking arches and load-bearing curves, operating according to a kind of fractal design ethos where the central structure is duplicated (with variations) at smaller and smaller scales. Take away one flying buttress, or one major keystone, and the structure is compromised. But the problem with trying to *build* a gothic arch is that the central keystone that "locks" the arch can't be put into place until the rest of the arch is complete, and the rest of the arch won't stand without the keystone. Our arch can't be assembled using simple sequential instructions, by placing one block at a time in the correct positions until the structure is complete. It will fall down. We either have to assemble our arch *in parallel*, moving all the component blocks into position *simultaneously*, or we have to use a system of temporary scaffolding to hold everything in place until the shape is complete.

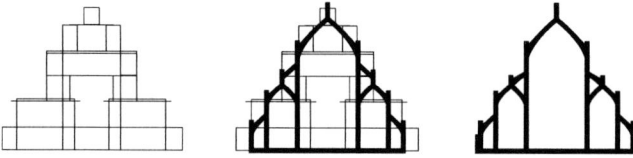

**Figure 15-1: Building a Cathedral:
Temporary and permanent features**

The shape of this scaffolding may be an approximate reflection of the shape of the cathedral, and it will share many of the same properties (alignment, approximate volume, and so on). But once the cathedral is built the scaffolding is discarded, and the final building will (hopefully) be freestanding and show no more traces of it. Although the construction of the scaffolding latticework might be a pragmatic step towards the construction of the cathedral, the final shape of the building doesn't *reduce* to the shape of the scaffolding as a limiting case.

This may be how things are with special relativity. Although Einstein's special theory may have been historically useful, and may well be a handy first approximation that let us establish some of the ground rules for relativity theory in a structure that obeys Euclidean logic (such as a weakening of the notion of "absolute time"), this doesn't mean that we're compelled to include special relativity as a limiting case of a final theory. It might be more like the skeletal framework of scaffolding that we use to build a cathedral: practical, and easier to build than the final structure, but ultimately disposable.

Rather than take an *incremental* approach, with all current theories expected to live on, embedded in the final structure as limiting cases, we could take an *iterative* approach and treat successful theories more cautiously, as *successive approximations* that don't have to be completely compatible with a final formulation of how Nature really operates. SR isn't guaranteed to be a limiting case of a final theory, and given some of the conflicts between its derivations and more general physics principles, perhaps it *shouldn't* be a limiting case, as a point of principle. Partial *principles* may well appear in the final structure as incomplete projections of broader law, but *theories based on them* don't have to be.

15.5: Does SR "do" acceleration?

Einstein's special theory doesn't model acceleration *relativistically*, but it lets us calculate what an inertial observers *might* see when observing an accelerating body, assuming that SR's velocity relationships are correct, and that any acceleration-related curvature effects that might appear in a more general theory can be ignored. Einstein's 1905 electrodynamics paper does discuss the dynamics of an accelerated electron, but it's a *slowly* accelerated electron.

This has led to disputes over whether SR can properly be said to handle acceleration or not. Some say that acceleration is technically outside SR's remit, that we ought to use GR for these problems – when SR then breaks down in an acceleration thought-experiment, it doesn't prove that the theory is wrong, it just shows that it's been applied outside its "proper" range.

Others will insist that SR clearly *does* deal with accelerations, point to the extreme accelerations that happen in modern particle accelerators, and say that if it *didn't* handle accelerating bodies, it wouldn't be much of a theory.

An intermediate position is that SR *can certainly be used* to *try* to model accelerating bodies, but that this isn't always guaranteed to generate exactly the right answers. The classic example of this was Einstein's suggestion in his 1905 electrodynamics paper that if the theory was correct, a clock at the (rotating) Earth's equator should tick more slowly than one placed at one of the poles: **Carroll Alley**'s 1977 experiment tested this and found no effect. Here's why ...

... centrifuge redshifts, the rotating Earth, and experiment

Special relativity's argument for centrifuge redshifts goes like this: the theory says that a passing object should show a Lorentz redshift (section 6.7), so if the object *circles* the observer, and is *constantly* moving at right angles to it, this *continual* redshift suggests that the object must *really* be aging more slowly (if we were feeling very brave, we might've also tried this same argument with the stronger transverse redshifts of NM). Einstein's argument was risky because it stepped away from the principle of relativity and applied its calculations *asymmetrically*, leaving us without an independent calculation to "sanity-check" the results for consistency. It also made an unsupported assumption that accelerations didn't matter, so it's probably safer to classify this as a *result* of the special theory rather than as a strict *prediction* ... "results" are allowed to be wrong without disproving the theory, "predictions" aren't.

We'll start by assuming that the statement is correct, and treat the Earth as a simple sphere, which, unbeknownst to its surface-dwellers, is then assigned the mysterious property of "rotation". Lets say that this effect is revealed to the Earth's inhabitants by clocks at the equator running more slowly, as suggested. What happens next? Well, if we slow the effective rate of timeflow in a region, we create an effective gravitational gradient that causes light and objects to deflect towards it (sections 1.9 and 5.6). This outward radial force pointing towards the equator will haul water and rock "downhill" into the equatorial region, causing a bulge that'll grow until its increased height gives an "uphill" slope that exactly cancels the effect out. We therefore expect sea level to only achieve a natural equilibrium when *all points on its surface have precisely the same clockrate* (forming an **equipotential surface** or **geoid**). So clocks *at sea-level* at the equator and poles should run at the *same* rate. We don't say that special relativity's *predictions* were wrong here, we say that the theory was applied too optimistically, and that these sorts of "complicated" situations need something more like GR.

Does special relativity a least earn some points for correctly predicting the existence of a centrifuge redshift (section 12.8)? Yes and no: SR's transverse predictions can encourage us to expect centrifuge time dilation effects, but so could the earlier, stronger, transverse redshifts predicted by Newtonian theory (section 6.4).

acceleration breakdowns

SR's carefully-constructed coordinate-systems don't work if the observer accelerates.

If we map out a set of events in Minkowski spacetime that special relativity could say are simultaneous for a given observer at a given moment, they mark out something called a **plane of simultaneity**. But if the observer then changes speed, their new "plane" will have a different tilt, and will now align with the block at a different angle (section 14.8). So ... if our inertial observer applies SR, briefly accelerates, and then applies SR a second time, there'll be a particular forward distance at which these "before" and "after" planes meet at an angle and then cut across each other.

Beyond this critical distance, the theory's method of assigning time-coordinates breaks down. Once the acceleration is over, a distant region of spacetime that would previously have been said by SR to lie in the observer's past gets assigned new time coordinates (for the observer's new inertial frame) that place it in the same observer's *future*. So special relativity's coordinates are only internally consistent if the observer doesn't change speed, otherwise they can do some very strange things indeed.

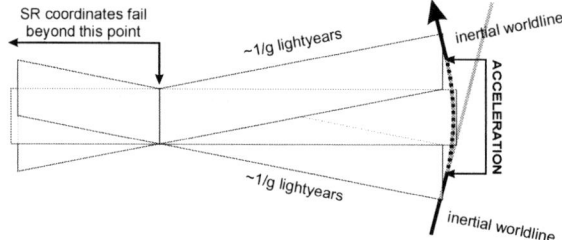

Figure 15-2: Constant acceleration produces an SR coordinate-system "pivot point" at approximately ~(1/geeforce) lightyears

This self-intersection distance is approximately **dist$_{BREAKDOWN}$ = ~1/acceleration**, with distances measured in lightyears and acceleration measured in Earth gravities (this relationship is a "happy accident", the universe doesn't *really* care about "Earth years" or "Earth gravities", the quantities just happen to match up nicely when we use this combination of units). So ... if a rocket ship accelerates at one gee, its SR-based coordinate systems break down at a forward distance of about one lightyear, accelerating at 10 gee gives a breakdown distance of about a *tenth* of a lightyear, and so on.

coordinate systems are artificial

We don't *really* think that leaping out of our chair makes time run backwards for distant stars, it just spoils the coordinate systems used by SR. Acceleration anywhere on an observer's timeline spoils the *event-labelling system* used by the special theory. But if these labels don't have a deeper physical significance, why did we feel that it was safe to construct a physical theory around them? If *physics* conforms to the principle of relativity, this doesn't mean that our artificial, idealised, *unphysical* sets of coordinates – an *artificial belief system* – must conform to the PoR in the same way that *physical reality* does ... and if we *do* manage to construct a relativistic implementation of those beliefs, the universe isn't automatically compelled to use the same interrelationships for real physical effects.

Some say that this shows that SR doesn't handle acceleration, but others say that SR deals with accelerating *objects* just fine, as long it's not applied to accelerating *observers*, or that perhaps it is *large distances* that give SR problems. This last interpretation says that although SR was derived as a **global theory**, it applies in practice as a purely **local theory**, and if we try to apply it over larger scales, any breakdowns are considered to be the fault of the user, not the theory.

15.6: Extensibility

Conventional Newtonian mechanics, though **incomplete**, did seem to be reasonably robust and **extensible**. If we start with textbook Newtonian theory (the good bits of Newton's model, with the mistaken inversions of section 5 corrected), it seems to be a decent basis for further work. Newtonian theory works quite happily with gravity, special relativity doesn't, Newtonian mechanics sits quite well with the idea of velocity-dependent light-dragging, special relativity requires a whole new layer of theory to put in effects that were assumed, *as a condition of its derivations*, not to exist.

Newtonian theory seems to pass the test of **extensibility**: Although "textbook NM" is presented as a flat-spacetime model, if we thought that spacetime *was* flat and absolute, a few basic calculations (such as Einstein's 1911 "gravitational time dilation" calculations) would quickly tell us otherwise, and we could then apply that knowledge to make our predictions more accurate without altering too many of the theory's core relationships. The general principle of relativity is missing from Newton's model, but adding it doesn't seem to require major structural corrections, and some of the general theory's other basic concepts don't seem too far removed from the ideas outlined in Newton's two main books. Newton wrote that gravity should bend light, and his arguments can *tell* us that space (and ultimately space*time*) ought to be curved, even if this wasn't our starting assumption.

But special relativity depends so strongly on the idea of flat spacetime that to allow curved spacetime or particulate behaviour seems to destroy the geometrical framework that we so carefully constructed for it. If we ask special relativity whether gravity exists, or whether the influence of moving matter should drag light, the theory either has nothing to say, or suggests the wrong answer. Similar problems afflict GR1915: although the theory's predictions regarding black holes seem flawed, the theory *itself* didn't tell us about the breakdown: we had to find it out independently using arguments from quantum theory. We *could* have used classical "cosmological horizon" arguments to point us in the direction of acoustic metrics (section 12.11) and towards a new version of the general theory, but the GR1915 worldview encouraged us to ignore these arguments and their implications.

15.7: Cumulative redshift effects ...

One of the things that gladdens that hearts of mathematicians when dealing with special relativity is the way that its frequency shifts and energy shifts cancel out over a round trip.

Let's suppose that we have a box containing a moving glass bead, and we shine a beam of light at the back of the bead, and watch the light emerging from the other side.

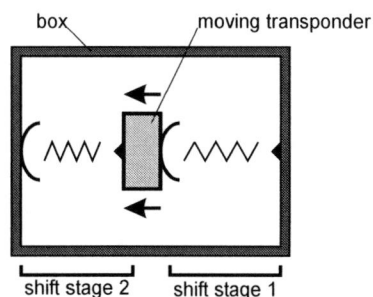

Figure 15-3: Round-trip redshift thought-experiment

Our light-source and our detector are bolted to the sides of the box, and the lightbeam's original source and final detector have no relative motion. What happens next?

SR calculation

Since the special theory assumes that light's characteristics are immune to the motion of any bodies in its path, it predicts that the beam should arrive at the far side of the box with precisely the same energy that it started out with, regardless of how the bead is moving.

And this is indeed what happens with special relativity's calculations: the recession redshift prediction is $f'/f = \sqrt{(c-v)/(c+v)}$, so flipping the +/- signs to get the equation for approach blueshifts gives $f'/f = \sqrt{(c+v)/(c-v)}$, and since the second prediction is the *inverse* of the first, it's guaranteed that when we multiply *two* of these consecutive shift stages together, with equal and opposite velocities, they'll cancel out perfectly.

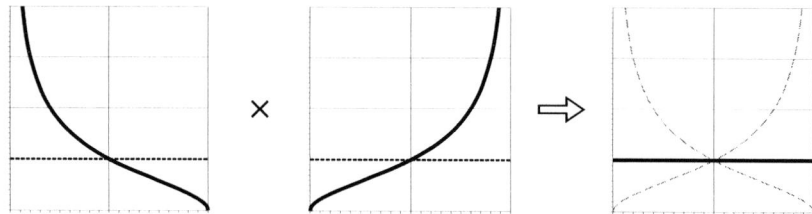

Figure 15-4: Special Relativity: Combining two equal and opposite velocity-shifts produces perfect cancellation

NM calculation

Under Newtonian theory we wouldn't have gotten this cancellation. With the "NM" Doppler relationships, the recession redshift would be $f'/f = (c-v)/c$, the approach blueshift would be $f'/f = (c+v)/c$, and if we multiply *these* two together, we end up with a round-trip redshift prediction, of $f'/f = 1 - v^2/c^2$. It's another "Lorentz-squared" relationship.

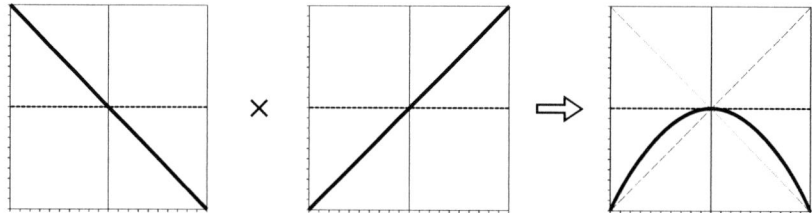

Figure 15-5: Newtonian relationships: Combining two equal and opposite velocity-shifts produces a Lorentz-squared redshift

The NM result is ... odd. With the old NM equations we predict a net energy loss, almost as if there is a time dilation effect in play that depends on the kinetic energy bound up inside the system, whereas with SR, which is more obviously *associated* with the idea of time dilation, there's no round-trip effect. The SR description maintains perfect flat lightbeam geometry, while the NM version doesn't work unless the relative velocity between bead and box is somehow distorting the region's geometry.

Since NM's velocity-dependent distortion result is more complicated, it's tempting to assume that the SR result "makes sense" and the NM one doesn't. If the NM account *was* right, we'd expect there to be a overall **thermal redshift** in signals sent through a material in which atoms were jiggling around – the greater the temperature, the greater the speed of the jiggling particles, and the greater the redshift. And this wasn't supposed to happen. So it was a little awkward that in the 1960s, when we were trying to carry out a new generation of ultra-sensitive gravity-shift tests, we found that our sensors were reporting shifts that were redder than expected, and decided that thermal redshifts were real.

15.8: ... Thermal redshifts

When we try to verify the Doppler predictions of special relativity by making very sensitive measurements of the frequency-shift on moving material, these tests have a nasty habit of reporting results that are "redder" than the SR predictions.

We often blame this on **recoil effects**: when an atom spits out a quantum of light, the associated reaction forces shove the atom back in the opposite direction, and since this recoil means that the atom is *backing away* from the light, the signal might show a certain amount of recession redshift. We can also reason that if the absorption site on a detector *also* recoils when it's hit by a photon, then perhaps the sensor's own recoil might cause it to report some additional excess redshift, too.

For experimenters trying to carry out precise frequency-shift measurements, this is annoying.

Rudolf Mössbauer's answer to the "recoil" problem was to use a material that not only had a very narrow "spike" in its absorption/emission spectrum, but which was also available in crystalline form (the usual material of choice being an **isotope** of iron, Fe^{57}). If individual atoms were held in a strong crystalline lattice, the recoil forces on one atom would be spread through the lattice, and the atom absorbing or emitting the signal wouldn't move nearly so much, and could be considered to be effectively "recoilless". Mössbauer's method of making "**recoil-free**" measurements caused a lot of interest, and the additional resolution provided by the **Mössbauer effect** encouraged researchers to try to measure very small frequency-changes that wouldn't otherwise have been practical, notably the 1959/1965 efforts by **Robert V. Pound, J.L. Snider** and **G.A. Rebka Jr.** to measure the gravitational shift across a university building only a few stories high.

the "SOD" effect

Some of the nuisances encountered in the Pound-Rebka-Snider tests are recounted in (Pound 2000). One of the more persistent problems was that even with the aid of the Mössbauer effect the experiment *still* insisted, inexplicably, on producing more of a redshift than the experimenters had expected to find. Pound attributed the extra redshift to thermal effects within the Fe^{57} material, and gave the new effect that was messing up his experiment the unfortunate (but perhaps appropriate) name, the "Second-Order Doppler" (or "**SOD**") effect.

The derivation of the SOD effect made some people uneasy. The "centrifuge redshift" reasoning of section 12.8 and 15.5 suggested that an oscillating, *accelerated* atom should age more slowly due to its acceleration, and if we then chose to apply the SR **clock hypothesis** we could reinterpret this physical redshift as a "speed" effect that was nothing to do with acceleration ... but our "bead-in-a-box" example, with its *unambiguously* constant pathlength and *genuine* lack of acceleration, gave a result of *no* SR round-trip redshift. It looked a little as if special relativity gave us a choice of *two* possible predictions, and allowed us to pick the one that we liked, retrospectively, to match the data.

Until this effect was discovered, no-one using SR seemed to have been *expecting* it to happen, and there didn't seem to be a record of anybody deriving the result from the SR relationships until *after* experiment had demonstrated that the effect was real.

stellar thermal redshifts: gravitation-equivalence?

The accidental (and late) discovery of the thermal redshift effect in the 1960's was also awkward for gravitational theory: we could argue from general principles that since energy has mass (section 2.7) the energy tied up in the thermal motion of a star's atmosphere should contribute to its gravitation according to the $E=mc^2$ law, and should contribute to its gravitational redshift ... but in the SOD example we were claiming that the energy tied up in oscillating atoms produced a redshift that was nothing to do with curvature.

15.9: ... Cosmological redshifts

These round-trip redshift arguments also seem to apply to *gravitational* shifts: if we send a lightsignal on a round-trip journey through a gravitational well, then, if we say that the gravitational field generates the same terminal velocity value for the uphill and downhill journeys, we can use Einstein's "falling clocks" argument (section 3.7) to say that the accompanying redshift and blueshift should be the same as the *velocity-shifts* that we'd see in objects that had followed a ballistic trajectory across the gravity-gradient (we'll meet some gravitomagnetic complications further down the page, but we'll start with the "simple" version of the "falling clocks" problem and see where it gets us).

> **With special relativity's "relativistic" Doppler equations** (without a retrofitted gravitomagnetic component), we find that when we calculate the frequency-shifts for the uphill and downhill journeys, they cancel each other out perfectly: light emerges at the far side of the gravity-well with the same energy that it started with.

> **With the Newtonian calculation** we don't get this tidy cancellation, and instead find ourselves with an overall Lorentz-squared redshift over a round trip.

These two results produce two different types of physics, with the "non-SR" description giving us a few additional effects:

1: energy loss from an oscillating system

With the NM relationships, an object oscillating back and forth across a gravitational cleft (see also "boomeranging", section 19.2) should seem to be constantly losing energy.

Where does the energy go? Well, current gravitational theory agrees that the oscillating mass should be radiating at least *some* energy, since it'll be throwing off gravitational waves ... so perhaps *some* degree of energy-loss isn't a totally outrageous idea.

2: spacetime route-dependence of gravitational shifts

A second odd thing happens with our dropped flask in section 2.3: if a portion of the light bouncing about inside the flask is bouncing *vertically*, then it's repeatedly moving up and down across a tiny piece of gravitational gradient inside the flask during the fall, and by the time the flask reaches the ground, the light hasn't just covered the distance that the flask has travelled, but also an additional round-trip distance. If a round-trip journey across a gradient involves a redshift, then the energy-change of the light inside the falling flask will depend not just on the initial and final locations of the flask *in space*, but also on the amount of time that the flask takes to move between them – the change in energy of the falling flask doesn't just depend on its route *through space*, but its route through space*time*.

The idea that the gravitational difference between two fixed locations depends on velocity sounds odd at first, and *seems* to undermine strict energy-conservation, but we came to a similar conclusion in section 9.4 when we found that the initial relative velocity of two objects seems to change "effective" gravitational potentials ... this led us to **gravitomagnetism** and velocity-dependent frame-dragging effects, which are "proper" outcomes in a full theory of relativity.

So while the route-*in*dependence of the SR mathematics might initially seem to be the superior result, it may be the wrong answer. We could try to correct the problem by deliberately *retrofitting* "Lorentzlike" gravitomagnetic effects to the description, but if we do this, its not obvious that our calculations are still using special relativity – if we were to assume that SR was "mathematically" correct but that gravitomagnetic effects produced an additional Lorentz redshift in problems involving *actual* matter, then what we'd actually have done would be to have recreated the earlier relationships of Newtonian mechanics by other means.

3: cumulative gravitational redshift (=cosmological redshift)

Similar effects appear in NM-based physics when we consider the case of a lightbeam crossing the universe. In the SR example the lightbeam underwent no overall energy-change and we could say that the universe was **quasi-Euclidean**: having lots of gravitational bumps and valleys at smaller scales that cancel out over larger scales to produce a shape with no large-scale curvature.

Figure 15-6: Cardboard eggtray: An example of quasi-Euclidean geometry

This sort of universe was favoured by Einstein when he originally constructed his general theory, and because one might expect gravitational attraction to add up cumulatively over large scales, Einstein invented a cancelling long-distance antigravitational effect to restore nice tidy Euclidean geometry at large scales: this cancelling antigravitational effect being the now-infamous "**Cosmological Constant**".

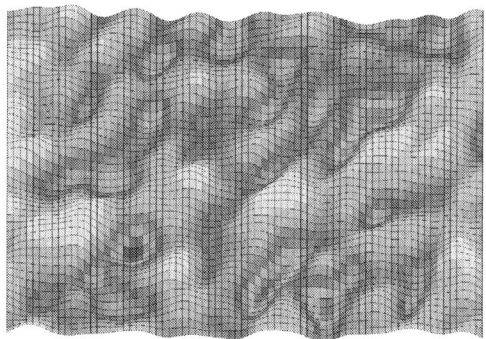

Figure 15-7: Large scale structure in a quasi-Euclidean universe

But these arguments for a static, Euclidean, constant-energy universe don't seem to apply in an NM universe, (at least, not in the original GR1915 sense). With gravity obeying the NM energy laws, our lightbeam progressively loses energy as it rides across the gravitational hills and valleys, and in a lumpy universe uniformly filled with stars and pockets of gas, the average amount of final redshift will be a function of distance travelled. If this distance-dependent redshift is then interpreted *gravitationally*, we have *distance-dependent curvature*, with the simplest geometry over large scales the being one in which the overall surface eventually curves right around on itself to form a hypersphere. If we then draw in lines at right angles to the hypersphere to represent worldlines of individual particles, these worldlines tend to splay apart by an amount that increases with time, meaning that for inhabitants of this universe, the distances between worldlines should on average be increasing or decreasing ... and since we know that this redshift should *increase* with time, this means that the arrows must be pointing *outwards* ... The NM equations suggest that the universe should be expanding.

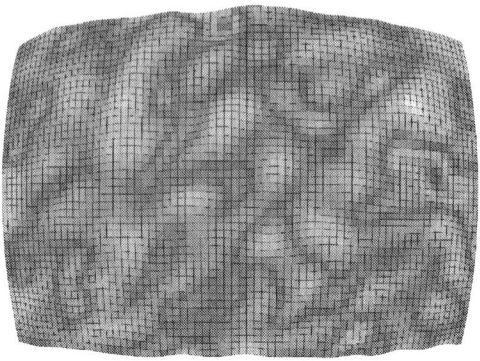

Figure 15-8: Large-scale structure without a cosmological constant

Hindsight is a funny thing. A century ago, a prediction of cosmological expansion might have been seen as a good reason to "bin" a model, but nowadays, predictions of cosmological redshifts and an expanding universe aren't so much "bugs" as "features", because this has become the "modern" description of how our universe is supposed to behave. It turns out that the untidiness of the NM equations, which might have seemed in the early Twentieth Century as a very good reason for getting rid of them in favour of the neater SR set, appear from the early *Twenty-First* Century to produce some very desirable behaviour. **Chaos** and **complexity** resulting from a model used to be considered as "bad design" but nowadays they seem essential tools for generating the sophistication of the universe that we inhabit.

15.10: Round-trip effects in general

When it comes to thermal redshifts and cumulative redshift effects, the track record of SR-based theory for predicting these things (in *advance*) doesn't seem to be particularly inspiring. Two effects that *should* have been predicted if we were using the NM relationships or more general statistical laws – thermal redshifts and cosmological expansion – were missed, apparently because they didn't correspond to the mathematical tidiness that we had with theory based on special relativity. These physical effects weren't "predicted" by the mainstream until *after* we'd accidentally stumbled across experimental evidence indicating that they were physically real.

If the usefulness of scientific theory to society is partly due to its ability to impart **foreknowledge** – to give us the ability to anticipate and predict and prepare for things *before* they happen (and to be able to uncover "new" effects more cost-effectively than by building giant multi-billion dollar accelerators in the hope that something "unexpected" turns up to surprise us), then we could suggest that perhaps special relativity and GR1915 mightn't be doing a very good job. Perhaps these subjects are just inherently "difficult" and naturally give our theorists trouble, but a more sceptical view might be that it would have been easier to predict these results in advance if we'd never heard of special relativity, and were still using the "redder" relationships of Newtonian mechanics.

... but before we assign too much significance to our tentative guesses as to what might *really* be going on in our experiments, we need to take a longer look at the historical experimental track record of special relativity, and we'll do this next ...

16
Experimental Evidence for Special Relativity

> *"Special relativity: Without a shadow of a doubt"*
> — title of the appendix of "Was Einstein Right?", Clifford M. Will (1986)

> *"You get used to one thing ..."*
> "Pleasantville", Gary Ross (1998)

> *"unlearn"*
> Picasso

> *"Common sense is the collection of prejudices acquired by age eighteen."*
> Albert Einstein

> *"This is the West, sir. When the legend becomes fact, print the legend."*
> "The Man Who Shot Liberty Valance", John Ford (1962)

> *"We conclude that free-float motion does not affect the structure or operation of clocks (or rods). If this is what you mean by reality, then there are **really** no such changes due to uniform motion"*
> "Spacetime Physics", Taylor and Wheeler, Box 3.4 ("Does a moving clock really 'run slow'?"),
> (the "realness" of special relativity's predicted time-dilation effects on inertial clocks depends on how we choose to define "real")

CERN

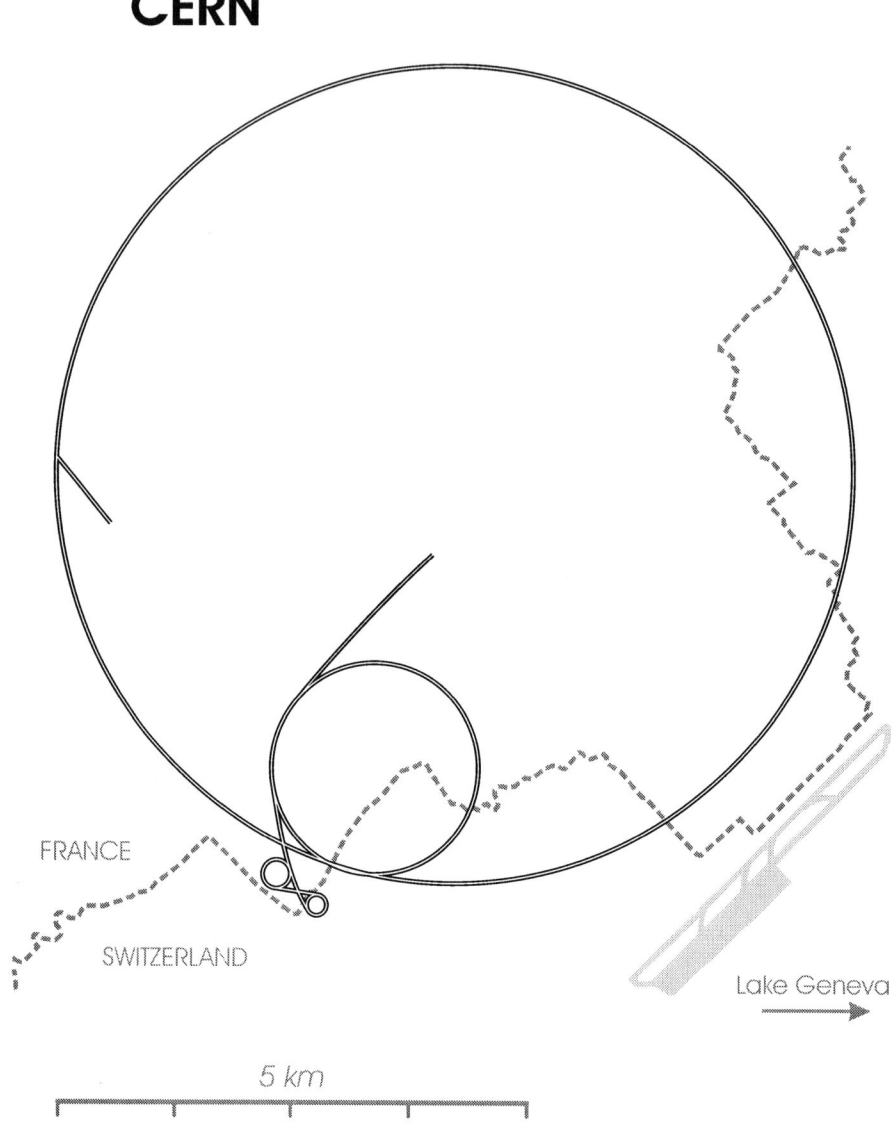

The CERN particle accelerator complex (originally
"Conseil Européen pour la Recherche Nucléaire") straddles
the border between France and Switzerland.
Its largest underground ring, the Large Hadron
Collider (LHC) is over 8 km in diameter.

16.1: Commonly-cited evidence for special relativity

We're told that the experimental evidence for special relativity is so strong as to be beyond reasonable doubt: are we *really*, *seriously* suggesting that *all* this evidence could be wrong?

Experimental results reckoned to support the special theory include:

- **E=mc²**
- **transverse redshifts**
- **longitudinal Doppler relationships**
- **the lightspeed limit in particle accelerators**
- **the searchlight effect (shared with dragged-light models and NM)**
- **"velocity addition" behaviour (shared with dragged-light models and NM)**
- **particle tracklengths**
- **muon detection**
- **particle lifetimes in accelerator storage rings / centrifuge time dilation / orbiting clocks**
- **the failure of competing theories**

… we'll be looking at all of these, along with a couple of important background issues.

16.2: … E=mc²

For a long time it seemed to be received wisdom that the **E=mc²** result was unique to special relativity. We were told that if special relativity wasn't true then nuclear bombs and nuclear weapons wouldn't work, and without SR's prediction of **E=mc²**, nuclear fusion wouldn't operate as it does. Without special relativity, the Sun wouldn't shine.

And while this was a good story to tell credulous schoolchildren, it was essentially **pseudoscience**. The idea that *E=mc²* "belongs" to SR doesn't hold up to basic mathematical analysis, and to Einstein's credit he went on to argue for the wider validity of the result by publishing further papers that derived the relationship (or a good approximation of it) from more general arguments outside special relativity. We also found in section 2.5 (with working supplied in the Appendices, Calculations 2), that **E=*mc*²** is an *exact* result of NM, if we ignore standard teaching and go directly to the core mathematics. Not only is the NM-based derivation of **E=mc²** reasonably straightforward, it's shorter than its SR counterpart, and it's also part of every hypothetical model in section 13.

While it's historically understandable that the equation wasn't widely recognised and embraced until Einstein came along, it's less clear why so many brilliant physicists with outstanding math skills continued to insist for so long that the equation somehow provides compelling evidence for the special theory. Since the math is so straightforward, how were so many clever physics people caught out? We might have expected that enough time had passed since 1905 for us to have checked the math dependencies, noticed the parallel compatibility with NM, and (in a respectable field of scientific study), made a high-profile retraction so that we didn't continue to pass misinformation onto students. But perhaps "**E=*mc*²** proves special relativity" was just too convenient a tale for people to want to give it up, regardless of what the mathematics really said.

RELATIVITY IN CURVED SPACETIME

16.3: "Classical Theory" vs. Special Relativity

When we read about experiments that compared the predictions of SR against those of "Classical Theory", we can away thinking that we've been told how SR's predictions stack up against most *earlier* theories (for instance, Newtonian theory).

This isn't usually the case. When we look at what's *meant* by "Classical Theory" in this context, we find that it's a sort of hybrid. It's a pairing of two sets of incompatible assumptions and math that have the advantage for experimenters of **(a)** being well known and standardised, and **(b)** making optical predictions that are so *exceptionally* bad that by comparison special relativity (and almost any other theory) looks very good indeed.

did "Classical Theory" ever really exist?

In the context of SR-testing, "Classical Theory" refers to a mixture of two sets of conflicting assumptions that didn't work together before SR/LET: "Classical Theory" uses Newtonian mechanics for the equations of motion for solid bodies, but for light, CT is equivalent to assuming an absolute, fixed, "flat" aether stationary in the laboratory frame. The energy and momentum relationships of these two different parts are, of course, irreconcilable ... NM requires the Doppler relationship to be $(c-v)/c$, but "Classical Theory" gives $c/(c+v)$. These aren't compatible. They never were. If they were, we wouldn't have needed special relativity.

There doesn't seem to be any *single* theory that attempted to combine these two predictions before LET/SR, or at least, there doesn't seem to have been anyone prepared to lend their name to one, and in a subject where people love having things named after themselves, this should make us suspicious. If "Classical Theory" doesn't mean "pre-SR theory", then where did it come from? The phrase appears in Einstein's explanations of the basis of special relativity, as a convenient form of words to refer to two apparently diverging predictions that special relativity then reconciled by applying Lorentz effects: to Einstein, "Classical Theory" represented incompatible aspects of earlier theories that didn't work together, but that could be reconciled using special relativity.

When we're look for a historical counterpart to Classical Theory there doesn't seem to be anything that would have made these optical predictions unless we go all the way back to pre-Galileo, pre-Newton times, and posit an absolute aether that permeates space and is locked to the state of a stationary Earth. That would give us the "Classical Theory" prediction of "no transverse redshift" for a laboratory stationary with respect to the Earth. But every other decrepit old theory that we can dig up seems to predict at least *some* sort of transverse redshift effect, sometimes weaker than SR, sometimes stronger than SR, and sometimes swinging wildly between the two depending on the Earth's motion. The one idea that *didn't* seem to be considered to be credible during the Eighteenth Century was the idea that lightspeed was fixed with respect to the observer, which is presumably why Michelson had so much grief with his colleagues over his "failed" aether-drift experiment.

SO, why do we persist in carrying out these "SR vs. Classical Theory" comparisons if they don't demonstrate very much? Well, to a cynic, Classical Theory is an excellent reference to test against, because its predictions are about as bad as we can get. If we set aside the theories that predicted time-variant effects, no other old predictions seem to be *quite* as bad as CT when it comes to predicting real Doppler shifts, and this makes "CT vs. Theory X" experiments very much easier to carry out and analyse. **Test theory** authors love CT because it meshes well with the chain of arguments that Einstein used when explaining the special theory, and experimenters design tests around the test theories that are available. It's a legitimate process – as long as we don't fool ourselves into thinking that that the results represent a realistic comparison of how special relativity's predictions *really* compared to those of its predecessors.

16.4: ... "Transverse" redshifts

Special relativity tells us that if an object moves through our laboratory, and we carefully point a highly-directional detector at right angles to its path (measured with a "laboratory" set-square), the signal that manages to register on the detector should be redshifted (section 6.7).

But the popular "educational" notion that this sort of redshift outcome is something unique to special relativity is as best misleading, and at worst ... it's simply wrong. The equations of Newtonian mechanics (or even the basic equations for audio, properly applied to the case of a stationary source) don't just predict redshifts in this situation, they'll often predict "**aberration redshifts**" that are *stronger* than their SR counterparts (section 6.4), so in a *physical* sense, the appearance of redshifts in this situation isn't just *not unique*, it's not even particularly *unusual*. In fact, the thing that *would* be unusual with this sort of experimental setup would be a theory that *didn't* predict some sort of redshift.

Although we tend to regard special relativity's transverse predictions as conceptually unique, experimenters have to know when supposed differences between theories generate *physically unambiguous* differences in the readings taken by actual hardware, and when the differences are more a matter of interpretation. This distinction isn't always obvious from the relativity literature.

Einstein's special theory requires these sorts of "pre-SR" redshifts to exist for its own internal consistency. The theory must predict the same physical outcome regardless of which inertial reference frame we choose to use for our calculations, so the *emitter* is entitled to claim that c is globally fixed for *them* (Einstein 1905, §7), and this means that they're entitled to claim that our relative motion makes *us* time-dilated, giving our view of the emitter's signal a *Lorentz blueshift* ... so in order for us to be able to instead see a Lorentz *red*shift, propagation-based effects in this situation – light moving at a constant speed in the emitter's frame, and arriving at us at an apparent 90 degrees – must, by default, generate a *Lorentz-squared redshift* to allow the same final SR outcome. This is the right answer (see *Calculations 3*).

So to fully understand the logical consistency of SR in this situation requires us to know that similar or stronger redshifts would appear in the same apparatus under other light-propagation models. Since different SR "views" can explain the same redshift component as the result of **(a)** conventional aberration effects, **(b)** time dilation, or **(c)** a combination of the two (we're allowed to try an infinite number of alternative views from intermediate reference frames), SR requires these two explanations to be *qualitatively indistinguishable*. Although expert sources may tell us that "transverse redshifts" are unique to SR, *the theory itself* tells us otherwise. We can distinguish SR's "transverse" predictions from those of other theories by their strength, but a redshift outcome in this situation doesn't automatically need SR.

the Hasselkamp test

We only seem to have one experiment that set out to measure the amount of redshift actually *seen* at 90 degrees to moving material (Hasselkamp *et. al.*, 1979), and it reported about twice the redshift predicted by SR, as we'd expect if the older NM equations were right. This result was nevertheless presented as supporting SR: the experimenters used a test theory that compared SR with "Classical Theory" (which predicted *no* redshift), and reasoned that the inexplicable excess redshift must have been due to an accidental detector misalignment. They were then able to use statistics to argue that, taking into account possible alignment errors, the "SR" prediction still made a significantly better match to the data than "CT" did.

But subsequent papers verifying that the presumed misalignment was *real*, or repeating the experiment (perhaps with the help of clever cancellation methods to eliminate the effects of these sorts of detector misalignments from further results), don't seem to have appeared. This makes it difficult to tell whether the result really supported the special theory, or invalidated it.

16.5: ... "Longitudinal" Doppler shifts

The Hasselkamp experiment was unusual – in practice, we don't normally try to measure SR's transverse redshift effect by *really* aiming a detector at the side of a moving particle beam – we find it easier to measure the forward and rearward Doppler-shifts, and then *calculate* the strength of the transverse effect by comparing them against each other.

This is a nice method ... because it compares *two* shifts, the technique makes it easier to cancel out various types of systemic error, known and unknown, and these "end-on" readings are less sensitive to the effect of small angular errors. By comparing the resulting three signals ("recession-redshifted", "approach-blueshifted", and an "unshifted" reference signal), we can derive a characteristic "signature" that lets us rule out certain relationships without having to commit to a theory-specific value for the exact velocity of the particle beam. We can select a theory, use *one* of the shift ratios to calculate what the velocity would have to have been according to that theory, use this hypothetical velocity value to "predict" the second shift ratio, and then compare this against the second set of figures to see how close we got to the real data.

Ives-Stilwell

The best-known of these "non-transverse" transverse tests is the early 1938 test by **Herbert Ives** and **G. R. Stilwell**, which set out to compare the predictions of Lorentz Ether Theory (and SR) against those of "Classical Theory". Ives and Stilwell's approach was simple: "Classical Theory" says that the two shifted signals (red and blue) should change in wavelength by precisely the same amount, so with all three wavelength values marked on a linear scale, we'd find *perfectly* even spacing between them. If the shift relationships obeyed the "redder" relationships of SR (or NM) there'd be an asymmetry.

Ives and Stilwell found a definite offset in the wavelength values. The simplicity of this experiment makes it tempting to reanalyse the data for a possible agreement or disagreement with NM, and when we do this we find that the stronger offset predicted by NM appears to lie *outside* the data range, by more than the experimenter's quoted experimental error. This seems to indicate that the SR predictions are significantly more accurate than NM.

further experiments

There've been several more experiments of this type published since Ives-Stilwell, using more advanced equipment, more complex optics and higher relative velocities, and these have supported the predictions of SR over "Classical Theory" with increasing confidence. However, when we try to use them to check how well they support SR over *NM*, we run into difficulties: with several of these tests, the more complex setup and calibration techniques make it dangerous to attempt a safe reanalysis for possibilities that weren't considered in the experimenters' setup procedures ... in others the quoted error margins seem rather similar to the margins that we'd need to be able to interpret an "NM" result as a "SR" result ... or extreme accuracy when making the comparison between SR and CT is achieved by a technique that makes it difficult to differentiate between SR and NM ... or "excess" redshifts are explained away as the result of mirror recoil .

It seems that even with this additional technological sophistication, our primary evidence for SR's superiority over NM is still that early Ives-Stilwell experiment. And since later experimenters have had trouble understanding how the test's accuracy could have been *quite* as good as the paper said (estimating accuracy can be difficult when using an experimental configuration for the first time), we don't yet seem to have a solid core of experimental results claiming that that the newer SR Doppler relationships really *are* more accurate than the NM set. Perhaps if our experiments had been devised with this comparison in mind from the beginning, we might by now have significant amounts of evidence to point us one way or the other ... but they weren't, and we don't.

16.6: ... The lightspeed upper limit in particle accelerators

Another of the results often trotted out as unambiguous evidence for the validity of special relativity is the fact that even our best particle accelerators can't persuade electrically charged particles to move faster than the background speed of light. As the speed of the particles approaches background lightspeed, it becomes progressively more difficult for the fixed accelerator coils to force them to move any faster. As the speed of a particle approaches background lightspeed, the energy that we have to pump through our coils to get an additional increase in speed seems to tend towards infinity.

Some commentators attach great significance to this result and argue that the outlandish scale and sheer brute force required by modern particle accelerators is an obvious indication that the special theory is correct. If we believed in the equations for light used by "Classical Theory" (section 16.3), we'd expect these machines to be able to accelerate particles to far higher speeds, but, in real life ... this quite clearly isn't the way that things work. Special relativity wins!

And certainly, special relativity wins when compared to CT. It just doesn't *necessarily* win when compared to other models. From the point of view of the *coils*, we can argue that the particle's resistance to acceleration (and its apparent inertial mass), goes to infinity as its speed through the accelerator approaches lightspeed, and we might blame this on the particle's additional **relativistic mass** at higher speeds. But the idea of relativistic mass isn't always fashionable amongst physicists, so it's handy to have another way of describing the situation, and we can do this by describing the experiment from the point of view of the particle.

coupling efficiency

Suppose that our "SR particle" is coasting through a straight section of accelerator tube at close to background lightspeed, and we throw more EM energy at it ... the particle sees the receding accelerator coils to be redshifted, reducing the frequency, energy, and radiation pressure of their signals. With the coils moving away *at* lightspeed, SR's Doppler relationships describe this energy and momentum of their fields disappearing altogether. So the **coupling efficiency** between the accelerator coils and the particle drops toward zero as their relative recession velocity approaches lightspeed, and with SR we therefore expect to be able to accelerate the particle *towards* the speed of light, but not *to* it or beyond it. This is what we see happening in our accelerators. SR wins!

... Except that, when we try a similar exercise with the Doppler relationships for other theories, similar things have a habit of happening. If we try the "Newtonian" Doppler relationships we find that with $f'/f = (c-v)/c$, setting the recession velocity to lightspeed once again gives a frequency (and energy, and coupling efficiency) of zero. When we *directly* accelerate a particle, the lightspeed limit that we usually think of as a validation of SR also shows up under Newtonian mechanics, and presumably also under a range of other theories.

indirect acceleration

This "direct acceleration" lightspeed barrier can have different characteristics under different models: in the NM version of the story, an unstable particle travelling at close to background lightspeed can fragment and throw off **daughter particles**, some of which might travel at *more* than background *c*. This effect is related to NM's support for classical indirect radiation effects ("semiclassical" **Hawking radiation**), and wouldn't seem to be possible under SR-based models. Unfortunately, when we start to deal with the more "particle-y" aspects of particle physics, quantum effects become relevant, allowing the appearance of particles in "impossible" situations to be explained away by ideas such as quantum tunnelling: even if we found something that looked like evidence of superluminal daughter particles, by classifying this as a quantum effect we could probably still get away with arguing that the result didn't threaten SR.

16.7: The "searchlight" effect

We met the **searchlight effect** in section 8.2: it's the tendency of moving bodies to throw more of their signal forwards rather than trailing it behind them. Special relativity and NM both apply the same "relativistic aberration" formula, and the effect also exists (to various degrees) in different dragged-light models.

This behaviour doesn't happen in the "Classical Theory" of section 16.3.

16.8: Velocity-addition

Special characteristics for "velocity addition" appear in a variety of models, including NM (section 14.8), and *usually* suggest that the propagation of signals is being affected by the motion of intermediate objects in the signal path. Although we usually choose to *interpret* the **Fizeau** and **Zeeman** results as supporting SR's **velocity-addition formula**, the special theory's match to the data isn't supposed to be any better than **Fresnel's** ancient dragged-light theory.

Again, this behaviour doesn't appear in the "Classical Theory" of section 16.3.

16.9: Particle tracklengths

Since we've brought up the subject of daughter particles, how do we test how fast they really go? Let's suppose that we have a particle that's only supposed to survive for a nanosecond, and we measure the length of straight-line distance that it covers between being created and blowing itself to bits. If we know the particle's "official" decay time, then surely we can measure the length of its track, and divide that by the time to get the speed? If this track length was *longer* than the distance that particle would travel at the background speed of light, wouldn't this mean that we'd shown that its velocity was **superluminal**, disproving SR? And if the particle tracks were always *shorter* than this, wouldn't this support special relativity?

But things aren't that easy. We're used to thinking of velocity as an unambiguous property, but since we can't be in two places at once, the property often has to be *interpreted*. Since special relativity redefines *all* of the properties associated with velocity – energy, momentum, distance and time – fair comparisons between SR and other theories can become quite convoluted, and this can make it difficult to tell, when we're using these agreed, *uninterpreted* quantities, whether there's *really* a physical difference between the SR and NM tracklength predictions.

Special relativity assigns greater energies and momenta to particles and signals than NM does, by a Lorentz factor:

	NM	SR
momentum	$p = mv$	$p = mv \times$ **gamma**
Doppler effect	$E'/E = (c-v)/c$	$E'/E = (c-v)/c \times$ **gamma**

, so ... for a high-energy particle moving along a straight line with constant speed, with a known energy and/or momentum, Newtonian theory and special relativity will be assigning *consistently different* velocity values to the same particle. The nominal "SR velocity" value ("vSR") will always be less than lightspeed, while the nominal "NM velocity" value ("vNM") will be larger than its SR counterpart by a Lorentz factor (calculated from vSR).

When we migrate from NM to special relativity, a particle's nominal velocity gets reduced by a Lorentz factor, *shortening* the distance that the particle would be expected to travel before decaying. But SR's "time dilation" effect then predicts an *extension* of the particle's lifetime by the *same* Lorentz factor thanks to time dilation, *lengthening* the particle's track by that same ratio. Because these two corrections exactly cancel, the particle's decay position as a function of its energy and momentum is precisely the same for both theories.

The results of both sets of calculations are necessarily identical.

16.10: Muon showers

Similar arguments apply when we try to assess evidence from "cosmic ray" detectors. High-energy cosmic rays hitting the upper parts of the Earth's atmosphere create showers of short-lived "daughter particles" that survive for an incredibly short amount of time before decaying – their lifetimes are *so* short that even if they were travelling at the speed of light, we might think that they still shouldn't be able to reach the Earth's surface before decaying.

But ground-based detectors *do* report the detection of muon showers, and there are two main ways that we can interpret this result:

SR-based interpretation

According to special relativity, we should explain the detectors' result by saying that since we "know" that nothing can travel faster than background lightspeed, the muons' ability to reach the ground shows that their decay-times must have been *extended*, and we interpret this as demonstrating that the special theory's time-dilation effects are physically real. We say that the muons move at a very high proportion of the speed of light and are time-dilated, and if it *wasn't* for this time-dilation effect, they wouldn't be able to reach the detectors.

Or ... we could adopt the muon's point of view, and suggest that the muon is stationary and the Earth is moving towards it at nearly the speed of light. In this second SR description, all of the approaching Earth's atmosphere is able to pass by the muon in time even though its speed is less than c, because the moving atmosphere's *depth* is *Lorentz-contracted*. These two different SR explanations (**length-contraction** and **time dilation**) are interchangeable.

NM-based interpretation

But is the success of the SR muon calculations *significant*? Is it *significantly different* to the calculations we'd have made using earlier theory? When we compare the tracklengths predicted by SR and NM, starting from theory-neutral properties, the final results seem to be identical (section 16.9): for a given agreed momentum, the muon's decay point according to SR would seem to be *precisely the same* as the NM prediction – the two models don't disagree on *where* the muon decays, they disagree as to whether it achieves that penetration by travelling at more or less than background lightspeed, which is more difficult to establish.

fast or ultrafast?

Muon bursts seem to be associated with **Cerenkov radiation** – the optical equivalent of a supersonic shockwave – but since lightspeed is slower in air than in a vacuum, using the **Cerenkov effect** to show that the muons are moving faster than *lightspeed in air* doesn't show that they're also moving faster than the official background speed of light, in a vacuum.

So how do we find the *real* speed of the muons, given that we don't have advance warning of when a cosmic ray is going to strike? With additional airborne muon detectors we can try to compare the detection times in the air and on the ground, but interpreting this data neutrally can be difficult: one such experiment seemed to indicate that the muons *were* travelling at more than c_{VACUUM} (Clay/Crouch 1974), but subsequent experiments seem to have supported the opposite position.

From here on, things get muddy. Given that we know that the record of SR-trained theorists trying to interpret non-SR theory isn't exactly faultless, it's difficult to know exactly how to treat this situation ... but there's one thing here that we *can* be sure of: When SR textbooks tell us that ground-level muon detection gives us unambiguous evidence for special relativity, and tell us that these muons *couldn't* reach the ground unless SR was correct, and *couldn't* have been predicted by earlier theories ... those statements are wrong.

16.11... Particle storage rings and centrifugal time dilation

If we apply strict SR protocols, we can't *in principle* isolate the SR time dilation effect in a particle moving at constant speed along a straight track, because if we could prove how much time dilation it *really* had, we'd have a measure of its true velocity compared to the speed of light – if other observers then had to accept our results, we'd have uniquely locked the speed of light to a *specific* inertial frame, which goes against SR's basic principles. So a constant-velocity "SR-compatible" result can be interpreted as including time dilation effects *or not*, depending on our chosen reference system. Part of special relativity's internal consistency is maintained by the theory's structure *preventing* us from being able to verify certain things.

particle storage rings

But as physicists, we really do *want* to be able to know how fast a particle is moving, and there's an unambiguous way of doing this: send the particle around a *circular* path, measure the length of that path in laboratory units, and then stand alongside the ring and use a local clock to measure how long it takes the particle to make a complete circuit. If we say that the circular track is one kilometre long, and the speed of light is just under ~300,000 km/s, then if the particle takes *more* than about ~1/300,000 of a second to make a lap, it *has* to be circling at less than background lightspeed. If the particle's usual decay time "at rest" is one *millionth* of a second, and we manage to send it all the way around the ring before it decays, then surely its ability to make a complete circuit is unambiguous proof that the particle is ageing more slowly, and surely this then *proves* the existence of SR's speed-based time dilation effect?

Unfortunately it doesn't. Although the particle now shows a *verifiable* time-dilation effect, changing to a *circular* track alters the physical parameters of the experiment. Our circling particle is now feeling a *physical acceleration*, and the reduction in its ageing rate can now be blamed on *acceleration* effects (section 12.8). By taking the geeforces as evidence of an apparent gravitational field, and calculating the amount of *gravitational* time dilation that should be associated with that field, we seem to get the right answers for the particle's time-dilation – so we can still predict the right result even if we've never heard of the special theory.

These simple particle accelerator-ring experiments don't prove that special relativity's time-dilation effect is real – they present us with a straight choice between interpreting the result as an acceleration effect or as a speed effect. We usually *choose* to accept SR's **clock hypothesis**, that acceleration has no effect at all ... but this choice isn't imposed by more general arguments, or by experimental data – we *choose* this interpretation in order to be able to continue using SR. Although the two types of calculation both work well in our circular particle-accelerator ring example, they don't necessarily agree in more complex situations.

Hafele-Keating, etc

In 1971, Joseph Hafele and Richard Keating performed a famous experiment in which they flew two sets of atomic clocks around the world, one set Eastward, and one set Westward. Seen from the frame of the background starfield, the "Eastward" set were circling *with* the Earth's rotation and had the greater speed (and acceleration), while the "Westward" set were moving *against* the Earth's rotation and circling more slowly. When both sets of clocks were brought back to their base, the set that had circled fastest seemed to have aged less.

But again, although this experiment (and more reliable repetitions of it) are touted as *proving* SR, it's another example of centrifuge time dilation, and the outcome should be calculable from more general gravitational and equivalence principles (as a consequence of the difference in accelerations) without using special relativity.

It seems to be that when the only explanation for time dilation is SR, then the effect can't be isolated: when it *can* be isolated, there's another, deeper explanation that makes SR redundant.

16.12: deSitter / Brecher disproof of simple emission theory

As we've already mentioned (sections 13.12 and section 13.6), the Newtonian equations of motion seem to become increasingly incompatible with flat spacetime at higher velocities. **Ballistic emission theory** had assumed that light was effectively "thrown" by a moving particle – the model was relativistic and used the NM relationships, but it didn't support the concept of local *c*-constancy and broke wave-theory rules. If light was emitted at $c_{EMITTER}$ and then *continued* at that speed without any sort of **gravitomagnetic** regulation, different signals could end up travelling along the same path at the same time with different speeds depending on the characteristics of their respective sources. As well as destroying the idea of an orderly light-metric, this gave some odd-looking side-effects:

1) The Doppler-*blue*shifted light generated in a star's atmosphere by atoms moving *towards* us would travel more quickly and reach us before redshifted light from other adjacent atoms moving away from us. If the star was also moving sideways across our field of view, we'd then expect to see it having a spread of positions and colours, as a rainbowed "streak" (which doesn't seem to happen).

2) If an object was "stationary" but being eclipsed, we might expect to see different colours disappear and reappear in sequence as the eclipse progressed (Newton seems to have made enquiries as to whether this effect had been seen – it hadn't).

3) For a double-star system, light emitted when a star was approaching would reach us more quickly than that given off earlier when the star was receding, and at sufficient distances, the star could seem to be occupying two drastically different parts of its orbit at the same moment.

The astronomer **Willem deSitter** realised that in this third case, the difference in signal flight-times between a star's "approaching" and "receding" images would increase linearly with distance until, at a sufficiently-great distance, the "fast" signals from the approaching star should be able to catch up with and overtake earlier signals thrown off by the same star when it was receding, as well as with other "slow" light emitted during earlier orbits. The signals would be "*all mixed up*" when we received them (Einstein: Shankland 1963), and our view of these distant circling stars would be hopelessly scrambled.

DeSitter's 1913 survey of known double-star systems didn't reveal these severe scrambling effects in any of the stars surveyed, and given the extreme distances of some of the stars and the extreme statistical unlikeness that *all* these stars might have their orbital planes facing us, deSitter concluded that either there was *no* (global) dependency between the speed of a signal and the speed of its source, or that any (fixed, proportional) dependency would have to be so *absurdly* small that the possibility really wasn't worth mentioning.

DeSitter's result is often said to prove that **the speed of light is independent of the source**, but we have to treat this statement carefully: showing that the speed of light isn't *wholly defined* by its original source isn't the same as saying that there are *no local dependencies* – double stars should throw off **gravitational waves** (Figure 4-10), so to prove the absence of *short-range* effects on signal flight times might be to disprove general relativity. The deSitter result indicates that lightspeeds aren't sensibly affected by the motion of their sources *over long distances*, but we should still expect short-range gravitomagnetic dependencies between the speed of a signal, the speed of the emitter, the speed of the detector, and the speeds of any other nearby objects that happened to be wandering through the region at the time.

We typically test the SR shift equations against the predictions of "Classical Theory" (section 16.3) rather than against NM. We justify this by quoting deSitter's result (replicated by **Kenneth Brecher** in 1977) and saying that since this rules out emission theory (the "historical" implementation of the NM relationships), and since we "know" that spacetime is flat (?), the NM shift relationships are therefore "known" to be unworkable and don't require testing.

16.13: "Domain of applicability" issues

We're told that special relativity produces good results, as long as it's used within its **domain of applicability**. It isn't compelled to produce good predictions when used outside this proper range. This sounds entirely reasonable.

But *what is* this proper range? How do we calculate it? It turns out that the correct range is determined *pragmatically* – it's the range of situations in which SR is already known (or thought) to produce good results. It's essentially an "engineering" definition, reducing our earlier grand statement to something more like: "Special relativity is known to work very well in situations where it is known to work very well. It does *not* have to work very well in situations where it is known *not* to work very well".

Knowing where these limits are tells where SR is a *useful* theory for engineering purposes, but their flexibility makes it more difficult for us to judge whether the theory has deeper validity. Flexible, *retrospectively-defined* domains encourage selectivity in how we evaluate evidence – cases of a good agreement between SR and the available data are taken as vindicating the theory, while disagreements are treated, not as evidence *against* the theory, but as showing that it's been applied inappropriately. This approach is great for engineering but rotten for experimental testing, because makes it more difficult to class a theory as **falsifiable**. We can end up insisting that the theory is "doing a damn fine job" within its (moveable) domain, and forgetting that the domain has been *preselected* as the range that *produces* that result. We can find ourselves saying that when an experiment disagrees with SR it's not the fault of the *theory*, but the fault of the *experiment* (or the experimenter).

SR vs. particulate matter: can a particle be an SR-style observer ?

The special theory tells us that if we have an array of observers in the same inertial frame (= all stationary with respect to each other) *they* should be able to claim that the speed of signals passing between them is fixed in *their* frame, and if *we* were to watch this array passing by, *we* should be able to explain the same results by declaring that the speed of the same signals is fixed in *our* frame. Reconciling these two views then gives us special relativity.

But this doesn't seem to be how things happen in real life. Suppose that our array of observers is real, and that the "observers" are water molecules. When we see these particles passing us, Fizeau's result (section 9.9) tells us that we should see light in the region to have a speed that's locally *offset* by the array's motion – so special relativity's reworking of inertial physics was derived to "explain" a counterintuitive result that doesn't quite agree with what we actually see – it explains how "the same signal has the same speed for all observers", even though what we *actually* detect is the signal being partially dragged along by the motion of the water.

A specialist can respond, well, *of course* SR's assumptions don't apply to light passed between water molecules, because light moving through water isn't the same as light moving through a vacuum, and SR doesn't claim to be valid for particulate media. But the *water molecule* could argue that its companion molecules are perfectly legitimate observers, and that the region between these molecule-observers *is* a perfect vacuum. Where do we draw the line between Einstein's arrays of rods and clocks exchanging synchronisation signals, which are supposed to obey the rules of special relativity, and arrays of *real* particles exchanging signals that are *not* supposed to obey the laws of special relativity? When does a group of particles count as a group of legitimate SR observers, and when does it count as a particulate medium? If our individual water molecules (in what is otherwise a vacuum) *do* count as legitimate SR observers then we have trouble explaining Fizeau's lightspeed offsets, and if they don't, then we have to ask, if a simple moving water molecule is too complex an "observer" to be correctly described by SR, whether it's valid to take the theory's predictions seriously for more complex compound objects such as spaceships and astronauts.

16: Experimental Evidence for SR

Textbooks sometimes explain the lightspeed offset in a moving particulate medium by invoking the **extinction theorem** – we're told that over an **extinction distance**, an incoming wavefront is absorbed (or "extinguished") by particles and is replaced with a new wavefront moving through the medium at a fixed speed with respect to that moving medium. This description presumably *works* (at least *reasonably* well, or it wouldn't be in the books) but it seems to be at odds with the story told in by SR that we "know" that lightspeed *isn't* affected by the movement of bodies. We might be told that this non-SR behaviour happens "because the particles act as transponders", but Einstein's hypothetical arrays of observers exchanging synchronisation signals *also* function as transponders. We seem to "know" different, mutually-incompatible things depending on the branch of physics that we happen to be studying.

dragging and kinetic energy – when is curvature "negligible"?

Concentrated energy warps spacetime, and a particle travelling at a "significant" fraction of the speed of light represents a "significant" concentration of kinetic energy, so we might expect high-energy particles to affect the geometry of their environment, and to warp a region's lightbeam geometry more strongly the faster they move through it. This would seem to be the logical extrapolation of Fizeau's experimental result (section 9.9). But if we accept this idea, special relativity's assumption of flat spacetime (and its resulting mathematical predictions) become progressively less reliable as the relative speed between particles approaches the background speed of light. To relegate SR to the status of a theory that holds for *less extreme* velocities would be unfortunate, since at lower velocities the theory isn't so easily distinguishable from Newtonian theory, and if we accept these dragging effects as important, then as the special theory's *theoretical significance* increases, its *theoretical validity* reduces.

How seriously should we treat these curvature effects for particles with ultra-relativistic speeds? Some particle physicists will tell us that SR is *entirely* capable of dealing with high-energy particles, that "curvature" corrections are unnecessary, and that the operation of our larger particle accelerators gives us ample proof of this ... but we're also told that in the next generation of particle accelerator tests at LHC Geneva, energy-densities are expected to be so high that they'll be creating microscopic black holes (which should then evaporate almost immediately thanks to the **Hawking radiation** process). Since black holes are *the most extreme examples* of spacetime curvature in our physics vocabulary, it would be odd to say that we know that experiments of this sort don't involve significant spacetime curvature.

Pressed further, particle-accelerator people may backtrack slightly and say that it's not so much SR that has the perfect track record in particle accelerator physics as **Quantum Electrodynamics**, or "**QED**"), which is a combination of SR and quantum mechanics. But since we know that "quantum" corrections can sometimes be used to *mimic* the effects of "**acoustic metric**"-style curvature, we might interpret the success of QED in different ways: it might suggest to us that since QED uses SR, this counts as a success for special relativity ... or it might suggest to us that some of QED's corrections may be inadvertently *recreating* the statistical results of the sort of velocity-dependent curvature effects that are missing from the SR description, and which we've been insisting don't happen.

We could also argue that if general relativity says that *on principle* energy densities are associated with curvature, and that *on principle* inertial mass is equivalent to gravitational mass, that a "general" theory *shouldn't* cleanly reduce to a theory that allows inertial mass to exist in the absence of gravitational effects, or allows arbitrarily high energy densities in the absence of any form of associated curvature. While special relativity is sometimes described as the limit to general relativity at which gravity is "switched off", perhaps a "complete" general theory of relativity wouldn't have such a limit, or would only have a *null* limit – in a general theory of relativity, the natural "medium" for gravitational effects is spacetime itself, so if we were to "switch gravity off", and remove *all* gravitational field-effects, a full theory arguably shouldn't turn into the physics of special relativity, but should disappear entirely.

16.14: Conclusions

Although we're told that the evidence for special relativity is beyond dispute, much of the supporting evidence and argument is *individually* so patchy that it wouldn't be taken seriously in other branches of physical science. Or at least, we should *hope* that this lack of sceptical scrutiny is unusual, because otherwise science in general would seem to be in a great deal of trouble. Almost every general argument for SR seems to have been missold in some way.

The $E=mc^2$ relationship *wasn't* unique to SR after all, neither were transverse redshifts, and the centrifuge redshifts that we'd been told had no other explanation *had* been predicted from more general gravitational arguments independently of SR. Although the experimenters may well have been scrupulously honest, some of the special theory's more active proponents seemed to be badly misrepresenting the available evidence and the mathematics, and their colleagues seemed to be allowing them to get away with it.

Since most of these mistakes can be found with a little basic critical analysis, this leaves us wondering whether the theory's proponents *genuinely* didn't realise that what they were saying was wrong or misleading (in which case the standard of cross-theory expertise is low), or whether they *knew* that evidence was being misrepresented, but chose not to raise the issue. Perhaps people thought that it wasn't so important if a few of these experiments were over-sold, because of the sheer breadth of other supporting evidence ... and that even if the SR-dependency of a few results had been hyped, that the exaggeration was harmless because *mathematics* told us that the theory was right ... but once a "casual" approach to scientific evidence is allowed to become widespread in a research subject, and once *everybody* starts to rely on the idea that the standards of evidence in individual cases don't matter so much, it allows the awful possibility that perhaps *every* piece of evidence used to support the theory might be similarly flawed. Mistakes will tend to cancel each other out in a diverse population, but in a **monoculture** they'll tend to reinforce one another. If *everyone believes* that the number of experiments provides a solid safety margin for their own work, and if *everyone depends* on the existence of that assumed safety margin, then it might be that the margin doesn't exist.

The experimental record may make a decent case for the *principle of relativity* being correct and also gives us strong evidence against a number of nonrelativistic models and against simple emission theory ... but when it comes to establishing whether SR is the *correct implementation* of the principle of relativity, things are less straightforward. If we believe that any relativistic model must reduce to SR *by definition*, we'll tend not to bother testing SR against other potential relativistic solutions, because we won't believe that they can exist.

The misrepresentation of the evidence for SR means that we're entitled to be suspicious, but it doesn't mean that special relativity's relationships are necessarily *wrong*. Definitive tests of "SR vs. NM" would seem to require direct tests of the Doppler relationships themselves, and in this case we seem to have two basic experiments, both slightly problematic – one apparently favouring SR against NM (Ives-Stilwell) and one apparently favouring NM against SR (Hasselkamp *et.al.*). If the "NM" Doppler relationships are correct, it seems incredible that we wouldn't have already noticed it, but if the SR set are really better, it *also* seems incredible that after a century of testing, we wouldn't yet have a body of results claiming to demonstrate it. It's hard to find *any* SR tests where experimenters claim to have compared the NM Doppler relationships against the SR set, and found the SR version better – it's just not something that people tend to do. If the SR set really *is* better, then the community really ought to have been able to find people able to verify it by now. A century should have been sufficient time.

Which of these relationships is better than the other at describing the universe we live in?

The honest answer seems to be: we still don't know.

Flip a coin.

PART VI

FUTURE PHYSICS

cosmology, wormholes and warp drives

RELATIVITY IN CURVED SPACETIME

17
Cosmologies

> " – Roll on, YE STARS! exult in youthful prime,
> Mark with bright curves the printless steps of Time;
> Near and more near your beamy cars approach,
> And lessening orbs on lessening orbs encroach; –
> Flowers of the sky! ye too to age must yield,
> Frail as your silken sisters of the field!
> Star after star from Heaven's high arch shall rush,
> Suns sink on suns, and systems systems crush,
> Headlong, extinct, to one dark centre fall,
> And Death and Night and Chaos mingle all!
> – Till o'er the wreck, emerging from the storm,
> Immortal NATURE lifts her changeful form,
> Mounts from her funeral pyre on wings of flame,
> And soars and shines, another and the same. "

description of an "oscillating" universe,
"The Botanic Garden", Erasmus Darwin, 1791

> " This was the untenable idea of Pascal when making perhaps the most successful attempt ever made, at periphrasing the conception for which we struggle in the word "Universe." "It is a sphere," he says, "of which the centre is everywhere, the circumference, nowhere." "

"Eureka", Edgar Allan Poe, 1848

> " ... the possible curvature and finite extent of space have been suggested by Zöllner as an escape from the reasoning of Olbers and Struve ... "

"Lectures and Essays", W.K. Clifford, 1879

> " The library is a sphere whose exact centre is any one of its hexagons, and whose circumference is inaccessible. "

"The Library of Babel", Jorge Luis Borges

> " I am very interested in the Universe: I am specializing in the Universe and all that surrounds it. "

Peter Cook

> " [economics] ...the last field in which a man can attain eminence without ever being right "

anon

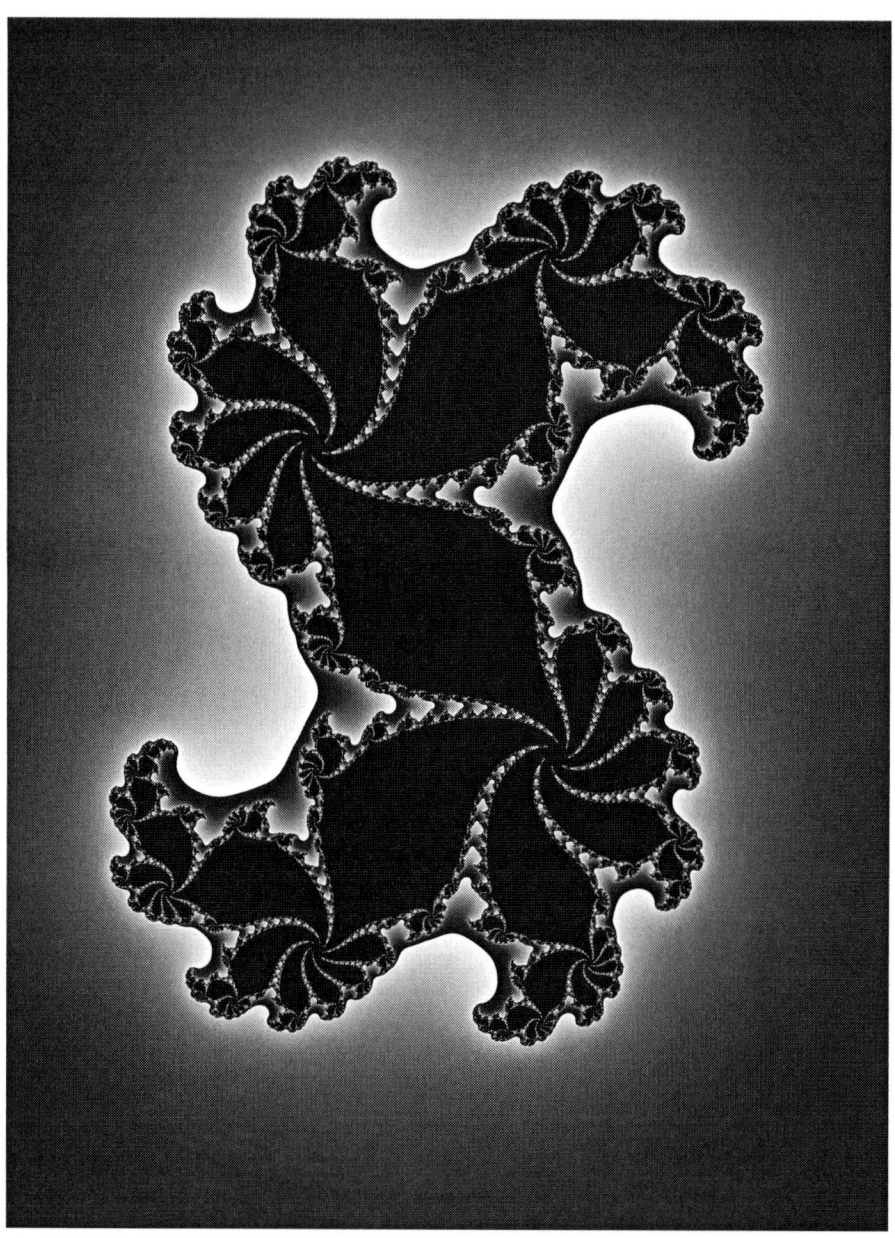

17.1: The expanding universe

As we've already mentioned (section 12.4), the universe appears to be expanding in size. This effect was reported by **Edwin Hubble** in the 1920's, after he surveyed the redshifts of a number of galaxies and found that, as a general trend, the further away objects are, the more redshifted they tend to be. This suggested that (as a generalisation), everything in the universe seemed to be receding from everything else – the universe seemed to be getting bigger. While there *are* other possible explanations for this distance-dependent redshift, expansion is generally considered to be the neatest and least complicated answer.

If we don't like the idea of expansion and try to suggest a different mechanism for the Hubble redshift, those alternative mechanisms tend to end up predicting expansion as soon as we try to introduce unification principles. For instance, what would happen if we argued that the distance-dependent redshift was instead due to gravitational effects? If we suggested that there was a gravitational field *in space* producing the redshift, then if objects more distant to us were being pulled away from us more strongly by this field, then, since they aren't anchored to anything, we'd calculate a distance-dependent recession effect as a result. But this *sounds* very much like "expansion" under another name.

Alternatively, we might consider a gravitational gradient existing *across time* – since we're seeing more distant objects as they were billions of years ago, we might suggest that perhaps the apparent redshift is a "temporal perspective" effect caused by our looking from a weak-gravity era at events that took place in a time of denser gravitation. If the background gravitational field was becoming progressively weaker with time, that might explain the Hubble redshift, but we'd also want to know *why* it was getting more rarefied, and if we set aside the idea that massenergy is disappearing, that leaves the idea that the field is becoming more diluted because the universe is getting larger. We're back to expansion again.

In the sort of unified (non-SR) approach taken in this book (12.13), these alternative explanations tend to merge together and become functionally equivalent, so that we should be able to describe the same redshift on a distant galaxy as being **(a)** a recession effect, **(b)** a gravitational effect, or **(c)** the effect of cosmological-scale curvature, with the resulting physical predictions (hopefully) being identical in each case.

In current, SR-based physics (GR1915), we can still sometimes apply some of these arguments as approximations, but *exact* physical equivalence won't quite work.

17.2: The "Big Bang"

If the universe *is* expanding, then if we "wind the clock back" and extrapolate our current expansion rate far enough into the distant past, we'll eventually arrive at a description of a universe with zero size. It's usual to refer to this extrapolated point in spacetime as the **Big Bang** event, and to a simple description of an expanding universe as a **Big Bang model**. This isn't quite the same as saying that the universe really *did* start with a point-singularity, only that that is the default *mathematical* extrapolation that we get before we start coming up with reasons why such a thing shouldn't be possible.

This is a bit like talking about a black hole's point-singularity under GR1915: since quantum mechanics hates sharply-defined objects, it might be more reasonable to expect the centre of a black hole in a "GR+QM" universe to instead be a sort of **"quantum fuzz"** (section 11.12), and we can apply similar arguments in cosmology: Information theory suggests a "fuzzy limit" to how far back we can safely extrapolate, and if we expect a cosmological horizon to shield the initial singular event from us (**"cosmic censorship"**), we can't be too sure whether or not there really *was* an initial singular event. A horizon tends to protect us from having to know, with any certainty, what "really" happened in the extremely early universe.

17.3: Spatial closure

Current theory also tends to assume that the universe is *spatially closed*, that is, that its positive curvature (section 3.3) is strong enough to make space curve right round on itself. This sort of universe is "edgeless" and self-contained – it doesn't expand like a conventional explosion (by flinging matter into a surrounding region of previously empty space), but by an increase in the *quantity of space* inside the universe. The usual analogy is to compare space in an expanding universe to the surface of an idealised inflating balloon, with points drawn on the surface finding themselves moving away from each other as the balloon expands.

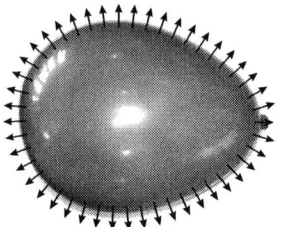

Figure 17-1: *Within* **the surface of a balloon, there is no special "direction" of expansion, and no "edge"**

If lines drawn *in* the two-dimensional surface of the balloon represents directions *in space*, then there's no edge "in space" to the universe, and no special "frontier" where the expansion happens – to two-dimensional creatures that can only exist within in the surface, the surface doesn't expand in any special direction *in space*, but in some mystical, "imaginary" direction that manages to be at right angles to surface-space, in a direction that they can't see or point towards. To *us*, the balloon's is simply expanding radially, but to our hypothetical balloon-surface dwellers, this set of radial lines don't point in any identifiable *spatial* direction, or originate at any spatial *location*, they are the directions that points in the surface travel in *over time*. What *we* see as the balloon radius is, to a surface-dweller, a particular sort of time axis.

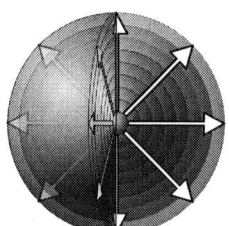

Figure 17-2: Radial time

The balloon analogy isn't totally satisfactory because it fails to suggest why the surface should be growing. We might instead use a "watchspring" analogy, and imagine a large quantity of springy wire wound up on itself to make a hollow ball superficially similar to the balloon – the wire's natural dislike of being bent will make the hollow ball's dimensions unstable, and it'll tend to enlarge as the wire finds ways to naturally unravel sections of itself. The unwinding ball continually increases in area as its local density reduces.

With a sufficiently powerful telescope, would it then be possible to peer into the furthest reaches of spacetime and find ourselves looking at the back of our own heads (Gamow 1940)? Probably not: we'd have to stay at our telescope for a billions of years to allow enough time for the light to completely circle the universe, and in some models a **cosmological horizon** could stop light from being able to make a complete round-trip.

observational limits

As we look across vast distances of intergalactic space, we're seeing light that was originally generated millions or billions of years ago, and which has only just been able to reach us. As we look further away, we're seeing images that correspond to earlier and earlier parts of the universe's history, and in most Big Bang models there's a certain limit at which we should be seeing either mangled remnants of the universe's origin (assuming that it has one), or a horizon separating us from the critical event(s). This **cosmological horizon** is usually taken to be an *absolute observational limit*, and even if we allow for the cosmological equivalent of **Hawking radiation**, the scrambling of information emitted through a horizon means that there's still liable to be a limit to *practical* observation. But it still seems to be an open question whether this limit of observation allows us to see the full *spatial* extent of the universe or only a small region of a much larger structure, and some astronomers are attempting to answer this by looking for possible correlations between large-scale structural patterns seen in different directions.

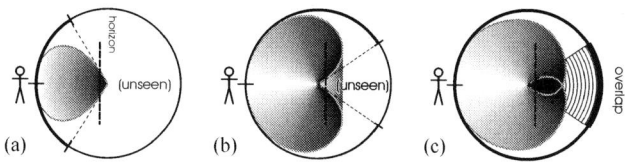

Figure 17-3: Alternative apparent universe shapes, depending on expansion profile

This effect is illustrated above. The large circle represents the *deduced* size of the universe at the moment of observation, and the inner shape represents the *apparent* dimensions of the universe, with the more distant regions appearing to have aged less and expanded less, and showing a correspondingly smaller apparent radius and a higher apparent density.

In case **(a)** an observer can only see a small part of the total universe, in **(b)** the majority of the universe is visible, and in **(c)** the universe visibly wraps around on itself, and the same ancient events and objects can be seen in different directions.

Olbers' paradox

The idea of a spatially-closed closed universe has some history to it: Einstein credited **Friedmann** with proposing the idea as a way of tidying up the cosmological side of general relativity and getting rid of the theory's Cosmological Constant, but **W.K. Clifford**, writing in 1879, credited **Johann Zöllner** as having already proposed a curved, closed universe as a way of getting rid of **Olbers' paradox**, and if we look really hard we can probably find earlier speculations and suggestions about "wraparound" universes.

Olbers' problem was that if the universe was infinitely old and infinitely large, and if it had looked pretty much the same during its entire history, then, since stars were continually pumping out radiant energy, by now there should be an infinite amount of radiation filling the cosmos. If something similar to our own local distribution of stars applied throughout an *infinite* universe, any direction that we could point in would lead inevitably to the surface of a star, and the night sky should look white: on the other hand, if light from the more distant stars was blocked by dust, then if the universe was also infinitely *old*, by now the dust should have heated up to a similar temperature and should be white-hot too. It seemed that the universe must have a finite age and/or spatial extent, or must be variable in some other way.

The idea can be tracked back further by the determined historian – the writer **Edgar Allen Poe** currently gets the credit for being the first English-language to mention this problem in print, but it's quite possible that earlier references may turn up.

RELATIVITY IN CURVED SPACETIME

Shapes existing in more than three dimensions

Suppose that we take a limited three-dimensional volume of space, and "close" it off to make it completely self-contained. We might do this by taking the different parts of the volume's perimeter and **associating** them, so that a lightbeam attempting to pass out of one region finds itself seamlessly re-entering the volume somewhere else. If the effect is completely symmetrical, an inhabitant of the volume might find it impossible to say where the original edges had been, and the "closed" volume's properties will be similar to the "closed" surface of a sphere, except that it has in one additional dimension. We can refer to higher-dimensional counterparts of a sphere as **hyperspheres**.

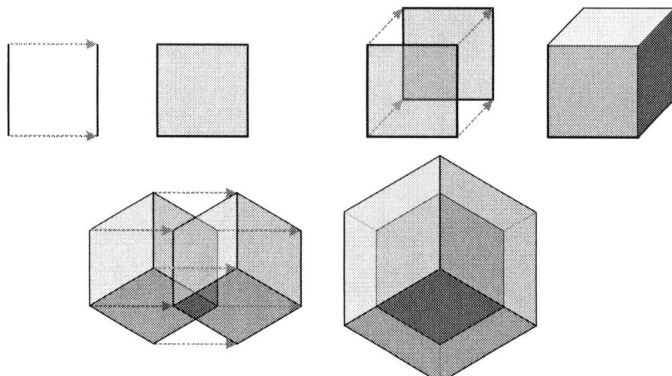

Figure 17-4: Sweeping a fixed-length line in space or time at right angles can produce a *square*: sweeping a square for an equivalent distance at right angles can produce a *cube*: a cube that appears abruptly in a fourth dimension, persists at constant size and suddenly disappears again can form a *hypercube* in four dimensions, sometimes called a *tesseract*

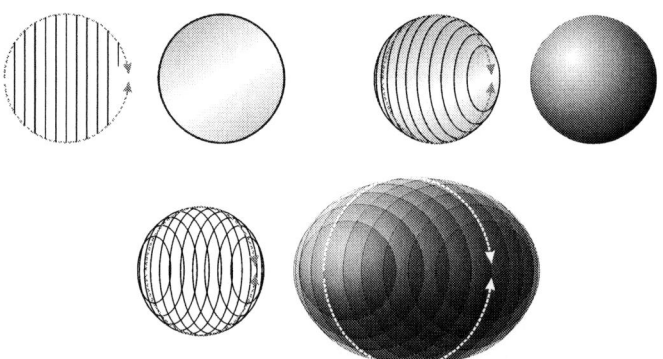

Figure 17-5: A swept line that smoothly inflates from zero size to maximum size and then deflates again with a particular profile can form a solid *circle*: if a swept circle inflates and deflates similarly it can mark out a *sphere*: A sphere that inflates and deflates in a similar way on an additional axis (such as a time axis) can be described as a *hypersphere*. We can also have hyperspheres that extend in more than four dimensions.

17.4: Expansion curves

Once we'd established that the overall amount of distance within the universe appeared to be increasing with time, the next obvious question was whether this expansion was constant. We might expect the combination of all the gravitational mass in the universe to produce a tendency to *collapse*, so (unless there was some counteracting force), perhaps this expansion should be slowing.

Before we started inventing additional expansionist parameters and effects (cosmological constants and the like), there seemed to be three main possibilities for the future of an expanding, decelerating universe:

If the universe was expanding faster than a critical rate, it would be able to overcome the gravitational attraction and continue expanding (at a gradually slowing rate), with no upper limit to its final size.

If it was expanding slower than the critical rate, the slowdown in expansion would eventually turn into a *contraction*, and that the universe would eventually start to collapse in on itself again.

The third option was a "knife-edge" solution between the first two, where the two effects cancel *exactly*, and the universe continues expanding at a progressively slower rate, so that there's a *maximum size* for the universe, that it always expands *towards*, but never quite reaches.

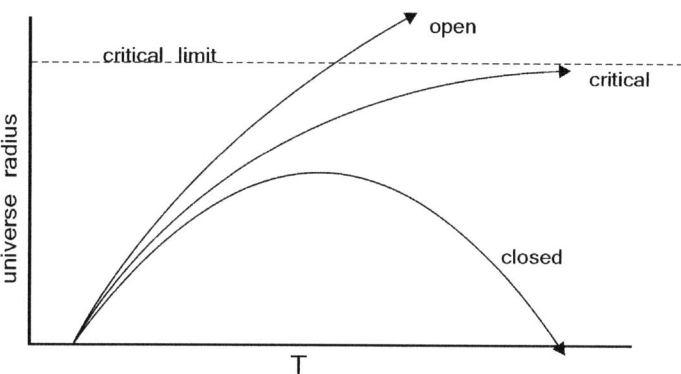

Figure 17-6: "Open", "closed" and "critical expansion" universes

In the upper part of the graph we have a possible universe in which expansion overpowers the attractive force of gravity. This universe expands forever, to infinite size. At the lower part of the graph we have a universe in which gravity dominates and the universe reaches a maximum size and then collapses again. Somewhere between the two there is the unstable "knife-edge" solution in which both effects balance perfectly, and the universe is always expanding, but the expansion rate is always decreasing, and there's a finite upper limit to the size that it will ever be able to attain.

Current measurements give a range of possible values that include this "critical" curve, but since the slightest deviation from the curve would seem to lead to runaway expansion or total collapse, these measurements don't yet tell us conclusively which of these three fates awaits us.

17.5: Cosmological time coordinates

Concepts of time, when we are talking about cosmological timescales, can become difficult.

Let's sketch a cross-section of a universe that's expanding with time, and call this time "**T**":

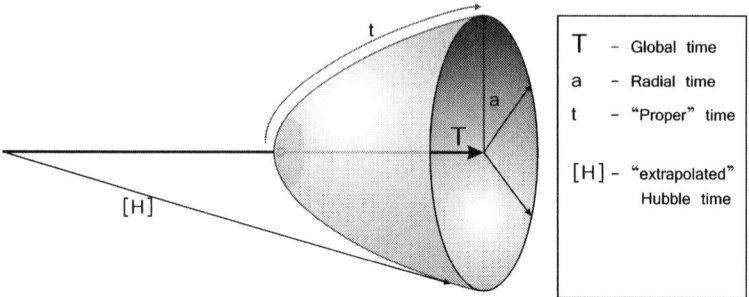

Figure 17-7: Different cosmological time coordinates

We then have a number of ways to assign nominal "time" values to points on this surface: the obvious thing to use would be our "**T**" values, but the *scaling* of this "global" **T** axis is rather arbitrary – different points in the surface will have aged at different rates according to local gravitational factors, so "**T**" won't necessarily correspond to everybody's (or anybody's) *experienced* time – it's an artificial, *imposed* reference value. Depending on how we choose to map "**T**-values" to local clock-times, the surface could end up as a cone, a sharp spike, a smooth hemisphere or paraboloid, or some other topologically-equivalent shape. In fact, this "**T**" seems to be suspiciously similar to the sort of universal, arbitrary, "**absolute time**" coordinate defined by Newton (section 5), in "Principia".

How can we improve on this and make our time coordinates more "physical"?

Well, we *might* choose to use real *elapsed* times (as experienced by local clocks), instead. This *physical* measurement of the time that passes for an object between two points on a **worldline** – "**t**" on our diagram – is known as "**proper time**". But the "proper time" between two points is **route-dependent**, and has different values depending on the shape of the chosen worldline. If we track the progress of a pair of particles created together shortly after the Big Bang, one particle might spend billions of years in deep space while the other sits in a strong-gravity region with a much slower local rate of timeflow. If they meet up again by sheer chance, one particle might insist that the universe has aged by twelve billion years since they last met, while the other could insist that only *ten* billion years have passed according to *their* worldline, and both claims would be valid. "Proper time" runs through the surface at different angles depending on the motion of particles, and using these *local time* coordinates, we can't assign a single value to the age of the universe, because adjacent particles with very different histories can assign different "absolute dates" to the same event. Another problem is that we don't know the exact "proper time" between here and the Big Bang even for a *single* particle, because we don't know how expansion rates and ageing rates may have varied in the past.

A third method (in an expanding universe) is to use the expansion itself as a form of time-reference: we can take the apparent radius of the universe as seen from a given location, and call this parameter "**radial time**" (MTW's "Gravitation" refers to it simply as "*a*").

A fourth method is to combine the *current* Hubble expansion rate and local rate of timeflow, and extrapolate these back to a Big Bang event: This distance ("**H**") is more artificial, but allows us to construct an arbitrary scale from local measurements. This sort of "**Hubble time**" value may not correspond exactly to "proper" times, but it's still useful as a reference-scale.

imaginary time

If we examine this diagram of alternative time-axes, we find that "**T**" (the time coordinate that we started out with) and "*a*" (time as a measure of radius) point at right angles to each other. This is a little weird: it means that if we define times according to **T**, the *a* values don't exist (or even project a shadow) anywhere on our number-line or time-line. In math-speak, we say that if **T** values are taken to be "real", then *a* values at right angles to it are referred to as "**imaginary**", and *vice versa* (Hawking, 1988, pp.155). This gives us the slightly perplexing term "**imaginary time**" for any measurements of timeflow that operate at right angles to whichever time axis we decided to use as our initial reference. If we'd instead decided to take *radial* time, "*a*" as our agreed "real" reference, it would be the *axial* time coordinates, **T**, that ended up lumbered with this horrible name, "imaginary time".

As we've mentioned, our choice of how to *scale* the **T** axis (which may depend on how fast we expect time to run at different periods in the universe's history) will influence whether the surface gets drawn as a straight-sided cone ending in a sharp point, or as a smoothly curved surface that passes through **T=0** without any nasty pointy bits. If the shape is to be "smooth", the universe might start with worldlines pointing exclusively in the direction of "radial time", and finish with them purely in the direction of with "axial time", arcing smoothly through an angle of 90 degrees as the universe evolves.

"arc" times

This gives us yet another measure that we can use for referring to cosmological timescales: after mapping the total surface to a conveniently curved shape (such as a hemisphere or paraboloid), we can refer to any point in the universe's evolution by the local angle of its surface. A more strictly-defined version of this parameter is sometimes referred to as "**arc-parameter time**", and for people who like Greek letters, it gets assigned the symbol η.

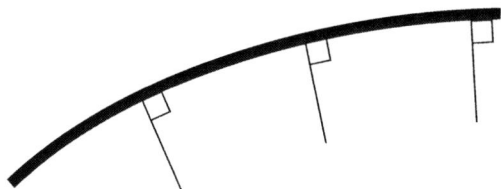

Figure 17-8: Cosmological ageing expressed as a change of angle

Our familiar, "local" conceptions of what time *is* (or what we think it *should* be) don't always work over these huge expanses of the spacetime landscape, and everyday language has a habit of breaking down when we try to talk sensibly about *cosmological* timescales. This is especially true when dealing with the earliest stage in our universe's existence, when energy-densities and temperatures are reckoned to have been so high that what we now think of as conventional matter shouldn't yet be able to exist. How confidently can we talk about "time" in the early universe, if its physics is too extreme for current means of *measuring* time to work ... when we can't use clocks or radioactivity, or the inertial mass of atoms or the electrons, because none of these things can be present? Is it meaningful to talk about the "passing of time" when there's no direct way to *measure* time in a way that we recognise, and no processes that we're familiar with to do the measuring?

Explanations of what happened in the first few fractions of a second of the universe can sound slightly incredible, and part of this is due to the application of familiar language to unfamiliar situations. We can explore what current theory predicts when a "young", "compact" universe has certain *energy-densities*, but common understanding of "time" might not be valid in such a region, and when we see "microsecond" time-values assigned to early events, perhaps we shouldn't expect those values to relate too strongly to what we *now* think of as "real" times.

17.6: The Hartle-Hawking "bubble universe"

Even the concepts of "before" and "after" may break down in the earliest region of the universe. On the Earth's surface, "North" and "South" are usually meaningful concepts, and when someone talks of walking "ten metres North" or "ten metres East", we usually understand what they mean. But as we approach one of the Earth's Poles, lines of **longitude** and **latitude** (which cross each other nicely at right angles at the equator), become progressively more distorted until, at the Earth's "poles", our system of describing directions fails. Where is the point "ten metres North" of the North Pole? If we're ten metres away from a Pole, and we aim ourselves at it and walk *twenty* metres, we find ourselves once again ten metres away from the pole, but on the opposite side of the globe.

In 1983, **James Hartle** and **Stephen Hawking** suggested this sort of geometry as a way of avoiding the nasty point-singularity that we'd otherwise tend to get at a "Big Bang" event. If we extrapolated the size of the universe back in time so that the surface appeared like the tip of a *cone*, then it was natural to ask what happened before the big bang, when all these converging worldlines met, and ... stopped dead in their tracks. This sort of sharp corner isn't really supposed to happen in a well-behaved field theory, so any method of getting rid of the resulting geometrical **singularity** was worth considering.

The Hartle-Hawking idea was to remap the shape so that its initial polar region was completely smooth. If we extrapolated a particle's worldline back to tiny values of a, and then extrapolated the line as a *purely mathematical* path through the fuzzy quantum-scale polar region, beyond the point where any physical particle could exist, then taking the line *even further back* than this didn't take us to a point "before" the big bang, it took us back out of the polar region and back into more conventional space, at the opposite side of the universe.

With this sort of geometry, there's no time "before the Big Bang", just as with Earth geometry there's no location "North of the North Pole", and questions about what happened "before" the Big Bang could be set aside. If we also allowed the universe to collapse again at the end of its life, then we could produce an exceptionally tidy "model universe" that existed as a completely self-contained hyperspherical "bubble".

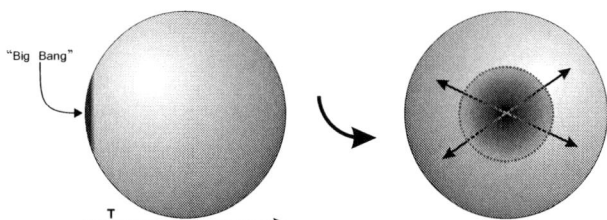

Figure 17-9: "Bubble" universe, "side" and "end" views

With this "bubble" geometry, the universe wouldn't physically support negative values for **T**, and there'd be no earlier event that "caused" the Big Bang: instead, the different parts of the bubble would be mutually self-supporting. When we look at the shape's polar region, one worldline exiting the region, extrapolated far enough back, becomes a *different* worldline exiting the same region somewhere else. The polar region becomes a quantum **particle-pair production** site, and if we look at the *full* surface, the borrowed massenergy needed for this to happen gets paid back at the second pole, where all the worldlines converge again.

The Hartle-Hawking idea describes the universe as a giant **quantum fluctuation**, with the initial and final singularities described not as nasty rule-breaking **geometrical singularities**, but only as artificial **coordinate singularities** – places where our usual labelling conventions and projections go crazy, but where the surface itself still obeys the usual geometrical rules.

17.7: Entropy, arrows of time, and the Big Crunch

If the universe *does* recontract into a mirror image of the Big Bang (a "**Big Crunch**"), we have to tackle some additional questions about timeflow. We normally identify the forward direction of time as being the direction in which things get messier and more disorganised, and this only seems to work properly if the universe is expanding. Our sense of "forwardsness" in time seems to correspond to an increase in disorder (or "**entropy**") that'd tend to be associated with an increase in universe size (our "radial time" parameter, "a").

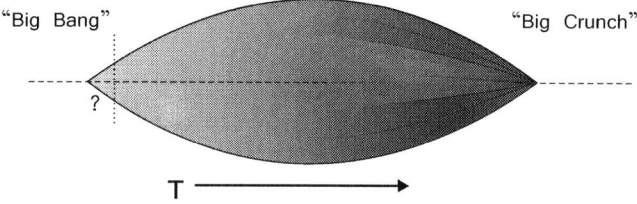

Figure 17-10: "Big Crunch"

So, if the universe decides to contract, and "a" starts to reduce again, does this mean that time should start to run backwards?

expansion and forward timeflow

Suppose that a set of gas molecules are positioned inside a box so that they spell out the word "START". As time runs forward, these molecules will tend to scatter and the order will be lost, and as they bounce repeatedly off the box walls, the pattern of the molecules will look more and more like random noise. But if the box walls are perfect reflectors and we wait long enough, our molecules should *eventually* cycle back to their initial positions, and again spell out the word "START". A series of **conserved operations** on a **closed system** should generate partial copies and "ghosts" of the original pattern as intermediate states, and as it cycles through these, the system will *eventually* return all the way back to its original state.

Where, then, is our concept of entropy as the "arrow of time"? Well, we can create a more persistent trend towards "proper" disorder by gradually enlarging the box – this makes it more difficult for the molecules to reform such tightly-defined structures, because there's always more space for them to spread out into. If an electron and antielectron meet each other inside the *original* box and mutually annihilate in a burst of energy, we'll expect the energy bouncing about the box to eventually refocus and reform the original electron-positron pair, and to keep flipping between "radiant energy" and "matter" states, forever. But if the box is *expanding*, the reflected, redshifted radiation won't have enough energy to recreate the original particles, and physics inside the box will be biased towards reactions that destroy matter and generate radiant energy, and this bias gives us our entropic arrow of time

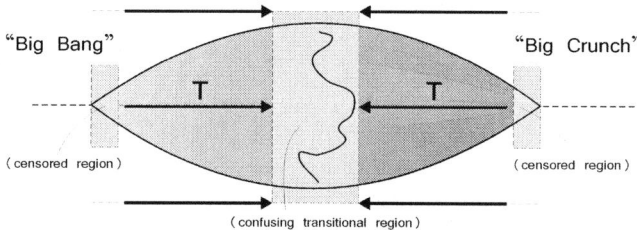

Figure 17-11: Entropic timeflow in a collapsing universe?

So ... an *expanding* universe will favour **exothermic** reactions (ones that give out energy) over **endothermic** ones (ones that absorb energy), and show increasing entropy and disorder.

But in a *contracting* universe (a "shrinking box"), this **thermodynamic** trend works backwards – instead of stars throwing out EM energy that gets redshifted away by expansion, we'd have a shrinking universe compacting and *blueshifting* EM energy, and trying to maintain equilibrium by *ordering* energy, to squeeze it back *into* objects more efficiently.

Does time *really* run backwards during the contraction phase? Do we have a double-universe, with a Big Bang at each end, and a nasty, ugly collision in the middle? If the universe has some structure and irregularity at its maximum size, and some parts start to collapse before others, does this equatorial region then contain a higgledy-piggledy jumble of time-arrows pointing in different directions? What would this look like to someone living through this phase, would they see time running forwards in some regions and backwards in others? And what happens when their own region starts to reverse its direction of timeflow? Many things in this scenario sound rather messy and unpleasant.

17.8: Extending the "bubble" model

One possible way around the time-reversal problem is to invent a new type of horizon that turns our "crumpled" transitional zone into something smoother.

Let's look at the problem again. If the universe has a maximum achievable size, then perhaps we might invoke conservation laws and suggest an upper limit to the quantity of space that can exist ... and since the expansion of space is also linked to a reduction in the energy of electromagnetic radiation in flight, perhaps we might solve this apparent "Hubble shift" breaking of energy conservation laws (section 12.5) by suggesting that the energy-loss isn't just *coincidentally* associated with expansion, but that energy and space share a larger *combined* conservation law (giving us a possible regulating mechanism that could give our universe a reason to comply with the condition of critical density in Figure 17-6).

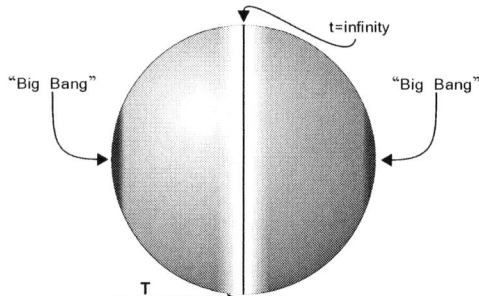

Figure 17-12: A variation on the Hawking bubble

If the maximum radius represents a time at which *all* available radiant energy has been converted, then we won't expect this to be able to happen after *all matter* has first been converted to radiant energy. So, in this model, matter becomes progressively less stable as the angle of "**proper time**", "t" curves around to align with **T**, with the hypersphere's equator representing a special region where the universe is effectively pure empty space and matter can't exist. Since nothing material can survive from one hemisphere to the other, this empty maximum-size universe is then free to contract again, with internal time "running backwards" in the second hemisphere, but without anyone being able to experience the transition.

Although this ultimate limit appears in Figure 17-12 to be reached in a finite amount of "T-time", reducing the stability of matter suggests a corresponding increase in the rate of timeflow for observers inside the universe. "Inertial" arguments relate reduced atomic stability to reduced inertial mass (and faster local timeflow), and Machian arguments support the idea that as the background energy-density shrinks towards zero, the rate of timeflow (a measure of inertia) should increase towards infinity, so at the **equatorial limit** (where matter has *zero* stability), the rate of timeflow should be *infinite*. As far as the universe's *inhabitants* are concerned, the deadly equatorial region lies *infinitely far in their future*, at $t=\infty$. If we're anxious about being "erased", we can comfort ourselves with knowing that in this model, the universe doesn't finally "fade to black" until we've had an infinite amount of user-time to enjoy it. In practice, the last proton would shuffle off its electromagnetic payload at some *finite* point in the future, but it would still be an awfully long time away.

This model combines the "closed" and "critical" solutions from Figure 17-6 – the model seems "closed" from the outside, but when viewed from the inside it is functionally identical to the forever-expanding "critical-density" solution. This rejigging of conservation laws would make the universe slightly more complex in some ways – for a finite expansion of space to be able to eliminate all radiant energy suggests that some conventional Doppler arguments might only be valid to a first approximation (analogous to the **Hooke limit**) – but it seems to be an interesting addition to our collection of possible expansion models.

17.9: Variable dimensionality?

And there are other ways of constructing cosmologies, depending on the questions we want to ask ... for instance, we might be troubled by the question of why the universe chooses to have three spatial dimensions (plus one conventional time dimension). A "3+1" universe does seem to have some nice properties necessary for life (or at least, for our sort of life) to exist, but there doesn't seem to be a *special reason* why the universe would choose to be so friendly. Some theoreticians invoke **anthropomorphic principles**, and suggest that there may be lots of other (less comfortable) universes out there, but that we happen to live in a "friendly" one for the simple reason that this is the one that allowed us to evolve. Reality hasn't selected this universe to be "friendly" to *us*, we've "selected" *it*, by appearing here.

But we could also imagine a slightly less arbitrary version of the anthropomorphic principle, in which our universe doesn't have a fixed number of dimensions, but is evolving downwards from a higher number of dimensions, unravelling as it goes. With this scenario, our familiar universe and its rules emerge when the major spatial dimensionality drops sufficiently close to "three", and then becomes more unstable as the dimensionality continues to fall. This way, our "3+1" universe becomes a fleeting transitional period within a more dynamic structure.

17.10: "Mirror" and "kaleidoscope" universes

If we *don't* use Hawking's idea, and *don't* smooth off the "peak" of our universe we're back to the old question: What happened before the Big Bang?

In a universe that starts with a singular point, we can extrapolate *through* the singularity, and if all information is conserved through the transition, this extrapolation will tend to generate an exact mirror-image of our own universe whose sense of **T** is exactly opposite to ours. The singularity then marks a particle-pair production point that sends one set of particles forward in time into our universe, and a second set in the opposite direction, which is "forward in time" for the mirror universe. We think that *their* universe collapsed to form *ours*, and they think that *our* universe collapsed to form *theirs*. If both universes are true mirror images, then we effectively have just one universe, mirrored.

Another configuration lets us take the identical contents of both cones and weld them together to form a single, fatter cone. This time when we plot a worldline through the singularity, it reflects off the "dead end" and emerges at the opposite side of our "fat" universe, whose details repeat every 180 degrees. It seems difficult to tell these two variants apart.

RELATIVITY IN CURVED SPACETIME

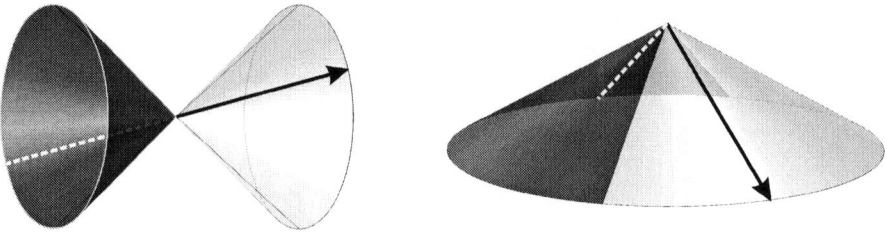

Figure 17-13: Mirroring and reflecting universes

In this second scenario we'd have an identical twin half a universe away, living on an identical Earth, as the end-result of a symmetrical pair-production process. We can track a strip of spacetime back through the polar region and argue that our two sets of worldlines are "*a-time*"-reversed copies of each other, or we can view the "fat" universe "side-on" and claim that there's a *spatial* reversal – either way, we can argue that our twin is made of **antimatter**.

However, if we were to somehow be able to make the journey around the universe to our twin's hometown, then since the trip involves rotating our local coordinate system through 180 degrees, we'd probably see it to be made of "normal" matter when we arrived. We'd never actually *meet* our identical twin, because on arrival we'd find that they'd set off on an identical mission to ours and was now visiting *our* home town. Unless there was some way to tell that the amount of curvature was half what it should have been, we might not have any physical way to tell that we hadn't simply done a round trip all the way around a "thinner" universe and arrived right back at our spatial starting-point.

near misses and partial transfers

But these "twin" and "kaleidoscope" games are only valid if every worldline passes through exactly the same point – if any worldlines veer off to one side or the other, then all bets are off. If the converging worldlines experience a "near miss", and universes connect through their polar regions by a *neck* rather than a true point-singularity, then we've avoided the problem of having a singularity, but if just one worldline fails to pass through the neck, or goes through at the wrong angle, the duplicate universe won't be a perfect copy.

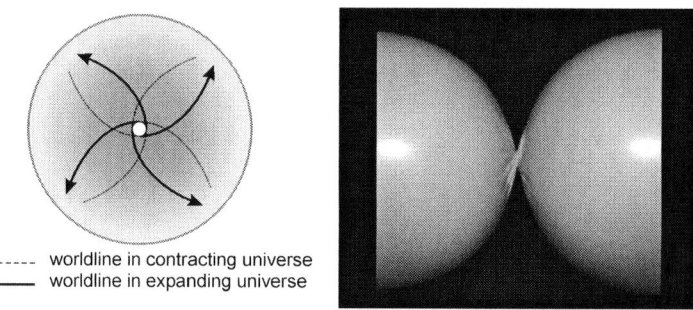

------ worldline in contracting universe
——— worldline in expanding universe

Figure 17-14: Worldlines narrowly avoiding a "Crunch" singularity

These "duplicate universe" extrapolations assume that *all* information from one side of a conjoined polar region emerges intact at the other side, but if we have a polar connection to a "prior" region of spacetime, and not all worldlines from that region arrive in ours – if theirs is a bigger universe, and ours is an incomplete offshoot – then we have no obvious way of calculating back to find the exact properties of that parent universe, because that information, by definition, didn't end up in ours. We could be confronted with a predictive dead end.

17.11: Oranges and raspberries

Gravitational **nonlinearity** can make it difficult to be sure of the reasonableness of our assumption that the universe is reasonably smooth (or "**homogenous**").

What exactly do we mean by *reasonably* smooth? How much "lumpiness" is allowed?

It's usual for us to think of the **"now" surface** of the universe, its spatial cross-section, as being a bit like an orange: approximately (hyper)spherical, but with its surface pitted by specks of matter, and with some larger-scale creases and dents due to clumps of galaxies.

Figure 17-15: "Orange" universe

Mach's principle tells us that timeflow ought to go more slowly in the creases and dimples than in the rest of the surface. But wait ... isn't the expansion of the universe *also* a temporal, entropic process? Shouldn't the "less dense" regions, which age faster, also be *expanding* faster? And shouldn't this make them even *more* rarefied, reducing the background field strength within them even further so that they expand faster again? This sort of universe sounds rather unstable. It's usually tidier to assume that the universe has an even background gravitational "floor" and to superimpose the effects of matter on top of this tidy hyperspherical background, but if we take the "expansion equals timeflow" idea seriously, we can get a rather different picture, of a universe in which the less-dense regions balloon out uncontrollably into lobes on the surface, and in which the background gravitational field strength within a "lobe" can drop arbitrarily far.

The resulting universe would look less like an orange and more like a raspberry, with galaxies tending to collect in the folds between lobes.

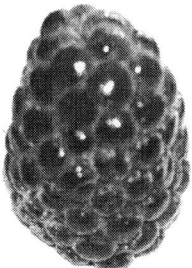

Figure 17-16: "Raspberry" universe

Plotting the apparent distribution of galaxies, we do see something that looks a *little* like this – galaxies *do* seem to collect into sheet-like structures (such as the "**Great Wall**"), surrounding vast voids, and this is at least *slightly* suggestive of the shapes we might expect from a universe with a tendency toward lobing.

RELATIVITY IN CURVED SPACETIME

A *heavily*-lobed universe has some interesting properties: the lobed regions might end up being described by observers in other regions as being occupied by an expansionist antigravitational field. In our "raspberry" model, there's nothing weird about this, we don't have to invent dark energy or other effects, we just take our existing idea that gravity slows time and take it in the opposite direction. Since a region of universe at the middle of a lobe would effectively be "older" than a corresponding region tucked away in a crack (where we live), this model allows us to see isolated objects in deep intergalactic space that appear to be older than what *we* think of as the age of the universe (as calculated from our own location). This illustrates the difficulty in assigning a definite age value to the universe: a "raspberry-flavoured" universe would have a range of ages at different locations, and since the less dense regions would tend to expel matter, "expelled" stars that were, say, fifteen billon years old could end up in a region that was nominally only twelve billion years old.

We also have to consider whether one can have "lobes within lobes". Is our entire observable universe simply a single mega-lobe of a larger surface, bounded by a real gravitational horizon? In other words, does our universe only *look* like a hyperspherical region, while actually being connected spatially to other regions with different background constants? This sort of multi-universe could support more "fractal"-looking properties, and allow more variation in constants between lobes.

17.12: A few Multiverses

Why is the universe the particular way that it is? If physics doesn't manage to get rid of all the fundamental constants of nature, then why do those remaining constants have the values that they do? And if all the fundamental constants *can* be gotten rid of, then how is it that, as the only possible universe that can exist, the *inevitable* laws of nature must end up specifying such bizarrely-specific events as the Declaration of Independence, "Roadrunner" cartoons, duck-billed platypuses and the entire story arc of Buffy the Vampire Slayer?

One way of trying to answer these questions (or trying to avoid them!) is to suggest that perhaps our visible universe is part of something much bigger:

Everett's "Many-Worlds" multiverse

With the **Many-Worlds Interpretation** ("**MWI**") of quantum mechanics, whenever the outcome of an experiment is uncertain, we can imagine different potential universes branching away from ours in which every possible outcome of the experiment really does happen.

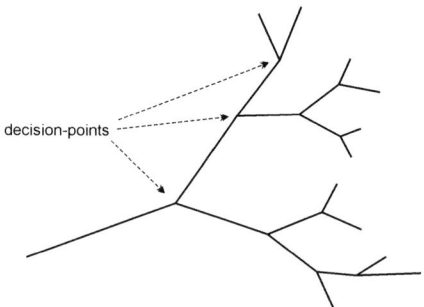

Figure 17-17: Simple universe worldline branching in possibility-space

With this approach to the two-slit experiment (section 10.4), we didn't have to worry about which slit the photon "really" went through, and we didn't have to invoke a separate probability wave passing through both slits: Instead, the electron genuinely *went through different slits in different universes*, and these different parallel universes interacted at the quantum level to produce the final interference pattern: the photon in our universe is still "**entangled**" with its counterparts in the other universes, and some information bleeds through as a result – our universe is affected by outcomes that we think never happened. With the MWI approach to probability, when we say that a future outcome has a 60% change of happening, we say that of all the universes that branch away from ours, 60% of them will give that result.

While science fiction writers have tended to embrace the MWI, mainstream physicists tend to be a little uncomfortable with it. Occam's Razor does favour the duplication of *quantities* of existing things over the duplication of numbers of *different* things (section 4.7), but the MWI does seem to be stretching this principle rather a long way, with the "many" in the title being something of an understatement. If *every* quantum event in *every* universe really spawns further multiple universes (and there are a lot of quantum events per universe!), then the MWI does seems to demand supporting infrastructure on a rather extravagant scale.

oscillating universes

In another variation, a collapsing universe (section 17.7) doesn't shrink away to a perfect point, and different components of its momentum and energy instead just miss each other and continue onwards to form a new expanding universe with a slightly different configuration, the end result being a sequential chain of universes that looks like a string of sausages.

Figure 17-18: "Chained" multiverse

Although this is nice and easy to draw on a lecture board, it is difficult to use it to make real predictions, and since each universe contains a "remix" of the information contained in its predecessors, this still doesn't tell us where the information originally came from, or why constants have particular values. If the universe is finite, and only has a certain fixed amount of "stuff" in it, and this conserved quantity of "stuff" has implications for the way that physics works, then who got to decide what this critical quantity is? Who got to pick the number?

baby universes

One attempted solution to the problem of where fundamental constants come from (and where information goes to that is "lost" in a GR1915 black hole) was the **baby universe**. This was quite an attractive idea, since it seemed to solve two problems at once. Information that disappears into a black hole (said the argument) *is* effectively lost to the outside universe, and when the black hole evaporates, its data *does* disappear without trace ... but it now exists in a universe of its own. When this little bubble of spacetime is cut off from ours, Machian arguments suggest that concepts of distance and time inside the bubble may change, leading to the bubble undergoing rapid inflation, so that it ends up looking rather like our own universe, its parent. The information that goes into the baby universe acts as "**seed data**" for the evolution of its other parameters, and if scientists evolve inside this new baby universe they may extrapolate back as far as they can and end up with the description of a Big Bang, or with one of the other descriptions that we've looked at. They couldn't extrapolate through the barrier to derive the exact properties of *our* universe, because they'd be working from incomplete information. What they *might* be able to do, though, is to make an educated guess about the approximate range of physical constants that the parent universe might have had, based on the mangled version of those constants that their universe had inherited.

Figure 17-19: **Black hole offspring universes**

Another nice aspect of the "baby universe" idea is that it involves a sort of Darwinian evolution of physical constants: a universe whose constants favour the creation of black holes will have more offspring, which in turn will tend to have constants weighted towards black hole creation. Since the combinations of constants that encourage black hole creation are supposed to be nicely set up for allowing life as we know it, the "baby universe" idea allows a multiverse to be continuously spawning "interesting" universes that are particularly well fine-tuned to the appearance of intelligent life, without having any sort of mysterious hidden agenda or bias. The black holes are the important thing, and do all the work, and we're just by-products.

Although the "baby universe" scenario is a compelling one, and may even allow us to make educated guesses about *some* aspects of how our observable universe should work, it depends on the existence of GR1915-style black holes, and it's not certain that these theoretical objects will survive the eventual solution of the **Black Hole Information Paradox** of section 11.18.

regionalisation

The last sort of multiverse that we'll look at has different physical constants spread over different regions of space. For instance, in our region, a thing called **symmetry breaking** makes "common" matter more stable than its mirror-image counterpart. There's no obvious reason why Nature chose to do this, and it's conceivable that parts of a large-scale matter-antimatter mix may be unstable and liable to collapse in one direction, which then becomes locally preferred. There may be other regions of the universe that have "collapsed" in the other direction, and matter in the boundary regions, with an even matter-antimatter mix, may be more liable to annihilate and radiate its massenergy into the other more strongly-biased regions, producing separated concentrated regions where either matter or antimatter dominate. Similar evolving separations may arise for other parameters, producing "mini-universe" lobes with different locally-stable configurations of physical laws.

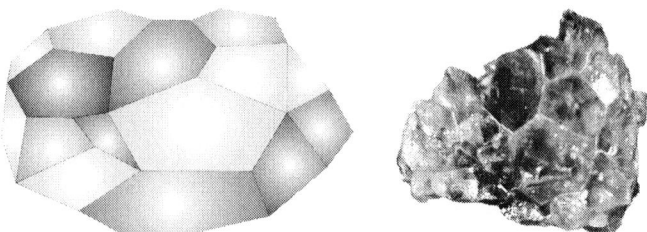

Figure 17-20: "Phase-change" multiverse, and monocrystalline zones in quartz

17.13: Fractal universe arguments

The **Mandelbrot Set** describes an artificial mathematical landscape (or, a set of landscapes) that show "**self-similarity**" – as we zoom in to higher and higher magnifications, we see recurring patterns that represent local variations and reiterations of a larger, "global" theme.

This set is intimately related to another type of fractal called a **Julia Set**, which again shows self-similarities at different scales. Each coordinate point on the Mandelbrot set represents the seed coordinates for a unique Julia set, as we sweep a test point across the Mandelbrot set, the Julia set that we obtain warps and swirls and shifts in character to match.

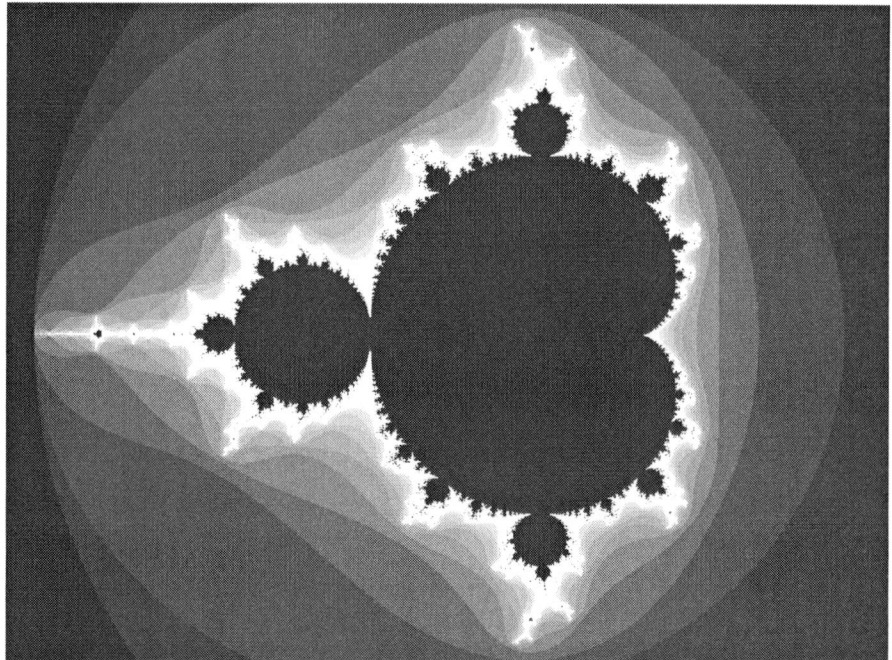

Figure 17-21: The infamous "Mandelbrot set" fractal

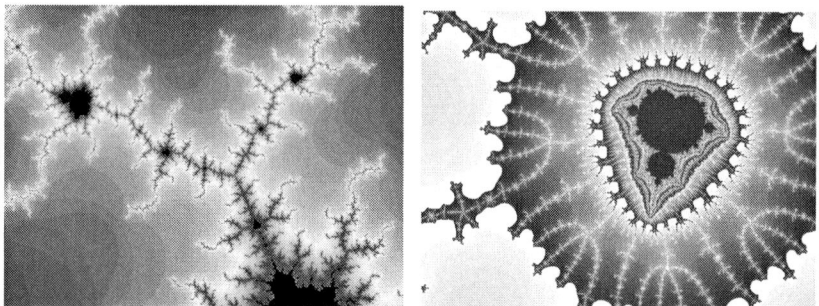

Figure 17-22: Zooming in ... and in ... and in ...

Since we can specify a point on the Mandelbrot set with infinite detail, we can obtain an infinite number of subtly different Julia sets.

17: Cosmology

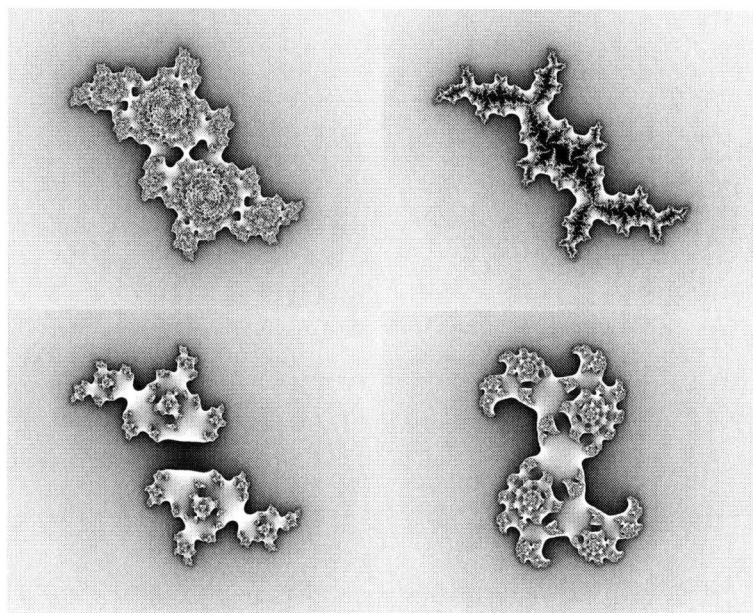

**Figure 17-23: Sensitivity to initial parameters:
Some Julia sets**

Julia sets (like the Mandelbrot set), *seem* to contain infinite detail, and we might think that they contain rather a lot of information. They look "organic" in the sense that they have a "feel" about them of something that has arisen naturally. And yet, other than the base equation, Julia sets only contain two unique numbers: a single pair of coordinates. An imperfect, partial image of one of these sets may well seem to contain a lot of data, but as we achieve more detail and a larger scope, the amount of **data redundancy** also increases, and when we have a *full* set represented in *infinite* detail, there's hardly any information there at all. This is reminiscent of the way that the **holographic principle** operates in theoretical physics.

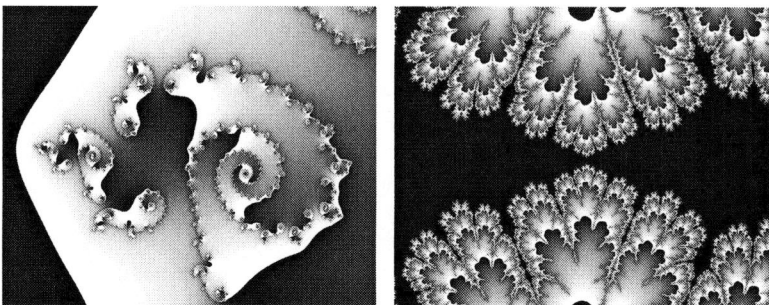

Figure 17-24: "Julia set" detail

Do these sets exist in any "real" sense? Certainly they exist as mathematical *possibilities* that we can explore and describe. But supposing that we could find a more complex multi-dimensional fractal, whose patterns and surfaces appear to generate patterns something like our laws of physics, would we be able to zoom in on this fractal and see counterparts of real galaxies and stars and planets, right down to single events in individuals' lives?

RELATIVITY IN CURVED SPACETIME

These fractal ponderings may not seem to get us anywhere practical: even if we did have some magical "equation of the universe" that was a counterpart of the Julia set, in order to get what we would consider to be "useful" fine detail would involve specifying the set of initial parameters – our fundamental constants – to such a ridiculously high precision that even if we were only dealing with a handful of numbers, the number of digits that those numbers would have to have would be unrealistically high, even if we had a computer powerful enough to process them. A vanishingly small initial error, multiplied by billions of years, might not give us the correct weather for next week: it might not even give us a recognisable solar system.

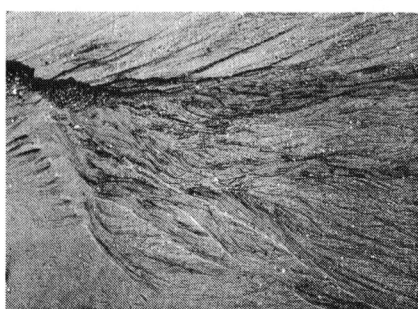 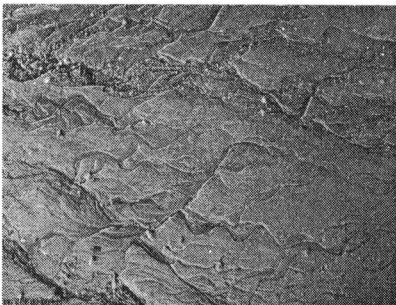

Figure 17-25: Chaotic evolution in Nature: Beach sand

What fractal arguments *do* achieve, though, is to illustrate a concept: It might seem silly that all the complexity in the universe could stem from some ultra-compact initial state and evolve into the baroque complexity of galaxies and rainforests and the internet, but to anyone who demands to know where this dazzling onslaught of detail comes from, we can now respond by pointing to one of these fractals and asking, well, where does all *that* complexity come from?

17.14: Why is the universe rational?

Let us suppose for a moment that the universe was *not* a rational place. Let's suppose that a larger **multiverse** or super-multiverse is a truly *random* soup where our usual rational laws don't apply. Any physical laws that we might be able to think of could exist somewhere in this soup, it might have no fixed number of dimensions, distances ... even the idea of distance or dimension might be alien to it. To a trained mind it might be absolutely the most horrid thing conceivable. And yet ... by even discussing it we are showing some conception of the thing, and we can start to see some emergent properties. For things to happen *truly at random* is very hard work, and we quickly run into a definitional paradox: if the soup is to avoid obeying *any rational laws at all*, even locally, then this is, in itself, a rational law. If the soup is to avoid giving any semblance of order, then certain sequences or combinations should be forbidden, and if they're forbidden, then the thing isn't random after all, and it *is* obeying a set of rules.

If a region or speck should then decide (randomly) to try a bit of order, and if the appearance of this little bit of order *isn't* forbidden, then we might expect it to act as a sort of "seed" to allow locally ordered laws to crystallise out of the soup. Once it has evolved a critical degree of order that allows it to become a **closed system**, it can be considered as a freestanding universe or multiverse regardless of what else might happen in the now-inaccessible soup outside. Now, suppose that our universe started in this sort of soup, and that some invisible and undetectable action in the soup snuffed out our entire universe, acausally and without warning, one hour from now. Suppose that our universe simply ... ceased. Would this have any consequences? Well, since an extrapolation of the universe's final uncorrupted data would still give the same future that the universe was originally going to have, it's not obvious that we'd notice. Outsiders might claim that our universe had ceased to exist, but its inhabitants wouldn't be obliged to agree, in fact they wouldn't be aware of the claim, because those outside events wouldn't exist for them. It would be difficult to argue that, for life inside the universe, the extrapolation forwards was any less physically real than an extrapolation backwards. Suppose that our universe had actually stopped *one hour ago*, and we were merely a mathematical extrapolated *possibility* that never happened – how would we tell? Should we even care?

Perhaps we shouldn't. Once a fledgling universe becomes a free-running self-extrapolating causal system, then any opinions of entities outside the system are probably irrelevant. If some outside … thing … tries to reach inside our universe acausally to mess with it, then if our causal system is strong enough, the attempted tinkering should have no effect. Taking a pair of scissors to a printout of a Julia Set doesn't affect the set itself or its mathematical relationships. It's a self-contained logical structure that can't be changed from outside.

Do we then have **free will**, or is everything predetermined? Again, it doesn't seem to be a useful question. The intricacy and sensitivity of deeply **chaotic** systems (such as human societies) mean that "classical" extrapolated predictions are hopelessly sensitive to initial errors. For instance: human reproduction involves the creation of sperm with "random" combinations of paternal DNA in a process that is totally beyond us to predict. It's about as close to being "random" as anything we know of, and the environmental coincidences that might then lead to a particular unique sperm (of *millions*), containing one particular selection of genes, winning its race to fertilise a human egg are beyond calculation. If you go to the desk of a male colleague and move his pencil one centimetre to the left, are you setting in motion a subtle chain of intricate events that may alter the genetic makeup of his future children, and therefore the path of future human history? Quite possibly. You'll never know.

Once we appreciate the exquisite sensitivity of the fine detail of our world to tiny changes, the older philosophical questions like "free will vs. determinism" look a bit naive. Even if our universe really *is* totally deterministic, **quantum mechanics** and **chaos theory** say that even if a precise future can *theoretically* be calculated in advance, in practice it's impossible. We could never collect all the necessary data with enough precision to make proper predictions without corrupting it. We can make educated guesses and statistical calculations based on system behaviour, but we can never *quite* know the future until we get there, and even then we won't be entirely sure about it. We can't even be *entirely* sure what's going on *now*.

past pasts

How real is the past? It sounds like a daft question, but in cosmology it can sometimes be a sensible thing to ask. We usually think of the future as being uncertain and the past as being fixed, but QM principles suggest that our knowledge of both is patchy and becomes less reliable the further we extrapolate away from the "known now" (Einstein, 1931).

It's also relevant to the earlier notion of a causal universe being "seeded" from acausal origins: if the emerging, evolving causal structure ends up being *perfectly* self-contained, then inhabitants will be able to extrapolate back through the "acausal" region to a false causal past that didn't actually take place. The newly-causal universe can evolve a new past for itself that "makes more sense" than what actually happened. And once all the details have ironed themselves out, it will become impossible for an inhabitant of that universe to say that the "false past" isn't genuine: according to the final causal structure, the false past *is* what really happened, with the original jagged, fuzzy, acausal entry point being erased from the historical record. This sort of universe may get to reinvent itself as a self-contained "Hawking bubble".

If randomness leads either to more randomness or to causality, but causality only leads to *more* causality, then if we live in a universe that seems to be at least *reasonably* causal, it shouldn't be a surprise to find that it seems to be *entirely* causal (albeit rather chaotic). In an **acausal superspace**, causality would seem to act as a sort of self-replicating disease. If we try very hard to extrapolate back to (or before) the very origin of the universe in order to find some deep truths about *why* our universe happened, we may be extrapolating back to a dead end. The *original* past, back in the pre-matter era, may have already self-erased.

17.15: The Drake Equation

Are we the only intelligent life here?

The **Drake Equation** attempts to calculate the likelihood of our being able to make contact with a living alien civilisation. If each civilisation gets to influence a region of spacetime, then in order to make contact, our region needs to overlap with someone else's. Two simultaneous civilisations at opposite ends of the galaxy are no good, two nearby civilisations are no good if one dies out before the other comes long. And even if a system has the ability to *support* interesting, intelligent, technologically-advanced life, there's no guarantee that it will *have* it: the Earth has been theoretically *capable* of supporting advanced life for billions of years, but we've only just started launching rockets and broadcasting radio signals in the last hundred.

And human life almost didn't make it this far. We barely survived the last Ice Age, we came worryingly close to wiping ourselves out with a nuclear exchange during the Cold War, and there's always the possibility that a supervirus or a meteor strike or rotten weather might set our civilisation back again. It only takes one big meteor every few hundred million years to make the evolution of technologically advanced life very difficult, and lifeforms on other planets may not have been quite as lucky in this respect as we've been.

We also don't know the likelihood that a planet that is capable of *supporting* interesting life can also *originate* it in a reasonable amount of time – we only have one example, and by definition, it was an example that worked. That doesn't mean that it always works. **Isaac Asimov** pointed out that the Earth-Moon-Sun system is a little strange: the apparent diameters of the Moon and Sun look improbably similar from the Earth, and the tidal forces are similar in strength. What if this was more than a statistical fluke?

What if the chemicals washing about in Earth's ancient rockpools needed to be stirred and deposited by a *matched pair* of tidal systems for the *efficient* production of the complex chains and sequences required to get organic chemistry kick-started quickly? Earth might be more unusual than we think. If this was the case (and it's just one possible example) our statistics on the likelihood of life appearing as quickly on other planets would take a nosedive: there might well be huge numbers of planets out there that could support *colonising* life, but fewer that would have convenient systems capable of building it from scratch as fast as ours did. Certainly, there's the **panspermia** idea (promoted by Hoyle and Wickramasinghe) that perhaps viruses and bacteria and simple proto-life components might drift through space from a seeding planet somewhere out there, and that perhaps *this* was the origin of life on Earth. But we can also look at this the other way around: perhaps life *did* appear on Earth from scratch, and perhaps the "originating region" for life in our section of universe is the planet that we're currently standing on. Perhaps Earth, with its unusual habitat and situation is Ground Zero for "early" life in our part of the universe.

Or perhaps it's not. But it's a slightly sobering thought that even if our universe is teeming with life overall *at some point*, there is no guarantee that this will be happening in *our* era. In a populated, colonised universe, it's true that for most of the lifeforms in it, they'll be able to point out to the stars and say, "There *must* be intelligent life out there, it can't only be us!". But for the early arrivals, it's almost guaranteed that for at least *some* of the civilisations saying "It can't only be us!" ... they'll be wrong.

Welcome to the universe. We might well have arrived at the cinema during the main feature, or we may be the first people here. But we can't yet assume that there's necessarily anyone else out there in our galaxy *within range*. Maybe they do exist, but aren't in the habit of wastefully broadcasting radio signals out into space. Maybe civilisations that are foolhardy enough to advertise their existence blindly to potential predators have a low life expectancy, and maybe the smarter civilisations keep damned quiet. Or maybe (just maybe) our galaxy will one day be colonised by intelligent life, but first we have to go out and seed it ourselves, by exploring and leaving a trail of unsterilised toenail clippings and dandruff and bacteria and mouldy TV dinners behind us that in a few billion years might evolve into something more interesting.

Perhaps this is our place in the great scheme of things, to wander an empty universe forgetting to clean up after ourselves. If so, it's a destiny that we might be quite good at.

17.16: Before Event Zero

It turns out that there are a number of possible answers to what happened before the Big Bang. If we try to extrapolate back through the cosmological horizon, we can use Hawking's "bubble" argument to say that as far as observer-time is concerned, "before" may not be a valid concept, or we can suppose that the horizon conceals some sort of "neck" that connects our universe to a previous incarnation, or to a larger parent universe. Or it might connect to a cluster of universes with smoothly or drastically varying fundamental constants, or to a more chaotic network of universe connections. We can have a patchwork universe, a lobed universe, a universe in which adjacent patches *become* lobes, or a more fractal universe in which lobes spawn other lobes.

The birth of our universe isn't necessarily all that *mysterious*. There are certainly a lot of critical things that we *don't know* about it, but that doesn't mean that there is a special mystery to *why* we don't know them. There are some quite good reasons why our knowledge is limited. Put another way, it's not that we don't have a clue how to model or explain the birth of our universe. We actually know quite a few different ways of doing it. We just don't yet know which of them (if any) is correct.

what is cosmology good for?

Enjoyable as these exercises are, playing mind-games with multiple universes can sometimes make us feel like kids playing in the sand on the beach – we feel that we've have learnt *something*, but it's difficult to pin down exactly what. How do we relate these abstract cosmological games to real, useful predictive physics?

One potential route to Usefulness is the way that some multiverse cosmologies modify our concept of space: although general relativity isn't supposed to assume a pre-existing metric (section 12.7) we can still fall into the trap of thinking of the gravitational field as a thing that is overlaid *on top* of a flat-spacetime background, a sort of standardised, absolute gravitational "floor" that the fields belonging to planets and stars and galaxies are imprinted on top of, and where we can't dip below this floor without involving "naughty" negative gravitational fields. But cosmological arguments can undermine this view. If *expansion itself* counts as an entropic process, and emptier regions have a tendency to "lobe", then the universe becomes a more intricate, finely-balanced (and more interesting) place. If time *is* expansion, then what we might previously have considered to be an "illegal" antigravitational field might just be the natural and reasonable tendency of a relatively less dense region to expand faster than our own.

With the "floor" approach, special relativity is a natural foundation-theory to use, and negative gravitational field components are difficult or impossible to produce. With a "no-floor" approach, acoustic metrics are arguably the more natural choice, and complex distortions of inertial and gravitational properties are not just *possible*, but are part of everyday physics. "Losing the floor" allows the existence of *relatively* negative field components, and opens up some intriguing possibilities for manipulating spacetime to our own ends.

So the "floor"/"no floor" distinction has important consequences for the subject of **metric engineering** and has implications for how far we're allowed to manipulate and control inertial/gravitational fields. To have a proper theory that could tell us whether warpdrives are possible or not, and (if so) how we can build them ... that sounds like a worthwhile goal.

But before we try to tackle these difficult "warp drive" issues, it's worth looking at a few related problems in wormhole theory ...

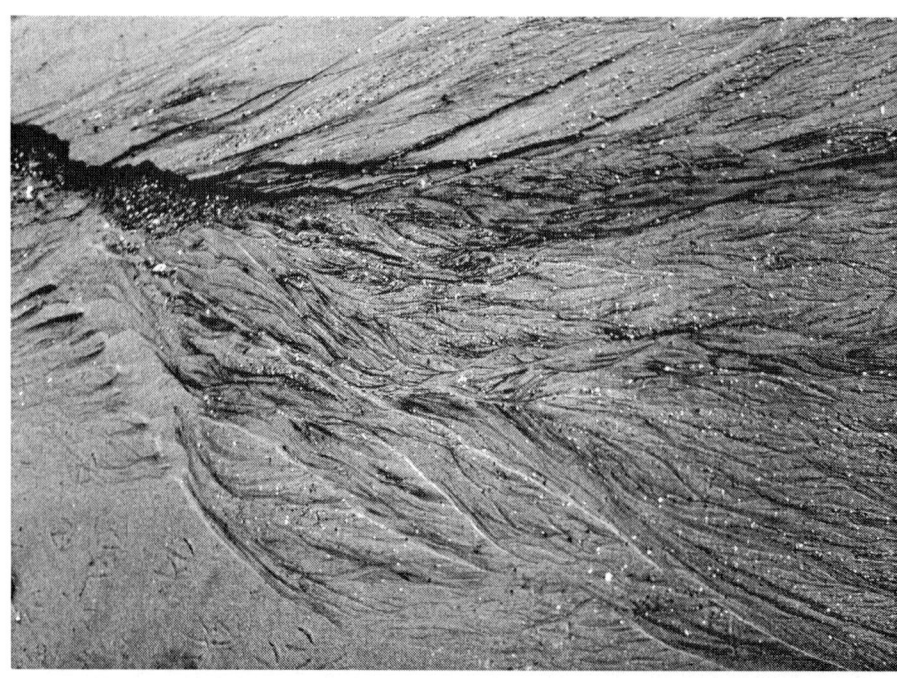
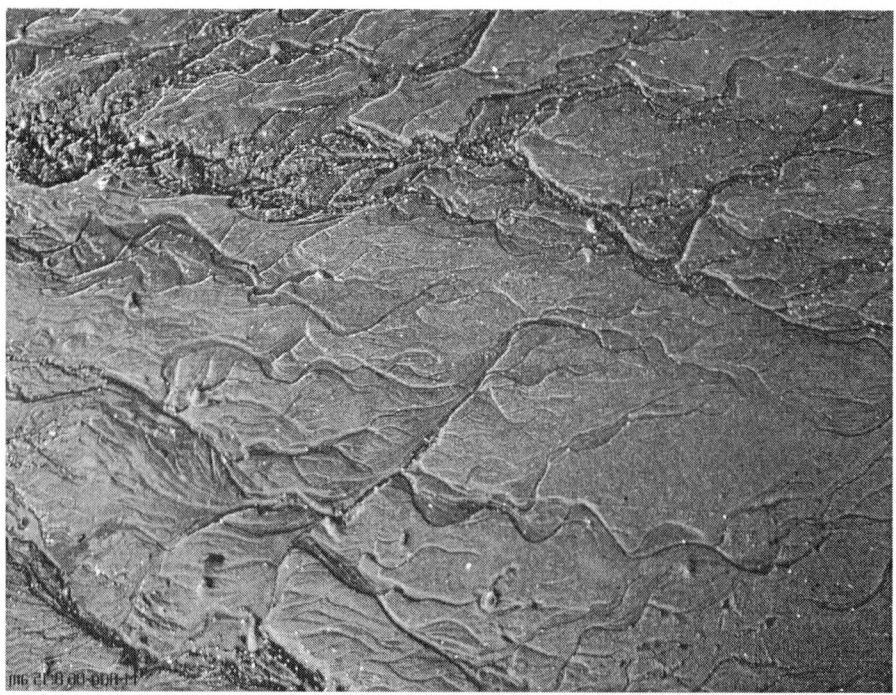

18
Trouble with Wormholes

> *" But his explanation involves the Fourth Dimension, and a dissertation on theoretical kinds of space. ... a kink in space ... two points might be a yard away on a sheet of paper, and yet be brought together by bending the paper round.*
>
> **"The Remarkable Case of Davidson's Eyes", H.G. Wells, 1895**

> *" Mathematical theorists tell us that the only way on which the right and left sides of a solid body can be changed is by taking that body clean out of space as we know it – taking it out of ordinary existence, that is – and turning it somewhere outside space. This is a little abstruse, no doubt, but anyone with a slight knowledge of mathematical theory will assure the reader of its truth. ... "*
>
> **"The Plattner Story", H.G. Wells, 1896**

> *" These solutions involve the mathematical representation of physical space by a space of two identical sheets, a particle being represented by a "bridge" connecting these sheets. "*
>
> **"The Particle Problem in the General Theory of Relativity"**
> **A. Einstein and N. Rosen, Phys. Rev. [48] 73-77 (1935)**

> *" One can consider a metric which on the whole is nearly flat except in two widely-separated regions, where a double-connectedness comes into evidence ... "*
>
> **"Geons", John Archibald Wheeler,**
> **Phys. Rev. [97] 511-536 (1955)**

> *" – 'I feel a draft '*
> *– 'We must be near the Sea of Holes.' "*
>
> **"Yellow Submarine" (animation), 1968**

RELATIVITY IN CURVED SPACETIME

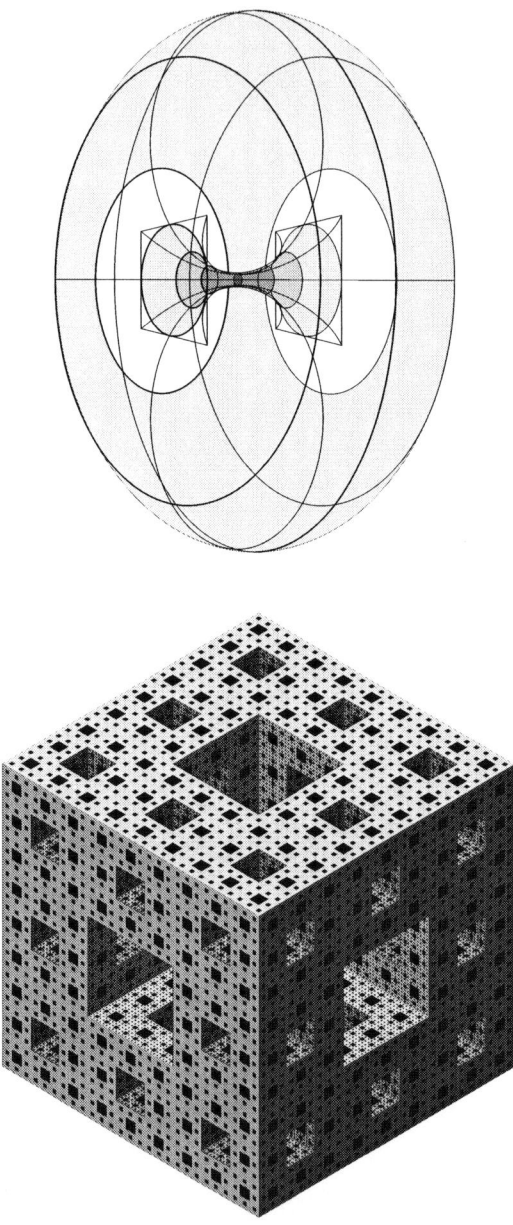

A doubly-connected surface and a fractal sponge

18.1: What is a wormhole?

A **wormhole** is a geometrical connection or short-circuit between two regions, and a universe that contains parallel wormhole connections can be said to be **multiply connected**. In pure geometry, connections between regions are referred to as **handles**.

In a universe that is only **singly connected**, any path connecting two points can be smoothly "morphed" or swept to become any other path with the same terminating points: but in a multiply-connected surface there are at least two *distinct* routes that a path can take. We can illustrate this with the example of two points drawn on a sphere. A sphere counts as a singly-connected surface, and if we stick two drawing-pins into it and link them with a section of rubber band, then our rubbery line can be stretched and distorted to become *any possible line* that we could draw on the sphere to connect those two points.

But if we try the same exercise on a "ring-doughnut" shape, (a "**torus**"), there are now at least two different ways of linking our two points: with lines that thread *through* the torus mouth, or with lines that don't. These two types of route are distinct and separate – a line that travels to its destination *through* a wormhole can't be "swept" to become a line that joins the same points *without* passing through the wormhole – it's trapped by the geometry.

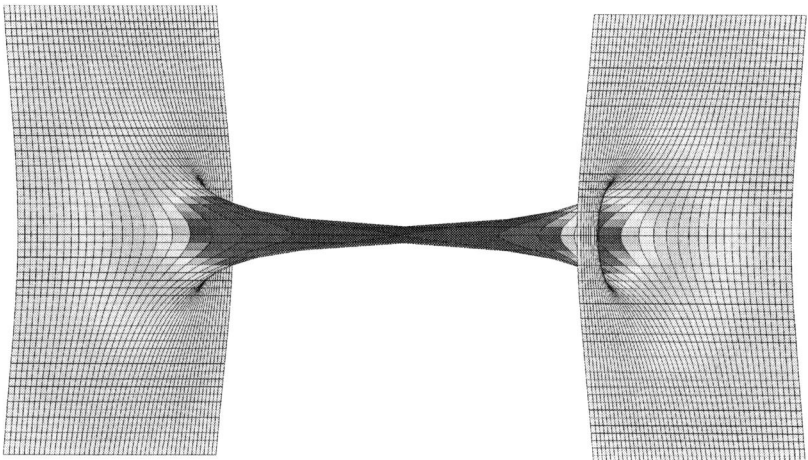

Figure 18-1: **Wormhole connection between two surfaces**

18.2: "Spacetime surgery" and simple optics

The easiest form of wormhole to visualise is a simple spatial "short-circuit". We can visualise the creation one of these creatures through (wholly illegal!) "**spacetime surgery**" (**Matt Visser**, 1996), by selecting two spherical regions of space, cutting them out of normal spacetime, and then connecting or associating the two resulting "bleeding edge" surfaces together. Anyone trying to enter one spherical surface then finds themself exiting the other, with a certain amount of wear and tear from the sharp tidal forces at the transitional region.

In order to visualise the optical properties of this kind of wormhole (what it *looks* like), we can replace each of the two spherical "edges" with a perfect spherical mirrors, and imagine the reflected image that we'd expect to see on each surface – then we swap them round: the sort of "reflected" image that we'd normally expect to see on mirrorball "**A**" instead passes *through* the connection and emerges at surface "**B**", and the reflection that we'd expect to see on "**B**" passes through the connection and presents itself at the "**A**" surface.

RELATIVITY IN CURVED SPACETIME

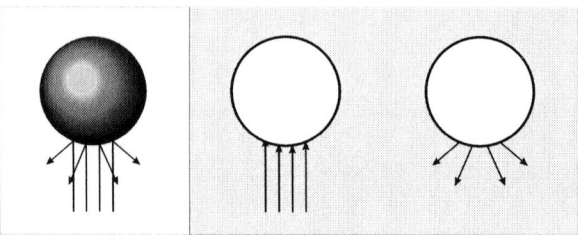

Figure 18-2: Simple wormhole optics: Reflecting mirror and wormhole

A pair of parallel rays hitting this spherical surface will be spreading (diverging) when they leave the other mouth, and since *positive* gravitational fields make rays *converge*, the ability to *spread* or *splay* light rays can seem like decent evidence of a *negative* gravitational field, and this is one argument that leads us to expect simple wormholes to be associated with negative gravitational effects.

Another argument is that since we've removed a certain amount of spacetime, the spatial density in the region should be naturally *lower*, the opposite of what we'd expect with a conventional gravitational field. However, the geometry becomes less straightforward when we deal with different mouth shapes, when the connection isn't so "simple", and/or when the throat connecting the two mouths has a significant length.

18.3: Wormhole instability?

We sometimes visualise the behaviour of "fluid" surfaces with soap-bubble films. A soap-bubble "sheet" will adjust its shape to form the minimal surface that links its perimeter points together, and if we take a soap bubble and create a new connection between two points on its surface, we'd normally expect this connection to either shrink and seal itself off, or to throw both its mouths at the same point on the nearest side of the bubble, and vanish that way. This *seems* to suggest that wormholes should be very unstable configurations that shouldn't stay open of their own accord, and which will tend to try to collapse down to quantum scales in a vanishingly small amount of time unless we somehow keep them prised open (with a negative gravitational field).

universe –sized wormholes

What happens if we take two distinct universes and link them with a *single* wormhole?

Our earlier "instability" arguments seem to say that the wormhole should collapse and leave the two regions isolated. But if we place two soap-bubbles next to each other and poke a tiny hole between them, the hole *doesnt* rapidly seal itself up, it rapidly *inflates*, so that the two smaller bubbles snap together to form a single larger bubble. So perhaps a wormhole connection isn't *necessarily* unstable, and perhaps their tendency to collapse in other situations might be a function of the global geometry and the wider geometrical context of the surfaces being connected.

Topology suggests us that a *single* connection between two regions shouldn't be unstable – or at least, shouldn't be any more unstable than conventional space – because topological arguments let us divide a "conventional" singly-connected region of space *arbitrarily* into as many "separate" regions as we like, linked by simple connections or "handles".

Since these "virtual" connections could be said to be everywhere, they obviously *aren't* unstable in the usual sense, and don't require an obvious negative field ... except in the sense that our *entire universe* might be considered unstable and permeated by a sort of negative gravitational field effect that describes its expansion.

18: Trouble with Wormholes

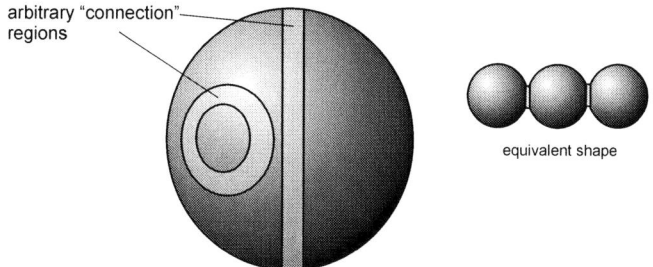

Figure 18-3: Simple connections between regions

Would these arguments then mean that *our universe* would then have to disappear in a vanishingly small amount of time? This depends on how we define "time": for someone inside the surface, a "sudden" collapse might still take a cosmological amount to time to complete, and "arrow of time" arguments (section 17.7) suggest that they might also consider this universe to be *expanding*. And on further consideration we realise that a spherical soap bubble's surface, *considered as a complete surface, is* naturally unstable, and that bubbles only manage to persist for a reasonable amount of time (instead of collapsing immediately into a blob) because of the air trapped inside them – soap-bubbles have a minimal surface because of the properties *of* that surface, but the thing that keeps them "propped open" is something that exists *outside* the surface.

We need to be cautious about wormhole disproofs based on "instability" arguments – if we're not careful, some of our over-enthusiastic proofs that *wormholes* can't exist might end up accidentally "proving" that our own universe can't exist either.

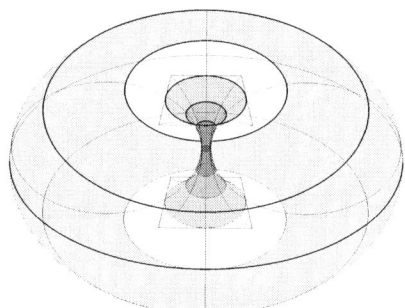

Figure 18-4: Doubly-connected universe

When we add a second connection to a single spherical soapbubble, surface tension tends to act to shrink and eliminate the smaller of the two connections, but if we consider the case of a universe that is doubly-connected on large scales, so that it forms the higher-dimensional equivalent of a torus, then we can invent a geometry (which isn't easy to draw!) in which all parts of the doubly-connected surface are equivalent, making it impossible to distinguish between "throat" and "non-throat" regions. In this sort of universe, it can be difficult or impossible to justify labelling one part of the universe as the "proper" region and another as being the "extra" wormhole connection, and if the universe itself can't tell which regions of space count as "normal" and "wormhole", then it's difficult to say that a "wormhole" section should be less stable than other regions. This suggests that some of our unusual rules regarding wormhole stability may be special cases, perhaps for wormholes that are very small compared to the regions being linked.

18.4: The distance problem

These "cosmological" and "collapse-time" issues highlight some potential difficulties with the idea of using "navigable" wormholes for interstellar travel. Supposing that we've managed to actually *build* a wormhole, and we ship one mouth in a conventional spaceship all the way to Sirius, and then carefully unpack it. If the wormhole mouths were initially only a few metres apart, we might *assume* that the length of the connection won't alter when the external distance between them is changed. But is this expectation realistic, bearing in mind that as an "artificially maintained" wormhole, many of our proofs regarding natural wormholes might not be valid? It'd be very annoying, having gone to all the trouble of pulling the mouths lightyears apart, to find that the "throat" distance linking the mouths had elongated by the same amount during the journey, and that the wormhole route was now just as long as the standard route! There's also the nasty possibility that by *artificially maintaining* the wormhole we might accidentally end up packing so much extra space into it that the wormhole connection ends up being *longer* than the conventional distance between the two region that it connects. This would not be helpful.

18.5: The ageing problem

To test that the distance through our interstellar wormhole connection really *is* as short as we'd hoped, we flash a Morse code signal through the mouth, and are gratified to receive a reply from Earth almost immediately. The distance through the throat *seems* to have stayed small. But our problems aren't automatically over: if the wormhole really does require a negative gravitational field to keep it propped open, then in "Machian" terms this amounts to the throat region being *partially isolated* from outside fields, and Mach's principle then suggests that the region inside the throat may have an increased rate of timeflow. We may have accidentally created an **entropic furnace**.

The default assumption that a wormhole mouth should have negative-looking gravitational properties gives the interior of the wormhole an accelerated ageing-rate. Even if lightpulses can pass between the two throats in a few milliseconds of "outsider time", this doesn't rule out the possibility that that if our travellers hopped into a small transport shuttle and tried to pop home through the wormhole for their victory celebrations, that their journey back might take tens, hundreds or even millions of years according to their clocks. It would be unfortunate if a traveller hopping into one mouth appeared a few hours later at the other as a pile of aged dust.

Humans might find it impossible to commute through this sort of wormhole at sublight speeds during a human lifetime, resupply through the wormhole might have to be limited to materials that could survive extreme ageing, and any signals passed through it might be badly degraded.

18.6: The "antihorizon" problem

The third nasty possibility that we face with an artificially maintained wormhole is that if the mouth region is enclosed by a sufficiently strong repulsive gravitational field, the wormhole may be **antihorizon-bounded**, and we might not be able to enter it at all, or even to send radio signals through it.

This sort of wormhole connection is conveniently "censored" from the outside world, and wouldn't necessarily appear on our maps of the region's lightbeam geometry unless we found some way to pierce the antihorizon and force the mouth open. Forcibly "lancing" the mouth would be difficult since the repulsive nature of the surrounding field would tend to make it super-slippery, deflecting anything we threw at it off to one side or another. We *might* be able to open the antihorizon with a focussed warp-field generator or warp-drive equipped spaceship – but if we had *that* sort of advanced warpfield technology at our disposal we might not have had so much need for the wormhole in the first place.

18.7: "Anti-wormholes" and spatial reversal

Suppose that some future laboratory cracks the problem of creating, maintaining and confining a pair of wormhole mouths. They take pairs of entangled particles and spread them over two spherical surfaces so that each set of particles maps out an exact copy of the other. The experimenters then throw a Magical Lever, each particle connects outside normal space to its partner, space "snaps" and reforms through the connection, and a wormhole forms. After the wormhole has been successfully maintained for a few days, and signals successfully passed through it, someone at the lab's celebration party decides to ceremonially throw a baseball through one mouth to their colleague at the other. The lab (and a few surrounding square miles of city) promptly detonates in something that looks like a nuclear explosion. What went wrong?

The problem in this scenario is the way that the surface maps were constructed. Visser's "spacetime surgery" exercise highlights a geometrical possibility that might make "true" wormhole creation dangerous: if we take our two bleeding edges (the two spherical surfaces in section 18.2), and start stitching them together outside of normal space, there's not just one but *two* distinct ways that a set of points on one surface can be associated with a corresponding set of points on the other:

Suppose that we have a navigable wormhole in deep space (which would probably be the safest place for it), and decide to fly a spaceship through it. Just before it enters the mouth, we ask the pilot to use a marker pen to write the letter "**p**" on the inside of our ship's front window. If the wormhole is a conventional one, an observer watching our spaceship emerge from the second mouth will be seeing this scrawl from the *other* side of the glass, reversed. This means that when the glass surface coincides with the wormhole boundary, a set of points on the surface that map out a "**p**" for an observer looking into one throat will appear from the other side as a *reversed* "**p**" (and should look more like the letter "**q**"). So, to make a "normal" wormhole, our two surface-maps need to be *mirror images* of each other.

But in our fictitious laboratory example we didn't consider this, and accidentally built a wormhole where the two point-maps were *identical*. What would happen if we flew our spaceship through one of these "misconnected" wormholes?

As we watch the spaceship emerge from the second wormhole mouth, the "**p**" that the pilot scrawled on the front window will now appear to be "**p**" to *us*, even though we're looking into the cockpit from the wrong side of the glass. If we jump into the ship's open airlock and peer at the writing from the *inside*, we should now see something that looks like a "**q**", and we might think that the pilot accidentally wrote the seventeenth letter of the alphabet on the glass rather than the sixteenth. But the pilot disagrees, insists that the letter really *is* a "**p**", and points out that the letter matches the corresponding "**p**"'s on all the ships' instrument displays and on their spacesuit's nametag. Looking around the ship we find that the pilot is right, and *everything* in the ship seems to be reversed – the pilot's heart appears to be beating on the wrong side of their chest, and every "handed" component in the ship, from the taps on the coffee machine to the screwthreads holding the craft together, now seem to be back-to-front. This includes the ship's navigation lights and markings, the twist of the pilot's DNA, and the "handedness" of organic chemicals in their bodies.

Worse, the electric fields generated by a currents flowing through the wires in the craft seem to be pointing the wrong way too, and after investigating we make a nasty discovery: The ship's electrical generators are spinning in the wrong direction, but still seem to be sending electrons along the wires in the same way. Except that these are no longer *electrons*, but **positrons** – the passage through the wormhole has twisted every atom and particle in the ship into its "evil twin" **antimatter** counterpart. Should we touch the walls of the craft or allow the airlock to start pumping the ship's air into the airlock, we'll be destroyed in a matter-antimatter explosion. Things are worse for the pilot: to them, their ship has emerged from the far side of the wormhole into a nightmarish "antimatter" version of their home universe, where anything they may encounter is liable to be deadly: their only hope of survival is to pass through the wormhole a second time. This is what caused the lab explosion: by using identical point-maps when assembling our wormhole, we accidentally built ourselves a **matter-antimatter converter**.

Möbius strips and Klein bottles

The ugly technical name for this sort of inverting connection is a **"nonorientable" wormhole** (Visser 1996, 20.3). To a topologist it would be called an **"Alice handle"**, and it's essentially a higher-dimensional version of the more familiar **Möbius strip**.

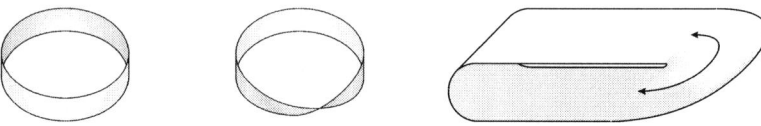

Figure 18-5: "Normal" and "Möbius" bands made from a cut-and-glued strip of paper, and a stylised representation

To make a Möbius strip, all we have to do is take a long strip of paper and join the ends together to form a loop: if the ends are joined "normally", we get a simple loop of paper with two sides (an "inside" and an "outside"), but if we accidentally join the ends with a half-twist, then when we trace our way around the loop we find that by making one complete circuit we return to our starting position, but on the "wrong" side of the paper. A second circuit returns us to our original starting point and orientation. We can no longer colour the two sides of the paper in different colours without causing a colour clash somewhere, and, technically, the paper loop now only has one side.

If we try to close the remaining edges of the paper around on themselves to form a single continuous surface, the result is a **Klein bottle**. We can visualise this shape as a closed, "sphere-like" surface, part of which extrudes outward into a tube that then (somehow) reconnects to the surface again from the inside (Figure 18-6). This gives us a *continuous* single-sided surface. We can't make a *true* Klein bottle in real life (using a three-dimensional **embedding space**), since the neck of the bottle has to pass through the wall of the bottle to reconnect from the inside, and we can't do this without interrupting the surface somewhere (unless we know a glassblower who can blow bottles in four dimensions). By passing a flat shape once around the surface of a Klein bottle (via the throat) it moves from the inside of the glass to the outside, and is "flipped over" when it returns to its original position.

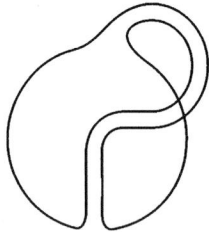

Figure 18-6: Klein bottle

By extending the "Klein bottle" idea by one more dimension, we get a shape that flips the "handedness" or **chirality** of solid three-dimensional objects that are passed around it. A left-handed glove made of "normal" matter passed through one of these connections emerges as a right-handed glove, made of antimatter.

The next two diagrams should give some sort of idea of how "normal" and "reversing" wormholes change the properties of a closed loop of space:

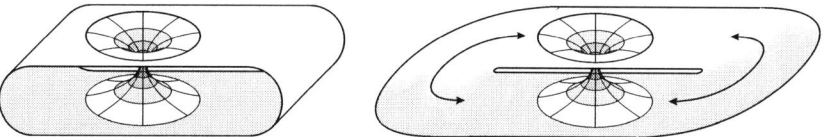

Figure 18-7: "Orientable" and "nonorientable" wormholes

With the left-hand diagram, a flat shape can be returned to its original position either by passing through normal space or through the wormhole, and it'll always end up with the same orientation: with the second diagram, whether the shape gets reversed or not depends on the route taken.

Alice universes

Antiwormholes are interesting theoretical toys, because it's not immediately obvious why (other than the **"boring physics conjecture"**, Visser 1996) they should be any less likely to exist than "conventional" wormholes. We can even imagine a variant on the toroidal universe (section 18.3) with "normal" and "reversing" routes between any two points. But in this case, the definition of which arm is "really" normal and which is "really" reversing becomes purely arbitrary, and we can no longer say definitively whether a given particle should "really" be counted as an electron or as a positron – the definition is *route-dependent*. We can still create useful *local* labels for whether a particle is positively or negatively charged (when compared side-by-side with a local reference charge), but we lose *global* definitions of charge and polarity. This sort of hypothetical universe is referred to as an **Alice Universe**, and electrical charge in such a universe is called **Cheshire Charge**, after Alice and the Cheshire Cat in Lewis Carroll's "Alice" books.

We don't normally take the possibility of antiwormholes very seriously, because the universe would have to be more complicated to allow them, and because they don't seem to solve any existing problems. Trying to be more scientific, we can say that since a "true" Klein bottle requires four embedding-space dimensions rather than three, a universe containing antiwormholes probably requires an embedding-space with one dimension more than a universe that only allows "conventional" wormholes, and we can then use **Occam's Razor** to argue that it would be bad practice to add extra dimensions to the universe just to allow antiwormholes to exist, when nothing else in physics seems to need them.

broken symmetry

On the other hand, non-orientable wormholes do provide a convenient way of looking at **symmetry-breaking**, which *does* seem to be a feature of our universe (Kaku & Thompson 1995, pp.183-). Certain nuclear reactions do seem to favour the production of certain forms of matter rather than its mirror-image, and since we'd normally expect physics to work symmetrically, this is a bit troubling. It suggests that either the whole universe's symmetry has at some point in the past collapsed into a state where our kind of protons and electrons are favoured over their anti-versions, or that different regions of the universe may have collapsed (more symmetrically) in different "chiral" directions. This would then imply a sort of physical "chirality field" that could vary in polarity from place to place, and our "antiwormhole" exercises are useful for studying the hypothetical properties of this sort of field. If the chirality of physical laws is a *locally democratic* effect, then perhaps when we start using our particle colliders to assemble significant amounts of antimatter, we'll have the chance to check whether or not proximity to antimatter has any effect on chiral "bias" in nearby nuclear reactions. There *might* be some real physics to be had, here, after all.

18.8: The Kerr wormhole

The circular singularity of a Kerr black hole (section 11.9) is interesting because some physics arguments (such as the idea that a singularity must always be hidden, or "censored") seem to break down for an observer inside the ring. It also *technically* seems to count as a slightly "odd" example of a wormhole, in that the ring singularity acts as a circular "cut" in spacetime that allows the existence of two distinct paths between regions: the paths linking two points *through* the ring and *outside* the ring can't be smoothly swept into each other without being swept through the singularity. The most basic **Kerr wormhole** seems at first sight to be an exceptionally *useless* sort of wormhole, in that it only links two regions that were already connected together anyway (!), but it still serves as an example of apparently multiply-connected spacetime that seems legal (and perhaps inevitable) under GR1915.

From here, we can go on to visualise a more advanced type of Kerr wormhole in which we have a matched *pair* of cross-linked Kerr black holes, where the two ring-singularities are actually the *same* object emerging in two separate parts of spacetime: If we were to claim that an object entering the disclike surface at **Kerr Hole A** emerges at the corresponding surface in matching **Kerr Hole B**, then no obvious laws of physics would seem to be broken (other than the more usual sorts involving black hole singularities).

This sort of "proper" Kerr wormhole would have some odd properties, not least the ability to exist in a sort of space where it can take 720 degrees of rotation to return to one's starting point: an explorer orbiting the singularity's arm would have to pass through the ring twice (2× 360 degrees) to get back to where they started. This has suggested to some people a similarity between Kerr wormholes and the properties of an electron (which also seem to have odd of "720-degree" properties), and there've been a number of "black hole electron" and "toroidal electron" models proposed by mainstream and fringe researchers.

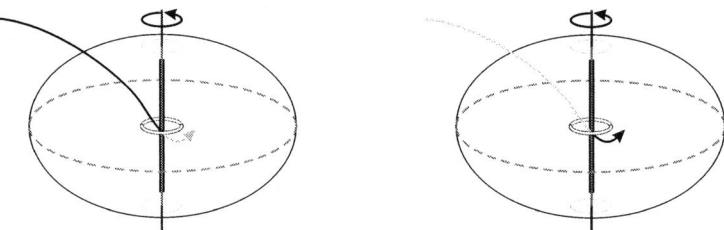

Figure 18-8: "Complex" Kerr wormhole

More of interest to us here is the fact that these (purely hypothetical) Kerr wormholes break one of the usual assumed rules of wormhole theory: since a Kerr wormhole is **planar** (flat), it only needs to be propped open in two dimensions rather than three, and the outward forces due to rotation are sufficient to do this (from the hole's point of view, the singularity is kept open by the pull of the outside universe rotating around it). **Centrifugal fields** aren't usually described as *properly* antigravitational because they only operate in two dimensions rather than three, but a Kerr wormhole only needs to be propped open in two dimensions, and for this, a centrifugal field is quite adequate. Wormholes may well still turn out to be illegal for other reasons, but the Kerr wormhole nicely disposes of the idea that they have to involve exotic matter.

Of course, "proper" Kerr wormholes still suffer from the more general wormhole problem that they don't provide us with an obvious mechanism for making a short-circuit connection between the two separated throats – while it might not be illegal to *own* one, there still seems to be no obvious legal mechanism that would let us *make* one. And, of course, we'd still have the problem of our destination being somewhere rather inconvenient, tucked away inside another black hole's event horizon.

18.9: The fieldline problem

We often visualise an object's associated *field* as a set of **field *lines*** emanating from it. How should these lines of electric or gravitational **flux** behave when confronted with a wormhole? The wormhole concept tends to undermine "fieldline" arguments, since any fieldlines that connect objects *through* a wormhole would seem to be trapped *inside* the wormhole mouth, and any fieldlines that connect *around* the wormhole would seem to be trapped *outside* it.

This causes problems. Suppose that we create a "nominal" wormhole connecting a near-Earth region to a region near Sirius, and attempted to jump through it. Before the wormhole's creation, the fieldlines that connect us to the rest of the universe obviously *don't* go through the wormhole, because it hasn't been created yet. But as we thread ourselves through the wormhole throat, all our previous fieldline connections will *still* point back through the throat and connect via the "Earth" region. If these fieldlines also represent *topological connectivity*, then, unless these lines can break and reform, we'll not be able to see out of the second mouth or interact with anything at that side of the wormhole. We might accept that we'd passed a long way *into* the wormhole, but the passageway might seem to us like a dead end.

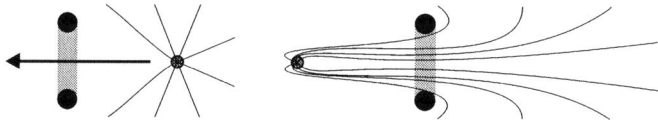

Figure 18-9: Wormhole fieldline incompatibility?

If the "fieldline" idea holds, then even if we could produce a *notional* change in the connectivity of the background metric, it's not obvious that the pre-existing web of connections that are overlaid on that background would be able to recognise and make use of the new connections, and this seems to rule out the possibility of "useful" wormholes.

John Wheeler has argued that *electric* fieldlines threaded through a wormhole could make the two mouths look like isolated positive and negative charges, so that the creation of a wormhole with these trapped fieldlines might be as an interesting model for the creation of a charged particle-antiparticle pair ("Geons", 1955). A weaker version of the "trapped fieldline" argument has been used to argue that a particle passing through a wormhole ought to deposit its mass at one mouth and remove a corresponding amount of mass from the other. These ideas are hindered (as always) by the persistent problem of how we're supposed to be able to create the wormhole in the first place without involving discontinuous changes in a region's topology. It's also conceivable that the fieldline concept itself may be unsafe or incomplete.

18.10: The gravity problem

"Connectivity" also raises the subject of how gravity might "connect up" through a wormhole. If we created a wormhole that linked our laboratory to the centre of a galaxy, would we, and our laboratory, and the rest of Planet Earth then risk being sucked *through* the wormhole into the higher-gravity region, with Earth's place in the solar system replaced by an anomalous orbiting wormhole gravity-source? If we accidentally connected to a star's interior, could the star balance the gravitational environments on both sides of the hole by squirting half its mass through the throat to produce a semi-star on our own side of the gateway, turning our solar system into a double-star system?

If the Earth accidentally got sucked through a wormhole into a higher-gravity region, it seems pretty unlikely that that region would be as conducive to human life as our own current orbit around the Sun. If we could *really* create wormholes, we'd have to bear these possibilities in mind before running the risk of doing something stupid and accidentally wiping ourselves out.

18.11: Wormhole politics

As well as the risk of accidental annihilation, there's also a risk of malicious interference.

If we take wormholes seriously as something that might be credible physics, then we have to take into account the possibility that somewhere out in the distant universe, there might be a civilisation older than us, that has already mastered the technology. If we suddenly open a wormhole mouth *here*, we might find ourselves unexpectedly patching into an existing network and finding ourselves face-to-face with some new neighbours, who may or may not be pleased to see us. An advanced, "happy" civilisation might decide that they like their civilisation they way it is, and that they'd really rather not be intruded on by outsiders who might be smelly, or warlike, or have incompatible concepts of law, ethics or religion, or might have deeply disgusting sexual habits or a compulsion to sell double glazing. They might not approve of societies that allow arms dealers or missionaries or corporate patent lawyers. They might *be* arms dealers, or missionaries or corporate patent lawyers. They might take the view that the best way to protect themselves from incompatible neighbours would be to "mine" wormhole-space, by setting up one half of a wormhole creation engine that would listen out for and connect to any tentative attempts by other alien civilisations to make a wormhole, then maintain the connection for just long enough to pop through a suitably-sized thermonuclear device to wipe out the alien's research lab, and possibly its whole species.

This strategy would allow a civilisation to automatically eliminate "stupid" civilisations at the point where they became a threat to the rest of the wormhole-using community: even if most wormhole-makers were friendly, it would only take one xenophobic alien civilisation to set up this sort of system and spoil things. With "mines" having limited destructive power, "brighter" and more empathetic emerging civilisations that were able to see things from other species' viewpoints might avoid destruction (by anticipating the risk and setting up their research in remote regions of space), and only the "incompetent" species that were dumb enough to use their own home planets as wormhole base-stations would be eliminated. If wormhole experiments really *were* feasible, this would be another good reason to ban any practical investigations unless they were performed off-world. Perhaps a *very long way* off-world.

18.12: The time-connection problem

Associated with the problem of how gravity "hooks up" through a wormhole is the issue of how *timeflow* connects across a wormhole "bypass". One argument against wormholes goes like this: if we take two connected wormhole mouths and put one into a strong-gravity region and the other into a weak-gravity region, time will run more slowly at one mouth than the other. The two mouths will age at different rates and will get progressively out of "synch". After a sufficient timelag has built up, a traveller stepping between the two mouths would be stepping backwards and forwards in time, with the wormhole functioning as a sort of time machine.

This scenario isn't necessarily correct: if a difference in timeflow *across* the wormhole also creates the usual associated gravitational gradient, *through* the wormhole, then instead of moving back and forth in time, we are just climbing in and out of a boring old conventional gravity-field: since travelling between these two locations in "normal" space doesn't produce reverse time travel, there's no immediate reason why taking an alternative route through a wormhole, *with similar gravitational gradients along the way* should follow different rules.

The "time machine" objection *might* apply if a traversable wormhole acts as a *discontinuous* stepping-stone between two regions of different gravitation ... if the wormhole connection and its surrounding transitional regions manage to be an exception to the rule that a difference in timeflow is associated with a gravitational gradient ... (spatial connection but no gravitational connection) ... but this doesn't seem to be a safe assumption.

18.13: Wormhole time travel?

Even if a wormhole connection *does* act as a short-circuit between our region of spacetime and a region that lies in the past, and we can *see* past events through its throat, it doesn't necessarily follow that we can use the wormhole for reverse time travel.

Let's take a simple wormhole whose mouths connect two different time periods at a single location, and draw the usual simplified embedding diagram.

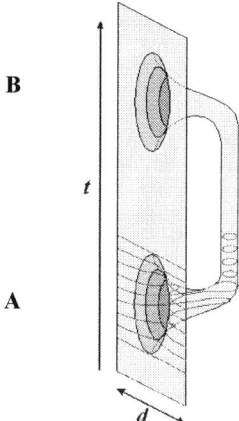

Figure 18-10: A "time-aligned" wormhole

Our sketch is rather like the usual "textbook" wormhole embedding diagram, but turned on its end (Figure 18-10 also nicely shows why mathematicians and topologists refer to these things as "handles"). The simplest interpretation of this diagram would seem to be: at time "**A**", our wormhole creation device causes a deviation from flat spacetime that grows larger and larger until it produces something like a "pocket" in space. As we follow the diagram upwards, the mouth of this pocket contracts and eventually seals off its throat to form a separate bubble of spacetime. This bubble then continues independently of the main region of universe until it reconnects at a later time "**B**", and then the second "pocketed" region flattens out again.

An observer at the second mouth can "see the past" though this second mouth when it opens, (the mouth can emit a burst of radiation that originally entered the first pocket some time earlier), but the observer *can't* use it to travel into the past. If they try to jump into the pocket, they end up being spat back into their own region as the pocket dissipates. As they stand on the threshold of the pocket, the fact that they can see signals from the past through the wormhole doesn't mean that time travel is possible: when they look back into their *normal* universe, this *also* carries images and signals from the past, but jumping back into "normal" space doesn't achieve reverse time travel. Topologically at least, "normal space" and "wormhole space" would seem to be supplying similar paths from the past into the future, and although this wormhole is traversable, it's only traversable in one direction ("forwards"). Trying to leap into the second mouth to arrive back in the past makes about as much sense as trying to do the same trick by jumping into a TV set playing reruns of old news programmes. Just because *you* can see *them*, it doesn't follow that *they* can see *you*.

Studying the diagram, we find that there *is* a particular closed line that we can draw that *does* seem to loop back on itself: a geodesic that perpetually "loops the loop" between the two mouths in normal space and through the wormhole throat. If we get onto this path, shouldn't we then be compelled to follow a worldline that takes us into the second mouth, down the throat, and then backwards in time?

RELATIVITY IN CURVED SPACETIME

There are a few problems with this idea: firstly, this central looped worldline or **closed timelike loop** may be an infinitely narrow, **chaotic**, unstable *mathematical* path, and there may be no realistic hope of following it – an object trying *very* hard to follow this geodesic might simply be pushed aside (or torn apart). Trying to jump onto this geodesic in order to follow it into the second mouth would seem to involve placing yourself at the exact position where the second wormhole mouth is expected to emerge – so the initial re-entry point for the mouth would then be somewhere inside your body. Rather than the wormhole allowing you to slide into its mouth and down its throat to the past, it seems more likely that the wormhole's repulsive properties would simply tear a hole inside you into which it could spit its contents before closing again. This does not sound like a fun exercise.

What if we "flatten" the wormhole's edge in the time direction so that a *parallel set* of worldlines point into the new mouth? Again, this idealisation (which ignores any curvature associated with real particles in the region) might not be physically meaningful, and a more sceptical guess as to what happens might be that by making the second mouth's emergence more abrupt, we might just be forcing any *physical* worldlines heading for the second mouth's coordinates to be wrenched aside more violently when that mouth forms.

18.14: Mistaken time machine behaviour

As the previous diagram indicated, simply connecting two different time periods together with a wormhole connection doesn't necessarily give us a time machine – our wormhole provided *an alternative path* between **A** and **B** from past to future, but it was still strictly a one-way connection. We can construct other forms of connection that *do* allow transfer of information between different time zones, but these don't work so easily as simple embedding diagrams, and we need to either include an additional "wormhole-like" connection *within the diagram* to make the ends join up, or to use a higher-dimensional embedding-space that allows more complex linkages between the two throats.

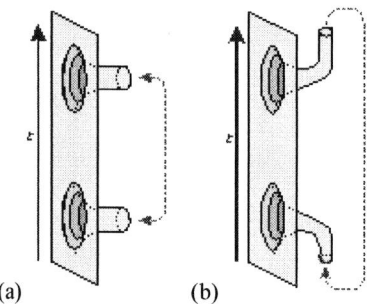

(a) *Bi-directional, and*

(b) *"reverse-only" timelike wormhole connections. Both require more complex embedding spaces than Figure 18-10, to allow the "pipe ends" to be directly connected without additional pipework. Occam's Razor then seems to frown on our inventing extra embedding dimensions just to allow reverse time travel, unless we can come up with some additional justification for the extra overhead.*

If we achieve the necessary reorientation by distorting time axes **outside and around** *the wormhole, then any physical closed causal loops are the fault of this external distortion, rather than the wormhole connection*

Figure 18-11: More complex geometries, suggesting reverse time travel

In an "orderly" universe, wormholes that *do* allow reverse time travel (Figure 18-11) seem to require an embedding space with higher dimensionality than wormholes that don't, unless local arrows of time in the surrounding metric do some very strange things: and if *that* happens, we can probably blame any resulting temporal oddities on the "weird" surrounding metric rather than on the innocent little wormhole embedded in it.

To sum up: "time machine" arguments are sometimes used to argue that wormholes can't exist, but wormhole connections don't create time machine paradoxes by default unless we make some very particular assumptions about how gravity can (or can't) hook up through a wormhole, or unless we supplement them with more twisted types of geometry that require a larger number of dimensions for its embedding space.

Route-dependent gravitational effects through and around a wormhole might allow an object to repeatedly fall "downhill" across the same gradient, gaining more energy each time, and this *sounds* like illegal behaviour. But route-dependent potentials also exist in conventional theory thanks to the dragging effects around rotating bodies, and these can be exploited to extract energy, quite legally. If route-dependent gravitational fields *aren't* illegal, we can't automatically use their presence in wormhole theory to claim breakdowns in conservation laws – we don't ban the relative rotation of matter by claiming that *it* has to produce similar energy-conservation issues and time travel paradoxes (although *see:* Tipler 1974).

18.15: Quantum foam

Classical spacetime geometry isn't supposed to allow discontinuous changes in topology, but it's been pointed out that according to quantum theory, spacetime at very small scales may look very odd indeed and may have an unfamiliar "frothy quality" to it that **John Wheeler** has referred to as **quantum foam**. The "quantum foam" argument comes with the usual QM declaration that the rules that we see operating at macroscopic scales don't necessarily apply in the quantum realm, and suggests that spacetime at small scales may be riddled with untold numbers of tiny wormholes that are too small to affect normal, "human-scale" physics.

Figure 18-12: "Quantum foam"

inflating quantum wormholes

One of the aggravating things about wormhole theory is that while field theory suggests that it may not be illegal to be *in possession* of a working wormhole, it might be illegal to actually *construct* one if the background metric isn't allowed to undergo discontinuous changes in topology. The idea of quantum foam has suggested an alternative approach: if we were to focus concentrated amounts of energy into a tiny space, to flex and distort the metric at small scales, we might be able to open up and inflate one of these pre-existing *quantum* wormholes to a macroscopic size, and then try to persuade it to do something useful.

We could then retry Plan A – try to stow one wormhole mouth in a holding pen in Earth orbit and tow the other to a neighbouring star system. The idea is still slightly fanciful, but it's not *too* many stages removed from known physics – the people interested in *making* wormholes from scratch have it much harder: they have work out how to create negatively gravitating regions, break spacetime, and persuade the two throats to successfully find each other and link up across superspace, which involves doing several things that are supposed to be quite impossible. "Tickling" a quantum wormhole that already exists and inflating it sounds *marginally* less impossible, and pumping energy into confined spaces is something that we have *some* experience of and expect to get better at as our particle-accelerator technology improves.

The problem with this plan (apart from the engineering problems of inflating, controlling and maintaining a wormhole, the energy requirements, and the need for a spaceship to tow it to its location), is that we still don't *really* know if quantum wormholes exist: even if "quantum foam" is real, we still don't seem to know for sure whether it should consist of *real* wormholes or whether it should be made of their more sneaky relatives, the **pseudowormholes**.

18.16: Scale-dependent topology

In Figure 18-13 we have a cute diagram of a **fractal solid**. If we built a model of this object one metre high, and tried to guess its shape by prodding it with a beachball, we'd probably conclude that it was a simple cube, or at least something fairly close to a cube. As far as our beachball is concerned, the solid has no navigable passageways, and the surface is **singly-connected**.

Figure 18-13: A fractal sponge

If we tried the same "poking" exercise with a smaller object (such as a tennisball), we'd become aware of the existence of tunnels passing through the shape, and as we tried smaller and smaller sensors, we'd begin to sense more and more of these additional connections. The *apparent topology* of the surface (its **rolling-ball geometry**) varies according to the way that we choose to make our measurements of it.

If we take a rubber ball that is slightly too big to enter one of these holes, the ball will "feel" the surface to have a pit or dent. If we then force the ball hard enough into this dent, the deformation of the ball (or of the shape) may allow the ball to suddenly pop through and appear at a second region that we'd mapped as another dent – although the *underlying* topology is constant, the *effective* topology and connectivity of the shape can change abruptly when we change the size of our sensor, or the force with which it is applied. It might *seem* that by using higher energies we were creating wormholes, but these connections would always have been part of the shape, even if they previously hadn't played any part in its larger-scale physics.

18.17: Pseudowormholes

Now we'll look at what happens when we try something similar with a differently-shaped solid, the three-pronged shape in Figure 18-14.

Figure 18-14: Open "cage"

When we roll our beachball over the shape, it appears singly-connected. Using smaller sensors reveals the existence of three large dents and one smaller dent, and as we apply more force, or use smaller sensors, these dents seem to deepen. At a particular size, our sensor is finally able to pass into the centre of the shape through one of the three larger gaps, and exit from any of the other two, and at this point we may be forgiven for thinking that we've shown that the solid contains a three-mouthed wormhole (or a central chamber linked to the outside world by three wormhole connections). We've demonstrated the existence of quantum foam! Hooray! Encouraged, we use successively smaller probes and discover that, with a smaller-sized probe, the fourth dent opens out into a passageway that joins the other three wormhole connections. At this point we may be quite convinced that the *underlying structure* of our spacetime model is riddled with wormholes.

 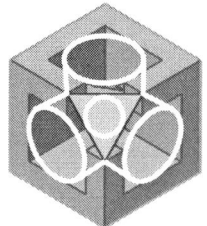 

Figure 18-15: Wormholes without wormholes

But as we probe with even more detail, we get a rude shock. Once we use a probe small enough to fit between the tips of any two adjacent prongs, we find that *all four wormhole mouths are the same mouth*. In fact, there is no *real* mouth in any meaningful sense, and the solid is (and always was) singly-connected and wormhole-free. Our "wormholes" weren't really wormholes at all, they were ***pseudo*wormholes** that disappeared and dissolved away completely when we tried to inspect them in more detail.

The next diagram represents an attempt to construct the "pseudowormhole" equivalent of our fractal sponge, by extending the previous three-pronged shape. Each cutout only extends into the solid far enough to join with its corresponding cutout from a different face, and none pass right through the solid. With this sort of shape, one size of probe might report the existence of a three-sided wormhole, a smaller probe may find an additional eighteen wormhole mouths the next size down, and so on, with more holes being discovered as we move to progressively smaller scales. It *seems* to be a fractal sponge. But if we stress this shape and allow it to flex, we find that each of the passageways stays connected to the outer surface along its full length – a cheese-wire threaded through any hole, with sufficient squishing and stretching of the solid, can be made to emerge through the front of the block without our letting go of the ends of the wire.

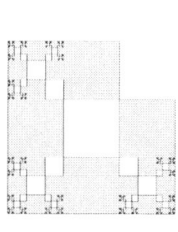 

Figure 18-16: Fractal "pseudosponge"

Wormhole-mining doesn't work in this sort of sponge.

18.18: Does quantum foam contain <u>only</u> pseudowormholes?

In our previous example, the apparent appearance of new topological features at the quantum level was an **artefact** of the measurement process.

So ... if we build a giant spaceborne particle-smasher and use it to create absurdly high energy concentrations, we might *hope* to open up a quantum wormhole to a sufficient size for it to be "catchable": then transport one half of the wormhole to a distant region of the universe and from then on, use the wormhole as a handy shortcut for rapid interstellar exploration ... but there's another possible outcome:

We build our hideously expensive atom-smashing wormhole-catcher, fire up the generators and find ... that it doesn't work. We poke spacetime about until we think we've found a dent, fire energy into its throat, it deepens and distort into a pit, but when we prise the walls of the pit apart, we find that it doesn't just dilate, but completely unfolds leaving us with no wormhole at all.

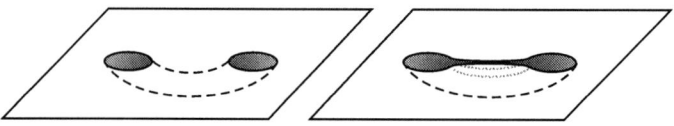

Figure 18-17: **Wormhole and mimicking pseudowormhole**

Certainly, we might expect that quantum foam should contain at least *some* pseudowormholes, but since pseudowormholes can recreate at least some of the "odd" behaviour of conventional small-scale quantum wormhole fluctuations, it seems reasonable to ask whether there is any need for "real" wormholes in our universe, even at quantum scales.

Perhaps the underlying quantum nature of our universe isn't *stranger* than we imagine, but simpler. Perhaps quantum foam is a statistical description of **virtual geometry** – the metric's *potential for creating pseudowormholes* once we start churning it up by firing high-energy particles into it. Perhaps even the pseudowormholes don't start to exist until we start hitting the metric with probes of sufficient energy to fold spacetime into clefts, as we try to find evidence for an underlying structure that wasn't there until we started looking for it.

If pseudowormhole geometry can be used to explain the theoretical behaviour of quantum foam (and this hasn't been established yet), and if quantum foam doesn't *need* true wormholes, then perhaps it doesn't actually have them.

18.19: Do wormholes exist at all?

So, if wormholes might not be real, does that mean that studying wormhole theory is a complete waste of time?

Not necessarily: even if wormholes don't really exist, these sorts of problems can be useful exercises in logic and geometry that we can practice on before tackling other slippery topics that might have a bit more to do with real-world physics.

By looking at some of these questions, we've started to explore issues that also crop up under more "reasonable" advanced physics, and if wormholes *don't* exist, well, at least that's one less way of blowing ourselves up. And some of the ideas that we've come across when wrestling with wormholes will pop up again when we start to look at warp drive problems.

We'll tackle these next.

19
Metric Engineering and Warp Drives

> *"Eppur si muove !"*
> (trans: " ... and yet it moves ")
> — **Galileo**

> *"What is to be expected along the line of Mach's thought?*
> 1. *the inertia of a body must increase when ponderable masses are piled up in the neighbourhood*
> 2. *a body must experience an accelerating force when neighbouring masses are accelerated, and, in fact, the force must be in the same direction as that acceleration.*
> 3. *a rotating body must generate inside of itself a Coriolis field, which deflects moving bodies in the sense of the rotation, and a radial centrifugal field as well.*
> *We shall now show that these three effects ... are actually present in our theory."*
> — **Albert Einstein, 1921 Princeton Lectures**

> *"A device which electromagnetic analogies predict might be able to create a non-Newtonian gravitational force that will accelerate a nonrotating, nonmoving body is one that contains accelerated masses whose mass flow is like the current flow in a wire wound torus."*
> — **Robert Forward, "Guidelines to Antigravity", American Journal of Physics [31] 166-170 (1963)**

> *"I cannae change the laws of physics!"*
> — **Chief Engineer Scott, "Star Trek"**

> "Matter tells space how to bend, space tells matter how to move"
> — **John Wheeler**

RELATIVITY IN CURVED SPACETIME

19.1: "Space bungees" and regenerative braking

Conventional wisdom says that to deliver a package to a distant location requires energy. First we need to expend energy to accelerate our payload up to a useful velocity, and then at the other end we need to expend more energy to slow it down again. This is the way that things tend to work with rocketry, and perhaps it's understandable that when it's suggested that the a (hypothetical) **warp-drive** might *not* require physics to work like this, that it might sound more like a bad case of wishful thinking than a "proper" theoretical argument.

But *in theory*, the net energy cost associated with the *displacement of an object's position* – getting it "from A to B" without worrying about what happens in between – is *zero* (plus some handling costs). Removing an object from one location and placing it at a second equivalent location *costs nothing*, because although *we* might consider useful work to have been done, as far as the universe is concerned, we have just flipped between two energetically-identical situations. We're acclimatised to the idea that transport costs are usually quite expensive, but this isn't because of the displacement *itself*, it's because of the secondary things that we normally have to do to *bring about* the displacement.

Conventional "double-ended" energy-expenditure is a practical rule of thumb when using certain types of transport (such as rockets), but it's not really a *law of physics*, and even as a rule of thumb there are some vehicle designs that don't obey it, not all of which are particularly exotic. It's not even obeyed by all public transport systems.

regenerative braking

Our "rule" is already broken by conventional electric and electric-hybrid vehicles that employ **regenerative braking**: An electric subway train spends a lot of time inefficiently accelerating and decelerating because of the large number of stations dotted along its route. These trains are fitted with large electrical motors to provide acceleration, and, to avoid wasting energy, when the next station approaches the train doesn't apply conventional friction brakes but uses its own electrical motors in reverse as generators. The train slows as the wheels drive these motor-generators to produce electrical power which is then fed back onto the subway's electrical grid, and when it's time to set off again, a similar amount of power is drawn from the grid to bring the train back up to speed. This sort of system is increasingly being designed into "hybrid" petrol/electric cars that are expected to spend a lot of time idling in traffic jams: the electric motor handles low-power accelerations, and braking involves using the car's electrical motors in reverse, generating electrical power that's used to recharge the batteries.

the space bungee / swinging on a star

Can we apply a similar concept to the theory of spaceflight? Energy-reuse in a vacuum is somewhat more difficult ... as a purely theoretical exercise, we *could* suggest tying a long bungee-rope to the end of our spaceship to reuse the outward trip's kinetic energy for the return journey, but this idea obviously can't be turned into serious engineering.

A *slightly* more practical solution might be to cut out the wasteful mid-way acceleration-deceleration stage for a round trip: A conventional rocket could be launched at a distant star, and when it reaches its destination some clever robotic piloting could thread it through the star's gravity-well so that it swings away from the encounter with its trajectory deflected towards another "interesting" star-system, and after a couple of these encounters it could finish by travelling back to us. This wouldn't leave a great deal of time for surveying systems, but perhaps lightweight probes could be dropped off (using the slingshot effect for deceleration) and their information scheduled for collection on a future flyby.

It's still a difficult plan to implement, and it still has drawbacks, but at least it's legal.

19.2: Boomeranging

The "swinging on a star" approach works when the craft's primary payload is *information*, because this can be picked up or delivered at a location without the craft having to stop. Energy recycling when a physical mass "stops and starts" is trickier to achieve. But it's still not illegal.

"**Boomeranging**" is a "proof of concept" of the idea that that simple spatial displacement is associated with **zero energy cost**. Taylor and Wheeler devote a chapter of their book "Spacetime Physics" (1990&1999, chapter 4) to the idea: boomeranging demonstrates that *in principle*, it's possible to pick up a package from one location, accelerate it to high speeds and then decelerate it again at its destination, without the package feeling any conventional accelerational forces, and without our being required to expend any energy at all.

Although this *sounds* rather unlikely, Taylor and Wheeler managed the trick without fancy warpfields or cleverly-knotted spacetime – they simply pointed out that if we bored a hole through the middle of the Earth, and dropped a package over one end of the hole, the package would naturally accelerate downwards (feeling no geeforces) until it reached the Earth's centre, and then decelerate again as it passed "uphill" back up to the surface at the other end. When it reached ground level at the far side it would momentarily slow to a halt at the highest point of its trajectory, at which point a bell could ring, a "package delivered" sign could illuminate, and a catcher-net could pop of the side of the wall to stop the package falling back into the hole.

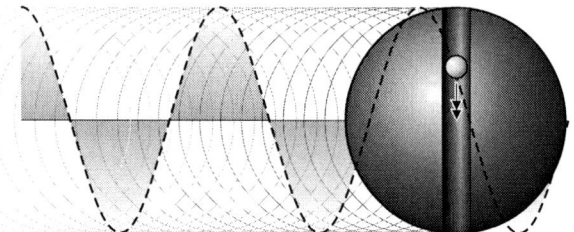

Figure 19-1: A free-falling object "boomeranging"

Zero transport cost.

T&W's book calculates a hypothetical peak speed for our package of a whopping 7.9 kilometres *per second* as it passes through the Earth's core, with the total time for the straight-line journey of more than twelve thousand kilometres being a paltry 42 minutes, using nothing but conventional gravitation. However, we probably can't use this simple "dropping" method to transfer packages faster than background lightspeed, since a basic gravitational field strong enough to produce a *lightspeed* acceleration will normally be expected to have its deepest region enclosed by an event horizon (section 19.5) … the package might well achieve arbitrarily high inward speeds, but never reappear at the other side.

Non-SR implementations

If we use "NM" velocity-shift relationships rather than SR, there *is* a certain amount of energy lost over the round trip as a function of peak velocity. In NM-based physics, the inward gravitational blueshift and the outward gravitational redshift don't cancel (section 15.6), and we have a round-trip "Lorentz-squared" energy-loss based on the peak velocity value. Our package, allowed to repeatedly swing back and forth through the shaft, will then be peaking at lower and lower heights each time. So … boomeranging with the NM equations isn't an *entirely* free trip because it effectively costs us a certain amount of gravitational potential.

Is boomeranging practical? Nope. But it's certainly *legal*, and it's a useful counterpoint to the claim that this sort of thing is impossible *on principle* in the context of warpdrive theory.

19.3: Exotic-matter drives

The concept of the **exotic-matter drive** was described in print by **Clifford Will** in the late 1980's (Davies, 1989, pp.31-32). The idea was that if we attach equal chunks of positive-mass and negative-mass materials together, the resulting unbalanced field should make the composite object try to scoot off in one direction, and since the overall mass of the object is *zero*, perhaps older SR objections (to infinite relativistic mass-dilation at $v=c$) might not apply.

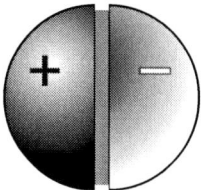

Figure 19-2: Early exotic-matter drive concept

This *seems* to give "free" acceleration without obviously breaking any laws of physics. The "catch" was where to get the negative mass, since negatively-gravitating material – **exotic matter** – is "what-if" physics, and isn't really supposed to be able to exist.

Miguel Alcubierre expanded on this idea in 1994. His paper famously described a metric in which "conventional" and "exotic" materials are cupped around a reasonably flat "payload" region, holding and moulding the perimeter geometry to form something that looked very like a science fiction "warp bubble". Alcubierre's paper produced a flurry of interest since it was using reasonably conventional mathematical tools to attack the problem, but again, we were still left with the question of where the necessary "exotic" material was supposed to come from (assuming that exotic matter was even possible).

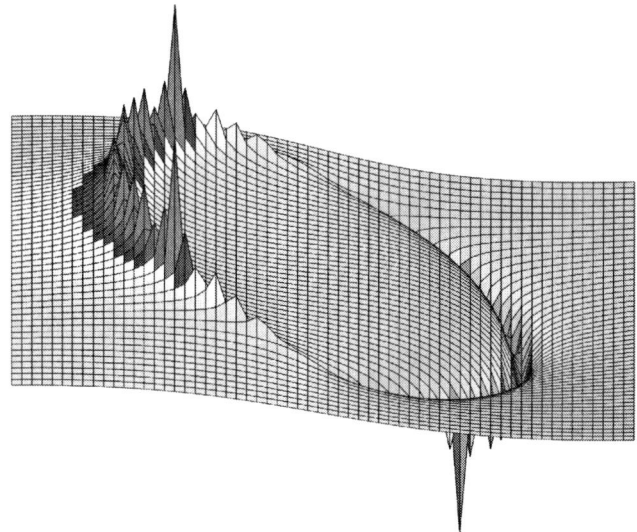

Figure 19-3: Cross-section through an Alcubierre-style warp bubble

Of course, to get the "acceleration" part of the argument to work, we don't need a *truly* negative field, we only need to unbalance the object's existing positive field. In other words, we require a **lightspeed offset**, or a deflection of the object's gravitational fieldlines, and thanks to the general principle of relativity (and possibly also to the arguments of section

RELATIVITY IN CURVED SPACETIME

9.12), we already have some ideas as to how to create this sort of distortion without using negatively-gravitating matter ... in fact, this sort of stable polarised distortion *might* be a feature of boring old everyday relative motion. While Alcubierre's distortion-field implies that an object placed within it should be moving, perhaps the argument goes both ways, and perhaps a conventional moving body produces a similar distortion-field – superimposing the perimeter field of Alcubierre's "tilted" shell onto a conventional gravity-well seems to produce something rather similar to the tilted gravity-wells of section 9.12.

19.4: The negative-field problem

If we want to move our spaceship through a region at more than the background speed of light, we usually say that this requires an increase in **the speed of light** in the region, and that **(a)** we don't know how to achieve this, **(b)** it would seem to require exotic matter, which isn't thought to exist, and **(c)** that the creation of an isolated negative gravitational field is probably illegal anyway, thanks to the **Positive Energy Theorem** (**PET**).

This approach to warpfield problems isn't really sensible. To understand why, we have to go back and examine what we *mean* by lightspeed, and remind ourselves why this term has become so prevalent. In physics there's a distinction between speed and *velocity* – "speed" is how fast something goes, and "velocity" is how fast it moves in a *specified direction* – physicists *usually* talk about "velocities" and try to avoid the word "speed" whenever possible.

However, Einstein's special theory bases its coordinate systems on the assumption that light must propagate at a rate that is fixed and **isotropic** (the same in all directions), so that the time that signals take to move between two objects along the path **A→B** is assumed to be precisely the same as for the opposite path, **B→A**. SR observers *define* the time taken by a signal to move between two points as the signal's "round-trip" flight-time divided by two, so "lightspeed" in SR-based physics really *is* a speed rather than a velocity.

If we accept these conventions, and we don't specify a preferred orientation and direction, then "to increase the speed of light" in a region suggests increasing the rate at which light through the region travels on *all three* spatial axes, and in *both directions* on each axis. *That* situation *is* difficult to produce and *does* seem to involve introducing a relatively-negative field component. But this situation isn't something that we'd need (or even want) for a warp drive.

A warpdrive-powered spaceship trying to move faster than conventional lightspeed only needs light's rapidity to be increased *in the direction that the ship travels*, which is a much less demanding condition. A matching increase in lightspeed in the opposite direction might be useful for allowing people back home to *see* the ship reaching its destination quickly, but isn't required for the warpdrive to function. In fact, a more "general" lightspeed increase isn't just more difficult, it's downright undesirable, because the associated increase in local timeflow would increase the journey-time for an occupant of the spaceship (sections 3.7, 18.5).

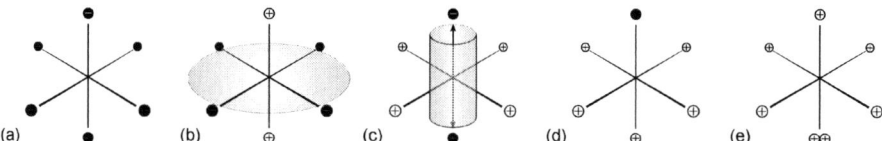

Figure 19-4: (a) full negative field, (b) apparent centrifugal field,
(c) bidirectional Krasnikov tube, (d) simple gravitational gradient
created with exotic matter, (e) simple gravitational gradient

To travel faster in one direction only requires that the velocity of light be increased in one direction (not six!), and if we don't mind the velocity of light in other directions being *reduced*, then to achieve something *like* our desired effect only seems to require a conventional gravitational field.

19.5: Ultrafast travel using simple gravity

Einstein's general theory already lets us travel faster than background lightspeed without any sort of "exotic" propulsion system – all we have to do is let ourselves fall into a black hole, and we'll then usually expect to find ourselves travelling (inward) faster than "background *c*" once we've crossed the event horizon. We don't overtake our *own* light, because it's falling in, too – we travel at more than the *background speed* of light without exceeding the *local velocity* of light. And, regardless of whether our object is really a GR1915 "black hole" or a Newtonian "dark star", this seems to be completely legal behaviour. But if we apply "round-trip" coordinate systems to the problem, we'll find ourselves saying that the *speed* of light in the stronger-gravity region is lower, because we're using round-trip measurements of how long it takes a lightsignal to pass into the region *and come out again*. So *technically*, our black hole example seems to be a case where one is able to travel faster than the official *speed* of light, without travelling faster than the local *velocity* of light.

Admittedly, a black hole's insides aren't somewhere that we'd normally want to visit, but if you don't mind a one-way trip, it's a legitimate way of breaking the human speed record … if you also don't mind that nobody back home is likely to see you doing it.

This distinction made in section 19.4 between a **velocity** (which has a specified direction) and a **speed** (which doesn't) can sometimes get a bit lost in arguments to do with warpdrive physics. It's natural for researchers to want to use nice tidy coordinate-systems based on the round-trip behaviour of signals because that makes it easier to generate results that can be proved mathematically, but those results may not always mean what we *think* they mean.

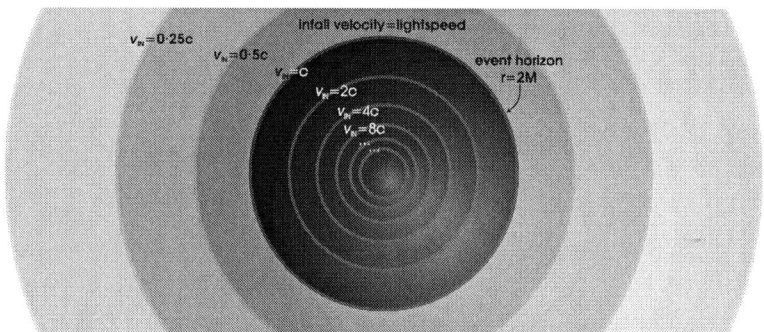

Figure 19-5: Ultrafast travel doesn't require negative energy densities

Because the phrase "speed of light" is so common in relativity theory, it's hard to stop ourselves habitually using it in situations where we should really be talking about a velocity, and this book is probably as guilty of this misuse of language as any other source. "The speed of light" is a nice phrase that trips off the tongue easily and *seems* not to require any further explanation. But when it comes to mathematical proofs involving *asymmetrical* underlying light-velocities of the sort that we expect in warpdrive problems, the phrase can sometimes turn tidy logic into a smouldering heap of ashes. While we often say that you can't travel faster than "the speed of light", our very basic black hole example involves travel faster than the *background speed of light*, faster than the *local speed of light*, and faster than the *background velocity of light* parallel to our path. It just doesn't involve travel faster than the *local velocity of light* in our particular direction of motion.

It all gets rather confusing, and as a result it seems that we can't always take strict mathematical disproofs of the possibility of "ultrafast" travel at face value. And now – having made our point – we'll go back to using the same inexact, "lightspeed" language used by everyone else …

19.6: The "cresting" problem

We've already seen how general relativity requires us to do away with the concept of global lightspeed constancy – but even with local lightspeeds that can vary from place to place, the idea of an external warp bubble that moves along *with* a spaceship (to allow it to continue travelling at more than the background lightspeed), still breaks our usual working definitions of what we *think* of as lightspeed.

Lets assume that a hypothetical warpfield-equipped spaceship exists, and is travelling at twice the usual speed of light. For the ship to continue moving, the warpfield has to move along at the same speed as the spaceship, or else the spaceship will quickly find itself overtaking its own warp bubble and entering normal space. But for the *exterior* of the warpbubble to move at *more* than background light velocity seems to involve contradictions.

- **If this warpfield can advance through the background at more than c**, then the leading edge of the warpbubble is moving faster than the velocity of light in the background metric, and we have something that moves through the region *outside* the warpbubble at more than that region's local velocity of light. The warpfield's leading edge would be an illegal **field discontinuity**, and if the warpfield caught up with "light in flight" from behind, that light could be expected to compact (illegally) to zero wavelength. Infinitely-short wavelengths suggest infinite energy concentrations, which shouldn't happen

- **But if each part of the warpfield moves at exactly the *local* velocity of light**, then the rear of the warpbubble's transitional region should advance faster than its front, so that the warpfield's leading edge quickly compacts to zero thickness, again creating a nasty field discontinuity, and again compacting the wavelength of any light caught in the transitional region down to zero. Which shouldn't happen either.

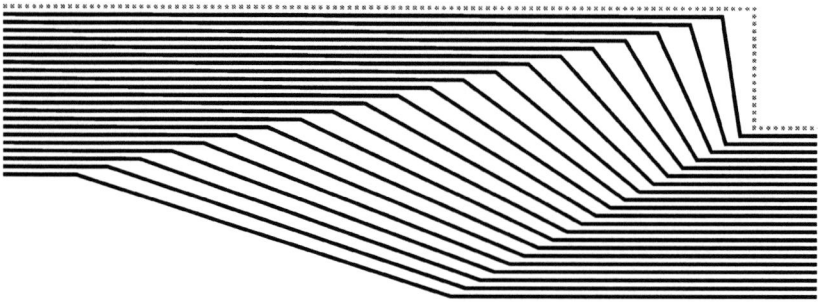

Figure 19-6: The "cresting" problem

So ... either "local c" is maintained but the warpfield advances at *less* than background lightspeed, or it advances at *more* than "background c" but violates our usual ideas about how signals propagate. Either way, we seem to get abrupt field singularities and a possible infinite energy requirement. Although the warpfield *payload* can move arbitrarily quickly, to get the warpfield *itself* to advance at the same speed, in a well-behaved manner, almost seems to require us to embed it inside another warpfield (Coule 1998)! The problem of how we're supposed to get the field *itself* to move at high speeds though its surrounding "unwarped" environment is perplexing, and for a budding warpfield designer this is a deeply annoying argument that makes it difficult to see how a self-contained warpdrive could possibly work.

Moving warpfields are complicated, so if we want to continue studying the subject we may want to start concentrating on the theoretical properties of warpfields that *don't* move, or at least, that don't move with respect to the background stars.

19.7: The Krasnikov tube

The simpler idea of a *stationary* enclosing warpfield is often referred to as a **Krasnikov tube**, and for the **bidirectional** version of the Krasnikov tube, we can visualise this "tube" as a low-density tract of spacetime that stretches all the way from the source location to the destination. We've actually met this creature before: it's a **pseudowormhole**, like the ones that we met in section 18.17.

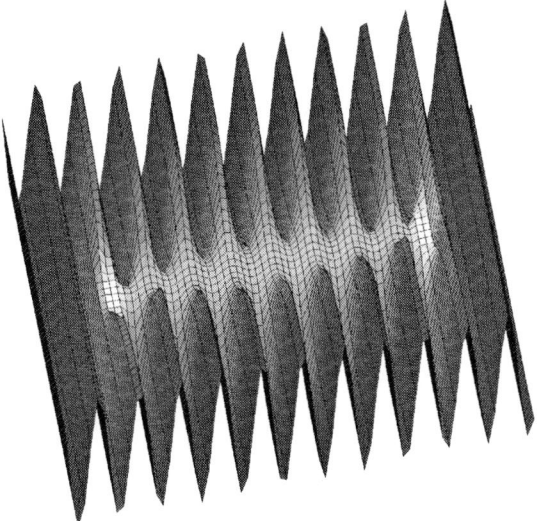

Figure 19-7: "Concertina-ed" spacetime diagram of a region containing a bidirectional Krasnikov tube

An interstellar Krasnikov tube doesn't immediately sound like a very practical device, since it seems to require warpfield-generating hardware extending over lightyears, and even if we only "switched on" sections of the tube while the spaceship was actually passing through (a "**peristaltic**" implementation of the tube – a moving warpfield generated by pre-synchronised stationary hardware), this still sounds like an awful lot of equipment.

But we might take some consolation from the hope that once the tube is switched on, the effective distance should be a lot shorter – after all, an effective contraction in space is part of what the tube is there to create. We might try to set up a small local "tube" and then drag the hardware that maintains one end of the tube across interstellar distances to the destination, so that the tube ends up "stretched", or we might set up two ends of the tube at remote locations, and then hope that they'd be able to latch onto each other and shrink the intervening distance after powerup. But although there have now been a few mainstream research papers on the hypothetical properties of Krasnikov tubes, this is still a deeply conceptual idea.

We don't need the tube to work in both directions at once, but we *do* need to remember that an **asymmetrical** tube (that increases lightspeeds in one direction but *decreases* it in the other) risks being described in common language as *reducing* the overall round-trip speed of light in the region, and this makes the resulting mathematical descriptions sound a bit peculiar.

Some SR-based arguments attempt to prove that ultrafast travel around a ring of Krasnikov tubes or wormholes inevitably leads to **causality violations** ... but since SR's geometry and conventions aren't valid in situations involving curvature or asymmetrical lightspeeds, these may not be reasonable – we have to distinguish between *real* geometrical breakdowns and cases where it's only a description (or an SR-based coordinate system) that fails.

19.8: Warpfield Hawking radiation?

But Krasnikov tubes may not turn out to be necessary.

If fields (including gravitational fields) generally shouldn't allow **singularities**, and shouldn't support discontinuous changes in field strength, or in other related parameters that could be considered as fields ("**law of continuity**"), we have two main ways to avoid this happening: we can suggest that attempts to produce a situation that suggests "cresting" aren't allowed, or we can suggest that perhaps something else happens to stop the process going to completion. If a waveform trying to "crest" would end up with infinite energy, then energy-conservation requires that something else has to give, and in the absence of any other solid candidate mechanism for what happens, we can fall back on our old standby, **quantum mechanics**.

A compacting gravitational wavefront will quickly produce an abrupt, arbitrarily-thin feature with arbitrarily-high energy – QM abhors clean, sharp edges and absolute barriers, and will act to dissipate, smooth and "fuzz" a region that seems to be trying to create one. The mechanism that we can associate with this is our old friend **Hawking radiation**, and since the sudden lightspeed differential across our wavefront can be reconsidered as a *gravitational* differential, the associated tidal forces may justify the appearance of significant Hawking radiation effects across the warpfield's leading edge, allowing at least part of the warpfield's signal to advance through the background metric faster than we'd otherwise expect.

This might solve our warpfield "cresting" problem: the information in the warpfield's leading edge might **quantum-tunnel** into the forward region of space to precondition it to accept the oncoming wavefront. Given a choice between breaking field theory and allowing quantum effects to intervene, quantum tunnelling seems to be the lesser of two evils.

Once we've tentatively established this principle, that an acceleration vector could be capable of propagating faster than the "standard" speed of light in the surrounding region, we're back to looking at "**acoustic**"-style metrics. Our arguments about the progress of the warpfield are then not so much about whether it *can* plough through the background metric faster than the conventional background speed of light, but about *how much* faster it might be able to move, whether a warpfield can "coast" or whether it requires additional energy to feed the tunnelling effect, and what the side-effects might be for the region of space being invaded by the warpfield. Does an ultrafast warpbubble advance *through* a region of spacetime in the normal way, or does it increasingly slipstream *between* coordinates, with the arrival of the warpfield appearing as a sudden inflationary bubble that inflates on the warpfield's arrival and collapses again once the spaceship has passed by? Would an ultrafast warpfield create the gravitational counterpart of the **Mach cone** and **sonic boom** associated with supersonic flight, or would it have the self-regenerating properties of our earlier "tilted gravitational well" configurations? These things are not immediately obvious.

Another approach is to try to apply the **democratic principle**, and to suggest that if lightspeeds inside the warpfield and lightspeeds outside it are creating a potential conflict in the intermediate transitional region, that this conflict shouldn't be resolved by saying that the outer region's concept of lightspeed *totally overrides* that of the inner region. We might apply a bit of topology to turn the problem inside out, and describe the warpfield interior as "normal space", and the external universe as being "inside" a warpbubble and moving past the spaceship at more than the *spaceship*'s local speed of light. While the democratic principle certainly allows one point of view to be *dominant*, there should be some sort of "give and take" between the concept of lightspeeds in the two regions where they meet together, and this suggests that "absolute" definitions used in *each* region should be damaged by the conflict.

This is the "brute force approach" to spacetime engineering: be nasty enough to a region of spacetime and it's not just spacetime that bends, but also the usual rules concerning how we might expect the bending to happen.

19.9: The "acoustics" analogue

Since the quantum-gravitational modelling of an advancing warpfield is fairly alien territory (not least because we don't yet have a working theory of **quantum gravity**), it's nice to be able to fall back on classical analogues. We did this earlier when we were facing the problem of **Hawking radiation** from black holes – in those earlier black hole horizon problems, the analogous case of an **acoustic horizon** was slightly obscure, but in our "travelling warpfield" problem, the "acoustic" counterpart is more familiar: supersonic flight.

Suppose that we have a jet aircraft moving at high speed. How fast can it move? If we start by assuming a particular speed of sound calculated from the mean air temperature and pressure, we can "prove" that there is no way for an aircraft with a sawn-off nose to move faster than the speed of sound. In our "naive" model, there's no way that the plane can move forward supersonically, because there's no way that the air can get out of its way fast enough.

The logic is essentially the same as in our warpdrive example in section 19.6: our supersonic aircraft can (in theory) drag or carry along with it a mass of air, so that the surface speed of sound around the aircraft is faster (in the plane's direction of travel) than the background speed of sound. And the plane's signals can move forwards through *this* air faster than the background speed of sound. If our definitions insist that air must move away from the nose at the speed of sound, then this forward speed is higher nearer the plane But in order for the plane to have a fresh supply of moving air to slide into, somehow the plane's motion has to cause air *in front of it* to react, and this signal has to move forwards *faster* than the normal speed of sound in the region. It's the wavefront "cresting" problem all over again: Stationary air in front of the plane has to somehow be able to "sense" that the plane is coming so that it can begin to move out of the way, but the signal telling it to start changing speed has to travel through the region faster than the "standard" speed of sound.

And yet ... *somehow* ... the plane manages it.

Once again, there are two ways that we can proceed:

ignoring theory

The usual way to deal with this situation is to say that our artificial definitions are stupid: since we know that there really *shouldn't* be an absolute sound barrier, idealised rules and conventions need to be modified (or ignored) in such extreme situations. A bullet or a cannonball, or even a simple unaerodynamic house brick *should* be able to move at many times the speed of sound, as long as we apply (and continue to apply) sufficient brute force to it.

Because we already know a fair bit about acoustics, we appreciate that the speed of sound *isn't* a primary, fundamental property: it's an *aggregate*, *averaged* consequence of more fundamental processes that can be modified in nonlinear ways. Individual air molecules aren't limited to the speed of sound: the speed of sound is the rate at which *group disturbances* usually travel though the medium, but the molecules in that medium can move faster if they want to (until they collide with another air molecule). Acoustics can sometimes appear give the impression of being a simple linear process, but as we examine acoustic processes in more detail, or start to look at more extreme cases, these nonlinear aspects become more important.

quantum acoustics

A second approach might be for us to continue to insist that the speed of sound really *is* a fundamental property, but to also retrofit QM effects to force our inappropriate "non-acoustic" metric to give the right "acoustic" answers: We could suggest that sound-information from the aeroplane travels as **phonons**, and then apply quantum arguments for phonon-phonon interference to argue that, faced with an apparent stationary barrier in the plane's frame, either **(a)** phonons quantum-tunnel forward across the barrier, or **(b)** that virtual phonon-antiphonon pairs forming ahead of the moving airplane are ripped away from each other by the shockwave – positive information appears somewhat ahead of the 'plane, preconditioning the air for the oncoming vehicle, while matching "anti-information" falls back and is absorbed by the 'plane, with "spooky" nonlocal effects conspiring to create the result that the information lost by the 'plane just happens to correspond to the information carried by the forward escaping phonon.

Of course, this second explanation sounds slightly insane. Nobody in their right minds would *seriously* attempt to explain how a supersonic aircraft moves by invoking quantum mechanics. It's a truly silly thing to do, except as an amusing mathematical exercise. And yet ... the QM explanation seems to be consistent with the physics, and produces the right answer.

It also tells us that, outside of GR1915, we already have at least one situation where quantum theory can allow a signal speed offset to advance faster than the "official" local signal speed of the background metric: in acoustics, a QM description of field-information tunnelling forwards though the shockwave gives us the correct answer. This may look like *crazy theory*, but the end result isn't *crazy physics*, because the resulting description does seem to agree with the behaviour of real physics. We've *seen* supersonic planes flying, and if that screws up someone's tidy mathematical definitions we don't really care, just as long as the math that we *can* still use allows us to do the essential engineering calculations that will let us design, build and maintain the things.

If a particular set of rigorous mathematical definitions makes this situation impossible then we need to use different mathematics, or different definitions, or change something else until theory fits reality. In physics or engineering it's less important to be rigorous than right. An exquisite proof of a wrong answer doesn't count.

spacetime is not air

Not everyone will like this analogy. Gravity is not acoustics, and treating distorted spacetime as if it was a conventional fluid sounds like a painfully misguided thing to do. But we should remember that when similarly dismissive things were said about Hawking radiation – that the idea was naive, a contradiction in terms, and that such a thing couldn't possibly happen – we got things badly wrong. Certainly, we shouldn't be too much in love with the idea that spacetime is a conventional fluid because clearly it's *not*, but we also have to admit certain *common patterns and themes* that must emerge within QM, fluid mechanics and gravitational field theory, because these three different subjects are all compelled to obey some common rules (thermodynamics, microcausality, and so on). If field theory can be used to describe transmission forces in a fluid, or gravitational effects, or quantum effects, then certain common characteristics and behaviours should emerge in all three areas as a result of these more general field principles, regardless of the particular details of their actual mechanical implementation. Quantum mechanics, treated as **meta-theory,** tells us that these common themes and patterns should appear in classical models thanks to deeper statistical principles and laws, regardless of how wildly different the mechanics and transmission mechanisms might be in each case.

So, does gravity follow the laws of acoustic metrics rather than SR/GR? It's probably still too early for the community to be able to arrive at a safe and solid agreement on this question, although perhaps the last quarter-century's work on the **Black Hole Information Paradox** has made it increasingly difficult to think of a different answer to how black holes are supposed to radiate. At the time of writing, experts are still coming to terms with the Hawking radiation problem, so perhaps five or ten years from now, if consensus is reached that gravitation really *does* seem to obey "acoustic metric" rules, we may see a sudden burst of activity in related warpfield research.

After all ... if boring old acoustics can avoid the "cresting" problem by using quantum arguments, then it's difficult to justify saying that quantum gravity applied to warpfield physics shouldn't be allowed to use the same trick.

19.10: Simple warpfield generators

linear accelerational dragging

If we surround a payload object with a hollow open-ended tube, and forcibly accelerate the tube along its axis, general relativity tells us that the accelerated mass of the tube should produce a gravitational field that will try to accelerate our payload in the same direction. This acceleration does not come for free: it shows up as an anomalous resistance to acceleration in the tube due to the gravitational coupling between the two objects.

This looks like a legal warpfield-based launcher!

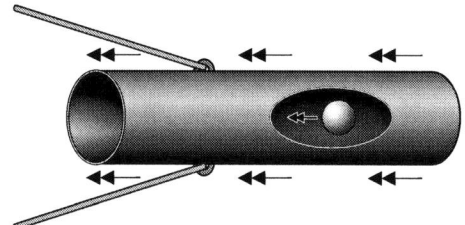

Figure 19-8: "Accelerator tube" principle

It's not, however, especially practical. To produce our accelerating field, we still need to accelerate the *tube* (using rockets, or some other conventional technology), and this rather defeats the purpose of the exercise. But our accelerated tube does provide a useful counter-example to arguments that these fields can't be produced at all – according to relativity's general principle, the association of these sorts of fields with a forcibly-accelerated mass is *compulsory* (section 4.10).

If we were to build a long series of tubes between here and a distant star, and insert our payload into one of the end-tubes, then by accelerating the tubes in sequence as the package passes through, we might be able to produce a gravitational wave that the package could ride (giving something that looks like a peristaltic variation on the **Krasnikov tube** idea). It might be very inefficient and absurdly expensive, but treated purely as a "proof of concept" exercise, the idea seems to be theoretically valid.

boring practicalities

There are a number of variations that we might try to make on the "launch tube" idea, but none of them seem very practical. Other than the reduction or elimination of the accelerational geeforces that would otherwise be felt by the accelerated payload, there seems to be no obvious reason why we'd want to go to all this trouble to accelerate a surrounding mass that then produces a weak coupling and change in speed in our payload, when we could have just accelerated the payload directly.

These ideas also don't correspond to the usual "science fiction" concept of a warpdrive because they aren't self-propelling. One of these launchers might experience a form of classical reaction force that pushes it in the opposite direction to its accelerated payload, but what we're really looking for in a "proper" warpdrive candidate would be something self-propelling – a device that could use its own field to push itself along *with* the payload, along the same route, without having to throw expendable propellant mass in the opposite direction.

We really want something that looks less like a glorified catapult and more like an *engine*.

19.11: Toroidal configurations

One of the problems with the simple "tube" configuration is that after a little while one tends to run out of tube. Is there any way we could *recycle* the material to produce a constant acceleration from a configuration of material that had a more persistent location? What if we continually removed "used" material from the front of the tube, swept it far away from the payload, then slid it around to the back of the tube again?

This would effectively turn our accelerating tube into an accelerating **torus**.

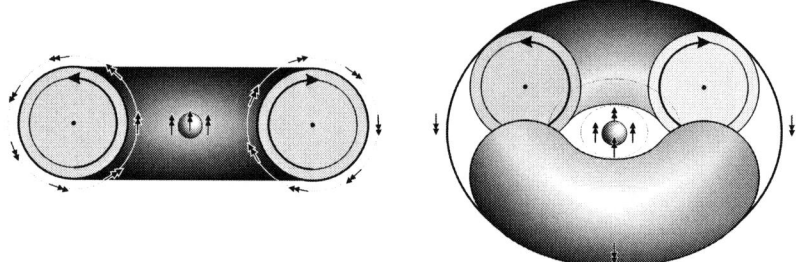

Figure 19-9: Rotational dragging through a flexible torus

The rotational dragging effects of section 9.7 suggest *some* sort of effect in this situation even with a *constant* rotation rate. If we're near the equator of a rotating star it should tend to pull us around with it, and if we're surrounded by an identical *ring* of stars rotating in the same sense as their neighbours (producing an approximation of our doughnut) there's no obvious reason why the effect should suddenly choose to stop working. So, perhaps our doughnut shape, constantly turning itself inside out, ought to accelerate material placed in its throat. The effect might be puny, but it *seems* that it should exist.

should doughnuts drift?

The next question is: if we build one of these hypothetical toroidal launchers in deep space, and its inner rim tends to push objects inside the throat forwards, shouldn't the outer rim be trying to push external masses backwards? Topologically ... yes it should. If our "pusher" is pushing surrounding matter backwards, doesn't this mean that the whole apparatus should have a tendency to push itself *forwards*? Have we accidentally created a self-propelling warpfield generator?

Well ... maybe, maybe not.

It might seem appropriate that *nearby particles* around the outer edge could be accelerated backward, and if we placed this torus in a weak atmosphere and blocked off the central passage, it might produce a weak form of propulsion by using the surrounding environmental particles as propellant ... but for a "proper" warp drive we want the unit to be capable of working in deep space, in a vacuum, in a homogenous gravitational environment.

This is where discussions of this sort of geometry tend to stop. We might argue that it's a basic precept of Newtonian mechanics that the equal and opposite reaction forces at the inner and outer edges must cancel out, *by definition*. We could break our torus up into a ring of cylinders rotating on the torus' circular **minor axis**, and argue that since each *individual* cylinder, modelled separately, will produce no sensible net reaction force of the sort that we're interested in, that the *connected ring* of cylinders can't produce a net reaction force either. The inner and outer rims have the same amount of material and the same nominal acceleration, so *if* forces add in a simple linear way *and* are entirely undistorted by the torus' body, then the combined reaction forces should cancel out if we're sufficiently far away. This result appears to be so obvious that no further argument against it seems possible.

But these are two big "ifs", and to check that cancellation is *really* the right answer, we need to investigate how **nonlinearity** and geometrical interactions might change the physics.

19.12: Cancellation and non-cancellation

We might well feel that the earlier "cancellation" result is so obviously, *trivially* correct that it doesn't need further proof, but as we try to make our argument watertight, we find that we've had to make a couple of important simplifying assumptions:

1. We assumed that the inner and outer reaction forces behave in exactly the same way, regardless of their different concentrations and background environments, and,

2. We assumed that gravitational fieldlines belonging to one part of the structure sweep through other sections of the torus at exactly the same speed as we'd expect if the region was simple empty space ("no gravitomagnetic refraction"). We've assumed no *interaction* between different parts of the torus, or between the gravitational distortions of its different parts ... we've treated our torus as if it has to obey the rules of **flat spacetime**, and we've assumed the absence of any complicating **nonlinearities** due to the different distortional effects combining in interesting ways (section 11.6).

Of course, if we start out by assuming that the rules of flat spacetime apply, then it shouldn't come as a great surprise when our exercise ends up "proving" that physics behaves as if it was a flat spacetime problem. We're simply getting out what we put in. Assuming "effectively-flat" spacetime and simple linear gravitational behaviour *will* get us to a simple answer, quickly, but it might it might not be the *right* answer, and if we look other field-manipulation devices, such as basic electric motors, we find that nonlinear effects can sometimes be important when trying to get these things to work efficiently. When we're looking at an unfamiliar field of physics, it's important not to jump to the "obvious" answers too quickly, until we've at least had a chance to find out what really counts as "obvious" in the new context.

What happens if we embrace the idea of spacetime curvature and *non*linear gravitation?

There are two immediate ways of ways of deliberately exaggerating these effects in a thought-experiment, to see what happens:

1. **We can create drastically different gravitational environments** for the inner and outer rims – we can spin the entire torus around its central axis to produce a radial **centrifugal field** across its structure (a **2-spin torus**), and,

2. **We can reduce the signal speed** in the region between the inner and outer rims by increasing the density of the material, until it ultimately becomes "singular", producing an (entirely hypothetical) **toroidal black hole**.

These two changes make the predicted behaviour of our torus much more interesting.

19.13: The 2-spin torus

If a rotation within a two-dimensional plane creates a field that acts *in* the rotation plane (our **centrifugal field**), can we give create an additional rotation, to distort this planar field and produce something more linear? If we take a body that rotates around one axis, and give it a simultaneous rotation around a *different* axis, do we get a more complex resulting field?

This might seem like a reasonable question, but it tends to prompt a short answer: that it's impossible. If we had *four* spatial dimensions, we might suggest having two totally independent rotations around two independent axes, but since we only have *three*, (plus time), this doesn't seem to be an option. We can't rotate a ball on two independent solid axes at the same time. If we drill two holes through a ball and fit two spindles, the ball is locked in position. So we're told that it can't be done. It's a naive, nonsensical question. End of exercise.

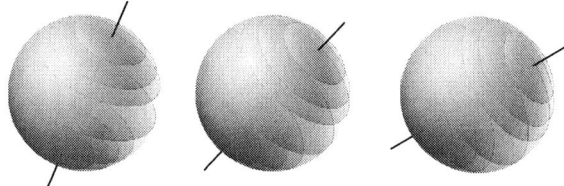

Figure 19-10: A solid, rigid, three-dimensional body can only rotate on one "persistent" axis at a time

But it we don't slavishly follow textbook conventions, and we refuse to let ourselves be fobbed off quite so easily, we find that there's a loophole in these arguments: if we don't limit ourselves to conventional "classical", rigid, solid bodies with simple surfaces, and allow the case of a *flexible torus*, this object can support two *distinct types* of rotation (before we start to confuse things further by warping geometries). A torus can have *two* rotational axes at right angles to each other, which don't intersect: a central **major axis** that passes through the torus mouth, and a **minor axis** that curves around to form a circle, runs inside the torus limb. Conventional rotation involves the *major* axis, and the torus' ability to turn itself inside out involves rotation around the *minor* axis.

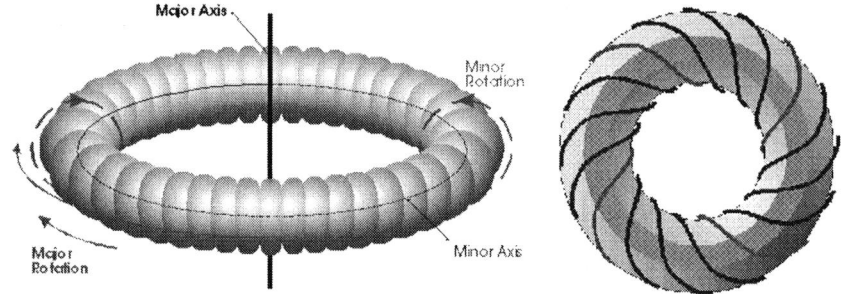

Figure 19-11: Combining two locally-perpendicular rotations

Our torus becomes a very strange object when it rotates around *both* axes at the same time: the surface vectors due to each of the two rotations are initially independent and lie at right angles to each other at each point on the surface, and the associated gravitomagnetic field becomes quite complicated. It's not now such a simple matter to say that the reaction forces due to rotation or acceleration on the minor axis *must* be the same at the outer and inner rims, because the context for these two motions is now different: the radial centrifugal field across the structure makes particles at the inner and outer rims appear to be at different depths in an apparent gravitational field, and experiencing different rates of timeflow ("centrifuge" time dilation, section 12.8). If we now try to "spin up" the minor-axis rotation rate, the two rims arguably ought to be associated with different inertial properties, so their reactions forces might be unbalanced too (unless clever higher-order effects cancel the effect). We are weaving matter in and out of this centrifugal field, through complex tidal variations that may be vastly more intense than anything we'd normally experience due to conventional gravity, and this may take us into new theoretical territory.

General arguments based on the principles of gravitomagnetism can produce what appear to be polarised gravitational field components even from simple combinations of rotating discs (Wald, 1972), so when we get to the more complex configuration represented by our 2-spin torus, perhaps we shouldn't be too surprised to find things becoming a little weird. Not only are the gravitational effects different at the inner and outer rims, the object has a distinctly polarised field pattern, with its two major-axis poles showing different properties, and this has already led **Robert L. Forward** (1963) to suggest the doubly-rotating torus as a possible candidate configuration for a (hypothetical) gravitational field generator.

This is a hellishly difficult configuration to model correctly in curved spacetime, especially if we aren't presupposing that the SR conventions are correct. If we thought that **gyroscopes** did odd things, then this configuration is like a gyroscope squared. The object's gravitational field is **chiral** – the object's field has a "twist" or "handedness" to it, creating a sort of "screwthread" distortion in spacetime. This isn't the sort of field that we're used to performing calculations on, and for spacetime along the object's major axis to be threaded like the interior of a rifle barrel strongly suggests the possibility of exchanging linear and angular momentum:

If a body connects to its environment through this sort of "rifled" spacetime, we could tentatively suggest that *perhaps*, if the object is then forcibly accelerated long its major axis, it should present an "anomalous" resistance to that acceleration that that will translate into a change in its rotation rate, and ... more intriguingly ... if we forcibly change its major-axis rotation rate, it *might* show an "anomalous" resistance to that imposed rotation that that translates into a change in its motion along the major axis, analogous to the result of spinning a nut that's been threaded onto a bolt.

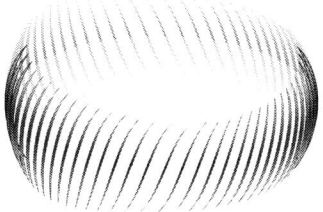

Figure 19-12: Perimeter lines of least resistance around a 2-spin torus

If we had a couple of identical toroidal units with the same minor-axis orientation, their major-axis rotations could be powered (in opposite directions) with a simple electric motor. Could we then get the device to accelerate and decelerate along its axis by spinning up and spinning down the tori? Wouldn't this count as a warp drive?

But before we get *too* carried away, we need to remind ourselves that this description is already beyond what is currently accepted as cutting-edge theory, and when we leave our tried and tested proofs behind, we increase the chances of overlooking some critical cancelling effect and Screwing The Whole Thing Up. Very Badly. These aren't easy questions (or rather, the *questions* might be easy but the answers aren't).

To say that we don't have a great deal of experience with solving dynamic problems in complex helically-twisted spacetime is rather an understatement, and relying on what intuition may *seem* to tell us about the result really isn't a safe option. When we involve complex gravitomagnetic distortions, it's no longer even obvious that forces that we'd naively declare to be at right angles to each other are really *still* at right angles. "Perpendicular" forces start to tip over and bleed through onto different axes, by different amounts and in different directions at different regions of the structure, and nice Euclidean proofs of what must or mustn't happen go out of the window.

19.14: Self-refraction and cross-refraction

The next thing we can do is to ask what happens to the gravitational field of our toroidal mass when it rotates, if those fieldlines take longer to sweep through the body of the torus than through the surrounding region of space (the gravitational counterpart of **optical refraction**, section 5.6). We might want to call this **gravitational refraction**, but this risks being confused with the gravitational refraction of light (gravitational lensing). What we're dealing with here is the more complicated nonlinear problem of how one set of gravitoelectromagnetic effects might refract another set (**self-refractive interference**).

Since we expect the speed of light to be slower in denser materials, it's not unreasonable to suppose that the speed of gravitational signals might be slowed as well, and if we take the *extremal* theoretical case of matter that's been compacted to form an *infinitely*-dense feature (a gravitational *singularity*), the speed at which gravitational fieldlines can be swept through that region should drop to zero. Singularities act as barriers to polite traffic. The "singular" version of a 2-spin torus is a **toroidal black hole**.

black hole tori

In our earlier case of a **Kerr black hole** under GR1915 (section 11.9), a body with conventional *major-axis* rotation collapsed to form a singularity, but in order to maintain the original angular momentum, the resulting singularity had to form a circle rather than a point. If we collapse a toroidal mass that is rotating on both its major *and* minor axes, then in order to also preserve the *minor-axis* angular momentum, a cross-section through the arm of the torus again has to produce circles instead of points – giving a singularity with a *toroidal surface*, with the angular momentum around the major axis expressed as the **major radius** of the torus, and the angular momentum of around the minor axis expressed as the thickness (or **minor radius**) of the torus.

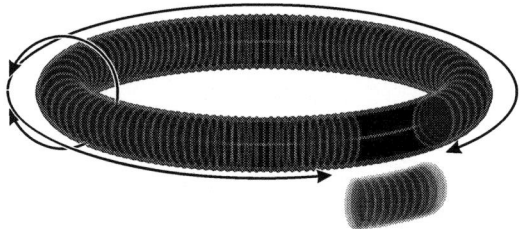

Figure 19-13: A Surface of a "toruslike" singularity

Black hole tori aren't studied much because we don't consider them to be real, even in a "perfect" GR1915 universe – they represent **extremal solutions** of GR1915, at the absolute limit at which the equations should break down. There are a few reasons for this, one being that minor-axis rotation takes the surface to the point of instability, and another being that this mathematical creature would do some very cruel and unusual things to gravitational fieldlines.

Fieldlines can't be swept through a singularity, so if the points at which fieldlines terminate are swept *through the mouth* of the torus by minor-axis rotation, their fieldlines become progressively wound up around the limb of the torus, like thread on a bobbin (this is closely related to the fieldline problem that we noticed in the wormhole chapter, section 18.9). After a very short period of time, all of this object's fieldlines would be expected to poke out from just one of its mouths (the one where rotation moves matter inwards), and we then might expect its gravitational field to be concentrated in the direction that this mouth is pointing towards. The object's gravitational field would be explicitly asymmetrical.

It's worth repeating here that torus-like black holes are *not* expected to form in real life under GR1915, quantum mechanics may prevent physical singularities from ever forming, and the "anomalous" linear momentum associated with our fieldline-winding effect may act to resist and perhaps eliminate and replace any minor-axis rotation. This angular momentum may be "used up" by the fieldline-winding effect. There are only a few published research papers that deal with on these objects, and these tend to treat black hole tori as interesting mathematical toys rather than as credible physics.

It seems that these objects should never *really* form. But they illustrate the extreme case of the *total* concentration of a body's fieldlines at one end thanks to a *totally* unintersectable body ... if our real distortions merely *slow* the sweep of fieldlines through the torus body objects only, we might expect a *partial* asymmetrical concentration or deflection of fieldlines.

If this slowing is a standard side-effect of conventional matter, we can start to tentatively suggest that a 2-spin torus might show an "anomalous" resistance to its minor-axis spin, and the fieldline deflection might create a corresponding "anomalous" amount of linear momentum along its major axis. What's not so obvious, though, is whether this effect and the similar sounding effects in section 19.13 couple together naturally, or whether they cancel each other out (like the arguments in section 9.11). We may have personal suspicions as to what the correct answer ought to be, but proving them is difficult.

19.15: General field-refraction issues

As we retreat from the "extremal" toroidal black hole solution to something more realistic, the possible effect of intervening matter on the apparent fieldline distribution of environmental matter raises some interesting questions:

If we go back to the "accelerated tube" example of section 19.10, we *might* reckon that since the usual gravitational influence of the tube is too weak to measure, that if we distort its gravitational field by accelerating it, the consequences should *also* be too weak to measure. But this idea is undermined by Mach's principle and the GPoR ... the equivalence of inertial and gravitational effects means that the tube's inertial mass should *also* be a field effect, and inertial mass is a very tangible thing indeed: if we talk about modifying an object's *inertial field* rather than its *gravitational field*, our expectations for the strength of the effect may be quite different (sections 3.4 and 4.16).

For simpler velocity-based effects, relativistic aberration effects (section 8.3) suggest that a signal passing through a wall moving at a constant velocity will emerge after two opposing aberration shifts with the same angle that it started out with ... but in more complex situations, cancellation probably isn't the correct outcome, and perhaps cancellation *shouldn't* occur when bodies are accelerated. So: does the payload mass within the tube see the distortion to be affecting just the tube's *own* geometry and fields, or does the tube's acceleration also warp the apparent geometry of the starfield behind it, and the *background* field?

spun glass

We might want to consider the case of a large rotating glass disc: the disc feels an outward centrifugal field and a rotational Coriolis field, and the effect of these *relative* fields extends through the disc's surface to produce frame-dragging effects around it (section 4.14). But we can't assume that the disc's own distortions sum linearly with the background gravitational field of the outside universe (section 4.16) – a significant coupling could strengthen the effect.

One way to try to model this would be with an "optical" approach: if we take the case of an observer sandwiched between two of these discs, their view of the starfield through the glass might end up *rotated* (Jones 1971) ... but a more interesting question would be whether lensing effects due to the rotational fields could also make these stars appear through the spinning glass to be offset *radially*. *If* that's a genuine result, and if the two spinning discs appear to change the environmental distribution of massenergy for a central nonrotating viewer (assuming that acceleration-shift and velocity-shift effects don't entirely cancel the effect, *e.g.* section 9.11), we might be manipulating the background field in a significant way.

But this involves going beyond SR and taking accelerational effects seriously. Nonlinear calculations are difficult, and this problem – of how strongly an intervening object's accelerational and rotational distortions affect any gravitational fields passing through it – still doesn't seem to be solved.

19.16: Momentum conversion

A warp drive based on momentum-conversion (section 19.13) is a pretty wild idea, so how do we prove that it's wrong?

The immediate objection to hypothetical devices of this type is that they violate the laws of momentum. Our hypothetical unit might not violate the *general* principle of momentum conservation, in that it's not giving us anything for free – we have to put in extra energy to force the thing to spin up – but it does violate the usual rule that angular and linear momentum are conserved *separately* and that one sort of momentum can't be converted into the other.

Is this separation reasonable? We used to have separate laws for the conservation of mass and the conservation of energy, but when $E=mc^2$ replaced them both with a combined law of **"mass-energy"** conservation, nobody seems to have been very upset. We also had separate laws of conservation for momentum and energy, which fused in later relativistic textbooks into a law of the conservation of **"momenergy"**, again, without any obvious rioting on the streets.

While we often pay "lip service" to this separation, we're usually happy to set it aside when it's inconvenient. For instance, in a steam-powered railway locomotive or a petrol-engined car, a burst of expanding steam or exploding petrol vapour pushes a piston in a straight line, this is coupled to a crankshaft or drive shaft that produces rotation, and then the rotation of the vehicle's wheels propels it forwards. At first sight, it *looks* as if we're converting back and forth between angular and linear momentum, and as the vehicle accelerates away, it *looks* as if that linear momentum has come from nowhere. We don't always *see* anything carrying an equal and opposite amount of momentum in the opposite direction.

The usual response is to say that if we studied each of these cases in sufficient detail, the two separate types of momentum *would* still be conserved once *all* the associated reaction forces are taken into account (if we include things like the angular momentum of the Earth being fractionally turned by the relative acceleration of the locomotive above it, the whorls of water leaving a ship's propeller, and so on). But unless we have an isolated system to study, the existence of some of these momentum components often can't be verified and has to be taken on trust: when we hop onto a bicycle and pedal off down the road, it isn't practical to check whether the Earth *really does* microscopically alter its motion through space beneath our wheels to compensate, but we don't lose sleep over whether our bicycle is really a "legal" device. Patent offices didn't refuse to patent bicycles and motor-cars on the grounds that the inventors couldn't *show* where the compensating momentum went. We understand the basic principles of the bicycle, and we assume that if momentum-conservation laws are correct, they'll be quite capable of looking after themselves, regardless of whether we're aware of all the details or not.

Eric Laithwaite produced a good example of this: a bullet is fired from a high-velocity rifle into one end of a wooden railway sleeper that is drifting in deep space. Before the impact, the bullet has simple linear momentum. After the off-centre impact, the "struck" sleeper will be rotating about its centre of gravity and will have a definite angular momentum. If we step away from clever definitional tricks and look at what *seems* to happen, it certainly *looks* as if one sort of momentum has been converted into the other.

If we want to argue that practical momentum conversion is *impossible*, we have to find a way to reconcile this with the Laithwaite's "sleeper" example. If we don't know how to do this, then the argument isn't a credible one. On the other hand if we *do* find a general method that explains how the "sleeper" example works without breaking these laws, then that method might also work for more involved cases, such as our warp drive examples.

Until we know how to make these laws obviously compatible with *current* mechanics, perhaps we shouldn't be too insistent to apply them to warpdrive calculations … perhaps we should worry less about the fine details of this accounting process and more about whether a unit should actually start moving when we power the thing up. If it does, we can argue about the accountancy details later.

19.17: "Reactionless" drives and deferred momentum

Warpdrives are sometimes referred to as **reactionless drives**, because they'd seem to involve "illegal" acceleration without any obvious Newtonian "equal and opposite" reaction forces pointing in the other direction. But since gravitational distortions can carry momentum, it might be that (in some warpdrive configurations) powering up one of these units creates a rearward-aimed **gravitational wave** that carries the missing reaction force – then, at some date in the distant future, when this wave strikes faraway stars and gives them each a tiny push in the opposite direction the "momentum sum" turns back into something more conventional-looking. In this scenario, a warpdrive's overall momentum-accounting would still balance out nicely even though it carried no conventional propellant … but it would still have to expend massenergy thanks to the energy carried away by the gravitational wave.

Ideally, a more advanced configuration might allow the craft to store this "reaction" momentum as a "momentum–debit" within a field-configuration that was carried along with it – if this is possible, the physical reaction forces would be **deferred** and used to brake the craft at its destination (the craft's own future deceleration acting as the "missing" reaction mass for it's initial acceleration). If momentum can be temporarily concealed within a field pattern whose exterior presents momenergy values to the outside world that aren't representative of its *total* field, then our reaction forces may be wound up inside a distortional kink that could be carried along with the warpdrive and released when the drive is shut down.

19.18: Can we build a working warp drive?

At the moment nobody really seems to know. Or at least, while there are plenty of definitive statements in print to the effect that such things are physically impossible, the supporting arguments are often fairly shallow.

We *can* produce a number of geometrical proofs of the impossibility of ultrafast travel by using logic based on special relativity, but since SR's internal consistency depends on the *absence* of curvature effects, invoking a "flat spacetime" theory to disprove a warp-drive idea is a little perverse: It's a bit like disproving the possibility of "heavier-than-air" flying machines or other aerodynamic effects by applying the simplifying assumption of total vacuum, or proving that fish can't swim by starting with the assumption that the oceans don't contain any water. This type of disproof isn't worth very much, no matter how much impressive-looking mathematical work may be bundled in with it. Warpdrives aren't compatible with special relativity? Well, neither is simple gravitation, but we don't use SR to "prove" that *gravity* is impossible. That would be silly. And we already know that in the case of objects simply falling into a black hole, conventional "proofs" of the impossibility of anything moving faster than the background speed of light can't be taken at face value.

extreme caution

It's tempting to get carried away with some of the arguments in this section and to start sketching out designs for two-stroke warpdrive configurations and the like, but we need to temper this enthusiasm with a certain degree of deliberate scepticism.

We don't yet know for certain whether any sort of warpdrive will ever be practical, and even if they *are* practical, we don't know if their applications might be severely limited, perhaps to sub-light, low-velocity long-term propulsion units for interplanetary probes, or for allowing satellites to make minor adjustments to their orbits without requiring disposable propellant. We might not be talking "Star Trek". And there may be other issues that make even these potential applications problematic or unrealistic. Does a given combination of fixed and variable rotations and accelerations on and around a given combination of axes translate into a final acceleration, or a velocity, or maybe just a displacement? We don't yet seem to know what the status of these devices might be, or what rules they follow.

If they work, then that's great. If we can prove that they *don't* work, that's great, too, because at the moment our knowledge of the subject is pretty awful. As of 2000AD, our current arguments as to why warpdrives *shouldn't* exist are horribly flaky, and if those are the *best* set of counter-arguments we can manage – well, there's room for a lot more work ... and that work may well be useful in other related situations where our limited knowledge is letting us down.

where next?

Warpdrive research within the context of GR1915 has been held back by the influence of special relativity, and the lack of proper tools for studying these problems in an explicitly curved-spacetime context, has contributed to a rate of progress that is dismally slow: we have the basic recognition of gravitomagnetic effects by Einstein and others, Robert Forward's key paper in the 1960's, some work by Wald in the 1970's, and not much else until we get to the more metric-based arguments of Alcubierre and Krasnikov in the 1990's (which deal more with the hypothetical properties of warped metrics than the question of how to actually create such things) ... and then a small flurry of secondary papers exploring the consequences of Alcubierre's and Krasnikov's ideas. Otherwise, we don't seem to have a lot of relevant published mainstream work to show for the last century.

Researchers can be wary of getting into a subject at the very early stages – it involves a high risk of failure, and the risk of being seen as someone involved in "fringe" research, or of being remembered for suggestions that turned out to be wrong. For now, even quality-assessed, peer-reviewed papers on the subject are often quite likely to be significantly wrong in some respects. But this is absolutely okay: even a "wrong" paper documents an argument whose rejection can act as a springboard to further work, as long as people don't take the rejections personally. "Blue sky" research often results in dead ends and false starts, but unless researchers are prepared to run the risk of being wrong, a subject can't progress. It's a process of exploration.

But with a few honourable exceptions, researchers have simply decided not to attempt these problems. This has left us in the humiliating position where the development of much of the subject's language, conceptual vocabulary and logic has been left instead to the science fiction community (some of whom are "moonlighting" physicists) ... this decades-long absence allowed physicists to avoid being called rude names by their colleagues, but there's a cost: the general opinion of the physics community on these matters now isn't worth so much.

But things aren't altogether hopeless. Since the "double-spin" torus seems to be a recurring theme when we try to attack these problems, and since it appears to be the only special-case geometrical configuration that we can think of that *might* allow this sort of effect (and since Forward already suggested it way back in 1963), we're not being confronted with a totally blank page, or overwhelmed by too many possibilities. If we want to try to prove or disprove the warpdrive concept rigorously, we have a good place to start.

PART VII

THE HUMAN FACTOR

RELATIVITY IN CURVED SPACETIME

20
Limitations of language and procedure

> **The 100% rule:** *Everything whose truth you are **totally** convinced of, is going to turn out to be wrong.*
>
> **The 10% rule:** *Ten percent of the other everyday stuff, that you just naturally <u>assume</u> to be true, is wrong, too.*

> *" The logic now in use serves rather to fix and give stability to the errors which have their foundation in commonly received notions than to help the search for truth. So it does more harm than good. "*
>
> — Francis Bacon, "Novum Organum" (1620)

> *" 'Treason never prospers'*
> *And aye, there's a reason*
> *For if it doth prosper*
> *None dare call it treason "*
>
> — Old English proverb

> *" 'This must be the wood,' she said thoughtfully to herself, 'where things have no names ... ' "*
>
> — "Through the Looking Glass", Lewis Carroll

> *" Computers are useless, they can only give you answers. "*
>
> — Pablo Picasso

> *" It's not the number. It's the meaning. It's the syntax. It's what's between the numbers. ... "*
>
> — "Pi", Darren Aronofsky (1998)

> *" Orthodoxy means not thinking – not needing to think. Orthodoxy is unconsciousness "*
>
> — "1984", George Orwell

> *" <u>**These**</u> go up to eleven! "*
>
> — "This Is Spinal Tap", Rob Reiner (1984)

> *" I'd notice the difference," said Arthur.*
> *" No you wouldn't," said Frankie mouse, "you'd be programmed not to."*
>
> — "Hitchhiker ... ", Douglas Adams

> *" Form a circle? But there's only two of us, that's really more of a line. "*
>
> — Giles, "Buffy the Vampire Slayer"

> *" You are not thinking. You are merely being logical. "*
>
> — Niels Bohr

RELATIVITY IN CURVED SPACETIME

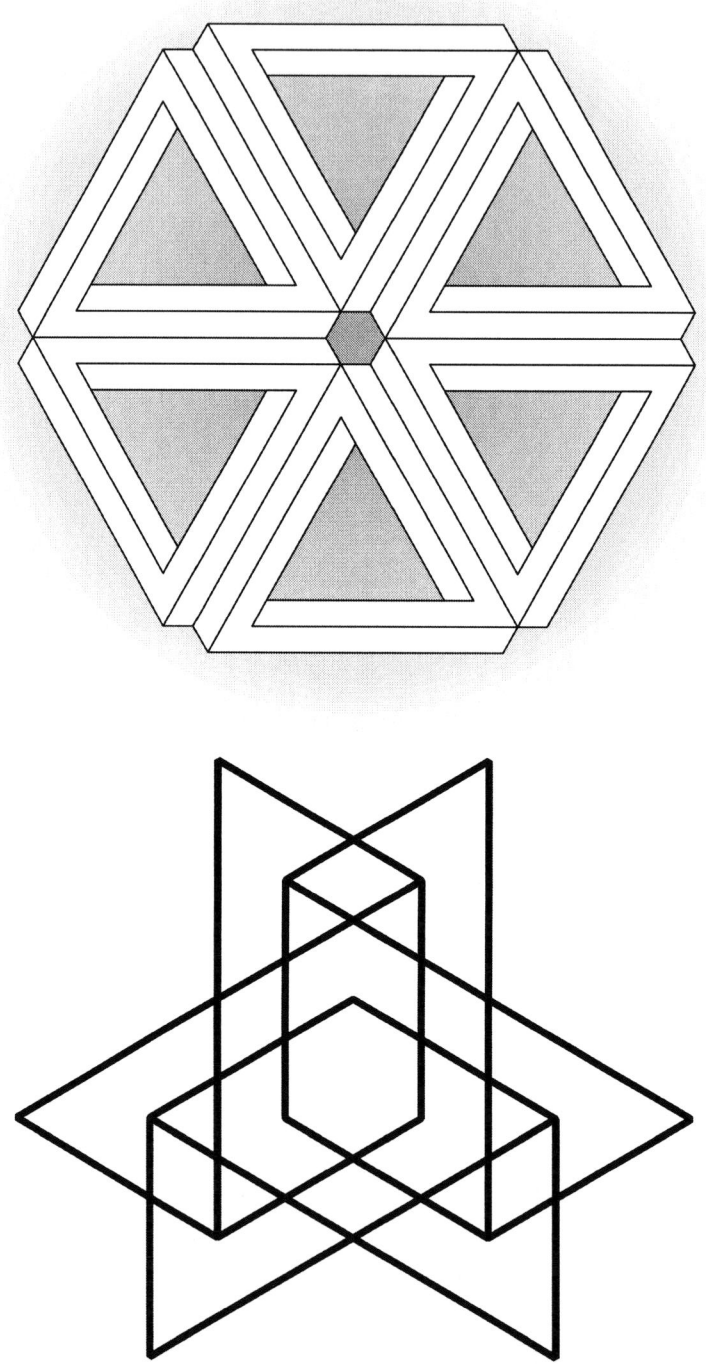

20.1: The order in which things are written

Definitional problems can show up not just inside theories, but also in our approach to *conceiving and constructing* theories in the first place. To study them we have to step outside the usual mathematical and logical structures of individual theories, and look at the psychology and thought-processes of the people who construct these theories in the first place. We can sometimes find examples where the inevitability or impossibility of a result seem mathematically *unavoidable*, but where deeper analysis shows that the conclusion may have grown from a convenient definitional decision made at an early stage of a theory's development, which may in turn depend on tiny changes in emphasis or on a particular use of language. A result that seems to be a "hardcore", rigorously-proved piece of physics can sometimes be tracked back to the psychological consequences of a historical accident.

"equals" signs and eternal damnation

Sometimes even the most innocent-seeming mathematical statements can carry a payload of assumptions and meanings that aren't always contained on the page, but are expected to be applied by the reader. In mathematical physics, there is often a great deal of "small print" that isn't just "small", but completely missing.

Let's take one of the simplest mathematical statements possible:

$$A = B$$

We might say that the statement tells us that **A** and **B** are equal. We might elaborate on this a little, and suggest that perhaps it means that **A** and **B** are carriers for *values* whose magnitudes are *numerically* equal. But this is not what the statement *says*. We didn't *write* "**A, B, (=)**", and if we read our three characters out loud, as a piece of sequential language, what the statement actually *says* is that "**A** *equals* **B**", which is a slightly different thing. It means that **B** is taken as our reference, and **A** has a value that has to match it. Considered as a statement in **sequential logic** or as a piece of computer code, or a statement used in physics to convey to cause-and-effect relationships, "**A=B**" and "**B=A**" can carry different operational meanings. If we say, "*This apple is green*", we accept the notion of "green" as an independent, pre-existing concept and assign it to the apple. If we say, "*Green is the colour of this apple*" (or, "*This apple IS green*"), we're using the colour of the apple as our reference property, and (re)defining our concept of "greenness" to agree with it.

In theoretical physics, the way that we choose to interpret these sorts of subtle linguistic nuances in a model's embryonic stages can dramatically influence the final structure of a theory, and can sometimes change the final physical predictions that it will end up making.

20.2: Lightspeed, velocity, and language traps

Just as we can't assume that the statements "**A=B**" and "**B=A**" always have the same function in sequential logic, we also can't depend on being able to freely exchange the two statements,

$$v < c$$

and

$$c > v$$

Both statements can be interpreted as saying that when we compare the values for *v* (the velocity of an object) and *c* (the speed of light), *c* will be found to carry the greater value. But the way that we choose to *write* this, as "*v is less than c*" or as "*c is greater than v*", can carry additional meaning about the relative priorities of the two properties being compared.

The different implied meanings of these two innocent-looking statements can be made clearer by rewriting them in full, and replacing "v" and "c" with phrases that signify the usual meanings of these two variables:

$v<c$, "The object's velocity is (always) less than the speed of light" implies that c should be taken as our reference speed, and that when we examine the velocity of a moving object we'll always find it to be less than this reference value. This statement tends to push us towards Einstein's special theory of relativity: we can find ourselves saying that because we *know* that lightspeed is constant, and that the velocity of an object is always less than this constant value, that the textbook speed of a lightbeam *in vacuo* represents an absolute upper limit to the speed that any object can ever attain. If all inertial observers then agree that a particular object *has* to be moving slower than "the speed of light" in their different frames, then we're led, perhaps inevitably, to the language, concepts and mathematics of special relativity. By referring to "*the* speed of light" as our reference, we also tend to assume that there can only be *one* speed of light, so that we use *global* lightspeed constancy in our derivations ... and if we want to reconcile *global* c-constancy with the principle of relativity, we seem to require the relationships of SR. When we then go on to build a more ambitious theory that can also "do gravity", we'll naturally assume that it should reduce to those SR equations, giving us GR1915 (or something very similar to it). So this first statement directs us towards special relativity, to GR1915's "perfect" and inescapable gravitational event horizons, and to a division between incompatible theories of general relativity and quantum mechanics.

$c>v$, "The speed of light is (always) more than the object's velocity" suggests a different sort of physics. If we say that no matter how great v is, c will always be *larger*, then if c isn't *infinite*, this implies that the speed of light should be modified somehow by an object's motion, to prevent the object from being able to move faster then its own light. Since this version of the statement doesn't put a limit on the legal values of v, if we wrote down a value for v that was greater than ~300,000 km/s, we'd require the rapidity of light leaving the object's nose and travelling forwards to be *greater* than ~300,000 km/s. The one-way velocity of light would then become a *variable* property influenced by the speed of objects. By reversing the order in which we write our simple three-character mathematical relationship, we're drawn (almost inevitably) to a different style of physics in which lightspeeds must be modified by the motion of nearby material. Velocity warps the spacetime metric, the mathematical relationships become those of an **acoustic metric** rather than those of special relativity, the basic equations of motion diverge from the textbook equations, the behaviour of gravitational horizons is different, GR1915's "perfect" black holes don't form, and there's no longer an obvious conflict between quantum theory and classical principles.

By starting with *either one* of these two statements, we can construct a chain of supporting arguments that *seem* to show that the type of physics theory that we ended up with was the only solution that was ever possible. If we define our technical language in the context of the prevailing theory, the "other" option (whichever it is) may end up not just seeming to be *impossible*, but might not be capable of being expressed using the standard technical terms – it might even seem logically *inconceivable*.

By emphasising incremental logic and standardisation, we can end up thinking that we've proved that the decisions taken in the past were necessarily correct, and we can shut ourselves off so completely from the wider range of solutions that had been available that eventually we lose the ability to see them altogether. Once we enter this sort of **logical black hole**, more rigorous mathematics usually won't help us to escape, but will just help us to dig ourselves even deeper. Strengthening our logical systems without querying the initial assumptions that we based them on just makes the trap stronger.

20.3: Fractured logic

Sometimes a line of reasoning can be derailed by a shift in the implied usage of words *partway though* an argument. Because this happens through a *failure* of language, it can be difficult to use language to prove that a mistake has happened, and perhaps it's easiest to illustrate the effect with a few examples.

For instance:

> "*The ancient Greeks thought that atoms were indivisible, but nowadays we know better – they can be split.*"

This statement is misleading: the ancient Greeks may have come up with the name and general concept of "an atom", but it was *us* that chose to apply that name to particular objects, only to find that it might not have been completely appropriate. The problem here (if there is one) isn't with the ancient Greeks but with our more modern application of the name "atoms" to things that aren't truly "atomistic". The "atoms" referred to be the Greeks and the "atoms" referred to by us are not *quite* the same things, and the "atoms" in the first half of the sentence are not *quite* the same as the "atoms" in the second half, even though the same word is used for both.

Let's try another example:

> "*Economists tell us that car ownership is a good indicator of the standard of living, But people in Germany don't drive "cars", but "autos", so "car" ownership in Germany is vanishingly small. We can deduce from this that the standard of living in Germany must be very low indeed.*"

No real economist would make this mistake, because obviously, a distinction that is purely linguistic (cars/autos) is being misapplied, and treated as if it carried a *physical* significance.

Now let's paraphrase a common misconception to do with special relativity:

> "*SR predicts **transverse redshifts**. The existence of these redshifts has been experimentally verified. When theories other than SR predict redshifts in the same situation, these should **not** properly be referred to as "transverse" effects, because the accompanying explanation is different – that wording should be reserved only for SR. Since SR is therefore the **only theory** that predicts transverse redshifts, it is the only theory that can explain the experimental data.*"

This argument gets abbreviated to something like:

> "*Transverse redshifts are unique to SR, so the experimental verification of this effect is compelling evidence for the correctness of the special theory of relativity.*"

Once we know what earlier theories *really* predicted, this second statement appears to be **junk science**. It takes a *linguistic* distinction between theories, and presents it as if it refers to a difference in the theories' *physically testable* predictions. Taken at face value, the statement is physically wrong, and if we allow enough ambiguity for it to be correct on a technicality, the statement becomes merely pointless (like declaring that "Germans don't own cars").

But this sort of logic trap is difficult to break out of, because the entrapped person can support every individual stage of their argument with reference to the literature. If they only think *linguistically*, it can be difficult to find a form of words that demonstrates that their conclusion is **pathological** unless we can eliminate enough linguistic and logical overhead. We might try to explain, for instance, that with a particular arrangement of hardware, special relativity will predict *this* number for a detector's redshift reading, and these older theories will predict *these* numbers, and that *all* these numbers are unambiguously greater than zero. Once people have been trained to accept a fractured line of reasoning, deprogramming them is hard work.

20.4: Logic traps and logical black holes

A **logical black hole** is the "logic" equivalent of a GR1915 black hole: it has logical paths or arguments leading into it, but none that lead out again. Once we accept a statement that leads us into a logical black hole, we descend into a system of mutually-self-supporting arguments and redefinitions that makes it impossible to leave.

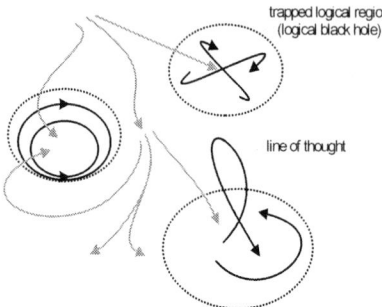

Figure 20-1: Logical black holes

How do we tell if we're trapped in a particularly effective logical black hole? Using conventional linear logic ... we can't. The hole's structure is *woven from* these linear logical strands, and following any strand that points away from the hole's innards takes us back into the hole. From the inside, it seems to be a self-contained, closed system, or at least a decent approximation of one.

circularity

One of the things that makes logical black holes so alluring (and so potentially dangerous) is that if we *were* ever to stumble across a recognisable "final truth", then perhaps it *ought* to be compelling and all-encompassing enough to count as a special-case logical black hole.

When a theory achieves internal consistency by relying too heavily on circular logic, we sometimes take this as evidence that the overall consistency of the theory has been achieved artificially ... we normally prefer to see a solid chain of sequential logic that takes us from where we are now into a particular belief system by travelling in a particular logical direction. However, if we were ever able to find a "true" final theory, a **Theory of Everything (TOE)** that explained *everything* within a self-contained system, with no arbitrary prior causes or initial assumptions or anomalous dead ends, then *its* definitions and relationships would probably be circular, too. This puts us in something of a quandary: the further we get to a Theory of Everything, the more difficult it is to use conventional logic to tell whether we're dealing with "real" truth or the result of an artificial **pathological belief system** that has trapped us somewhere other than our intended location.

sanity-testing and the search for a better hole

If conventional linear logic won't tell us whether there is another, potentially better world outside our own hole, how are we supposed to judge whether our current beliefs represent the best of all possible holes or just a local pothole? How do we break out of our logic-trap?

The usual method is to momentarily set aside sequential reasoning and create a deliberate logical short-circuit that lets us jump *discontinuously* to a different region of logic-space, to see what the scenery looks like from outside. We call this game, "what if". If we currently believe that certain things are inescapably true, how possible is it that we could still have ended up believing in these things even if they were *not* true? Do all paths lead to our original belief system, or do similar numbers of paths lead to other systems that are comparable to ours, but which have certain differences? Does the wider range of possibilities (even if we consider them misguided) lead to a number of alternative logical black holes, and is our originating system automatically the *best-looking* hole, or just a location that ensnared us because of an accidental choice of initial path?

This deliberate use of discontinuities or "random" data to break out of a pre-existing logical system is sometimes referred to as **sanity testing**, and can be particularly important in testing data analysis software. If we trained a program to look at images and spot the location of fish, and we wanted to know how far the program's output could be trusted, it could be useful to feed the program a few "wildcard" images that *didn't* contain any fish, to see whether the program correctly identified and reported a problem, or whether it continued to happily "see the fish" and tell us where they are when shown images of the Eiffel Tower, or Heathrow Airport, or randomly generated noise.

Another nasty possibility is that perhaps a final theory might be so *totally* circular and so *perfectly* internally consistent that it might be completely sealed off from the surrounding logic-space. In this case, we might not be able to enter the system using a conventional incremental route. We might still be able to use guesswork and a sense of design principles to work towards something that progressively approximated the theory, as a *parallel* attack on the problem – in the hope that once we had the solution completely surrounded, we could tighten our net and see everything finally snap into place – but having found our final theory, we might not necessarily be able to come up with a conventional sequential mathematical derivation ("*A* therefore *B* therefore *C*") of a route that someone else could follow to reach it.

If conventional derivations start from known foundations and build upwards, and if a final theory has no conventional hierarchy or foundations as such, but a set of emergent properties which all mutually define one other, then there might be *no* conventional starting point that we can use for a rigorous linear derivation. In this scenario, if a "final-theory" candidate was *correct* it wouldn't be provable by conventional means, and if it *was* conventionally provable, it wouldn't be true.

theories as logical black holes

It's sometimes difficult to tell from *inside* a theory whether it's become inconsistent in a wider sense, because once we've come to believe that a theory is correct, and have committed ourselves to it, any subsequent discoveries of mismatches between the theory and reality can be seen not as evidence that the theory has made *wrong* predictions, but as showing that the theory has helped us to discover "new" physical effects (section 20.6).

With an inappropriate model we may find ourselves continually discovering "new" effects and adding them as corrections, much as the **Ptolemaic system** (in which the Earth was considered to be immovably fixed at the centre of the Universe) required a seemingly limitless series of correcting **epicycles** to bring the theory's predictions about planetary orbits into line with observation. Does the Hay-Schild controversy (section 12.8) tell us that GR1915 is *structurally flawed*, or does it usefully tell us that we need to adjust and downgrade our concept of the importance of equivalence principles? Does the "dark matter" prediction demonstrate that the universe *really* contains some new sort of invisible material, or does it show that GR1915's equations just don't work properly for large-scale gravitation?

It can become difficult to tell whether these things mean that a theory is giving us valuable new information, or whether they simply demonstrate that it's broken. Once we reach the point that we believe in a theory so strongly that we're *always* willing to restructure other information around it, it's probably more proper to class the theory as a **belief system** than as a proper, fully-**falsifiable** scientific theory.

20.5: More examples of circular thinking

conspiracy theories

A **conspiracy theory** is good example of a logical black hole. Anything that supports the theory can be taken as supporting evidence that the theory is right, while anything that contradicts the theory can be taken as disinformation by the conspirators, and as again providing evidence that the theory is right. When two or more conspiracy theories contradict each other, showing that at least *one* of them must be wrong, this can be taken as evidence of an even larger conspiracy that constructs false conspiracies to prevent people from seeing the *real* conspiracy.

Once we've taken this path, there may be nothing that anyone can tell us to show us that we're wrong. We're sceptical of conspiracy theories because they can usually be used to explain almost *any* evidence retrospectively – ("*dinosaur bones exist in rocks because people put them there*") – but they aren't so good at making predictions.

If we can *always* construct a conspiracy theory to explain any situation, then an agreement between one of these theories and the known facts isn't so impressive (and won't usually be taken to be particularly significant) unless there's some particular set of circumstances that makes a particular *sort* of conspiracy seem especially likely to happen.

religious conviction

Some established religions have had time to evolve "logical black hole" patterns of thought that make it difficult for converts to conceive of other ways of thinking. In these systems, anything that might support the "truth" of the system is typically taken as positive evidence, while any indications to the contrary are taken as evidence that the Creator is working in mysterious ways, or (in more extreme circumstances) is actively supplying disinformation in order to test us ("*dinosaur bones exist in rocks because God put them there*"). Evidence, lack of evidence and contra-evidence are all taken as supporting the system.

Adherents may get the chance to see their own belief system more objectively when different religions come into contact – if they can see "the other guy's religion" as a *false* logical black hole, this may suggest the possibility that they could be in a similar situation themselves. To avoid these unwanted revelations, some mainstream religions have evolved the concept of **heresy** – ideas that are considered too dangerous to be considered, documented or even spoken out loud. Considering alternative views may be considered sinful or even **"evil"**, and people holding those views may be said to be working for Satan or some other powerful figure that God allows to exist, again for the purpose of testing us (a *theological* conspiracy theory). There may be social penalties for writing or allowing others to be exposed to these "heretical" views, and the propagation of "harmful" arguments contrary to the established doctrine may be taken as a malicious deed intended to hurt a worthy institution, demanding punishment. In the case of **Giordano Bruno**, who dared to disagree with Church teaching on cosmology (amongst other things), his punishment was being burned alive at the stake.

The more heavily a belief-system relies on extreme countermeasures, the more we may begin to suspect that perhaps its view may not be entirely optimal – perhaps *real* truth would be sufficiently robust to not require such levels of protection from dissenting voices, and perhaps an embedded belief-system supported by these sorts of defences would tend to *continue* being successful even if it was pretty rotten. So the enduring popularity of a system that is defended by extreme mechanisms for exerting emotional and social pressure doesn't necessarily signify that the system being defended is *really* any good. The success of a system that manages to become popular while still allowing adherents freedom of choice and openly allowing challenges – the way that science is supposed to work – will be more impressive.

military intelligence

One of the telltale signs that we may have entered a delusional state is when *every* outcome that might indicate that we're wrong becomes interpreted as further evidence that we're right.

A classic example happened in the runup to the 2003 invasion of Iraq, which was said to have an ongoing **"Weapons of Mass Destruction"** (**"WMD"**) development programme. As supporting evidence for the existence of the programme obstinately refused to materialise, it began to be argued that this *absence* of evidence didn't mean that we were wrong, but showed that the programme was instead bigger and more extensive than we'd initially thought.

The reasoning went like this: since we "*knew*" that the programme existed, our inability to find it told us that the country must be *actively concealing* the programme from us, and the harder we looked without success, the more insidious the concealment programme seemed to be. Eventually, we'd looked *so* hard (still finding nothing significant) that we concluded that for the country to be *this* successful at hiding the programme, it must be allocating significant national resources to a huge concealment operation, and this effort would only be worthwhile if they considered the project to be strategically important. As time passed, and the absence of evidence became increasingly marked, the perceived threat continued to inflate. Experts who questioned the existence of the programme were ridiculed and discredited, and to query whether the programme really existed was to look naive, or worse, to look like a dupe of an enemy power who was intent on attacking us.

Shortly before the invasion, the British Prime Minister explained on television that if only Iraq was prepared to come clean about the existence of its WMD programme, even at this late stage, the invasion wouldn't have to happen. But if the programme was so important that they were prepared to lie about it even when faced with imminent attack, it meant that they must have *strategic intent* to use it against us. Faced with this degree of blatant deception over a threat to our national security, we had no option but to invade.

After the invasion, it seemed that the reason for the scarcity of evidence was more prosaic: we hadn't found an ongoing WMD programme in Iraq because ... there wasn't one.

We'd hallucinated it.

compatibility "by definition"

We met the suggestion in sections 12.8-12.17 that there might be deep incompatibilities between the definitions of GR1915 and those of special relativity. Where SR models the actions and transfer of inertia and energy in the absence of fields or distortional effects, the general principle of relativity seems to make such a thing a contradiction in terms. Is GR1915 then *incompatible* with SR?

The "stock response" seems to be that GR1915 is compatible with SR "*by definition*".

Definitional consistency is a good thing during the development of a theory, but once the theory's structure has been set, compatibility should be a natural consequence of the chosen structure. If we have to continue to make adjustments and suspend principles from time to time in order to *force* compatibility, then any statements that we might make that the theory is known to be consistent won't carry a great deal of weight. We could take a theory whose components we believe to be *in*compatible, and by artificially writing "compatibility" into its specification, magically upgrade every fudge and bodge that would be required to keep that theory on track to the status of legal derivations: to say that such a theory is then consistent *by definition* isn't so impressive, *defining* something as true isn't the same as demonstrating that it *should* be true according to deeper laws or principles. When we say that GR1915 is compatible with SR *by definition*, and then use this *declared* compatibility to say that GR1915 is known to have no internal conflicts, we've entered a logical black hole by way of a logical fracture.

20.6: Is consistency all it's cracked up to be?

A more difficult problem shows up when we try to define what we *mean* by a consistent theory. Suppose, for the sake of argument, that we think that Einstein's general theory is inconsistent in its definitions and its approach, and we say so. A mainstream theorist can tell us that our opinion isn't worth much until we can make our case properly. We go away and find what seems to us to be a *definite* problem with, say, the relationship between GR1915 and its SR-based foundations. The theorist tells us that hand-waving arguments aren't enough, and that we now have to *prove* that these problems are *definite and unavoidable*. We go away and prove an *unavoidable definitional incompatibility* between some components of GR, and show that GR is *definitely* inconsistent. Have we won our argument?

We haven't. Because by showing that a theory is *consistently inconsistent*, we've unwittingly made it consistent again. We've generated an *extension* to the core theory. If we were to demonstrate *conclusively* that, say, equivalence principles don't work properly under GR1915, our theorist could respond that *of course* these principles are *now known* not to work consistently, but that GR1915 earns extra credibility by being able to *show* us the breakdown. What we originally identified as a "bug" has magically transmuted itself into a "feature". GR1915 is now considered to be an *even better* theory because of our work, because it's now apparently telling us new things about the universe that we didn't know before.

Each *consistent failing* that we find in a theory can be presented as the starting point for a *consistent development* of it. When something fractures in the deep structure of a model and the community invents an ugly bodge or workaround, they can initially say that they're not doing anything wrong because this is just a temporary "engineering fix". But once we *prove* that the theory would crash if we didn't use this process, our proof inadvertently elevates the *ad hoc* fixup to the status of a proper legally-derived procedure. Once the fudge can be proved to be *necessary*, fans of the model can say: since this fix-up is now known to be *unavoidable* in the context of the theory, it can now be legitimately absorbed into the theory as part of the theory's "modern" implementation.

This is a recurring problem with mathematically sophisticated models – almost any model, no matter how bad, can be shown to be consistent if we're allowed to apply enough mathematical analysis to the problem. If our theory says that a particle possesses one value on Mondays and another on Tuesdays, we might try to argue that the theory is inconsistent in its application of that value, but a theorist can respond that, actually, the theory's behaviour may be strange, but that it is entirely *consistent* once we factor days of the week into the equations.

But surely consistency is a good thing to aim for? It certainly is in mathematics, but in physics the situation isn't so clear. If a theory is *incomplete* in the sense that it's not pretending to be a final theory, then perhaps it ought to contain holes that are waiting to be filled by the things that the theory is missing. Perhaps for an incomplete "theory of the universe" to be credible, it may have to contain a small degree of inconsistency in order to have a chance of being correct. If a theory that is *too perfect* in its implementation of an incomplete set of phenomena, it might not just be incomplete but wrong.

This turns some of our older certainties on their heads. If "good" theories must be "bad", and only "bad" theories might be "good", then how do we judge whether the apparent breakdowns in GR1915 make it a "bad" theory for being inconsistent (before we patch it up) or a "good" theory for being (correctly) incomplete?

Perhaps part of the answer involves walking away from strict mathematical proofs and applying old-fashioned physics criteria. Is a theory true to its founding principles? Is it efficient? Does it make testable predictions? And are they correct? Instead of asking whether it's "true", perhaps we should be asking: *Is it useful?*

20.7: "First answer" syndrome

Humans have an understandable tendency to decide that once we have *a* solution to a problem, that the work is finished. It's solved. From here, it's a small step to assume that since the answer that we found seemed to be correct, that it's "*the*" correct solution", with the word "the" suggesting that there can only be one. But sometimes a problem has multiple potential solutions, and sometimes different solutions can be good for different things.

Let's suppose that we're given an exam question designed to test our understanding of basic geometry, and it asks us: "What regular polygon is obtained by cutting a cube in half?"

The obvious answer is: "a square", and that's probably the answer that most people will write down. But it's not the only answer, because by cutting the cube at a different angle, we get a perfect regular hexagon.

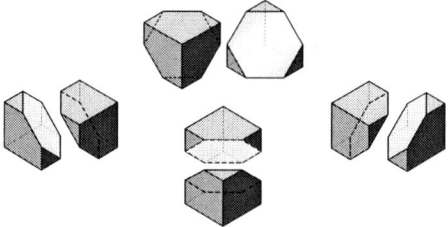

Figure 20-2: Seeing the hexagon: Cross-sections of a cube

So, there are at least *two* possible solutions to the question, not one. When the question asked us *what shape* we get, it implicitly told us that there was *one* solution, and that there was *a* correct answer (as opposed to: "*name one of the shapes* ..."). We're not likely to be rewarded for writing down something other than the obvious answer, and risk being losing marks for giving an answer that an examiner might consider to be not just *wrong* but *pathetically* wrong. Why risk spoiling your exam score just to make a point? We know which answer we're *expected* to produce, and after a while (if we're good at exams) we become accomplished in analysing the psychology and social background of examiners, and second-guessing which answers they'll consider the most pleasing. It's a game: we learn to tell authority-figures what they expect to hear, and in return, we get suitably rewarded.

And, of course, to us, the "hexagon" answer looks awkward. A square is surely a much more convenient thing? But this may betray a cultural bias in our way of thinking about geometry: as land-dwellers, we naturally find it easier to position a cube by resting it on a flat surface and then cutting at right angles to the surface, and we "like" right-angled blocks because of their usefulness for constructing buildings in a surface-based gravitational environment. But to a zero-gravity creature this advantage may not be so obvious: they may find it easier to hold the cube by its opposing corners, and to cut at right angles to the line joining those corners, especially if the creature's vision makes it easier to sense corners and edges than flat faces, or if it analyses shapes by feeling them. If the creature has three-fold symmetry rather than two, it may find the "hexagon" answer more aesthetically pleasing, and might argue that this provides *four* distinct ways of halving the cube, whereas the "square" solution only allows three. A starfish or octopus that spends a lot of its time feeling shells for corners and edges to prise open might consider the "hexagon" answer to be more intelligent. But *we* don't see the "hexagon" possibility because we've been trained to see a cube in terms of squares.

"First answer syndrome" seems to be common in SR literature ... because SR predicts something (such as $E=mc^2$ or transverse redshifts), we assume that SR is the *only* way to generate these effects. But we don't have a huge incentive to look for alternative solutions, and if we *find* one, we're liable to be told off for having done something stupid.

20.8: Life, Death, and the Square Root of Two

The Pythagorean Brotherhood

In the Sixth Century BC, **Pythagoras** founded what was meant to be the perfect society, where pure logic and pure reason would hold sway over petty political squabbling and more mundane considerations. The **Pythagorean Brotherhood** had its own laws and conventions, and was founded on the belief that Nature operated according to the rules of perfect harmony and order. This was supposed to be self-evident. So, when one of their brethren called **Hippasus** suggested that some numbers might *not* fit into this orderly scheme and – worse – that that a ratio that ought to be one of the most beautiful numbers in Nature, *the square root of two*, could be one of these numbers, the Brotherhood reacted in just the way that we would expect of any perfect society that considered itself to be composed of rational seekers after Truth.

They killed him.

an infinite number of monkeys

In science it's common to assume that one can arrive at an absolute truth by starting with an orderly set of basic rules and conventional logic and, by applying a sequence of precise logical steps, to either arrive at the correct final answer, or to be able to at least snuggle up close enough to it for some clever cancellation trick or mind-flip to take us all the way there. This is the usual way that "industrialised" science research operates ... it assumes that if enough people work diligently and logically, like the components of a vast computer, eventually all the gaps in our knowledge will be filled in. With enough worker-bees calculating additional decimal places, finally we hope to come to a full stop where the sequence ends.

"**Root 2**" kicks this idea in the teeth. It says that for some problems, successive approximation and curve-fitting will never quite get us to a final truth, if the language that we use isn't adequate for the job. No matter how big a computer you use, or what number-base, we'll never be able to write "√2" in full as a conventional stream of digits, or as a perfect ratio between two integers. For certain problems, the "successive approximation" approach isn't only *not guaranteed* to get us to a final answer, it's *guaranteed not* to get us to a final answer.

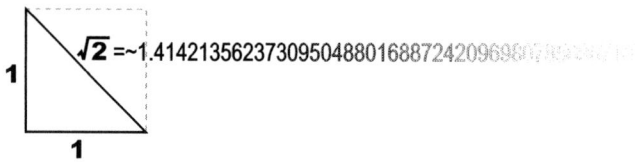

So, Hippasus earns his place in history for three reasons: Firstly he recognised the concept of the **irrational number**, a horrible creature that simply didn't exist in the orderly number-bases and perfect integer proportions of the Pythagorean research project. The second aspect of his discovery was that the square root of two was such a number. It's difficult to imagine how much of an affront this must have been to the Brotherhood's sensibilities: to persist in contemplating even the *existence* of these perversely indefinable beasts may have seemed like the action of a warped mind, deliberately setting out to poison everything that the Brotherhood stood for: but Hippasus was going even further than this, and claiming that the simplest right-angled triangle possible, the one that we get by cutting a square in half diagonally, with sides of 1, 1, and (by **Pythagoras' theorem**) "root two", birthed one of these monsters.

The reaction of the Brotherhood gives Hippasus his third historical distinction. We're told (Singh, 1997) that he was executed by drowning and that Pythagoras himself is supposed to have signed the death warrant. This makes his demise possibly the first instance on record of death by scientific peer review.

20.9: The story of Pi

Part of the reputation that mathematics and geometry have for being definite, respectable, *fundamental* subjects is that they appear to be beyond cultural bias. Okay, so we scored a few points for social bias in the "square vs. cube" problem, and we found that some *supposed* geometrical proofs associated with special relativity turned out not to hold up to serious inspection (sections 6.4, 16) but there are surely *some* fundamental truths to be had that are definitely not open to debate, and the two basic results that people tend to remember from school geometry are **Pythagoras' Theorem**, and **Pi**.

Pi-i-i-i-i-i-i-i-i-i-i-i-i-i-i-...

Everybody knows what Pi is. Represented by the Greek letter "π", Pi is the ratio between the circumference and diameter of a circle (the ratio between the circle's perimeter distance and its width). Like the square root of two (section 20.8), Pi is an **irrational number**, and can't be represented as the ratio between any two whole numbers.

If we lay a conventional square coordinate-grid on top of a circle, the perfect curve of the circle's perimeter never *quite* matches up with the grid. We can *approximate* the circle's area by colouring in and counting the squares inside the circle, or perhaps counting the squares inside and those outside and splitting the difference, and by using greater and greater grid magnifications to make sections of the line approximate more and more closely to straight-line sections ... but no matter how far we zoom in, our grid-counting result still won't be exact as long as the circumference still has a slight curve to it, and there seems to be no clever error-cancelling method that can produce a nice whole-number ratio. The rules that Pi operates under are not the rules of conventional integer-ratio mathematics, and if we want to express Pi as a conventional number, we need an infinite number of digits to do it, as the value weaves in and out of our carefully constructed number system without ever hitting it exactly. This is true whether we construct our grid in "base ten", "base three", or *any* other conventional number base ... no matter what equably-constructed grid we apply, under whatever magnification, Pi can never quite be pinned down. We can describe Pi as an infinite *sequence* of operations or *series of terms* that will allow us to churn out a numerical value to an arbitrarily high number of digits, but when using a conventional number-base, the stream of digits never ends.

~3.   1415926535 8979323846 2643383279 5028841971 6939937510 5820974944
      5923078164 0628620899 8628034825 3421170679 8214808651 3282306647 0938446095 5058223172 5359408128 4811174502 8410270193 8521105559 6446229489 5493038196

It's the three!

The second reason why π has such an occult attraction, the reason why people write books about it and obsessively plot it to thousands of decimal places ... is because it starts with a three. As we've pointed out, the square root of two is *also* an irrational number, but it begins with a "one", and hardly anyone seems to lie awake at night worrying about *it*.

As a general rule-of-thumb, fundamentally significant numbers close to unity ("=one") tend to be expressed as a value *greater* than one, and they do *not* start with threes. They start with ones. "Ones" are straightforward and need no further explanation. "Twos" are a bit disturbing, and if they exist can usually be gotten rid of by finding some way to justify dividing the thing by two to get back to a ratio of one point something. But a *three* ... that's downright freaky. So for centuries, numerologists have been spending their lives studying these dancing digit sequences, trying to decode the hidden message. *What does the three mean?*

As it happens, there *is* an interpretation of the digits of Pi that ascribes a deep meaning and significance to the leading number three. That message is: guys, you've been spending all these years looking at the wrong number.

RELATIVITY IN CURVED SPACETIME

If we look at the properties of a circle, one of the fun things that we can do (if we have a mathematician's concept of "fun") is to mark two points on the circumference and compare the difference between the shortest *curved* path and the shortest *straight-line* path between them. This number is always one-point-something, and varies between being vanishingly close to one (when the quantity of curve is vanishingly small), all the way up to a maximum of about 1.57... (when the two selected points are at opposite sides of the circle). The number is a useful measure of the "amount of circle" that we're looking at and the degree of curvature involved. We can generate an interesting sequence of irrational numbers by dividing the circle into equal-sized segments (by drawing a **regular polygon** inside the circle), and reading off the ratios, and for a six-sided figure, where the hexagon's sides happen to be the same length as the circle's *radius*, our special ratio is π/3, or ~1.047197551... .

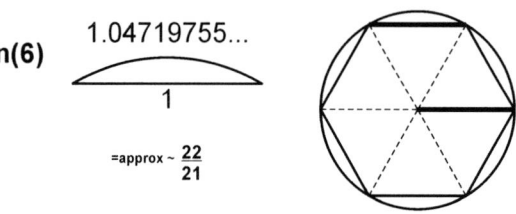

Figure 20-3: Pi-alternatives

For a *square* drawn inside the circle, the ratio is 1.5 × n(6)/√2 = ~1.1107... , for the limiting (borderline-legal) case of a "flat" two sided figure, it's 1.5 × n(6) = ~1.570796...), and so on.

But π itself does not appear as part of this sequence

why Pi?

Geometrically, our choice of the "circumference-to-diameter" ratio for Pi doesn't seem to make a lot of sense. A circle's *circumference* or perimeter is a round-trip, "there and back" measurement, while "diameter" is an open, strictly one-way distance. With π, we aren't comparing like with like, and the result therefore isn't likely to be a truly fundamental number. π/2 and π/3 *are* fundamental in this sense, but π itself doesn't seem to be in the same league. So instead of making schoolkids memorise **π=3.14...** and "the circumference of a circle equals 2πr" (~"six-and-a-bit"), perhaps we should have been showing them the "hexagon" diagram, telling them that the "hex" ratio is about ~1.04719... and then having them remember that the circumference of a circle is six times *that* times *r*. Much less mysterious.

So where did the convention of using π come from? Perhaps it's to do with stonemasonry. Presented with a stone pillar, you can't easily measure the radius, but the circumference is easily measured by wrapping a piece of string around the pillar, and the diameter can be measured with callipers, so perhaps the mysticism surrounding Pi is less to do with *geometry* than with conventions used in the construction industry a couple of thousand years ago.

The moral of this story is that even when presented with the most neutral-looking, apparently *unarguable* fundamental building block, that seems to have *no possible* social bias, it can still come with built-in assumptions or less-than-optimal implications that can survive unchecked for hundreds (or in this case thousands) of years. Certainly, Pi freaks some people out, and certainly it mystifies schoolkids, and certainly, if we were inventing mathematics again from scratch, we'd probably be better off without it. But making the change *now* would require some effort, and people who've grown up with Pi probably aren't that bothered to change things. If it's an ugly evolutionary leftover like the human appendix, well, we don't go to the trouble of taking people's appendixes out until they become life-threatening. Even in mathematics, efficiency is sometimes considered less important than following established conventions.

20.10: Pi and global extinction.

Pi may not be harmless. Some people think that we should be broadcasting Pi into space in the hope that some alien civilisation may pick up the signals, decode them, and realise what great guys we are. But a broadcast incorporating Pi might carry more information than we'd intended to share with alien civilisations, and perhaps it would be a message that we shouldn't want anyone to hear ...

the "hidden meaning" interpretation

How would a hypothetical alien signal-analyst treat our broadcast?

To begin with they may have trouble understanding what it is, since our number may not be on their hotlist of "special" numbers that other civilisations might want to broadcast, but let's suppose that they realise that we're broadcasting something that *relates* to part of a fundamental sequence.

They might think it vanishingly unlikely that a civilisation would go to the trouble of beaming a signal out into space to be intercepted by other species, and then screw up the choice of numbers. They may say, it's *inconceivable* that a technologically advanced civilisation could be this dumb ... *obviously*, our perverse choice of number sequence is meant to communicate some very deep and fundamental message.

It would have to be something guessable, but what? In a gestural language, "Pi" might represent the drawing of a circle and then a single line striking through it. If it represents a "slashed circle", the slash might be taken to symbolise death or destruction, or division – an attack on a disc representing a star or a planet, or (more generally) the destruction of the perfect order represented by a circle. To a society that respects the circle as a universal symbol of perfection, our broadcast may be alarming. The "Pi" sequence may function as a distress call or quarantine warning, and might translate as "death star" or "death planet", and it might be that by broadcasting 3.14159...., we're inadvertently broadcasting an **ideogram** that carries an implied threat of violence or danger, and could even be taken as an open declaration of war.

This might not be a good thing.

the "incompetent species" interpretation

Alternatively, a "Pi" broadcast might suggest that although we were technically advanced enough to be sending the signal, and valued our own sense of mathematics, we hadn't really gotten the hang of this "alien communication" business yet and – even more alarmingly – didn't seem to realise it. Pi is a **stupid number**. We'd be advertising the idea that we have an exaggerated sense of self-worth, sometimes make rotten decisions, and sometimes have real trouble understanding other cultures' points of view (to the extent that we might not always realise that there's an alternative point of view to be understood).

The last point is troublesome. Our signals analyst might conclude that the last thing their species needs is an encounter with creatures that are technologically advanced and keen to explore but don't have a clue about social interaction even at the most basic mathematical level. How could one make sensible diplomatic agreements with a species like that? An encounter with the Earth-creatures sounds like a nightmarish prospect, and our analyst might recommend that perhaps, since the species in question seems keen to meet others without understanding the possible consequences, and might even be *psychologically incapable* of dealing gracefully with any result of any encounters that *did* happen (with potentially disastrous consequences for all concerned), that the safest thing to do might be to lob a cheap global extinction device in the direction of the signal before these idiots started building proper spaceships and starting wars, and generally screwing things up for everyone else.

20.11: Naming rituals, binary logic and Giant Pandas

To a shaman, a word has power, and to know a thing's *name* is to have command over it.

And naming does give psychological and practical advantages. If an unknown creature roars in the dark woods, we may feel insecure and scared – when someone hears the roar and asks us what it is, we have to admit that we don't know. Not knowing things about that affect us is scary. On the other hand, if we *name* the unknown creature (perhaps with some Latin name that means "unseen thing in the woods that roars"), then we can give our friend a more scientific-sounding answer, and sound knowledgeable. Having named the thing, we feel that have exerted some power over it, and forced *it* to fit *our* laws. Naming is a form of control.

And naming does give us power: it allows us to organise gangs to go looking for the beast, and it lets communities exchange and pool information about the creature, and to pass that information down. The name provides a *persistent identity* to which we can attach further information. To admit that our existing gravitational models fail to predict 90% of the universe's gravity is embarrassing – to guess that there might be an unseen "something" out there supplying the other 90%, and admit that we don't have the faintest idea what it is, this is awkward. But as soon as we give the "unknown" a name, and start confidently talking about "dark matter", it sounds as if we have *some* idea of the thing that we're talking about, even if we don't. It's still an unknown, but it's a *specific* unknown, with a catchy, standardised name.

The danger of naming is that we can begin to assign a physical reality to our naming system as if its rules are necessarily the same as those of the universe outside. Does dark matter really *exist*? Are we even sure what the question *means*? Classification systems attempt to *classify* reality, but can drift into attempts to *define* it – classifications can sometimes be dubious, and can sometimes impose strict, artificial, nominal boundaries that aren't reflected by reality.

Kids tend to think of giant pandas as bears. For years, some adults have taken great satisfaction in pointing out that pandas are *not* bears, but are more closely related to racoons, so as adults we now know: pandas aren't bears. Except that genetic research now suggests that perhaps pandas *are* related more closely to bears than we'd thought. So are pandas bears or aren't they? Well, it depends on our definition of a "bear", and if we have this much trouble deciding, then perhaps our "bear" category is pretty loosely defined, and perhaps we shouldn't be insisting on a simple yes/no answer to the question. Perhaps that level of specificity isn't appropriate. Perhaps pandas are simply *pandas*, and if we want to think of them as bears, that's up to us.

While imposing strict categories can simplify a worldview, it can also have harmful consequences. People are complex, and in the modern world with its increased mobility and communications, the fact that some people may share a particular surname or ethnicity carries no guarantee that they'll share beliefs or upbringing or worldview or skills. When we impose externally-defined categories onto people, and invent particular words to "pigeonhole" them, these nouns are often considered insulting. To accept that one is "X-ish" is to accept a *description*, but to accept the statement "that person is an X" may be to accept a *defining* classification system that lumps one together with other "X"'s, and may amount to an implicit acceptance that "X"'s are *inherently different* to "Y"'s or "Z"'s. If someone has been taught at a young age to classify people according to, say, skin colour, then their inbuilt pattern-recognition abilities will tend to fixate on details that agree with the stereotype as "significant", and to ignore those that disagree with it for want of a category. This is a simple matter of data-compression – information tends to be discarded when it can't be filed efficiently. Even smart, well-meaning people who understand intellectually that certain categorisations are pathological can still find themselves unable to entirely break away from the classification schemes that they were brought up with, and through which they first began to perceive the world. Conditioning through *linguistic* definitions of reality can be difficult to escape, and to a someone with a linguistically-oriented mind, these categorisations can *become* reality. They will demand to know, well, is a panda a bear or not? Yes or no?

20.12: Intransitive logics

In section 20.1 we met the idea that serial languages can sometimes impose implied **rankings** between properties that we might otherwise have considered to be of equal importance, sometimes without our noticing. Another complication is that sometimes we have *real* rankings that are **intransitive** – they only function locally, and can't be applied globally.

paper-scissors-stone

A good example is the "paper-scissors-stone" game. This game is often used as a way of breaking decisional deadlocks: the two players each pick an option, ("paper", "scissors" or "stone"), make the appropriate symbol behind their back, and then reveal them to each other at the same moment. The ranking of the two symbols decides who wins. "Paper" outranks "stone" ("**Paper wraps stone**"), "stone" outranks "scissors" ("**Stone blunts scissors**"), and "scissors" outranks "paper" ("**Scissors cut paper**").

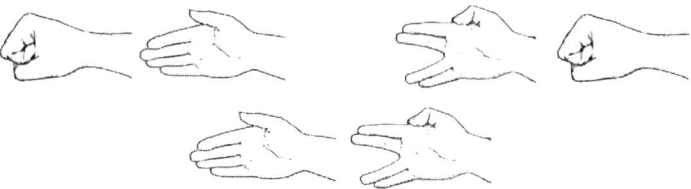

Figure 20-4: "Paper-scissors-stone" game

The ranking of these symbols is *circular*. The exercise is similar to taking three properties, **a**, **b** and **c**, and saying **a>b**, **b>c**, and **c>a**. Algebraically, any *pair* of these statements looks reasonable, but the combination of *all three* seems nonsensical: there are no simple numerical values that we can assign to **a**, **b** and **c** that will let this system work.

other examples

In road traffic networks, a particular sort of junction may give one motorist right of way: but the network may be organised so that when the roads meet again, the other motorist has priority. Priority then becomes a matter of **local geometry**, not global superiority.

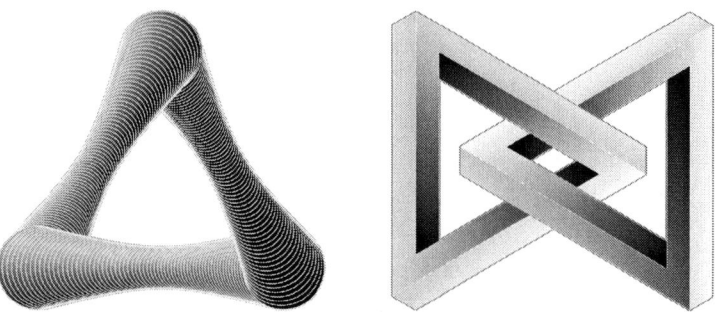

Figure 20-5: "Impossible" tribars

Logical systems in which some ranking parameters only exist as local, *relative* relationships have sometimes inspired graphic design, notably by **Oscar Reutersvärd** and **M.C. Escher**. In the **impossible tribar** drawn by Reutersvärd, each of the figure's three corners looks like a conventional right-angled joint between two rectangular-cross-section beams, and we might expect four of these angled turns to be able to make a sort of picture-frame shape. But the figure produces closure with only three right-angled corners. The figure has no *locatable*

RELATIVITY IN CURVED SPACETIME

flaw: each region of the figure is *locally* consistent, and if we cut the figure at any point it becomes a diagram of a "conventional" three-dimensional shape. It's only when we come to look at the entire shape *globally* that our ability to visualise the shape fails.

20.13: Complex logical spaces

Logical relationships are sometimes expressed using **set theory**, where a **set** is a well-defined collection of items, and where sets can overlap or nest inside other sets, depending on the particular interrelationships that we want to express. The conventional way of representing relationships in "naive" set theory was the **Venn diagram**, which neatly packaged the relationships between sets and subsets of items and properties into graphical form.

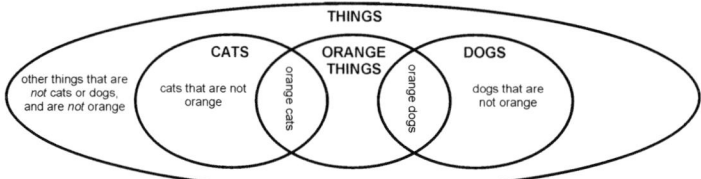

Figure 20-6: Example of a Venn diagram

Sets can refer to physical objects, properties, or abstract concepts, for instance, the set "*Even Numbers*" is a **subset** of "*Integers*" which is in turn a subset of "*Sets To Do With Numbers*". Mathematical infinities (such as the quantity of numbers that can exist) can be tidied up by putting these things into their own sets, and then grouping them into larger sets such as "*The Set Of All Number-Related Sets*".

But the idea of **self-referential** set definitions then seemed to defeat conventional logic. The *collection of all possible sets* (let's call it "S") is *itself* a set. But since "S" is a possible set, it has to contain *itself*. We seemed to end up with an infinite series of nested copies of "S", which implied that every other set seemed to have to be copied an infinite number of times too. And if we looked *outside* "S", although there shouldn't be any sets outside S, if we were looking at a nested copy, we'd find every other set that existed, *outside* S's borders. Self-nesting not only seemed to involve infinities, it seemed to destroy the very concept of "inside" and "outside", and things got even worse when we started to define *self-exclusive* sets.

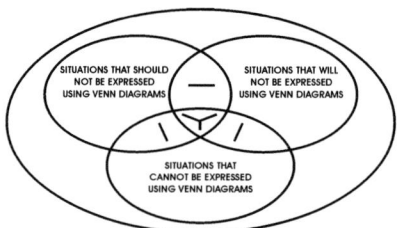

Figure 20-7: An example of a Venn diagram

This class of self-referential logic problem was famously illustrated by **Bertrand Russell** in his **Barber Paradox** (sometimes called the **Russell paradox**): a barber shaves everyone in the village who doesn't shave themself. So who shaves the barber?

We can go some way to tackling this breakdown by rethinking the *concept* of a logical boundary. It turns out that we *can* represent "S" with a Venn diagram drawn on a finite surface ... but only if that surface isn't a simple flat sheet of paper. If we drew a circle representing **S** on the top surface of a *torus*, by going deeper into the centre of S and into the centre of the torus, we eventually end up outside S again. S (and all the other sets within it)

20: Limitations of Language and Procedure

are only drawn once, S surrounds the entire contents of the surface, and it surrounds itself. Imposing a one-way system on how we track across the surface then completes the model.

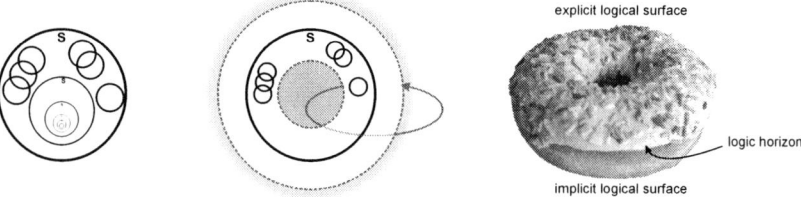

Figure 20-8: Infinite regress:
Logic within a multiply-connected surface

Self-referential sets highlighted several problems in older logics: **(1)**, it had been assumed that for the simple boundary of a Venn diagram's set, the concept of "inside" and "outside" was self-explanatory: but it wasn't enough to identify the boundary, we also had to identify a **polarisation** across the boundary, or label a distinct region on one side of it, **(2)**, it was also assumed that the natural underlying logic-surface for a Venn diagram was *simply-connected* and that relationships were absolute, which isn't true in all problems, and **(3)**, it was assumed that distinct regions *existed* and that boundaries were necessarily *closed*, and our experience of problems in acoustic metrics tells us that this isn't necessarily the case, either. Even these modifications aren't enough, for S to be "inside itself" but not "outside itself", we'll probably also want the extra connecting topological surface to only be navigable in one direction.

polarisation

If a boundary divides two regions with a distinct relationship to one another, then the boundary can be reconsidered as a *polarisation*: items hugging the outside of the boundary have distinct properties that make them different to items on the other side. But in our studies of acoustic horizons we found that although this sense of direction may be locally very strong (perhaps strong enough to make it *seem* as if conventional binary logics were operating), the degree of polarisation could vary from place to place along the horizon, and in some cases, could fade away altogether.

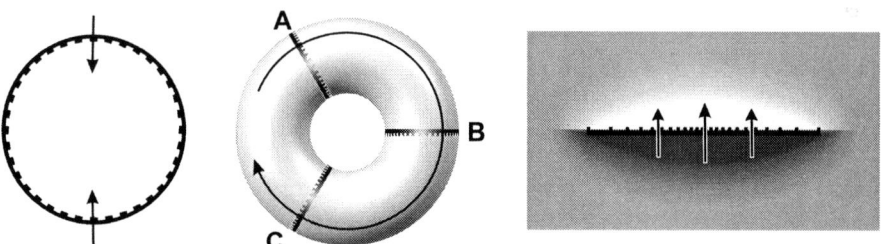

Figure 20-9: (i) Explicitly polarised boundary, (ii) $A \supset B$, $B \supset C$, $C \supset A$,
(iii) polarised boundary that only exists locally

In these cases, the polarised concept of "inside" and "not inside" might seem solid *locally*, but globally it doesn't exist – logical relationships become **relative** and **route-dependent**. We can respond by replacing our boundary with a wall of little arrows signifying local rankings, and these arrows can *emulate* the effect of a boundary or surface, but they can also support more complex logics in which our arrows fade away, or swirl, or self-layer in ways that look more like **field theory**. If we start with a simple binary logic based on the values "true" and "false", we'll find situations where it can't cope, and where these *relative* logics are required.

East is East and West is West

A familiar example of intransitive ordering shows up in geography when we look at our everyday concept of "East" and "West".

Does New York lie "East" or "West" of Los Angeles? If we're looking *locally* at a "**patch**" comprising the United States then it's well understood what we mean when we say that LA is somewhere to the West of NY. But technically, we can also go "East" from NY, and reach LA, the long way around, so *globally*, LA is also to the *East* of NY.

This isn't *just* an abstract geometrical problem caused by an inappropriate coordinate system, because in real life, there do seem to be some "intransitive" properties that operate like this, an obvious example (in physics) being the gravitational field of a moving or rotating object.

20.14: Intransitive ordering and gravitation

In our East/West example, we can draw little arrows of local "westwardness" around the surface of a globe, and the arrows will be locally consistent with their immediate neighbours except at the poles, where the system either "fades away" (with the sizes of the arrows representing "west-yness" shrinking dropping to zero length), or it breaks down.

This sort of "circular" ordering isn't just a "mathematical" labelling problem, it also has a physical manifestation: Lets consider the problem of what we mean by the word, "height". We could take it to mean "height in a gravitational field", and we could assign relative "height" values to different positions depending on whether energy is required or released by a change in height between the two locations. We might then think that the problem is solved, and that we can draw in surfaces of equal gravitational potential around the Earth that represent particular agreed heights. Every point then has a globally-assigned "height" value, and by comparing the numbers assigned to different locations, we can say whether one location is "uphill" or "downhill" of another. At first sight, there doesn't seem to be anything wrong with this idea.

But the general principle of relativity teaches us that this approach doesn't quite work. We might naively expect that if we made the Earth's surface perfectly smooth, and made all points on the Earth's equator *exactly equivalent*, that these points would *obviously* all be at the same height, by definition. We'll expect a ball placed at the equator to sense that it's resting on a surface of **equipotential**, and to stay put. We'll expect this surface to be at right angles to the gravitational fieldlines pointing through it.

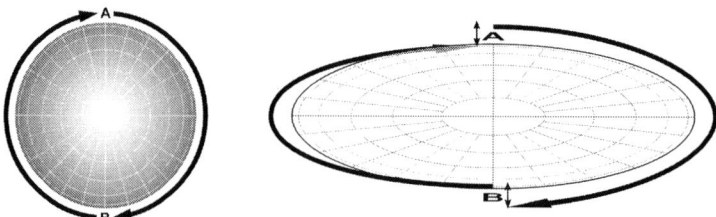

Figure 20-10: Two points A, and B, both simultaneously "uphill" and "downhill" from each other

But because the Earth rotates relative to its environment, these rules are broken. An object placed at our "idealised" equator will feel a gravitational dragging effect that pulls it in the direction of the stars moving overhead, East-to-West, and so it should have a slight tendency to roll "downhill" to the West. The distortion of geometry due to relative rotation means that the perfect equatorial surface still manages to feel a slight tilt, to the West. Taking points on two opposite sides of the Earth's equator, we might say that "**A is higher than B**" when we measure around the equator in one direction, but "**B is higher than A**" when carry out our measurements along the other route. For a fully-circular measurement path, we get the apparently-nonsensical result that **A** and **B** are each higher and lower than *themselves*.

Figure 20-11: "Intransitive" stairwell: Repeatedly climbing "uphill" or "downhill" around this staircase will return you to your starting point: "Downhill" is clockwise, "uphill" anticlockwise

The effective gravitational potential between two equivalent points isn't absolute, but ***route-dependent***. This means that if we try to label the Earth's surface consistently using contour-lines, those lines won't always correspond to the physical relative heights between locations, and if we try to draw our lines so that they *do* correspond to relative heights, then the global mapping system breaks down. We can still use *relative* mapping: we can cover the surface with arrows and labels giving the height differentials between locations along given lines … but we can't apply a global system of *absolute* physical values and expect its properties to correspond perfectly to the region's real physical properties.

In **M.C. Escher**'s lithograph, "Waterfall", water is seen flowing downhill around a closed circuit. This is usually considered to be a clever draughtsman's trick, similar to the graphical sleight-of-hand used in Figure 20-5: but what's less well appreciated is that sometimes Nature really *does* operate like this. For instance, the moving gravitational "humps" caused by the apparent passage of the Sun and Moon across our sky encourage water to try to flow East-to-West, in a dramatically exaggerated counterpart of the effects that we expect to exist due to the gravitomagnetism of a "smoother" rotating background, and when these flows are blocked by landmasses, and water is forced to deflect sideways, the resulting circulating ocean currents – clockwise in the Northern hemisphere, anticlockwise in the Southern – help to determine our planet's weather patterns. So while it may seem ludicrous to suggest that we could use the Sun's and Moon's gravity to generate power, when we set up hydroelectric systems to harness *tidal* power, that's exactly what we're doing – using the Earth's oceans as a medium for collecting and channelling these forces to create the end-result of water flowing through our generator turbines.

Figure 20-12: Two shapes that mutually surround and are surrounded by each other.

20.15: "Certainty" parameters

losing boundaries

Let's look back at our "East-West" example: if we take a globe representing the Earth, and draw little arrows all over it representing the direction "West", then these arrows will line up perfectly with their immediate neighbours everywhere on the surface except at the poles. We may say that our system of arrows produces a **coordinate singularity** at the poles.

But gravitomagnetic effects give us a clue as to how to *dissolve* this singularity – a west-pointing arrow indicates the horizontal direction of dragging, and since *Nature* manages to avoid *physical* singularities at the poles, perhaps we can steal the method.

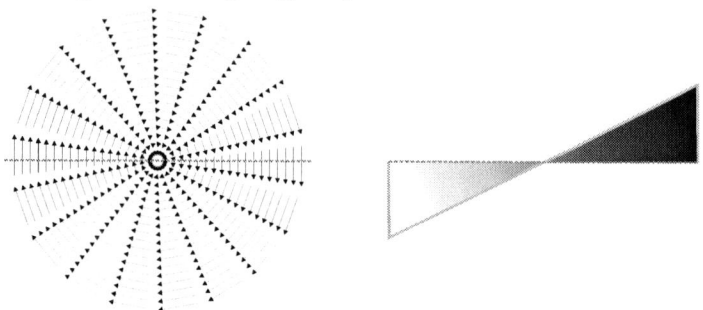

Figure 20-13: Direction and *strength* of the concept of "westerliness" over the Earth's northern hemisphere.

If we take the direction of the dragging effect as our *physical definition* of "westwardness", things become easier. We find that since the dragging effect is strongest at the equator and grows weaker at more polar latitudes, where the Earth's rotation makes surface objects mark out smaller circles. As we cross a pole, our directional dragging effect doesn't suddenly "snap" from one direction to another ... its *polarity* flips, but the strength of the effect fades down to zero at the pole and then grows again, pointing in the opposite direction. So by making each arrow's length proportional to the strength of the dragging effect at that location, with the longest arrows at the equator and progressively shorter arrows towards the poles, the field (and our definition of "westward") fades politely to zero there. The arrow-lengths indicate how *emphatically* westward a direction is, and when we want to describe the local relationship between two points, we can use *three* parameters, direction, distance, and *emphasis* or *certainty*. When the last parameter is zero, the coordinate system evaporates. Similar arguments can be used to remove singularities at the centre of other sorts of field "whorls", to allow more complex, purely *relativistic*, logical structures.

Our first step in trying to tackle a problem is often to try to impose some sort of global order onto it. This is understandable. We like to be able to apply fixed grids and coordinates and locations, and to be able to label the parts of a problem with absolute numbers and coordinate values. We hope that by using these numbering systems, we can break the problem into a more mechanical, more easily-solved problem of *ordering*. Faced with the concept of gravitational horizons, we decided that it would be best if they were considered as boundaries, with a clear sense of who was outside a boundary and who was inside. We decided that these boundaries had to be perfect and edgeless ("*a boundary has no boundary*"), and we proceeded in the case of black hole theory to go on to prove a number of striking results ... which were quite wrong.

Reality doesn't always respect absolute rankings. Sometimes properties can't be expressed using simple "absolute" numbering systems, and instead their relationships split and disappear and swirl and circle and knot, like the patterns on fingerprints or the eddy-currents in a stream.

20.16: Living with uncertainty

Certainty parameters can be used within conventional logic systems and can play a useful part in fault-tolerant analytical systems, and in parallel-processing computer networks.

If we send a robotic spacecraft on a long mission, then eventually its processors and sensors will inevitably start to fail, and the system will increasingly have to make decisions based on incomplete and unreliable information. We can deal with this by adopting the idea of having an additional parameter that says how strongly a value is believed to be correct. When two values are multiplied, the confidence of the result is the product of the confidence values (in the range 0→1) for the two numbers. We can build a parallel set of logic to automatically generate "confidence" figures for all our calculations, to suggest the impact of degraded or missing data: if a sensor fails completely and its datastream bears no relationship to reality, then by setting that stream's confidence tags to zero, the confidence values quoted in our final answers should *automatically* degrade according to the importance of that data to the results.

We can add various additional layers of abstraction to this process: "default value" registers, registers to hold values for our *confidence* in the "confidence" values, and so on. The fact that we're dealing with situations that appear to "go singular" with conventional "dumb" processing methods doesn't mean that we can't deal with them using *reasonably* conventional tools. Our approaches may need to change, but the new logics aren't difficult to learn.

why computers crash

The inability to deal properly with uncertainty has been a historical failing of traditional computer systems, partly because the clever people who devised these systems and languages and their data-structures wanted computers that could produce *definite* answers. Suppose that we have an old "eight-bit" computer, and we want to use just a single eight-bit **byte** to represent whole numbers. Conventional computers only use ones and zeroes, so each **bi**nary digi*t* (or "**bit**") can have two states. An eight-bit sequence gives us $2^8 = 2 \times 2 \times 2 \times 2 \times 2 \times 2 \times 2 \times 2$ possibilities, so if we used up the first bit as a "±" sign, the remaining seven bits will let us count from zero to 127, giving us a total range of "−127" to "+127", and including a redundant value of "minus zero". So as not to waste this "extra" code value, the designers used "-0000000" to represent "-128".

With hindsight, this was a rotten decision: the asymmetrical number range meant that by taking a "legal" range of numbers and swapping the plus for a minus, we could almost guarantee that one of the numbers would give an "out of range" result, crashing the computer. As well as having been lumbered with asymmetrical number ranges, these systems had a more serious problem, in that they had no concept of a "wrong" number: when we tried to do something illegal, like dividing a number by zero (which gives an "infinite" result), since there was no "correct" response to this result, the computer let us know that it had a problem either by crashing (after throwing up an error message), or, if we were lucky, by catching the "impossible" calculation, and shutting down the program that caused it (which would no longer have a "legal" way to continue obeying its chain of instructions).

What we desperately needed was a code for "nothing", or "invalid", or "**null**", and this is how we should have used the spare code-symbol. Any data-field whose contents weren't known could have been set to "null", and any calculation that gave an illegal or unknown result could have generated "null". Instead of crashing our program and/or our operating system, programs would have been able to continue working, and would have politely offered up a final value of "null" for any parts of the calculation that had gone badly wrong. "Null" times another number would again equal "null".

The lack of an adequate way of expressing uncertainty within computer software was a major part of the infamous "**Year 2000**" or "**Y2K**" problem: since the original programmers hadn't been able to use "null" as a value, they'd had to hijack existing *proper* numbers as private personal codes within their programmes, so for instance, if an operating system didn't allow "illegal" dates, and had no way of expressing a date as "unknown", programmers would use something memorable like **"December 31st 1999"**, or **"01/01/00"** (**"1st January 2000"**) to symbolise a "blank" date. Decades later, when these dates started to appear within the software for *real*, systems started to fail in unpredictable ways as "real" data was treated as "invalid", and vice versa.

Some systems would also fail because they only supported "two-digit" year-values, and could only cope with dates up until the end of 1999. Instead of failing gracefully and *explicitly* (**"99+1=null"**) they would fail *inappropriately* (**"99+1=00"**), the danger being that when somebody tried to hire a car for a week, instead of printing an invoice, the software might issue a refund cheque for several million dollars, representing a negative hire charge, for "- 99" years. Even when we expanded these two-digit date fields to cope with four-digit dates, the original "dummy-handling" code could still damage our data and records, and when this ancient code activated and corrupted our calculations and our data, the fact that the software was still putting out real-looking dates and numbers rather than "nulls" made it difficult for us to know that we'd identified and fixed all of a system's date-bugs. This made Y2K certification more difficult, and many non-Y2K systems and subsystems had to be expensively rewritten from scratch.

In some more modern languages, we've finally started retrofitting additional number registers that can carry error codes associated with a number (such as the much-needed "null"), but where we *do* now have this facility, it tends to be retrofitted rather than being a piece of deep design: we preserve the quirks of the older systems (such as asymmetrical number ranges) in order to guarantee that older "legacy" software will run on newer hardware and platforms in exactly the same way, and that code components can be transferred back and forth between older and newer systems. "Null" processing and certainty registers should have been built into processor chip designs right from the very beginning, but unfortunately, the brilliant minds who wrote the logical templates that became the basis of the design of Twentieth Century computer processors and languages didn't seem to think that a computer should be able to respond by saying "I'm sorry, I can't calculate this properly", "I'm sorry, the figures I was given weren't complete", or "I'm afraid that this result only has a given confidence value".

Our craving for *certainty* … even when certainty was *inappropriate* … led to bad design decisions that crippled a generation of computer systems, and which still have harmful "legacy" effects in more modern systems.

20.17: Conclusions

Logic is good. Logic helps us to get answers to problems, and the working assumption that we live in a logical universe seems entirely sensible. But before we start dictating the laws that we expect Nature to obey, we should ask what *variety* of logic is appropriate to a given situation.

Acoustic metrics supply us with a wealth of hypothetical situations in which conventional rankings and methods fail, and in which the underlying logic-surfaces are more complex and more subtle than those we're used to dealing with. In GR1915 we assume that an event is either inside or outside a horizon, and we construct proofs and networks of relationships around this idea. But Nature doesn't seem to be compelled to play along with these simplistic logical systems, and in the case of simple acoustics, supplies us with other sorts of horizon that break these rules.

When we try to construct or assess theories, we often bring along a lot of historical and mathematical "baggage" to the task: we typically define whether theories are "credible" or not according to how well they reduce to simple Euclidean or Minkowskian metrics, and theories that show more subtle behaviours are liable to be dismissed as *logically wrong*.

But it may instead be that we are applying the wrong logic.

21
The Perils of Experimentation

> *"The human understanding when it has once adopted an opinion (either as being the received opinion or as being agreeable to itself) draws all things else to support and agree with it. And though there be a greater number and weight of instances to be found on the other side, yet these it either neglects and despises, or else by some distinction sets aside and rejects, in order that by this great and pernicious predetermination the authority of its former conclusions may remain inviolate."*
>
> **Francis Bacon, Aphorism 46**

> *"If you know the answer, it's not an experiment"*
>
> **anon.**

> *"Then Eddington would go mad, and you would have to come home alone."*
>
> **Frank Dyson, on what would happen if Eddington's experiment should produce *twice* the GR1915 prediction. Related in (Bernstein, 1991, pp.116-117)**

> *"... then I would have been sorry for the dear Lord: the theory is correct."*
>
> **Einstein, when asked how he would have reacted if Eddington's expedition had not confirmed his prediction**

> *"It is a good rule not to put over much confidence in a theory until it has been confirmed by observation. I hope I shall not shock the experimental physicists too much if I add that it is also a good rule not to put overmuch confidence in the observational results that are put forward until they have been confirmed by theory."*
>
> **Arthur Stanley Eddington**

> *"Never make a calculation until you know the answer."*
>
> **"Wheeler's First Moral Principle", John Wheeler**

> "I thought you said that your dog does not bite?"
> "That is not my dog"
>
> **Peter Sellers, as Inspector Clouseau**

> *"The technician ... added in a moment of unscientific enthusiasm 'But we'll keep on repeating the experiment until we get the right result'."*
>
> **E.B. Mpemba and D.G. Osborne "Cool?" Phys. Educ. [14] 410-413 (1979)**

> *"The first principle is that you must not fool yourself ... and you're the easiest person to fool."*
>
> **Richard Feynman**

> *"With a new theory, nobody believes it except the theorist: with a new experimental result, everyone believes it apart from the person who actually carried it out."*
>
> **anon.**

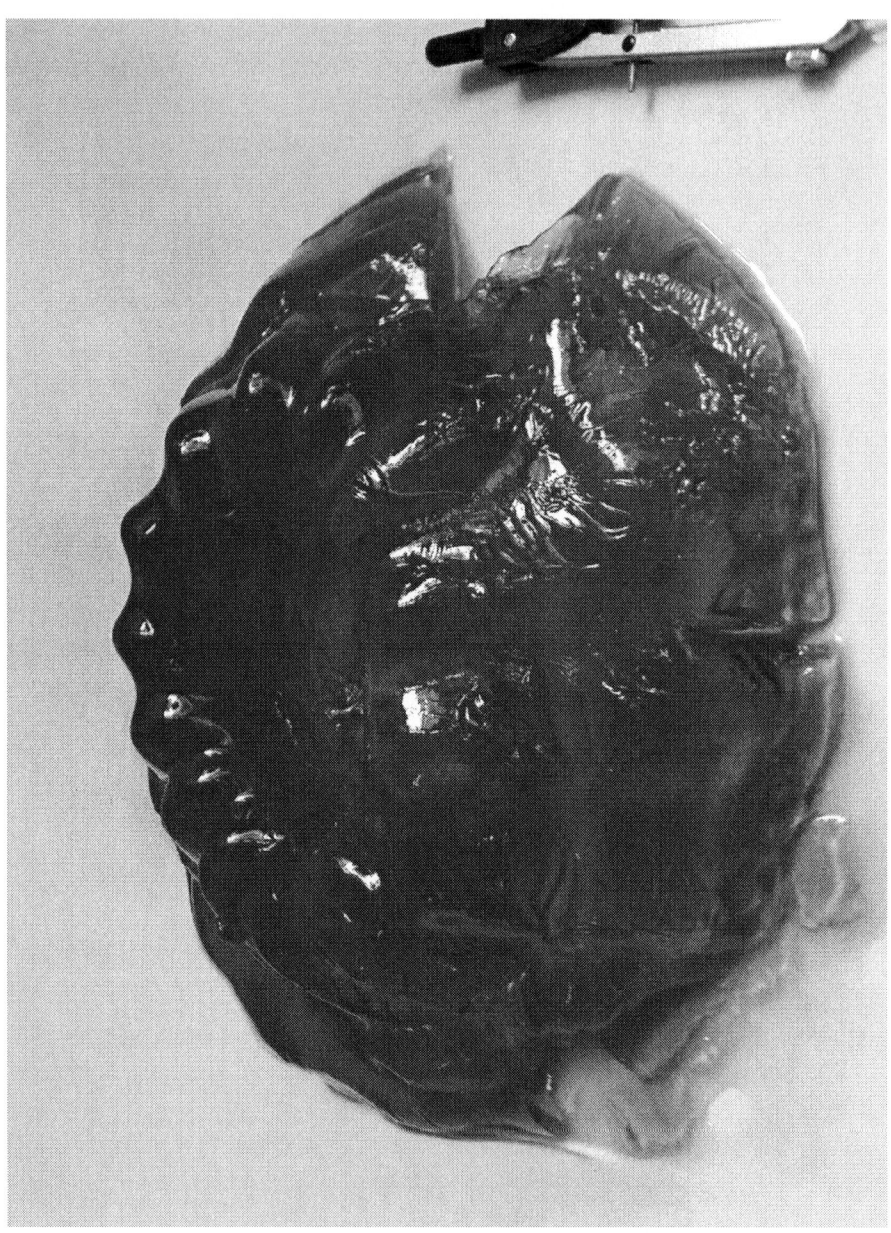

Trying to measure a jelly

21.1: Evaluating science neutrality

Science is a positive-feedback process. We grope about in the dark and tentatively try things, and when something seems to work we pat ourselves on the back, chalk it up as a possible success, and see what else we might build on it. It's a process of exploration, and it involves creating a balance between maintaining a healthy scientific scepticism, and the optimistic working assumption that our work has at least *some* prospect of success.

People who present science as a cold, dispassionate business can be fond of saying that it only takes one experiment to disprove a theory. This isn't how physics operates in the world outside the classroom: in practice, a large helping of subjectivity helps to decide whether experiment overrides theory or vice versa. In the *real* world, we evaluate the significance of experimental data based on what we believe to be the **balance of probability** that a result is what it appears to be – it's a human process where experience, expertise, precedent and consensus view all make contributions, and we don't always get it right the first time around.

Experiments have a habit of being ... well ... *experimental*, and the experiments that become famous as "world firsts" have a natural tendency to be more "experimental" than others. And as a result, when we go back and look at classic pioneering experiments decades later, we sometimes find that we've invested a great deal of credibility in results that we *liked*, and that we *agreed* with, but that with hindsight perhaps weren't particularly reliable.

Newton

Isaac Newton's early career was distinguished by his outrageous (but correct) claim that light didn't come in three, or five, or seven distinct colours, but in a *continuous range* of wavelengths. This contradicted some expert opinion at the time, and Newton had a hard time of it. Newton claimed to have demonstrated this continuous-colour theory of light by splitting light with a prism, and finding that when we isolated a narrow colour range and fed it through a *second* prism, no new colours were generated. The prism didn't *create* new colours (said Newton), it only *separated out* colours that had already been present in white light, as a mixture. After Newton described his experiment, other researchers on the continent had trouble replicating his result, and Newton was accused of fraud. Disgusted by the response, Newton vowed to abandon natural philosophy and didn't take up the subject again publicly until years later, when asked to help out with an astronomical problem (after which he seems to have cheered up and went on to write "Principia" and "Opticks"). But thanks to the wonders of peer review, "Newton the theorist" was nearly lost to us.

How do we assess Newton's case? With the benefit of hindsight we now know that Newton's "continuous colours" theory was right, and that the mainstream critics were wrong. But this leaves us in a difficult situation, because it was claimed that Newton's result wasn't **replicable**: if Newton's results were made up, then according to modern quality control standards, one of the greatest theoretical physicists of all time should probably have been kicked out of the group and not allowed back in. The later "Opticks" and "Principia", which helped lay the groundwork for the Industrial Revolution, might not have been written. On the other hand, if Newton's claimed experiment was *genuine*, it suggests that peer review showed **pathological scepticism**, wrongly dismissed one of the most significant physics experiments ever, and nearly deprived us of one of the greatest physics minds that ever lived – Newton stomped off in a huff and spent the next few years studying Bible chronology instead.

Either of these interpretations is a little unnerving. Newton often built his own equipment using tools that he'd made himself, so were his opponents unable to replicate his results because their hardware was inferior? Or were they embracing their failure to see Newton's result, because they didn't like it and had already convinced themselves that it *ought* to be wrong?

Einstein (special relativity)

Moving forward to 1905 and Einstein's special theory, the first paper to cite Einstein's famous electrodynamics paper is generally reckoned to have been the write-up of a 1905 experiment by **Walter Kaufmann** explaining that the special theory *wasn't* a good match to experiment. That paper isn't now taken too seriously, but it demonstrates that if we *really* accepted the idea that a single conflicting experimental result "kills" a theory, then special relativity would have been considered "dead" almost as soon as it appeared (which, clearly, didn't happen). As **Jeremy Bernstein** (1993, pp.21-23) points out, one of the main predictions in Einstein's 1905 electrodynamics paper, that clocks at the Earth's equator would run more slowly, later turned out to be *genuinely* at odds with experiment and with more advanced theory (section 15.5, pp.201). We now recognise that this was perhaps more a *result* of SR than a strict *prediction*, a case of the over-enthusiastic *application* of the theory rather than a problem with the theory itself ... but if we'd been less inclined to be charitable, we could have used this an excuse to declare the 1905 theory disproved. Embarrassment at an "Einstein 1905" result being wrong seems to have kept this subject rather obscure for decades, but the issue gets documented in a peer-reviewed physics paper (a century after Einstein's original piece appeared), in Harvey/Schucking 2005.

Einstein (general relativity)

With **Einstein's general theory**, we had another case of slightly over-enthusiastic application: Einstein managed to "explain" why there'd be no distance-dependent redshift shortly before Hubble's measurements showed that the redshift was real. We decided that Hubble's observations didn't invalidate Einstein's general theory, they just showed that Einstein's subsequent use of an arbitrary cosmological constant to cancel the redshift effect had been a mistake. We liked the theory and gave it the benefit of the doubt, but had we been especially predisposed to *dis*like it, we would have had an excuse to declare *it* falsified, too.

Although we may be fond of the idea that experimental tests of theories are strictly deterministic things, the processes that lead up to them and the way that we treat the results afterwards are very much human affairs. Dubious experiments are sometimes accepted and made famous because they say what we want to hear, and other experiments are disbelieved and set aside because they produced what we reckon to be "wrong" results.

There's a certain amount of interpretation involved.

Michelson and Morley

Sometimes an experiment moves from one category to the other.

If we look at the case of the 1887 **Michelson-Morley experiment**, modern teaching says that this was a critical experiment that changed the face of modern physics, and that helped pave the way for special relativity. However, the M&M experiment doesn't seem to have generated *quite* such a wave of interest at the time. Einstein was rather noncommittal about how strongly the M&M experiment had influenced his 1905 paper (or, indeed, if he'd even heard of it at the time), and **R.S. Shankland**'s interview with Einstein suggests one possible reason why: according to Shankland, Michelson's own colleagues had considered the experiment to have been a failure, with Michelson to be pitied not only for having constructed a test that *didn't work*, but for then having been dumb enough to broadcast his own incompetence by publishing the non-result. Shankland's piece suggests that amongst Michelson's immediate peer group, the feeling may have been that if your experiment gave an "inexplicable" result, the *smart* thing to do was to keep damned quiet about it.

The community would have had many reasonable-looking grounds for ignoring the "anomalous" M&M result: Michelson was working at an US institution at a time when most "respected" work on physics was being conducted in Europe: he published in English when "the language of physics" was predominantly German, Lorentz wrote in 1895 that Michelson's analysis of his earlier experiment had been fatally flawed (overestimating the effect by a factor of two), and subsequent researchers seemed to have difficulty reproducing the "null" result to the degree of certainty claimed in the second test. Einstein later suggested that Michelson's ability to wring results from equipment that others couldn't match might have been due to his "artistic" approach and his exceptionally keen eyesight, but generally, when an experimenter produces a result that doesn't fit current theory, and which others can't "see", we tend to make some fairly brutal assumptions about the experimenter's abilities. With hindsight we consider Michelson's experiment to have been brilliant, but to a contemporary researcher "playing the averages" it may well have seemed safe to dismiss the M&M result as an irreproducible and inexplicable result from an experimenter working outside the major institutions, who had already gotten this sort of experiment wrong once before.

We don't seem to have any reason to suspect that there *was* anything wrong with the M&M experiment ... but in a research world supposedly founded on objectivity and repeatability, our favourable assessment of the M&M experiment (as with Newton's experiment) seems to be partly based on a combination of reputation, social factors and on the fact that the experimenter was reporting a result that we now believe to be correct. But we can't examine the M&M hardware to evaluate whether it could *really* have worked as well as claimed, because it was scrapped so long ago that we no longer know what happened to the parts.

So ... although Michelson's experiment may have seemed dubious when it *didn't* correspond to our expectations, after our expectations had changed (and the experiment had become a convenient validation of our new beliefs), we decided retrospectively that it had been a very fine experiment after all, and that, in fact, it had marked a decisive moment in physics history. But this change in attitude wasn't caused by a change in the information available to us about the test – it wasn't the data that changed, it was us.

21.2: Perception filters

Suppose that we obtain an experimental finding that disagrees with current theory. What happens next? In the "educational" account of how science works, we go back and double-check our results, send them off to a major journal, get published, and then step up to receive a Nobel Prize.

In practice it doesn't always work like that. The first thing that we'll do is suspect that our experiment has somehow gone wrong. We'll go back and check all our sensors, and check that they're working properly. Then we'll run the experiment again. Perhaps our next experimental run produces something closer to the official answer. Good – we're ironing the bugs out of the system. We tweak everything and run it again. Along the way we may discover some previously-undiscovered effects that are throwing our results out (such as Pound's "SOD" effect, section 15.8). Eventually we may be able to get the experiment to produce the "right" answer, and declare the result a success.

If we can only get into the *neighbourhood* of what we consider to be the right answer, we may go back and reassess the accuracy of our equipment, and publish our result as *basically* agreeing with expectations, but with the hardware demonstrating a disappointing lack of accuracy. If other researchers try the same experiment and have the same difficulties, they can say, well, we *know* that this technique has accuracy problems, because "experimenter X" already ran into the same issue. If our experiment refuses to agree with the theory, or refuses to *disagree* with it in a way that makes sense to us, we'll tend to think that there must be something badly amiss with the experiment. And sometimes this will be true.

What if, after a great deal of additional effort, we *still* can't persuade the figures to fit? Well, we still might not believe that our own experiment worked properly. We might be using equipment in unfamiliar situations, perhaps outside its usual operating range. The experiment might be so new that nobody really knows yet what the reliable operating characteristics of the equipment actually ought to be. We may have to cut our losses, and publish our procedures with some description of how far we got, as a "preliminary" paper, and hope to get further funding to tide us over until we can figure out what went wrong. If we don't ever get the experiment to work "properly", and funding dries up, that may be as far as things go.

Let's say that our "rogue" result is correct, that the experiment seems to report it unambiguously, and that after spending an additional couple of years trying to find an identifiable fault with the equipment, we come to believe that perhaps our equipment is telling us the truth (or at least, seems to indicate that an interesting unknown factor may be at work). Having checked everything one last time, we send off a full paper to one of the major journals for publication.

But the journal isn't obliged to publish the result. When a peer-reviewed journal publishes a research paper, it puts its own stamp of authority on it – it says that the journal considers the paper interesting, significant, and free from identifiable error. But in an *experimental* paper, the journal and its anonymous referees aren't in a position to vouch for what actually happened during the experiment, because they weren't there. They may well consider the *result itself* to be evidence of an unspecified but "identifiable" error. And this may even be a reasonable reaction – after all, if even the *experimenter themself* didn't entirely believe the result until they had actually seen it with their own eyes and rechecked all their equipment, it's difficult to expect a reviewer to take the same result as correct on pure trust, and "sign off" on it on the assumption that the researcher *definitely* knew what they were doing. The anonymous reviewer may say, well this result can't *possibly* be right, and trot off a list of supposable reasons why a piece of equipment might give that answer in error (which the experimenter may or may not have already thought of, and checked themselves). Even if every single eventuality can be dismissed (which may not be practical), the reviewer may still decide, on the balance of probabilities, that the result is still most likely due to equipment malfunction or experimenter error. The result might never be published, and the dispirited experimentalist may give up and decide that in future they'll work on something else. Once their test has been repeatedly rejected as "not suitable for publication", they may prefer not to discuss the experiment any more. Since the acceptance or rejection of future papers may depend on the experimenter's presumed credibility, they may not want to stake their precious reputation on support for a result that they might not understand and may not have *wanted* to find in the first place. If they become known for submitting "dubious" results, they may find that future papers meet with more scepticism and have more trouble getting published, and they might find it more difficult to find funding for other work. Persistence may damage their credibility, and affect potential funding for other research that they consider more important.

There are many layers of deterrence that can prevent an "anomalous" result getting into print, and for some of the rejected papers this might seem to be the right outcome. But the peer review system has the potential to block and deter the reporting of "inexplicable" experimental results without *really* knowing whether they're good or bad, based on how well they fit our existing expectations.

If we *all* believe strongly enough in a particular result, "quality control" mechanisms can conveniently eliminate any conflicting evidence that might have suggested that our beliefs are flawed ... and the more persistent this absence of opposing evidence becomes, the more strongly we believe. This is a **positive feedback** process – the more that we trust something, the more the (published) physical evidence will be selected for agreement with it, and the more that our trust will seem to be supported by the accumulating evidence.

Beyond a certain point, positive feedback in a system can achieve **resonance**: after this, the system or network's state becomes fixed and self-sustaining regardless of the input data.

At least ... up until the point at which the whole system crashes.

21.3: System bias and "v1.0" syndrome

Now lets look at our last hypothetical experiment again, but suppose that it had instead produced a result that *agreed* with a popular theory, first time. We'd be happy. We'd be much more likely to believe that the experiment was successful, and we might not feel the same urge to go back and rip the equipment apart, piece by piece, looking for the tiniest flaw. A result that we already *expect* to be true won't tend to be subject to quite the same critical scrutiny as one that we really didn't want. We may be quite enthusiastic about posting our paper to a journal, and the journal may be quite enthusiastic about publishing it. The peer reviewers may not feel obliged to spend quite so much time inventing hypothetical scenarios to explain how the experiment *might* have produced the result by mistake. The experimenter wants to be the first person to publish the result, and the journal wants the prestige of having a "world first" result in their pages.

The result is published, the experimenter becomes famous, and lecturers and writers cite it approvingly as the first verification of the effect. And then, when it finally turns out decades later that the experiment was badly flawed, it's almost too late to do anything about it. As long as subsequent experiments agree that the effect was real, as far as the community is concerned, it's now less relevant whether the original experiment was *really* any good.

With hindsight, it now seems that some of the best-known experiments associated with relativity theory mayn't *quite* have deserved the accolades that we heaped upon them at the time. There are some quite innocent reasons for this: if an experiment was the first of its type, then the experimenters may have been working at the "bleeding edge" of contemporary technology, using materials or methods that weren't fully understood at the time. A certain degree of optimism also seems to be required in these situations, and sometimes "optimistic" assessments of the data can turn out afterwards to have been a little *too* optimistic. If an experiment fails to "work" for six months while we tinker with it, and then finally kicks into life, and produces the result that we were looking for, it's tempting to publish the one "good" result and overlook all the others.

natural and unnatural selection

Sometimes this enthusiastic selection of "good" data goes horribly wrong. A classic example seemed to happen in 1915, when two researchers (one very well known), set out to test the value of a constant, and reported a figure that agreed well with their calculations. The calculations were subsequently found to have been incomplete and their predictions out by a factor of two, which begged the question of how the heck they had managed to come up with an *experimental verification* of those wrong numbers (in experimentation "two" is often **the demon constant**, a factor that represents the ratio between the answer you *wanted* to get and what you actually found). One possible explanation that eventually dribbled out was that perhaps the team had actually performed *two* experimental runs that had produced different answers – the run that agreed with the *correct* (but unexpected) numbers being discarded as obviously faulty, and the run giving the *wrong* (but expected) numbers being seized upon as "good", and duly published. According to Jeng (2006), things then went from bad to worse: another experimenter who'd tried a similar experiment and gotten a rather *better* result, on hearing how the first experimenters' figures had turned out, decided to go back and repeat their own test ... and then reported a new outcome that was more in line with the first researchers' *wrong* answer. An understandable desire to publish only "proper" results led to the inadvertent suppression of the "good" data and the publication of "bad" data, without the experimenters realising until afterwards that they'd done anything wrong.

"Selection" is a procedural grey area where we often have to trust researchers to make a multitude of sensible decisions that may not be mentioned in their papers. Journal space is limited, and papers are expected to be concise. An experienced experimenter isn't expected to give every tiny detail of what the equipment did on days when it wasn't working properly, and if they *did* supply this detail, their readers probably wouldn't thank them for it. On the other hand, an experimenter who *deliberately* selects data for publication to fit a particular answer may be accused of fraud. Our requirement that experimenters only select "significant" data for publication, but also don't show any human "bias" can put them in an impossible situation, and in recognition of this, we tend to give experimenters the benefit of the doubt over their motivations when things are subsequently found to have gone wrong.

Or at least, we tend to give them the benefit of the doubt if the results that they reported in error *agree with the results that we want*. If an experimenter accidentally produces an "unreliable" result supporting a mainstream theory, we look on it benignly. We say, well, no real harm was done, research efforts weren't wasted trying to replicate bad results, and the researcher meant well. We encouraged them to publish, and it's not their fault if occasionally a cutting-edge experiment later turns out to have been a little over-ambitious. Experimenters do a crucial job, and mustn't be subjected to an atmosphere of recrimination. We need these people, and they need our support.

But if the researcher makes exactly the same degree of error in accidentally reporting an result that *disagrees* with mainstream data, then the community is less charitable. The experimenter is liable to be accused of incompetence, fakery, or of having some sort of emotional disorder, or a sinister financial motive … if they persist in publishing, they're liable to be accused of any or all of these things even if there is no real evidence to indicate that they've done anything wrong.

This is a little rough if the poor soul *did* conduct the experiment properly, and was simply trying to leave an objective record of how the experiment then played out, as scientific methodology tells them they ought. Experimenters learn that they're expected to apply a certain degree of self-censorship to inexplicable results, or risk degrading their reputations.

first past the post

When experimenters *really* know what they're doing, it's quite likely that they won't be the very first people to be doing it. Almost by definition, the famous "world's first" experiments tend to be a bit flaky, because if your experiment is *genuinely* straightforward to carry out, with established, well-understood and easily available equipment, the chances are that someone else out there will have already done it before you – one of the occupational hazards of pushing back the frontiers of human knowledge is that you don't always know exactly what you are doing. If your equipment has never been built before, it won't come with a handy user-manual.

And the pressures on experimenters can be intense. If two research teams are racing to be the first to announce a particular result, then even if each team is entirely scrupulous and ethical in their dealing with data, it will tend to be the more "enthusiastic" team that'll win the race and end up in the history books: the team that decides that their result is "good enough" and publishes first will get the credit, whereas the team that decides to give the experiment another few months to be *absolutely* sure of their facts will tend not to be remembered. The science community loves news, and the educational sector loves "historic dates". Reporting the same result as another team, more safely, some time later, doesn't tend to win one fame and fortune – natural selection favours the more "optimistic" experimenter.

games with error bars

Experimenters can soften the impact of an "unwanted" result by reassessing the supposed degree of experimental error. If an experiment hits slap bang in the middle of other results, it's tempting to consider it a success and contract the quoted error margins, and if the figure seems to be *incompatible* with the group consensus, its natural to go back and find excuses to *increase* the quoted error margins until they overlap with what you believe to be the correct figures. This means that "reading" error margins isn't always straightforward.

21.4: Safety in numbers

getting the right answer

When we look at the history of lightspeed measurement we find an odd pattern – we normally expect experimental results to show a broad initial "scatter" in reported values, with wide quoted error-margins that then progressively reduce as the experiments become more accurate and the methods become better understood, and we expect these values to converge with increasing confidence on a particular range that contains the correct value. And in this case, this is almost exactly what happened, apart from one detail: independent experimenters somehow managed to converge, as a group, on a range that *didn't* contain the right value.

There doesn't seem to be an obvious explanation for how this could have happened other than **social filtering** (and more specifically, the **"bandwagon effect"**). What *usually* happens is that our understanding of a result evolves through **positive feedback** ... a rough agreement forms as to which values seem to be the most believable, values in the "right" range are then increasingly believed and published more enthusiastically than those that aren't, and eventually this emerging consensus makes it more difficult for researchers to maintain credibility while announcing a figure outside this common range. The more strongly-favoured results are encouraged, and the less-believable ones discouraged, and this positive feedback accelerates the rate at which we can arrive at a definitive-looking answer. But occasionally this system can work *so* strongly towards an emerging consensus that it can become self-steering, and can veer away from the underlying reality – experimenters' *expectations* of what a figure should be can become more important than how their equipment actually behaved (**"expectation bias"**, Jeng 2006). Researchers may select figures for release with an eye on other researchers results, or they may select their calculation methodology for better agreement with what their colleagues consider to be the "right" figures. Experimenters are liable to consider their own results to be more reliable when they agree with those from other experimenters ... but if everyone is playing the same game, reported results may drift *as a self-sufficient block* away from what is later found to be the right result. In lightspeed results after around 1950, reality intrudes: the historical graph of reported lightspeed figures suddenly "jumps" outside the previous converging range, and then dots about reasonably consistently in what we now consider to be the "proper" range, before becoming *defined* as 299,792,458 m/s.

We expect this sort of **consensus bias** to happen as part of the scientific community's natural process of exploration, and it usually shows a healthy process of successive approximation converging on an agreed and reliable truth, backed up by multiple independent researchers – but this case suggests that sometimes social dynamics can act to mimic the *appearance* of scientific advance, and generate a convergence in reported results towards a core truth that may not actually exist. Social pressures and beliefs can persuade a group of scientists to generate figures that converge on *wrong* numbers with increasing confidence. Legend says that when physicist **Louis Essen** produced a figure for the speed of light that really *was* in line with our "modern" value, his boss commiserated with him and reassured him that he shouldn't worry, and that he'd be able to get a "good" value once he'd became more familiar with the equipment. Essen stubbornly insisted that his figure was right, and that the "community" figure was bad, but he might have had an easier ride socially if he'd simply selected less accurate experimental data that agreed better with the community answer, published that, and joined the club of experimenters who were able to declare that they'd been able to get the same answer.

The "lightspeed" example shows one problem with the socialisation of physics: while group consensus takes the judgement load off individual experimenters and spreads it over a group as a way of increasing the reliability of our assessments, there's a danger that instead of consensus being an enhanced route towards objective truths, consensus can start to take over and impose its own reality. Social mechanisms in science are supposed to be a means to an end, but they may sometimes become an end in themselves.

pathological quality control

The awkwardness of the "lightspeed" example is that every individual concerned seems to have believed that they were doing the right thing. Many of the people concerned were well-known and well-respected experimenters, with acknowledged skills and good track records.

For a lone experimenter refusing to allow social pressure to influence their reported results, things may have been tough: a sceptic would be entitled to ask, why do you think that *your* results are right and everybody else's are wrong? Are you impugning the abilities of everyone else in the field? Are you claiming that your data is *that* much more reliable than everybody else's results put together? And how do you explain how everybody else's equipment is giving figures that independently converge on the same result? Are you seriously suggesting that the great and the good of physics are all colluding and are all involved in some sort of conspiracy to fraudulently fool people into accepting the wrong numbers? What possible reason would they have for doing this?

The idea sounds absurd, and it'd seem that the appropriate reaction of a well-behaved researcher is to set aside their own "anomalous" findings and accept the consensus view. To persist in claiming that everybody else is reporting "wrong numbers" for no apparent reason doesn't sound rational, and to suggest that a form of **group hallucination** has taken over a part of the research community sounds like the raving of a conspiracy theorist. It may seem like act of a delusional individual who can't come to terms with the more likely explanation, that their *own* experimental expertise isn't quite as good as they'd like to believe.

And yet … in the lightspeed case, the community consensus figures, reproduced and certified independently by experimenters using different equipment in different countries, *were wrong*. An excess of enthusiasm for the common goal – nailing down the speed of light – destroyed the scientific basis of the work, and turned it into an accidental exercise in collaborative fiction.

21.5: Accident reporting

"Mission-critical" professions tend to develop special error-reporting procedures to allow the anonymous logging of failures, such as aircraft near misses or surgical mistakes. It's assumed that honest accidents will inevitably happen from time to time, and that the importance of gathering and collating accurate statistics on how often they happen and when outweighs the importance of being able to make disapproving noises at those involved. We decide that we're prepared to live with a situation in which mistakes aren't *always* fully reported and acted on, just as long as we have reasonable statistics to reassure us that these incidents are rare enough to be counted as unforeseeable and unavoidable random error, and don't resolve into a recognisable pattern of **systemic error** that would require our intervention. If the data showed that aircraft near-misses usually happened at the start of a particular work-shift, or in certain conditions, then this'd be useful information: we'd want to know about it and then we'd want to try to find out why, in the hope of being able to do something to rectify the situation.

But there doesn't seem to be an comparable reporting or "logging" procedure in experimental physics for keeping track of experiments that fail to work as expected. A researcher can attempt to publish the results of a failed attempt, and perhaps make it more "publishable" by presenting it as a preliminary paper, describing a method or technique that hasn't yet produced good results… but generally it's not considered sensible to announce to the world that you haven't been sufficiently skilled to get an experiment to work, and journals (whose space is limited), might not consider the details of a failed experiment to be worth publishing.

But if we're trying to review a subject and are looking for evidence of possible departures from standard theory, it's often these "quiet failures" that are the most important. If this information can't be found in journal papers we can be forced to resort to a more risky and interpretational approach of trying to analyse gaps and silences in the experimental record.

21.6: Quantum sociology?

How well does science work?

Social groups can be delicate and chaotic systems, and groups that traffic in information or perform decision-making on behalf of a larger groups can provide the most extreme examples, thanks to the higher complexity of networks that have inbuilt feedback mechanisms operating at high bandwidths. If we want to model some of the basic behaviours of these delicate and deeply chaotic systems, then it seems sensible to try to apply the rules of acoustic metrics, or of information theory, or, more familiarly, of quantum mechanics.

crash-prediction

One of the first things that we learn about these systems is that they have a tendency to crash, and the greater the degree of positive feedback, the larger and more determinedly single-minded these crashes will be ... at least, unless there are countermeasures in place. Brilliant success has a habit of leading to extinction, as systems that are able to tune themselves too perfectly to a particular situation shed the "unnecessary" abilities that would allow them to adapt to different circumstances. In general, "control-freak" systems tend to fail the fastest, and the more "perfect" the society or system, the more disastrously it tends to self-destruct (and the worse the final crash turns out to be). We see this story told repeatedly in stock market charts and in the fossil record.

Titanic syndrome

One of our more useful rules is that when something *can't* be known for sure, the more strongly a source insists that it *is* known, the more likely it is that the thing isn't just uncertain but actually *wrong* – the classic example being the **RMS Titanic**, the ship that was reckoned to be "unsinkable", and which obligingly sank on its very first commercial voyage in 1912. This is pretty much par for the course. If you're lying in the road after a car accident and a passer-by comments that you look pretty rough, then that's probably a fair assessment of your condition. If they hold your hand and stare into your eyes and insist, "You are going to be ALL RIGHT", then you may be in dire need of specialist medical help. If they scream at you "YOU ARE NOT GOING TO DIE!!! YOU ARE NOT GOING TO DIE!!!", then you probably are. Very soon. With the added annoyance of having to do it while someone screams at you.

If a government public announcement tells you that there is no reason to panic, there probably is. If a policeman tells you that there is nothing to see, there probably is. If your airline pilot suddenly feels an inexplicable urge to announce over the plane's intercom system that there is absolutely nothing to worry about and that the plane is definitely *not* going to crash, definitely not, no sir-ree, then it's time to start eyeing up the emergency exits.

As individuals, there are many areas where our own direct experience is limited, forcing us to rely on *indirect* perception – we use other individuals and groups closer to the raw information than ourselves as extensions of our own senses. This doesn't necessarily mean that we believe what they have to say – those closest to the data may lack perspective – but we watch how specialist groups interact with data and then use our inbuilt and intuitive knowledge of how groups behave to judge what underlying data would be most likely to produce those responses.

We regularly apply general, "instinctive" value judgements like this in our daily lives: in the early days of the "global warming" debate, many people reacted to early claims that the effect was *definitely real* by deciding that it was probably a load of bunk ... until government sources started insisting that there was *definitely not* anything to worry about, which is when we began to take the thing seriously. In order to assess what might really be going on, we apply some quite sophisticated rules based on how we expect groups of our fellow humans to process the information that they're exposed to, sometimes in quite perverse ways.

21.7: Pattern Recognition and group decisionmaking

emergent behaviours

Societies (especially those dealing with the categorisation and creations of knowledge) can start to show global characteristics that one might associate with the apparent **emergence** of an effective **collective consciousness**, or a group subconscious. These groups can *appear* to be producing deliberate, consciously-designed responses to outside threats or actions that impinge on the group, without there being any controlling mastermind or group of conspirators, just as there's no centralised consciousness behind the flocking and wheeling behaviour of groups of birds: **distributed decision-making** can generate *resonant* behaviours reminiscent of some of the patterns and behaviours that we see emerging from quantum theory.

These forms of **herd behaviour** can be powerful. As a single buffalo in the middle of a charging herd, your options are quite limited – charging wholeheartedly with the herd makes you almost invincible, safe from almost any outside predator, but not running means that you'll be trampled underfoot by your neighbours. You might decide to take a compromise position and try to run slightly more *slowly*, or try to veer in a slightly different *direction* in the hope that if enough other buffalo do the same, the herd will eventually slow down and stop, or change course ... but trapped in the centre of the herd, your ability to make these sorts of decisions is sorely limited. You can't see the distant landscape, or the lie of the land. You can't even see the ground a few feet away, all you can see around you is the other charging buffalo and the dust cloud that's been kicked up by the group.

In the right circumstances, this sort of group behaviour can be very, very successful ... but occasionally a group of stampeding buffalo will charge straight over the edge of a cliff, either through sheer bad luck or because some sneaky predators have realised that it's possible to use the herd's own defence mechanisms against it. The buffalo at the back can't see the cliff, and continue charging, and the buffalo at the very front don't have the option of stopping even if they want to. This sort of group catastrophe shows us the importance of valuing diverse input in any group decision-making process – a more successful herd strategy may involve "scout buffalo" scattered around the periphery who can signal to the herd of any outside danger, but if the entire population is packed into a central block, or if the group consensus is so strong as to automatically obliterate outside views, then entirely disastrous decisions can be made in which the whole herd is lost.

nightmare states

Another lesson that we learn from the "statistical mechanics" part of quantum theory is that although the emergent behaviour of a group will usually be in the group's self-interest (at least in the short term), sometimes resonances can build up that appear designed to destroy it. If we take a parallel-processing computer network designed for image recognition that's been trained to extract particular types of image from a pixel field, and then disconnect its inputs, it can lapse into a quasi-resonant state in which its image-starved pattern-recognition abilities start to organise random input noise into the sorts of shapes that the system has been taught to identify – the system "sees" familiar patterns in the background noise and enters something similar to a **dream state**. It starts to hallucinate. But a system that has been taught to look out for and recognise certain shapes and patterns as precursors to "unpleasant" stimuli can find itself imagining these, too. An "anxious" system can be prone to nightmares, and an over-trained, "traumatised" system can find itself in a "waking dream" state where it can hallucinate and project "nightmare scenarios" even when real information is present.

Problems due to over-sensitive threat recognition were a factor in "Cold War" decision-making, before the breakup of the old Soviet Union (McNamara 2004), but "resonant" behaviour also seems to have appeared in US intelligence and decision-making after the September 11th bombings, where we see repeated instances of information apparently being generated spontaneously from background noise by a highly-energised system trained to find it. In the post-9-11 environment, the instruction to be aware of, say, any ongoing "weapons of mass destruction" programmes in Iraq was so strong that any tiny scraps of information that *might* have been interpretable as "evidence for" tended to be passed up the command-chain, while any that contraindicated the belief system were more likely to be dismissed or ignored.

Eventually the level of resonance became *so* strong (with multiple intelligence agencies sharing and amplifying each other's information and disinformation in a feedback loop), that expectations and convictions overruled more "rational" evidence assessment. If we have a sufficiently-strong need to believe in something, physical evidence will tend to appear that will support our belief, even if that evidence shouldn't rightly exist: we don't have to consciously manufacture false evidence ourselves: sufficiently-strong reward systems will encourage other people to do it for us, either inadvertently or deliberately.

self-fulfilling prophesies and threat-addiction

As we look at the behaviours of complex systems, another idea emerges that has social implications. There are no perfectly-isolated systems in our universe, or at least, if there are, we can't see them (because if we *could* see them they wouldn't be isolated), so **cause and effect** within a processing subsystem tends (to some extent) to be smudged out into its surrounding systems, and to some degree even outside its official processing environment. Even in pure computing, the fact that a computer programmer sees a program's output means that the human designer is influenced by the program's behaviour and may make subsequent changes to the code differently as a result: we can say that the program exists as part of the programmer's feedback loop, but with a sufficiently complex "smart" systems, the programmer (and their colleagues and friends and superiors, and the wider society) can increasingly also become part of the feedback loop of the computer program. As we couple a "thinking system" to the outside world, events at the periphery of the system that are within its realm of physical influence begin to be increasingly co-opted as part of the system's decision-making matrix, and although in theory this will increase the influence of the outside environment on the system, a sufficiently "overtrained" or "stressed" system, with a well-defined and isolated region of direct influence, may start behaving in ways that provoke or subtly encourage the real-world events that it's expecting to see. The system's *expectations* of particular events in the outside world may start to affect the probability of those events actually happening, (or being reported as happening), because self-tuning parallel-processing systems will tend to adjust their own behaviour to pick out resonant relationships between outputs and inputs that best satisfy their internal reward systems, even if the exact cause-and-effect relationship for how they achieve these effects isn't obvious. The internal reward system of an "anxious" network strongly encourages the identification of threats, because it'll assume that threats really do exist whether it sees them or not. Since these systems may be happier in a crisis than in peacetime (because a "known" threat allows a more computable response), their need to identify threats can become overriding, even if this means that they have to create those threats and crises themselves. A delusional system may unconsciously find ways of making its own nightmares physically real.

Putting these last two effects together, we can end up with a rather alarming scenario: a complex analytical network or *organisation* that has some control over its immediate environment, and which enters a "monomaniacal" state, may inadvertently behave in such a way as to recreate what it considers to be "nightmare scenarios" not just in its *perception* of how reality works, but in the real physical world outside. Not only can it unconsciously promote the collection and creation of data that supports its own assumptions, but a "stressed" system can evolve responses that may almost seem to be consciously *willing* a disaster to happen. Changing the weightings of a reward system can change how results that are reported, and if those reports have a significant effect on the world outside, a feedback loop can be created in which changing a few lines of analytical code in an important subsystem can end up changing subsequent real-world events in ways that we may not be able to analyse.

"seeding" disasters

Quantum principles warn us that we need to be on the lookout for systems and organisations for whom a catastrophic accidental mistake would result in increased power and influence for the offending body. Although nobody in one of these organisations may deliberately engineer a social disaster in order to benefit their group, quantum theory suggests that perhaps when we have a structure that rewards "accidents", "suitable" accidents may have a tendency to happen "all by themselves", thanks to the unconscious and apparently uncoordinated actions of disparate members of the group. If there's a system potential that would reward a particular action, then we should expect that action to have a tendency to happen even if there is no direct, planned, cause-and-effect relationship to bring the thing about.

For example a catastrophic accident in the biotechnology industry (such as the inadvertent creation and release of a lethal virus) might well lead to that industry becoming more powerful and better funded, as resources are diverted towards it so that its expertise can be used to clear up the mess, and toward the end of the Twentieth Century, it sometimes seemed that the industry was taking an improbable number of odd decisions with safety implications (some of which, were identified after and after a gene-therapy death prompted the medical side of the industry to undergo a ritual period of painful self-examination). Its initial enthusiasm for **xenotransplantation**, for instance, seemed likely to create opportunities for viruses to cross the species barrier to humans, and if and when this sort of accident eventually happens, the industry may make a lot of money from it.

These pathological behaviours can usually be disrupted by setting up damping fields and counter-potentials, and one of the most effective of these is social disapproval. Social disapproval also has the advantage of costing almost nothing to implement. If it's made clear to members of a community that they'll be stripped of wider social respect and advantage if an accidental mistake happens, then the risk tends to recede.

But fear of *outside* social disapproval is less effective when a group has its own self-contained social structures. An immediate example of a high-risk network would seem to be a country's national security intelligence infrastructure. This seems to have all the necessary qualities: it deals with "delicate" information, its workings are so complex and confidential that nobody really understands quite how it works, its staff operate as distinct cultural subgroup of the larger population, it has significant leverage and control over some aspects of its immediate environment, and it has pronounced sense of what it wants to happen or not to happen. It will also be highly trained to recognise certain "nightmare scenarios", and could gain increased funding and control in the aftermath of one of these disasters becoming real.

Another aspect of these organisations is that they consider the job they do to be so important that the their own continuing access to resources can find itself being promoted to the status of a *primary* goal. They show an instinct for self-preservation and self-advancement. If an organisation's funding and influence depends on the existence of outside threats, and this threat wanes, then, consciously or unconsciously, members of the organisation may find themselves acting together in such a way as to make those threats more credible.

From what we know of quantum theory and complex system behaviour, we therefore need to mindful that one of the major threats to a country's national security may be the behaviour of its own well-resourced national intelligence systems. It may seem fanciful to suggest that the people tasked with protecting us from terrorism may be inadvertently working to encourage threats against us, but if "unfortunate incidents" increase the resources available to an organisation, it'll tend to evolve behaviours that may encourage – perhaps unconsciously – those incidents to happen.

A sufficiently-energised system might find itself creating wave after wave of new threats as a way of earning rewards more efficiently, unless it has sufficient self-awareness or governance to recognise that *this behaviour itself* poses a threat, and to implement countermeasures.

21.8: Market Forces

The idea that action-reaction principles apply across chaotic social structures – that by setting up a system of reward and deterrence forces at the periphery of a complex system, the system will automatically respond and reconfigure itself to best harness the potentials applied to it – isn't new. It's essentially the idea proposed by **Adam Smith** in his 1776 treatise on capitalism, "*...The Wealth of Nations*". According to Smith, the most efficient way to meet a need was not for a government to impose a solution, but to allow the natural **market forces** of supply and demand to allow an economy to self-organise and spontaneously generate the most efficient response to a problem. Under Smith's scheme, a "**virtual conspiracy**" emerges spontaneously to solve economic problems without the need for central planning. Smith's philosophy has been attractive to generations of governmental advisors, partly because it has a strong element of truth to it, and partly because it seems to say that the job of government is to get out of the way and allow the economy to do pretty much what it likes.

There are a couple of difficulties with the idea of an *entirely* free market: firstly, our experience of complex, chaotic systems tells us that sometimes the natural behaviour of a complex system leads to a partial or complete extinction of the system itself, and even if we can argue that the occasional catastrophic economic crash (and the resulting war, pestilence, and so on) is an entirely "natural" occurrence, most of us would really rather prefer that these things could be avoided if at all possible. The second argument against a full-blown "free market" implementation of economic policy is that perhaps the appearance of democratic interventionist governments is *itself* the result of market forces, as a response to the population's demand for there to be people with some degree of responsibility for how things are run, and who can be called to account when they go badly. Understanding the operation of market forces is obviously critical to understanding the behaviour of economies, and anyone who chooses to build an economy by ignoring them may be as likely to succeed as an architect who chooses to design a building while ignoring the effect of gravity. But market forces, like gravity or thermodynamics, makes for a poor religion: We routinely manipulate and exploit gravity, thermodynamics, electromagnetism and so on without feeling that we're somehow corrupting a pure force of nature. The idea that water flows downhill is useful, and it would be foolhardy to build housing on a flood plane without taking this into account, but on the other hand we don't declare that pumping water uphill or building sewage processing systems is morally wrong on the grounds that tampering with the natural order is somehow automatically bad.

In order to understand how a system or group behaves, it helps to understand something about the workings of its internal potentials and reward structures (its **internal market**). With an artificially-engineered system, these quantum principles can be critical to understanding and avoiding system failures, because one of the features of complex free-running systems is that they have a nasty tendency to find the most efficient way to exploit a potential, sometimes in ways that we didn't plan. A sufficiently-strong potential will tend to find a way to "short-circuit" or "quantum-tunnel" to its desired state, and with sufficiently-strong incentives, money can leave a system in ways that we didn't plan for. The behaviour of "market forces" has an affinity with thermodynamics: "engineered" markets (such as the "artificial" markets devised to operate within newly-privatised industries) tend to find ways of extracting maximum rewards for minimum work – if market traders get to extract more money from a market with an oscillating exchange rate, then exchange rates may show a mysterious tendency to oscillate.

While it's sometimes said that a free market leads to maximum consumer choice, our experience at the start of the Twentieth Century was that some entrepreneurs realised that the most efficient way to make money from the system was to *eliminate* choice, by gaining **monopoly control** of a sector and then using that control to eliminate competition. This is why we have strong anti-trust laws, to stop people playing these "market forces" games *too* well.

Enron, reward-tunneling and system failure

The energy utility supply company **Enron** (which got its name in 1985,) was considered the "poster-child" for the deregulation of energy markets, and the company's success was *supposed* to make the supply of energy more efficient and cheaper for the consumer. What actually happened was that once the rewards system had been set up, market forces found that the most efficient way of making profits was to take the rewards *without* producing the benefits, and some Enron employees ended up deliberately creating artificial energy *shortages* in order to (efficiently) inflate prices. The company ended up using its ability to "turn off the power" to extort money from regional governments in a way that we'd normally only expect from organised crime. Enron's founders didn't seem to *set out* to create a company that depended on criminality, but after one disastrous investment went bad, the temptation to "spin" the facts and indulge in some creative accounting snowballed into a cycle of dependency as the company's executives dug themselves deeper and deeper into a hole.

Nobody inside Enron seems to have been *instructed* to do anything illegal, but by setting up a cascading reward structure whereby employees who bent the rules in the company's favour were told in a tangible (and spendable) way that what they were doing was "good", and by turning a blind eye to profitable activities that might not have been entirely legal, Enron developed an efficient system of distributed criminal decision-making that turned the company into an empty financial "bubble", and when that bubble burst finally burst in 2001, it exposed one of the greatest business disasters of the Twentieth Century. Enron's institutionalised criminality doesn't seem to have involved a conventional conspiracy, but an "emergent" effect that arose naturally from what we might call "cultural problems" within the company's internal community. Loyalty to the company overrode loyalty to wider society (after all it was Enron that was paying the wages and the nice bonuses), and the prevailing ethical system seemed to drift towards the idea that since the company was so great, anything that helped the organisation out of a bind must be "righteous" behaviour. Inside Enron's organisation, people acted in isolation according to self-interest and the rewards system set up by the company, and when they saw their colleagues doing the same, the social disincentives to commit what *outsiders* might consider to be misbehaviour disappeared: it became a local cultural norm.

Ultimately, Enron collapsed because misleading information designed to better the company's profile in wider society was fed back into the company's own decision-making networks. Key employees needed to believe that the company was doing well in order to keep taking their bonuses, and the painful readjustments that were necessary to bring the company back on track didn't happen. The company didn't just "con" the markets, it conned itself.

The Enron experience is instructive: excessive loyalty to a group or institution ("*my country, right or wrong*") can sow the seeds for that organisation's destruction further down the line, and while we might *expect* that an organisation will always act to ensure its own survival, this isn't always the case. Overemphasis on meeting short-term goals can lead to **"reward junkie"** behaviour: if a company's CEO decides to earn extra bonus payments by maximising short-term profits at the expense of long-term sustainability ... this is effectively what the company's reward system is *asking* them to do.

The appearance of these sorts of **emergent behaviours** in complex systems links aspects of quantum theory, field theory and information theory, with the emergence of consciousness and personality, and the appearance of apparent group psychological characteristics in isolated and semi-isolated social groups. Group psychology and group behaviours (conscious and unconscious) can play a critical role in decision-making, and the tendency of these systems to ultimately crash has implications for human society as we increasingly move towards a single, increasingly-interdependent system of economics and decision-making.

These rules seem to apply most pointedly to groups whose business is the acquisition and processing of information, and which apply strong positive feedback ... and this would seem to include the physics community.

21.9: Physics nightmares

It's sometimes said that physics is a unique profession because a physicist would never (knowingly) commit fraud – any physicist faking their data would *inevitably* be found out, and publicly excommunicated. The idea that physics is a special, incorruptible activity is a comforting one if one is a physicist, but unfortunately we've learnt from experience that whenever we hear it said that people of "**profession X**" would never do "**Y**", it usually turns out that not only *are* **X**'s **Y**-ing, but the sheer "unthinkability" of their activities means that their colleagues may find it difficult to accept that such a thing is happening ... which may in turn mean that **Y** is *more* likely to happen, unreported, in a profession in which it has the greatest **taboo**, and in which even the vaguest *suggestion* of **Y**-ing is liable to cause most offence.

whistle-blowing

A behaviour may be taboo because of the extreme loss of credibility that a group would suffer if such a thing happened, but the unwillingness to believe that such a thing *could* happen can make it easier for offenders to hide their actions, and the group's self-preservation instinct can help offenders to avoid a public unmasking after the thing is discovered. The group suffers obvious short-term damage when the wider public realise what has happened, and many guilds have a "catch-all" offence of "**bringing the profession into disrepute**" ... while this *seems* to be aimed at offenders, it's sometimes applied less to the competent wrongdoer (who is, after all, trying *very hard* to keep their activities secret) and more to the person trying to unmask them. Where problems exist, *to publicise the existence of those problems* can sometimes be considered as serious professional misconduct.

occupational hazards

Unchecked, "nightmare" behaviours *do* have a nasty habit of coming about: mathematicians (who live by logic), seem to have a notable incidence of insanity, America's most prolific arsonist turned out to be a member of the fire service, and Britain's most prolific serial killer turned out to be a doctor.

We recognise this tendency in other professions, and put in (often unpopular) safety measures to guard against, say, police officers being in the pay of criminal organisations, or intelligence officers working for enemy states, because we understand that the "taboos" in these professions are accompanied by a greater ease of means and opportunity, and that recruitment may also attract people with a higher than normal motivation to be offenders.

This isn't surprising: we know that the fire service is likely to attract a disproportionate number of people who find fire interesting, and a statistically lower number of people who have a natural aversion to the idea of running into burning buildings. To someone already obsessed with fires, it's a natural career choice. While some individuals already have a propensity to stray when they join a profession, others can be led astray by the pressures and psychology of the job: a person who initially has no inclination to fire-starting may find that once they've been trained to fight fires, and learned to do it well, and have been trained in the arcane business of how fires start and spread, they may find themselves becoming restless between callouts, and itching to have more chances to put their skills into practice. If fighting fires is considered a Good Thing, then this requires there to be fires out there to fight, and sometimes a keen specialist may feel that they have to start a few themself.

social personality and "Monsters from the Id"

So these things do tend to happen, and sometimes it's the *group's* behaviour that goes wrong rather than that of individuals. Once individuals are deeply **embedded** in a system, emergent behaviours and responses can act themselves out through its unwitting members, and members of a community may find themselves participating in aberrant behaviour without being aware of it. The personality of a *group* often isn't the same as the personalities of its members, and the drives and reward systems of a group may be substantially different to (or even completely at odds with) the goals and ideals that the system was originally meant to promote.

Once a methodology has taken hold, it can take priority over a group's founding principles. The "lightspeed" case in section 21.4 serves as an example: researchers' laudable enthusiasm to isolate and replicate a consistent value for the speed of light resulted in a value that diverged from reality. The motivations and rewards favoured the isolation and agreement on *a* value for the speed of light ... but didn't implement sufficient safeguards to make sure that the value was a *correct* value. As long as everyone *agreed* on a value, it was assumed to be right.

We see some of this attitude emerging in statements that a thing is "not scientific" unless it has been subjected to peer review, or that science is defined as being, simply, "what scientists do". These sorts of statements promote a parody of "hard" science. There's a distinction between science as an ideal, and science as it is commonly executed, and to turn the *implementation* – with all its inevitable flaws – into a *definition* means that when an enterprise goes off the rails, we no longer have an external yardstick to judge it by. If scientists had a fashion for wearing yellow hats, it would not mean that wearing a yellow hat was essential to scientific research (except in a social sense), nor would it mean that the possession of such a hat automatically made one a scientist (except in a social sense).

Our implementation of social scientific structures is meant to be a means to an end, a way of ensuring that science gets done. These structures aren't meant to become an end in themselves, and the *purpose* of science isn't supposed to be their perpetuation. They're not supposed to *define* science but to serve it. When they don't, they quickly become pathological.

so, how safe is physics?

Since physicists tend to believe so strongly that their subject is based solidly on "hard" experimental fact and in the benefits of peer review, a "nightmare scenario" for physics might be to find that a fundamental fact that had been widely believed turned out to be a mere social construct, and that peer review had acted to hide the fact rather than to reveal it ... another might be the failure of a core theory, and the realisation that ignorant outsiders who'd been criticising the thing for years based on dumb intuition had been right all along. These, to many physicists would be Unthinkably Awful Things, and they may comfort themselves with the idea that they could never, ever happen. After all, about the only physics theory that physicists think *can't* be wrong is special relativity, and its chances of failure are reckoned to be zero ... other theories could fail, certainly, but not the special theory. It's been *proved*, the mathematicians have shown that it *must* be right, and the experimenters have repeatedly shown that its unique predictions are validated. How could *all* of this possibly be wrong?

But physics has gotten so big that it's difficult to be in full possession of all the facts. Experimenters typically test SR against "Classical Theory" (section 16.3), because that's what the "test theory" guys suggest. The theorists work on information received from the mathematical community about available options and how to distinguish between them ... but the mathematicians only provide options based on SR, because they've been told by the physicists that since SR's correctness has been shown *experimentally*, only SR-compatible solutions need to be considered. The experimenters think that SR's status as the only relativistic solution that requires testing has come from the math community, and the math community believes that it's been established beyond doubt by the experimenters. The theoretical physicists, who sit in the middle and who are the one group that should be able to see the potential for a breakdown in communication, think that *both* other groups have come to the same conclusion, independently. But while everyone in this network seems to believe that another group has done the necessary background work, it seems that nobody has.

In a very small research enterprise, things get missed because there aren't enough people to do all the work. In a very large enterprise things get missed because everyone thinks that someone else has already done them.

22
Conclusions

> "It would be an unsound fancy and self-contradictory to expect that things which have never yet been done can be done except by means which have never yet been tried."
>
> **Francis Bacon, "The New Organon" 1620**

> "... This is the reason why all attempts to obtain a deeper knowledge of the foundations of physics seem doomed to me unless the basic concepts are in accordance with general relativity from the beginning. This situation makes it difficult to use our empirical knowledge, however comprehensive, in looking for the fundamental concepts and relations of physics, and it forces us to apply free speculation to a much greater extent than is presently assumed by most physicists. I do not see any reason to assume that the heuristic significance of the principle of general relativity is restricted to gravitation and that the rest of physics can be dealt with separately on the basis of special relativity, with the hope that later on the whole may be fitted consistently into a general relativistic scheme. I do not think that such an attitude, although historically understandable, can be objectively justified. The comparative smallness of what we know today as gravitational effects is not a conclusive reason for ignoring the principle of general relativity in theoretical investigations of a fundamental character. In other words, I do not believe that it is justifiable to ask: What would physics look like without gravitation?"
>
> **Albert Einstein, Scientific American, April 1950**

> "You take the blue pill – the story ends, you wake up in your bed and believe whatever you want to believe. You take the red pill - you stay in Wonderland..."
>
> **The Matrix, Wachowski Bros. (1999)**

A geometrically-ambiguous surface

22.1: SR-based or NM-based physics?

In the previous sections of this book we found that it's possible to make apparent progress in some deep problems in relativistic physics without having to depend on advanced mathematics. We can take some basic principles and methods and reuse them in unfamiliar situations, and generate answers that are consistent with some of the better respects of modern theory, and at least *reasonably* consistent with published experimental data. This approach also seems to allow agreement with some additional equivalence principles that aren't implemented in current theory.

The disadvantage of our approach, and the reason why some of these details aren't already documented in the mainstream literature, is that it appears to be fundamentally incompatible with Einstein's special theory of relativity.

To those whose first exposure to special relativity prompted an immediate instinctive loathing, this may not seem altogether a bad thing ... but to established physicists who want to play by the existing textbook rules, a solution that doesn't agree with SR is a disaster. To even suggest this sort of possibility is to risk being labelled as a crazy person (and not in a good way).

So which of the two approaches is right?

testing

The ultimate measure of these ideas, of course, is how strongly they appear to correspond to physical reality. Do Doppler shifts, tested in an appropriate context, give results that are significantly more strongly redshifted than "barebones" SR?

We *should* be able to test for the value of our curvature factor, "©", using current equipment, but this is a potentially dangerous process: a "false negative" for "©=+1" might unfairly discredit the idea, but so might a "false positive" that was subsequently found to be untrustworthy. A "*true* negative" would tend to be believed but would probably be considered unimportant and unexciting by people who never thought that there was a real prospect of any other outcome, and a "true positive" is unlikely to be believed. So there's little motivation for an experimenter to carry out these sorts of tests unless they think that "©=+1" has a reasonable chance of being right, and in this case, the mainstream is liable to reckon that the person is too emotionally invested in getting a non-standard outcome to be a proper person to be trusted to do the work properly. Any experimenter motivated enough to carry out the test is liable to be accused of personal bias if the results don't go the way the community expects.

This leaves us at something of an impasse – in the last 50-100 years of experimental testing it's difficult to find a record of *anyone* having tried to compare the "©=0" and "©=+1" relationships against each other and against real experimental data, regardless of outcome.

Attempts to reanalyse existing experiments are complicated by the fact that **test theory** usually only tells experimenters to try to make a fair comparison between the predictions of "©=-1" and "©=0", and because any *positive* result for © in this context strongly rules out the minus one result of classical theory, it risks being interpreted as strong evidence for special relativity. We can invent explanations of why we get too much redshift, and bring the remaining redshift down into the SR range and then say that even with these generous corrections, we are still nowhere near the Classical Theory predictions. So, in some circumstances, a *more* redshifted, *non*-SR result may make it easier for us to claim that we've validated the predictions of the special theory.

Some physicists may argue that this scenario is far-fetched – that it's inconceivable that redshifts significantly *stronger* than the SR predictions could be reported without their potential significance being noticed and widely commented upon – but with the Hasselkamp test, this is exactly what happened. We found *twice as much* transverse redshift as SR had predicted (in line with "©=+1"), and instead of announcing the result as potentially undermining the case for SR, we used statistical analysis to present it as a result that emphatically disproved classical theory and therefore confirmed SR's predictions. Regardless of whether we choose to interpret the result as due to non-SR physics or unexpected experimental error, the lack of any obviously insightful response shows that the community has been presented with an apparent experimental disproof of SR *at least once*, and failed to see the potential significance of what they were looking at. Our "inconceivable" scenario, in which stronger NM-compatible redshifts are found but presented as evidence supporting SR isn't just *possible*, it's already happened.

minimalism

If we don't yet have a sufficient body of experimental evidence to guide us, we might try to use other methods of guessing at the probability of SR being right or wrong.

Part of SR's claim to fundamental correctness is its simplicity: we can say that for Nature to allow us a solution like this *and not use it* would be perverse. Why would this exquisite "flat" solution exist if it wasn't right?

But we have *two* apparent solutions here: SR for flat spacetime and NM (or something like it) for curved spacetime, and at least *one* of them *has* to be wrong. If a "curved" solution is correct, then the "flat" version can be explained away as a sort of partial, artificial "wireframe" projection of some of the correct properties onto a flat background ... but if the "flat" SR solution is *physically real*, there seems to be no reason for Nature to also have this additional "curvature-based" solution, available and unused.

22.2: The fork in the road

Another way of assessing probability is "summing histories": if the special theory is correct, this will explain why we have come to believe in it so strongly ... but might we have still been able to arrive at similar levels of belief even if the theory wasn't right? How strongly did the course of physics history depend on whether © equalled zero or plus one?

Well, when we look back at our subject's history, there seems to be a **decision fork** at the end of the Nineteenth Century and the beginning of the Twentieth, leading to two different types of physics, SR and non-SR.

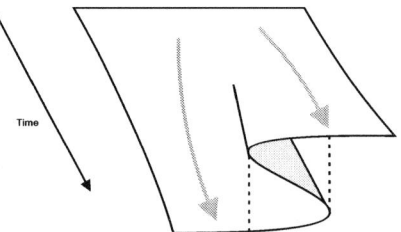

Figure 22-1: Catastrophe Theory:
"You can't get there from here "

If we'd recognised the idea of gravitational time dilation *before* the beginning of the Twentieth Century (and, by rights, we really should have), then the mathematicians trying (unsuccessfully) to apply **Gauss**'s methods to Newtonian physics in three dimensions would have realised that their basic approach was wrong, and that they should instead have been applying curvature in *four* dimensions. This could have given them a "GR-like" theory before the beginning of the Twentieth Century (Rindler, 1994), based on the idea of local lightspeed constancy, regulated by curvature effects. With *space itself* now considered as the medium in which light travels, and with Mach's ideas, we'd probably then have ended up with an *NM-based* general theory. The idea of trying to make *global* lightspeed constancy relativistic might then not have occurred to anyone, and when quantum theory appeared, acoustic metrics would probably be a logical way of explaining some of their stranger behaviours.

Regardless of what the actual underlying relationships actually are, if history had developed along this alternative path, the questions that led us to special relativity might not ever have been asked, and the theory might never have been developed. But our delay in realising the inevitable effect of gravity on timeflow meant that the story unfolded differently: we didn't have the missing piece of the curved-spacetime puzzle until Einstein pointed it out to us, and by this time, our Nineteenth-century "space-crumplers" had become dispirited (or had died), "crude" aether theories had a chance to become popular, Lorentzian Ether Theory had turned into special relativity, and Einstein was already assuming that his special theory was right We were then on the "SR track" to GR1915, "perfect" black holes and a "deep" incompatibility with quantum theory.

Given that these two solutions are more difficult to tell apart than the usual SR/non-SR comparisons that we test for, it seems possible that we might have ended up on *either* of these two tracks, not because of anything in the underlying physics, but because of an accident of human history. The path that we take seems to be dictated not by any experimental result, but by the particular *timing* of our (theoretical) discovery of gravitational time dilation. Whichever of these two tracks we take, the resulting arguments and evidence will seem to reassure us that our track is the right one, and that anything else is inconceivable. In each of these two hypothetical histories, the physics community might be quite sure that they had gotten physics right: but only one of them really *would* be right.

This exercise doesn't tell us which of the two solutions is the more likely, but it warns us that our current confidence in a current track doesn't have to mean that it really does correspond to the underlying reality. We might be just as confident if we were wrong.

22.3: Warning signs

If there are *historical* reasons for our taking the route that we did, the next thing to consider is whether, if we *had* accidentally taken a wrong turning, we'd have been canny enough to have realised it by now. Is our confidence in the route that we took founded on real expertise? Have we assessed both routes *properly* before committing to our present path?

Given that we have a decent-sized community of clever people who spend their lives thinking about these sorts of problems, how likely is it that this entire community could have collectively made a wrong turn without anybody raising the alarm up until now? To a trained physicist the idea may seem palpably absurd, but perhaps the strength of this conviction is itself a warning sign that our community might be showing some of the signs that we associate with group failure. When the "unthinkable" happens, it sometimes goes unrecognised because people don't think that it *could* have happened, even when the evidence is staring them in the face.

Is it *at all* possible that we could have messed up?

From our earlier ponderings on how systems tend to fail, the SR section of the physics community does seem to be showing at least some of the warning signs that we'd expect from a community that was in the middle of a Really Big Mistake:

Titanic syndrome – as we've pointed out, when a thing is said to be *probably* right, it often is, but when it's said to be *definitely* true, it's usually wrong. Science is supposed to allow us to question things, and usually it does: it's okay to query aspects of Newtonian mechanics, there's an increasing recognition that GR1915 may not be a final answer, and if you have almost *any* gripe with quantum mechanics, many physicists might agree that you have a point. You can take liberties with energy-conservation, thermodynamics, and almost any branch of physics and still be considered respectable, provided that you do it with some grace. But the one physics subject where we find instructions that conformance with a theory is *non-negotiable*, is special relativity, and if a scientist dislikes SR, to actually say so would seem to be career suicide. Purely as a matter of *social* system behaviour, if we were looking for a potential disaster in theoretical physics, special relativity would be our candidate.

Overconfidence – physics has a bullish culture, where debating styles sometimes verge on the gladiatorial. The "university style" of adversarial debate, where players each take a position and try to argue the other into the ground, is sometimes reckoned to provide a sort of Darwinian proving ground for physics theories, where only the truth survives. But when we see the same style used in *political* debates, we *don't* tend to associate it with a deep search for underlying truth – the object of the exercise isn't to arrive at the truth, it's to win, and when we see these sorts of debates happening in politics, we don't always come away with a lot of respect for them, the conclusions reached, or the people involved.

Things tend to be different in mathematics: a mathematician will often start a major presentation by saying something like, "I know that this is probably all wrong, but ..." – in physics, this level of caution would tend to be seen as a sign of weakness, and in an environment where debates are won by whoever manages to stand their ground the longest, to admit a honest uncertainty is often to lose the game.

Polarisation – The adversarial approach encourages physicists to commit to particular "fixed" positions on given subjects, and sometimes makes it more difficult for us to admit when we don't know something. Discussions of special relativity are usually heavily polarised – we have a mainstream that insists that the principle of relativity makes SR unavoidably correct, and we have a small minority opposition (largely outside physics) who take the position that, since we're told that you can't have relativity theory without SR, and since some aspects of SR seem dubious, that *the whole subject of relativity* must be one big con. To defend their subject from these critics, some proponents of SR have been making inflated claims in order to try to convince people that SR *must* be true, and the more overstated the claims have become, the more the theory's sceptics have gotten the idea that perhaps the whole field is somehow dishonest.

Neither side of the debate seems to have much incentive to explore the middle ground, that perhaps the *principle of relativity* might be reasonable, but that that special relativity might not be the appropriate implementation of it.

Inadequate fact-checking – physicists have a huge amount to material to deal with, and although they're supposed to retain a healthy scientific scepticism about what they're told, in practice they tend not to have enough time to check everything. Consequently they have to learn to take a great deal on trust. This trust isn't always justified.

It takes a very long time to be trained as a theoretical physicist, and someone who has gone through this lengthy process may be understandably unhappy to be told that a critical part of their training may be wrong. The easiest way to extricate themselves from this discomfort is to decide that it *can't* be wrong. So, although physics is supposed to be the most deeply-grounded of the physical sciences, in practice, a few of its most fundamental concepts still seem to be based on our willingness to accept slogans that we were taught in school.

We say that we *know* that SR must be right, because we *know* that lightspeed is constant and we know that its not affected by the motion of nearby material. It doesn't matter so much if these ideas seem to conflict with the experimental evidence, we still feel that we ***know*** these things to be true, because we *have* to – they're foundation-beliefs that are necessary for a continued acceptance of special relativity.

Ignorance of previous failures – in most communities, the occupational hazards of the job are well-known and steps are taken to avoid them. This level of self-awareness helps us to avoid making the sorts of major mistakes that would otherwise be almost inevitable. In physics we aren't so good at learning from past historical mistakes, partly because the embarrassment of earlier generations and their effectiveness at tidying away the evidence means that we may be blissfully unaware that some of these mistakes ever happened.

Although we make a big deal about mistakes that were spotted *before* they became widely accepted (such as **N-rays**), we keep rather quieter about mistakes that happened *within* the established scientific mainstream, and which can have more serious potential consequences. We're taught that people who produce theories at odds with the mainstream sometimes make fools of themselves: we aren't taught that sometimes the mainstream folk do it as well.

Cross-contamination – in our "Iraq" and "Enron" examples in section 21, misleading information intended for an outside target market ended up influencing key decision-makers *inside* the organisation: Information that had been skewed to "sell" an idea to outsiders was fed back into the organisations' own decision-making processes, and organisations were partly fooled by their own hype.

In physics, this sort of cross-contamination can happen between information intended for the "educational" and "research" communities. The two groups have different approaches to dealing with information: in research, factual accuracy is supposed to be sacrosanct, whereas in education, it's considered okay to take a few liberties with the facts in order to "help" students to accept and become proficient in a theory and get their qualifications. The hope is that by the time someone becomes a *professional*, they'll know which facts are "real" and which are merely "educational". But since it now takes so long to train someone to be a cutting-edge physicist … and they have to spend so long within the educational system … the distinction can become blurred.

Overreliance on social approval and peer review – we sometimes hear that a piece of work can't be said to be scientific unless it has gone through the peer review process. But this is a social definition of science, not a scientific one: if physics is supposed to be unambiguous "hardcore science", then a result or a calculation should either be "right" or "wrong" regardless of whether or not one's colleagues at the time happen to like it. Certainly, we all make mistakes, and we may not feel that it's *likely* that one person could disagree with a thousand trained physicists and win … but we have to acknowledge that when science has advanced in the past, it's often been by the questioning of generally-accepted knowledge, and this questioning sometimes has to start as an extreme minority view. It's happened in the past, and it'll happen again.

For incremental mainstream work, **peer review** acts as a useful stamp of certification that a piece has been independently proofread and error-checked, and that its representation of the existing body of knowledge on a subject corresponds to current opinion. For work that is trying *not* to fit existing theory, this last feature is almost useless. When physicists talk of "proper" science being "peer reviewed" science, they sometimes forget to make a distinction between science as a *collaborative social enterprise* and science as a *fundamental discipline*. Peer review is almost essential to the first, and almost irrelevant to the second. If we're involved in a search for fundamental truths, then an argument's *intrinsic scientific worth* (if such a thing can be said to exist) isn't affected in the slightest degree by whether many people happen to believe in it or not, or whether the idea happens to be unfashionable at one particular point in history.

If physics becomes too strongly defined by popularity and by politics (and by the need to keep up appearances) then it risks becoming a glorified branch of the social sciences. If we allow mass belief to overcome logic – if we insist that certain things must be true because they're written in our holy reference books, and because our scientific "priests" declare them to be true – then physics becomes a religion. Science is cumulative, and it's right to put a lot of weight in the careful efforts of our predecessors, but as scientists we must never quite surrender that last piece of scepticism.

Failure rate – for some unknown reason, the reliability record of statements supporting SR is bad. In fact it's *really* bad. In fact, when we try to list the most popular reasons why SR is supposed to be "known" to be correct – exclusivity of $E=mc^2$, exclusivity of transverse redshifts, absence of other explanations for centrifuge redshifts and so on – we find that the majority of them seem to be historically, geometrically and mathematically wrong.

The general student-level understanding of SR *does* seem to have improved in the last couple of decades, but this may be more to do with the unregulated exchange of information over the internet than through the use of official channels and "mainstream" quality control. It sometimes seems that when these issues are finally addressed, it's often *in spite* of the community's quality control procedures, rather than because of them. This sort of failure rate is not supposed to happen in science, and especially not in physics.

Bad error-correction – one of the best ways of judging the veracity of an information-source is to see how willing that source is to issue retractions and corrections when its information is found to be faulty. To have to admit that you've been wrong is embarrassing and unpleasant, but the mark of a scientist is that when we recognise a mistake, we'll grit our teeth and do it anyway. It's a mark of scientific discipline. The annoyance of having to do this (and the embarrassment if one has to do it too often) is part of what keeps us on our toes. Scientific reputations are valuable things, but if our reputations ever become *too* precious to us, to the extent that we make factual accuracy secondary to protecting the *image* of science or scientists – if we put appearances and political considerations above truth – we may not me entitled to those reputations. We *have* to be prepared to acknowledge and admit mistakes when they happen.

Looking back at the record of the last sixty or so years, error-reporting in any subject related to special relativity seems to have been *exceptionally* bad. The story of the **Terrell Effect** shows us that special relativity is a subject in which – for decades – the theory's proponents somehow managed to get the theory's predictions wrong for the appearance of moving objects. The theory made a *deterministic* set of predictions, and yet when put on the spot, physicists presented what they believed the theory *should* say, rather than what the numbers actually stated. When the issue was finally addressed in 1959, it wasn't fixed out by any of the established mainstream physicists, but by **James Terrell** (an "undergraduate who had problems getting his result through peer review) and **Roger Penrose**, a mathematician interested in physics (who bypassed peer review and had his result published as a letter in a small local mathematical journal). "Proper" physicists either weren't capable of seeing the mistake or weren't willing to discuss it in print. Even after the flurry of papers prompted by the Terrell's paper, one could still find academics forty years later insisting that the faulty results that were taught "pre-Terrell" were correct and were *known* to be correct.

Special relativity seems to be a subject where bad predictions are able to persist uncorrected by experts for decades, the corrections seem to have to be forced through by outsiders, and even after mistakes are finally acknowledged, it can take decades for the news to filter down through the community's information systems to the "grass roots" physicists.

These aren't encouraging signs.

Overactive defence mechanisms – does the scientific community actively *deter* dissent over special relativity? We're told that scientists don't behave like this, or at least, while there may have been documented instances in the "bad old days" under various repressive regimes of "inconvenient" research being marginalized, that this sort of thing doesn't happen in modern, free societies. The physics community believes in academic freedom and doesn't intimidate its own members. In theory.

So it's instructive to look at the case of **A.G. Kelly**, who in 1996 published a paper saying that the potential sources of experimental error in the 1971 **Hafele-Keating** test appeared to be at about the same level as the collected data (making the result unsafe), and that using more recent statistical procedures, perhaps the results wouldn't have been regarded as acceptable. Kelly's paper didn't say that *special relativity* was wrong, or even that *the result itself* was wrong, only that the experiment shouldn't be regarded as a reliable verification of it.

Kelly broke a taboo in criticising the methodology of a famous experiment used to support special relativity, and ... what made it worse ... it looked as if he might have had a point. Instead of the community making a glum face and thanking him for his contribution, Kelly quickly became a hate figure on the internet forums, and was roundly criticised for almost everything one could think of ... apart from the actual contents of his paper, which most of the critics didn't seem to have read.

The message that the community was sending seemed to be seemed to be – this is how we dispose of troublesome people who criticise the status quo, and if you find a possible problem with our arguments, the same may happen to you. After seeing the community's response to Kelly, any mainstream researcher with similar doubts about another mainstream SR experiment might well decide that the most sensible thing to do is to keep their misgivings to themselves.

So, although we're told that special relativity *can't* be wrong, when we look at the ways in which the SR community operates, it seems to have all the characteristics that we'd expect from a pathological enterprise (when we're told that a horse "can't possibly lose" a race, it usually does). These problems don't necessarily mean that the theory *is* wrong, but if SR *was* wrong, we'd at least have an explanation for why these things keep happening.

the least worst option?

The scheme we've outlined here is probably the simplest possible alternative to the existing structure. It may or may not be right. But we should probably hope that it *is* right, because the alternative – that the real structure of physics lies in another direction that we haven't yet been able to see – suggests that our chances of finding it may not be good. Our scheme is reasonably intuitive, only uses components that we're already familiar with, and suggests relationships that were already in use before special relativity ... and yet we somehow still managed to miss it.

If a structure *this straightforward* could still be overlooked, the chances of our spotting a more obscure solution that is *genuinely* counter-intuitive and alien to us don't seem high. If *this* alternative doesn't pan out, we might have nowhere else to go.

22.4: Mathematical "truth" vs. relevance

The internal validity of an abstract mathematical model doesn't always translate into practical geometry, or at least, it doesn't always do it in ways that are useful or expected.

For instance, here are two popular ways of stacking objects:

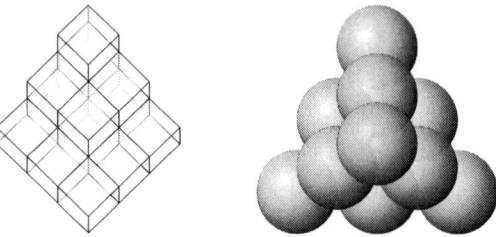

Figure 22-2: Cube-stacking and sphere-stacking

On the left we have simple cubes, which stack regularly and which completely fill space, and on the right we have a stack of spheres, which seem to follow different rules, arrangements and geometries. Each layer of spheres forms a triangular or hexagonal grid, each layer is offset from its neighbours, and, the pattern repeats every two or three layers, depending on how we decide to stack. The rules, conventions philosophies and intuitions that we might devise from our cube-stacking experience don't obviously carry over to a "sphere" case – they might, or they might not. If we're modelling close-packed spheres it doesn't matter whether the properties of the "cube" system are mathematically true in their *own* geometrical world, because these are two *different systems*.

Certainly we can try to stack spheres as if they are cubes, but the result isn't guaranteed to be optimal. Some themes and patterns *will* carry over between the two systems, but the way that they do this might be quite subtle.

In Figure 20-2 we found that we could slice a cube diagonally to produce hexagonal faces, and below we find that eight of these half-cubes will reassemble into a semi-regular 14-sided polyhedron (with eight hexagonal faces and six smaller square faces). A collection of these polyhedra will then stack to produce a perfect space-filling array. If we squint at our new stack, we find that each pair of touching hexagonal faces represents one of our original "cut" cubes, and the stack can be cut up and reconsidered as a simple array of these smaller cubes.

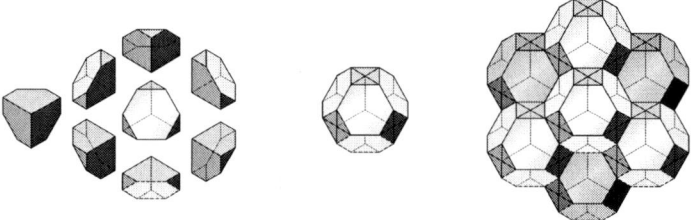

Figure 22-3: Another space-filling method

At this point, a cube-enthusiast may say, "See, I told you, its nothing but cubes! Cubes are fundamental for perfect space-filling! Nothing else can work!"

But although our space-filling array may have some geometrical properties in common with a cube array, and may in some sense contain *aspects* of the cube-array, the uncut 14-sided polyhedra are *not* solid cubes, and aren't stacked like cubes, and someone brought up in "cubeworld", where every angle must be a right-angle, might never be able to reach this solution. A deep training in cube-stacking may be counterproductive. Some of the more *abstract* geometrical properties of the array may correspond to those of a cube array, but we can't produce our new system by assuming that solid cubes are an essential, *indivisible* building block of other complex shapes. Our new unit-polyhedron can't be made by gluing cubes together, and no matter how long we play with a set of wooden cubes, we'll never manage to assemble them into our new array, or our alternative solid. We need a hacksaw. In the new context, proofs based on the essentialness of cubes may not just *not give* us our new stacking system, they may seem to prove to us that such a system is *geometrically impossible*.

A similar relationship may hold between special relativity and a final theory of physics. The fact that SR's geometry is "mathematically true", and appears self-sufficient and inevitable in its own context, doesn't mean that a final theory must use SR as a building block, and must be a full logical superset of the special theory. We may need a hacksaw. Ghostly echoes of *some* of the SR patterns and relationships may well live on within the structure of the new theory, but the theory isn't compelled to reduce *physically* to special relativity. It might be that the final theory's relationship to special relativity will only become apparent with hindsight, and that the larger theory can't be produced *in principle* by starting with SR and building outwards.

This happens in our "©=1" model. If this model should turn out to correspond to reality, we may like to be nice to our SR-trained brethren by agreeing that Minkowski spacetime is a very fine mathematical model of a hypothetical, empty, unphysical relativistic space, and that Lorentz relationships are *indeed* fundamental and inevitable. We may say that, in fact, what we need to do to produce a more modern theory is to apply these very desirable Lorentz relationships not once, but *twice* – once to create the "empty space" relationships of Minkowski's geometry, and then a second time to reproduce the missing gravitomagnetic distortions and warpages associated with the interactions of real particulate matter. Special relativity, multiplied by this *additional* Lorentz redshift, gives us the relationships of "©=1".

But if our colleague gets too excited and starts to jump about and say that this just shows that SR was right all along, we must quietly point out that the new model was *not* developed using SR, that it was only conceivable if we assumed that SR was *physically wrong*, and that SR's main contribution to the project was in fact to hold back this development for half a century by claiming to have proved, mathematically, that no such theory could ever be constructed.

Special relativity represents one way of assembling *some* of the properties that we want in a final theory, to produce a system. But there are other desirable components that aren't used in the SR system, and aren't compatible with it. The optimal shape that we obtain by using up *all* the parts is not necessarily going to be "SR with extra accessories". We do not necessarily have to go *through* SR to reach the final system, and SR is not required to be a full *subset* of the final system, except in some invented, abstract, unphysical sense.

22.5: Alternative alternatives

Of course, our arguments for a ©=1 solution aren't watertight. Although positive values for © are attractive, and ©=1 appears to be the simplest solution, we haven't produced what a mathematician might consider to be a fully-worked-up proof. We've essentially said that if the model *can't* be ©=0, and has to reduce to Newtonian mechanics in the simplest way possible, that *if* the NM set can work in curved spacetime with no modification, then this should be our default solution unless anybody can justify building-in additional overhead. Minimalism and **Occam's Razor** gives us ©=1 until we have reason to suspect another solution.

These arguments aren't exhaustive, and a mainstream physicist may argue that our ellipse arguments (convenient though they are) don't explicitly take into account spacetime curvature. Might there be some additional Lorentz-factor trick here that turns our NM-based calculations back into SR? If light-signals leave the emitter at $c_{EMITTER}$, and reach the observer at $c_{OBSERVER}$, then isn't it reasonable that the shifts that we see should be some sort of hybrid of the two associated Doppler predictions, potentially giving us the "root product" averaged outcome of special relativity?

Perhaps. But in order to assess that idea we need something more concrete to work with, and the SR community doesn't yet seem to have been able to get a "curved" implementation of SR to work. If *our* arguments in the context of curved spacetime are somewhat skimpy, the SR's community's efforts so far to prove *their* equations in the context of curved spacetime ... with hordes of available researchers, and a century to do it ... seem to be missing.

Ultimately, our decision as to the correct set of relationships and principles will have to be established by experiment rather than by theorising, but it would be sensible to do at least *some* preliminary work on possible alternatives before the results arrive.

the "bad universe" possibility

Some of the "curvature-based" arguments that we've been looking at seem to be very efficient, and we may find it comforting to think that Nature doesn't miss tricks like this, and that we live in a rational, minimal, efficiently-organised universe where these sorts of arguments must be right. Perhaps a sense of human design aesthetics favours a fully "curvature-based" solution over SR-based models, and perhaps this might be how *we'd* design a universe. But this presumes that Nature will necessarily always use what *we* think of as the most efficient solution.

This is a good working assumption when devising laws of physics, but we also have to consider the annoying possibility that, when we look at multi-universe cosmologies, even if we can devise what *we* consider to be the "best" laws of physics possible, it isn't guaranteed that we occupy a section of multiverse in which those "best" laws apply.

So even if we think SR-based physics *sucks*, this doesn't guarantee that it's wrong. We may inhabit a slightly shoddy "reject" universe where the laws of physics are a bit substandard. That's a slightly depressing possibility for a theorist, but it's a difficult one to rule out completely. Our ultimate theory could describe a "Universe 2.0", while we're stuck in a "Universe 1.9". The physical system that we come up with might not apply to here, or to now.

However, if we're to be optimistic (which is pretty much an essential attitude for theoretical physics), we have to assume (or hope!) that if we ever do produce an ultimate theory, reality will turn out to be not too different.

22.6: Life after special relativity

We seem to be living in interesting times. If future experiments were to show conclusively that the SR predictions were seriously wrong ... how should we react?

If we are scientists, there is only one way *to* react – to be scrupulously honest about it. We might have to acknowledge that we messed up, but we'd be able to point out that the difference between science and religion or politics is that when things are found to be wrong, *we* admit it. Certainly, if ©=1 turns out to be right, it would be tempting to retreat behind some flimsy argument that *of course* special relativity was never claimed to be valid for real matter, and that ©=1 represents a logical, gravitomagnetic *extension* to special relativity ... but this wouldn't be honest. Learning curves can be painful, but the process incentivises us to try to identify and avoid future failures. To ignore or rewrite the history of SR – to "spin" this sort of non-SR result as a logical *progression* of SR to avoid short-term embarrassment – would be a further failure, and an indication that perhaps we still haven't learnt our lesson. Hopefully, if the situation arises, we'll do the right thing.

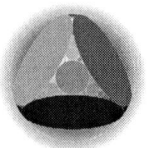

PART VIII

CALCULATIONS, REFERENCES AND INDEX

RELATIVITY IN CURVED SPACETIME

Calculations 1: Doppler shifts

Calculating the emission-theory Doppler relationship

We've already worked out the audio Doppler-shift for an object moving with velocity v through a "stationary" medium (in section 6.1) as *freq'/freq* $= c/(c+v)$. How do we find the Doppler prediction for an *object* that is "stationary" and an *observer* that is "moving"?

Well, there *are* methods for tracking or graphing the progress of lightsignals against the location of the objects and observers, but a quicker and more "cheaty" way is to invent a situation that involves *both* equations and has to produce a known result.

We'll take a (hypothetical) idealised box, beam a signal from one side of it to the other, and place a signal transponder in the middle of the signal path, to absorb and then retransmit the signal (a small glass bead would be adequate). We'll also declare that in our thought-experiment, the speed of the signal is *absolutely constant* with respect to the box. Since we're *insisting* that the bead's motion *doesn't* affect how the signal moves, the signal has to arrive at the far side of the box with the same frequency that it started out with.

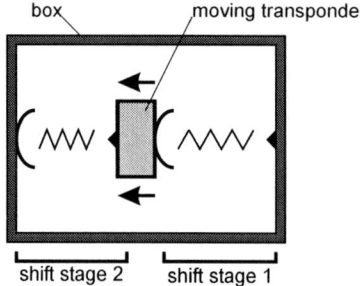

The far wall of the box should see the bead's signals to be Doppler-shifted by the frequency ratio *freq'/freq* $= c/(c+v)$, because this involves a "moving source" and a "stationary observer", as in section 6.1. This is "**shift stage 2**" in our diagram.

We now have all the information that we need to work out the corresponding prediction for the earlier "**shift stage 1**" – the shift that the *bead sees the box* to have. This relationship is the one that we're looking for, since it assumes a lightspeed that's supposed to be fixed in the *emitter's* frame (in *this* case it's the *box* that is the emitter). We'll call this unknown ratio, "**X**".

We know that ratio "**X**" and the "other" Doppler shift, multiplied together, need to exactly cancel each other out, so

$$\frac{c}{c+[v]} \times X = 1$$

which means that our "unknown" Doppler relationship has to be

$$X = \frac{c+[v]}{c}$$

Our last step is to acknowledge that, for the signal crossing the box, if the transponder is *leaving* the front wall, it must be *approaching* the rear wall, so if we're to quote both Doppler relationships consistently, using the same "recession velocity" format, we'll have to reverse our new equation's ± sign.

This then gives us our finalised Doppler prediction for a signal moving at a fixed speed with respect to its *source*, of

$$\frac{freq'}{freq} = \frac{c-v}{c}$$

, where v is recession velocity. Cross-checked against other methods, this turns out to be the right answer. This was the historical prediction of NM-based **ballistic emission theory**, and could also be calculated from the NM momentum law. It's a "redder" result than SR.

The "box" under special relativity

Special relativity has no identifiably "special" frame for the propagation of light (although by convention we usually use the observer's frame as a convenient reference), so both shift stages are forced to use the *same* Doppler formula in order to satisfy the principle of relativity. SR also requires that spacetime remains flat, even when objects move past each other with "ultrarelativistic" velocities.

To meet these two conditions, we use a different Doppler relationship, $f'/f = \sqrt{(c-v)/(c+v)}$.

Combining two of these stages with opposite velocities, we notice that changing the "velocity" signs flips one half of the equation upside down, so that the two shifts will automatically cancel each other out, without us having to do any calculations:

$$\frac{freq'}{freq} = \sqrt{\frac{c-v}{c+v}} \times \sqrt{\frac{c+v}{c-v}} = 1$$

This is the correct result for special relativity: SR assumes that the motion of an object has no physical effect on the behaviour of light, so the "rebroadcast" light reaches the far side of the box with the same characteristics regardless of whether or not the moving transponder is there.

The "box" under Newtonian theory

If we apply Newtonian relativity, both observers once again need to see a single shift law in operation, but *now* this'll be the NM Doppler relationship, $f'/f = (c-v)/c$, for *both* observers. This gives us a predicted signal frequency received at the far side of the box, of

$$\frac{freq'}{freq} = \frac{c-v}{c} \times \frac{c+v}{c} = 1 - \frac{v^2}{c^2}$$

In other words, with "pure" NM relationships, the signal sent via the moving transponder somehow manages to arrive with a Lorentz-squared redshift. Pulses arrive at the far end of the box more slowly than they're sent, and would seem to be "stacking up" in the region of spacetime inside the box interior, as we might expect if the region contained a gravity-source.

This is a more complicated result, but might be reasonable in a model based on gravitomagnetic effects ... when the transponder-object moves more quickly inside the box, the "transponder-box" system has more recoverable kinetic energy, and the region between the "moving" object and its enclosing "stationary" box is more strongly warped as a result. According to the classification scheme used in section 13, this sort of "odd-looking" round-trip redshift effect would seem to have to appear in any relativistic model that includes a degree of positive velocity-dependent curvature (values of © greater than zero).

RELATIVITY IN CURVED SPACETIME

Calculations 2:
$E=mc^2$ from Newtonian mechanics

Assumptions used

- Momentum of particle, $\quad p = mv$
- Momentum of light-pressure, $\quad p = E/c$
- Energy of light taken as being proportional to its frequency
- Newton's law of the conservation of momentum

and,

- The default "Newtonian" Doppler shift relationship, $\quad freq'/freq = (c-v)/c$

Non-radiating particle

A particle travels forwards at **v** m/s inside a nominally-stationary box.

When the particle embeds itself in the rear wall of the box, it contributes to the box a forward momentum of *mv*. We'll make the mass of the particle extremely small compared to the mass of the box, so that the change in box-velocity caused by the impact can be arbitrarily small.

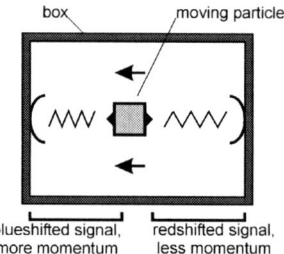

blueshifted signal, more momentum   redshifted signal, less momentum

Radiating particle

Now we'll repeat the experiment, but this time the particle will try to emit a burst of energy **E** before it hits the box, as two plane waves each with energy **E/2**, one aimed "forwards" and one aimed "backwards" along the particle's path. The particle shouldn't change speed when it ejects this light, because the particle sees the reaction forces from both waves to be equal and opposite.

However, seen by a *box* observer, the two waves *don't* have the same energy or momentum. The frequency of the *rearward*-aimed wave is redshifted by

$$freq' / freq = (c-v) / c$$

which gives a *reduction* in energy and momentum, and the frequency of the *forward*-aimed wave is blueshifted by

$$freq' / freq = (c+v) / c$$

, which is an *increase* – so the two waves strike the box with different amounts of force.

If the particle reckons that the momentum carried by each wave is $E/2c$, then for a rearward observer standing in the box, the overall momentum of the two opposing waves hitting the opposite sides of the box (one wave redshifted, the other blueshifted), becomes:

$$E/2c \times (c-v)/c \ + \ E/2c \times (c+v)/c$$
$$= E \ [\ (c-v) - (c+v) \ ] \ / \ 2c^2$$
$$= -Ev/c^2$$

, so, as far as the box is concerned, these two pulses of light, added together, should convey Ev/c^2 of total combined forward momentum to whatever they hit.

Conservation laws say that this momentum has to have come from *somewhere*, so to balance the books we're forced to conclude that when the moving particle ejected the light, it also supplied it with some of its own momentum – the it's remaining momentum then has to *reduce* by this same amount, Ev/c^2. But since we've already said that the particle didn't change speed, so to lose momentum, it must have lost some of its mass. To work out how much, we can write

$$mass = momentum \ / \ velocity$$

, and we can then see how much mass has to disappear to match the amount of overall momentum carried by the Doppler-shifted light.

$$mass_2 - mass_1 = (mom_2 - mom_1) \ / \ v$$
$$mass_{LOST} = (Ev/c^2) \ / \ v$$
$$mass_{LOST} = E \ / \ c^2$$

The (unspecified) velocity value cancels out nicely: when the particle radiates an amount of energy E (measured in its own frame) and loses an amount of mass m, (measured in its own frame) the relationship between the two quantities must be

$$E = mc^2$$

Since this argument works for any velocity, *no matter how small*, we can go on to argue that a particle should still lose this amount of rest mass even if it appears to us to be *stationary*, and its relative velocity to us is too small to be detectable. The calculations still give $E=mc^2$.

Special relativity

For the corresponding calculations under SR, we can rewrite the energy and momentum laws for mass to include a Lorentz-factor increase, and also increase the energy of the light by a Lorentz factor compared to NM. These two changes to the math cancel out, so special relativity shares the result (as does every other potential relativistic theory addressed in section 13).

---

light at other angles

For a pair of plane waves aimed *at 90 degrees* to the direction of motion, we get the same final result thanks to relativistic **aberration** effects (section 8). Once again, the emitting particle doesn't change speed when it emits the light because it sees the reactions forces to be balanced ... but the box sees both beams to be tilted forward due to aberration effects, and therefore sees them to be carrying a forward momentum, that must have been stolen from the emitting particle. Once we have the $E=mc^2$ result for both the longitudinal and transverse cases, we have it for balanced groups of rays emitted at *any* combination of angles, or for an omnidirectional, spherical pulse of light that throws rays in all directions.

RELATIVITY IN CURVED SPACETIME

Calculations 3:
Non-SR transverse Doppler effect / "Aberration redshift"

Some physicists might not like us saying that "physical" transverse redshifts can happen outside SR, and might declare that a ray arriving at the observer at 90 degrees *can't possibly* show a redshift under any theory other than special relativity. They may reject our ellipse arguments (section 13.6), and our SR-based arguments (section 16.4) as misguided, on the grounds that these give answers that they consider impossible. To address this scepticism, and to demonstrate the result in a more physicist-friendly way, we'll derive the thing yet again, using a longer and more boring method, this time using tedious high-school trigonometry.

Working:

This calculation is for the redshift that we'd expect if the speed of light was absolutely fixed with respect to the emitter. This sort of redshift also appears in everyday audio problems, if we take c to be the speed of sound and assume "still air" for the sound-source ("stationary source, moving observer"). It's also the correct "optical" prediction for an absolute light-aether moving along with the source, and for simple Newtonian ballistic emission theory.

We can break the calculation into three stages: **(a)** the aberration angle for a ray received at 90 degrees in the lab frame (which tells us the original angle at which the emitter *attempted* to aim the signal), **(b)** the longitudinal velocity component for the ray at that original angle (how much recession velocity should be associated with the original ray), and **(c)** the amount of "conventional" Doppler shift that should be associated with that amount of recession.

(a) *Aberration angle*

Working in the emitter's frame: An observer moves past us in a straight line at constant speed. Coincidentally, at the precise moment that they're closest to us, we happen to give out a burst of light **(i)**, and shortly afterwards this light intercepts the observer and is seen **(ii)**.

Since the ray of light that successfully intercepts the observer will necessarily have been moving sideways at *exactly the same speed* as the observer, the observer will reckon that this light has *no* sideways motion, and that it hits them side-on, at a perfect 90 degrees. We can now work out how far this ray deviated from 90 degrees according to *us*.

If the speed of light in our diagrams is c metres per second, and the detector is moving at v metres per second, then, if we were to say (for convenience) that the light happened to take one second to reach our moving detector, we'd have a diagonal path for our aimed ray of length c metres, and the baseline of the triangle parallel to the direction of relative motion would have length v metres (the distance that the detector moved in that time). Since trigonometry tells us that the "sine" of an angle is the ratio of its "opposite" side to the hypotenuse diagonal, this gives the "**relativistic aberration**" formula:

$$\text{SIN A} = v/c$$

(If we prefer to quote angles relative to the object's *direction of motion*, we'll use the triangle's *other* angle and write "**COS A**", with a plus or a minus depending on the scheme we're using.)

(b) Longitudinal velocity component

By the time that the signal has reached the detector, the detector has moved on from its point of closest approach, and is now beginning to move away from us with an increasing recession velocity. Our next challenge is to work out the rate of recession at the *exact moment* that it is struck by the ray (NB: because the recession speed is increasing, this "momentary" value will be larger than the total amount of recession that occurred during the previous second).

Looking along the line of the light-ray, the velocity v can be broken into two perpendicular velocities (iii), the recession component w, and a sideways component that we don't care about.

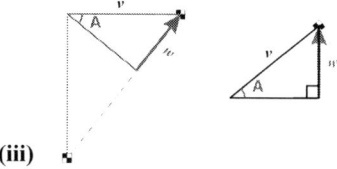

(iii)

These create a smaller right-angled triangle that (for "trig" reasons) has exactly the same proportions and angles as its parent – the angle **A** in (iii) is identical to the angle **A** in (ii).

In this new diagram the "opposite" and "hypotenuse" lengths are w and v, so **SIN A** = w/v, and the instantaneous recession velocity w can be written as

$$w = v \times \text{SIN A}$$

But since we've already established that **(SIN A)** equals v/c, we can write v/c instead of the trigonometric part, so that the recession velocity component of v becomes just

$$w = v \times v/c$$
$$w = v^2/c$$

(c) Recessional Doppler shift

We've now established that according the emitter (us), the observer was moving away from us when the signal struck it side-on, and that the observer should therefore see the signal to be redshifted ... all we have to do now is to work out the effect's exact strength.

For light moving at constant speed with respect to the emitter, we already know that the appropriate Doppler relationship is $f'/f = (c-v)/c$ (*see:* Calculations 1), so all we have to do now is take this relationship and write in w as our recession velocity. This tells us that when the observer sees the ray appearing to strike them at 90 degrees, it should still be redshifted by the amount:

$$f'/f = (c-w)/c$$
$$= 1 - w/c$$
$$= 1 - v^2/c^2$$

SR aficionados will immediately recognise this as a **Lorentz-*squared*** redshift – square-rooting it gives SR's more familiar, weaker, Lorentz redshift prediction of $\sqrt{1-v^2/c^2}$. So, if the speed of light is locked to the observer we have no redshift, if it's locked to the emitter we have a Lorentz-*squared* redshift, and if it was locked so some intermediate frame, we'd expect an intermediate amount of redshift somewhere between these two extremes. When we receive a signal at what *seems to us* to be 90 degrees from a moving object, aberration effects will cause *most* theories to predict that we should see the signal to be redshifted.

Calculations 4:
The "Box of Frogs" depiction of classical Hawking radiation

Let's suppose that you own a damp basement and decide that it is a good place to keep frogs. In order to stop the frogs escaping, you sink a vertical-sided box into the centre of the basement, higher than a frog can jump. Let's say that by studying a "test frog" you establish that it is impossible for these frogs to jump more than one metre into the air, so by making the walls of the pit 1.2 metres high, you are sure that the box is frog-escape-proof.

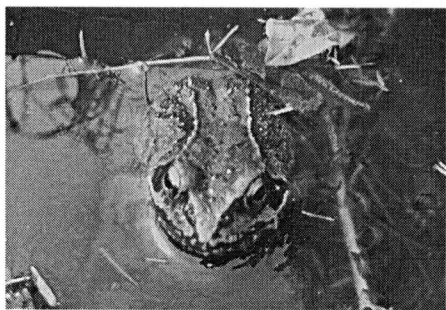

A froggy creature

That night you put a load of frogs into the box, make sure that they have sufficient food and water, and go to bed. During the night, something unexpected happens. As the frogs leap about, eventually two frogs collide in mid-air, and one of the frogs then kicks against the other, giving it enough additional speed to clear the wall.

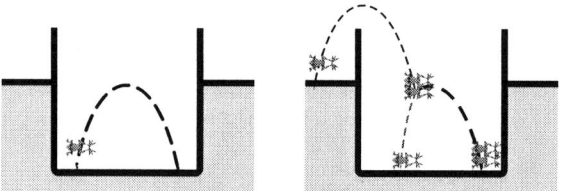

Lone frog, trapped ... "leapfrog" escape mode

The group behaviour of a **box of frogs** is clearly not quite the same as the behaviour we'd expect by modelling how a lone frog acts in isolation, and multiplying this by the number of frogs. There is additional **system-level** behaviour involved.

In the morning you come down to check the basement and find a frog hopping about, quite impossibly, outside the pit. How would you try to explain this supernatural occurrence?

"Froggy" quantum tunnelling

Modern particle physics gives us several ways to model the escaping frogs. By making observations over a long period of time, we can establish that the faster the frogs jump, and the more frogs there are, the more often a frog escapes. We can say that even though the box wall presents a classically impenetrable barrier, given frogs that want to escape badly enough, and given sufficient time, we can expect a frog to magically appear on the wrong side of the barrier every so often.

When we try to summarise the results of these very abstract statistical arguments, we can invent a story in which an unwatched frog has *quantum-tunnelled* through the barrier during the night.

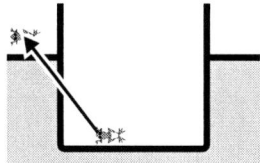

"Quantum-tunnelling" interpretation

The story sounds impossible, but it does predict the appearance of the escaping frogs.

"Froggy" statistical fluctuations

A second approach would be to say that the frog's ability to jump one metre is unchanged, but that the height of the walls or the height of the floor, or a combination of the two undergoes a *statistical fluctuation*. At certain moments in time, a leaping frog coincides with a window of opportunity when the barrier is leapable, and at these moments, a frog will escape.

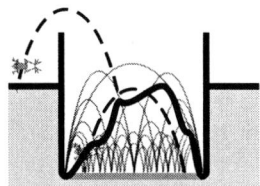

Statistical description ("fluctuating floor")

In some ways, this description is close to what we think *really* happens in our box – we can say that since the *effective* floor of the box is a complicated surface consisting of the actual box floor and the backs of any frogs leaping around above it, the "effective" box floor really *is* constantly fluctuating. Since the mid-air collision-rate of frogs depends critically on the fact that frogs have a given size (zero-radius "pinpoint" frogs would always miss each other and couldn't use this escape method), it *might* even be a legitimate use of quantum statistics.

"Froggy" pair production

If quantum statistics really *can* correctly describe our frog problem, we can attempt a third, more involved QM-style interpretation. If we accept that frogs are landing outside the box, from a height that they shouldn't really be capable of jumping to, then we can reinterpret the same statistics as describing a **pair production** process. The new story that we manufacture to explain the frogs then goes something like this: We could try to extrapolate the path of the escaping frog all the way back to the box floor, but it fails – we *know* that frog couldn't possibly have jumped all this way. So we extrapolate back as far as we legally can, and say that this is where the frog must have originated, magically popping into existence somewhere above the floor. Since we're not allowed to create a single frog, its "anti-frog" twin must also have been created, with equal and opposite momentum, moving downwards along the remainder of the path. The creation of the frog-pair causes a two-frog energy deficit, with one frog's path leading out of the box and the other leading back into it. Once inside the box, the "antimatter" frog lands on and annihilates another (identical) frog in a burst of energy that is immediately absorbed by the region to pay back the initial two-frog deficit. The final outcome all of this is that the box loses a frog, and an identical frog appears outside the wall, *as if* it had escaped.

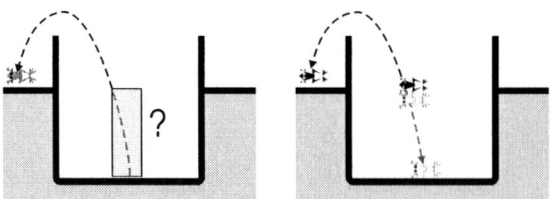

illegal extrapolation ... frog-antifrog pair

At this point, a physicist may feel entitled to break into the conversation and complain that the comparison between our "frog" exercise and black hole radiation isn't fair. A frog represents a very complex and sophisticated lump of massenergy, and while it wouldn't be *impossible* for a black hole to emit a frog as Hawking radiation, this might not be any more likely than for it to emit, say, a wombat or a pineapple. The frog-basement description is unfair and unrealistic because the basement *only* emits frogs, and not similarly-improbable objects such as wombats or pineapples or DVD players. If a black hole *did* emit any frogs (says our imaginary physicist), then it would have to emit truly *random* frogs, whereas if we stocked our basement with orange frogs, then the frogs that magically appeared outside the enclosure would tend to be orange, too.

So if we believe that gravitational horizons are truly inescapable, and that the "box of frogs" explanation of Hawking radiation is *wrong*, then we'll tend to believe that the information streaming away from a black hole must have *no causal relationship* with the information that previously fell into the hole. But further work on information theory has tended to undermine this view of the "GR+QM" black hole – when a black hole emits a particle (or a frog!), its entropy changes, it loses massenergy, and its information content drops. In the particle-pair description, when the antiparticle enters the hole with a mirror copy of the information held in the escaping particle, we might expect the total information in the hole to go *up*, but instead it goes down. This is what we'd expect to happen if the information held in the infalling antiparticle conveniently cancelled a matching amount of *identical* information already contained within the hole, as would happen if it met its exact mirror-double travelling towards it inside the hole. But this would suggest that the particle that's annihilated *inside* the hole by the infalling twin must carry the same information as the *escaping* particle – it would mean that there was a correspondence between the objects that disappeared from the hole's innards and the objects that magically appeared outside as Hawking radiation. If this correspondence is real, then the escaping particles' characteristics *weren't* random, and the simplest explanation for the match would be that the "disappeared" particles and the "escaped" particles were one and the same, despite the fact that GR1915 says that this is impossible.

Recent work on the black hole information paradox has tended to concentrate on explaining how the information that leaves a hole might be *based* on the information that entered, without *explicitly* contradicting GR1915 – for instance we could suggest that perhaps the event horizon takes a copy of any information falling through it, and then uses this information to condition and dictate the exact form of subsequent Hawking radiation emitted by the hole.

But this sounds suspiciously like saying that the characteristics of a black hole's Hawking radiation depend on what previously fell into the hole. Fill a black hole with frogs, and its random emissions will look suspiciously like squished frog detritus. Throw coffeebeans into it and the Hawking emissions should have a pleasant aroma of freshly roasted coffee.

At the time of writing (2007), classical mechanisms are reckoned *not* to be the underlying reason for QM's Hawking radiation from back holes, because they don't work under GR1915. But so far, nobody seems to have managed to demonstrate that the "quantum" and "non-GR1915 classical" explanations really are any different.

Calculations 5: Comparison table

"Classical Theory"

Recession redshift: $freq'/freq = \dfrac{c}{c+v}$

"Transverse" redshift: *none*

Momentum laws: *different for light and for matter*

Special Relativity

Recession redshift:
$$freq'/freq = \sqrt{\dfrac{c-v}{c+v}}$$
$$= \dfrac{c}{c+v} \times \sqrt{1-v^2/c^2}$$
$$= \dfrac{c-v}{c} \div \sqrt{1-v^2/c^2}$$

"Transverse" redshift: $freq'/freq = \sqrt{1-v^2/c^2}$

Momentum law: $p = \dfrac{mv}{\sqrt{1-v^2/c^2}}$

"re-calculated" Newtonian

Recession redshift: $freq'/freq = \dfrac{c-v}{c}$

"Transverse" redshift: $freq'/freq = 1 - v^2/c^2$

Momentum law: $p = mv$

MAJOR PLAYERS

Pythagoras	~582 - ~497 B.C.	Roland von Eötvös	1848-1919
Zeno	~450 - ~425 B.C.	Oliver Joseph Lodge	1851-1940
		George Francis Fitzgerald	1851-1901
Aristotle	384 - 322 B.C.	Albert Abraham Michelson	1852-1931
Aristarchus of Samos	310 BC - ~230 BC	Hendrik Antoon Lorentz [*1904*]	1853-1928
William of Ockham	~1280-1349	Jules Henri Poincaré	1854-1912
Nicolaus Copernicus	1473 - 1543	Max Planck	1858-1947
Giordano Bruno	1548-1600	Alfred North Whitehead	1861-1947
Galileo Galilei	1564-1642	Hermann Minkowski	1864-1909
Isaac Newton [1687] [1704]	1642-1727	Willem de Sitter	1872-1934
		Karl Schwarzchild	1873-1916
Ole Rømer (/Roemer)	1644-1710	Albert Einstein [1905] [1911] [1915]	1879-1955
James (Bishop) Berkeley	1685-1783		
James Bradley	1693-1762	Arthur Stanley Eddington [1919]	1882-1944
John Michell [1783]	1724-1793	Niels Bohr	1885-1962
Henry Cavendish	1731-1810	Alexander Alexandrovich Friedmann (/Friedman)	1888-1925
Johann K.F. Gauss	1777-1855	Edwin Powell Hubble	1889-1953
Augustin Jean Fresnel	1788-1827	(Abbe) Georges Lemaître	1894-1966
Gaspard de Coriolis (/Koriolis)	1792-1843	Pavel Alekseyevich Cerenkov (/Cherenkov)	1904-1990
Christian J. Doppler	1803-1853	John Archibald Wheeler	1911-
Armand H.L. Fizeau	1819-1896	Robert L. Forward	1932-2002
Jean B.L. Foucault	1819-1868	Kip Thorne	1940-
Georg F.B. Riemann	1826-1866	Stephen Hawking [1974]	1942-
James Clerk Maxwell	1831-1879		
Simon Newcombe	1835-1909	William Unruh	1945-
Ernst Mach	1838-1916	Matt Visser	1956-
William Kingdon Clifford	1845-1879		

Topic References

Dark stars, Hawking radiation, pair production and acoustic metrics (→ 2005)

1784 John Michell "On the Means of discovering the Distance, Magnitude, &c., of the Fixed Stars ... " Proceedings of the Royal Society, 1784
— *John Michell's 1783 paper to Henry Cavendish published in 1784, calculating the critical "r=2M" radius and suggesting using prisms to check for spectral shifts due to the gravitational weakening of light*

1799 Translation of a piece by **Peter Simon Laplace**, as pp. 365-368 of

    Hawking and Ellis, **The large scale structure of spacetime** (CUP, Cambridge, 1973)

1939 Albert Einstein "On a stationary system with spherical symmetry consisting of many gravitating masses" Annals of Mathematics **40** 922-936 — *argues that total gravitational collapse should not be allowed*

1960 Carlton W. Berenda "Temporal Reversal of Events in Restricted Relativity" Am. J. Phys **28**, 799-801

1966 Clyde R. Hardin "The scientific work of the Reverend John Michell" Annals of Science **22** 27-47

1968 Russell McCormmach "John Michell and Henry Cavendish: Weighing the stars" British Journal for the History of Science **4** 126-155

1970 Gerald Feinberg "Particles that go faster than light" Scientific American **222** 68-77

1972 Stephen Hawking "Black Holes in General Relativity" Commun. Math. Phys. **25** 152-166
— *GR1915 black holes can collide but not split*

— the "Hawking radiation" concept begins to emerge —

1972 Y. B. Zeldovich and A. A. Starobinsky "Particle production and vacuum polarization in an anisotropic gravitational field" Sov. Phys. JETP **34** 1159

1973 Jacob. D. Bekenstein "Black Holes and Entropy" Phys. Rev. **D 7** 2333-2346

1974 S.W. Hawking "Black Hole explosions?" Nature **248** 30

1975 S.W. Hawking "Particle creation by black holes" Comm. Math.Phys. **43** 199-220

1975 P.C.W. Davies "Scalar particle production in Schwarzchild and Rindler metrics" J. Phys. A **8** 609-616

1976 S. W. Hawking "Breakdown of predictability in gravitational collapse" Phys. Rev. **D 14** 2460-2473

1976 Don N. Page "Particle emission rates from a black hole: Massless particles from an uncharged, nonrotating hole" Physical Review **D 13** 198-

1976 Don N. Page "Particle emission rates from a black hole. II. Massless particles from a rotating hole" Physical Review **D 14** 3260-

1976 W.G. Unruh "Notes on black-hole evaporation" Physical Review **D 14** 870-891

1979 Simon Schaffer "John Michell and Black Holes" Journal for the History of Astronomy **10**, 42-43

1979 Gary Gibbons "The man who invented black holes: His work emerges out of the dark after two centuries" New Scientist (28 June 1979), 1101

1981 W. G. Unruh "Experimental Black-Hole Evaporation?" Phys. Rev. Lett. **46** 1351-1353

1982 William G. Unruh "Acceleration radiation and the generalized second law of thermodynamics" Physical Review **D 25** 942-958

1984 William G. Unruh and Robert M. Wald "What happens when a accelerating observer detects a Rindler particle" Phys. Rev. D **29** 1047-1056

1986 Kip S. Thorne, Wojciech H. Zurek and Richard H. Price "The Thermal Atmosphere of a Black Hole" - Published as Chapter VIII of
    Black Holes: The Membrane Paradigm (1986). See also other papers in the same compilation 281-340 — *Black hole atmospheres, history, mining and dumping*

RELATIVITY IN CURVED SPACETIME

1986   Thorne, Price & Macdonald (eds.) **Black holes: The membrane paradigm** (Yale, New Haven, 1986)
– *A compilation of papers treating the exterior of a black hole as a conventional radiating surface, as an "engineering approach" to modelling Hawking radiation and related QM effects*

1991   Theodore Jackson "Black-hole evaporation and ultrashort distances" Physical Review **D 44** 1731-1739

1991   J. Eisenstaedt "De L'influence de la gravitation sur la propagation de la lumière en théorie Newtonienne. L'archéologie des trous noirs." Arch. Hist. Exact Sci. **42** 315-386
– *Good historical reference review article on the history of dark stars. In French, but with many useful references to other relevant works in English*

1992   John Preskill, "Do Black Holes Destroy Information?" hep-th/9209058 , in **1992 International Symposium on Black Holes, Membranes, Wormholes and Superstrings** eds. Sunny Kalahara and D. V. Nanopoulos (World Scientific Pub. Co., Singapore, 1993), pp.22-39.
– *" I review the information loss paradox that was first formulated by Hawking, and discuss possible ways of resolving it. All proposed solutions have serious drawbacks. I conclude that the information loss paradox may well presage a revolution in fundamental physics. "*

1993   Ulf H. Danielsson and Marcelo Schiffer "Quantum mechanics, common sense, and the black hole information paradox" Physical Review **D 48** 4779-4784 gr-qc/9305012 – *clear exposition of the black hole information paradox and possible methods of attack. Quantum copyrights and information duplication*

1993   Jacob D Bekenstein "How Fast Does Information Leak Out from a Black Hole?" Phys.Rev.Lett. **70** 3680-3683 hep-th/9301058 – *" The initial information may ... gradually leak out ... "*

1993   Matt Visser, "Acoustic propagation in fluids: an unexpected example of Lorentzian geometry" gr-qc/9712016

1994   H.C. Rosu "Hawking-like effects and Unruh-like effects: toward experiments?" Grav. Cosmol. **7** 1-17 gr-qc/9406012

1994   John Preskill "On Hawking's Concession" (24 July 1994)
http://www.theory.caltech.edu/~preskill/jp_24jul04.html

1994   John Preskill: "Black holes and information: A crisis in quantum physics" (Caltech theory seminar, October 1994) http://www.theory.caltech.edu/~preskill/talks/blackholes.pdf

1994   Kip Thorne, **Black Holes and Time Warps: Einstein's Outrageous Legacy** (W.W. Norton & Company, 1994) ISBN 0393312763 – *Discussion of Hawking radiation, the conflict between GR and QM, the history of dark stars and black holes, and most other black-hole-related topics. Recommended resource.*

1995   G. 't Hooft "Black holes, Hawking radiation, and the information paradox" Nucl.Phys. B (Proc. Suppl.) **43** 1-11

1995   T. Banks "Lectures on Black Holes and Information Loss" Nucl. Phys. (Proc. Suppl.) **41** 24-65
– *hypothetical black hole remnants ("cornucopions")*

1995   Larus Thorlacius "Black Hole Evolution" Nucl. Phys. (Proc. Suppl.) **41** 245-275

1995   W. G. Unruh, "Sonic analogue of black holes and the effects of high frequencies on black hole evaporation" Phys.Rev. **D 51** 2827-2838

1995   G. 'tHooft "Black holes, Hawking radiation, and the information paradox" Nucl. Phys. B (Proc. Suppl.) **43** 1-11 – *Information paradox and the possible options*

1996   Jeremy Bernstein "The Reluctant Father of Black Holes" Scientific American (June 1996) 80-85
– *" Einstein sought to prove that black holes – celestial objects so dense that their gravity prevents even light from escaping – were impossible. "*

1996   Stephen W. Hawking and Roger Penrose "The Nature of Space and Time" Scientific American (July 1996) 44-49
– *"Hawking and Penrose both believe that when a black hole radiates, it loses the information it held. But Hawking insists that the loss is irretrievable, whereas Penrose that the loss is balanced by spontaneous measurements of quantum states ..."*

1996   Ronald Gautreau "Cosmological Schwarzschild radii and Newtonian gravitational theory" Am.J.Phys. **64** 1457-1467 *cosmological event horizons*

1997   Leonard Susskind, "Black Holes and the Information Paradox" Scientific American **276** p.40-45 (April 1997)

References

1997 Matt Visser "Acoustic black holes: horizons, ergospheres, and Hawking radiation" Classical and Quantum Gravity **15** 1767-1791 gr-qc/9712010

1997 Leonard Susskind "Black Holes and the Information Paradox" Scientific American **276** pp.40-45 (April 1997).

1998 Matt Visser "Hawking radiation without black hole entropy" Phys.Rev.Lett. **80** 3436-3439 gr-qc/9712016

1998 Matt Visser "Acoustic black holes: horizons, ergospheres and Hawking radiation" Class. Quantum Grav. **15** 1767-1791 gr-qc/9712010

1999 Matt Visser "Acoustic black holes" gr-qc/9901047

2000 Matt Visser, ("Matters of Gravity" report): "Optical black holes?" **MOG-15** gr-qc/0002027

2000 Stefano Liberati, Sebastiano Sonego and Matt Visser "Unexpectedly large surface gravities for acoustic horizons?" Class.Quant.Grav. **17** 2903-2923 gr-qc/0003105

2002 Ralf Schützhold, Günter Plunien, Gerhard Soff "Dielectric black hole analogues" Phys.Rev.Lett. **88** 061101 quant-ph/0104121

2002 U. Leonhardt, T. Kiss, P. Ohberg "Bogoliubov theory of the Hawking effect in Bose-Einstein condensates" cond-mat/0211464

2003 Matt Visser "Essential and inessential features of Hawking radiation" Int.J.Mod.Phys. **D12** 649-661 (2003) gr-qc/0408022

2003 Uwe R. Fischer, Matt Visser "Warped space-time for phonons moving in a perfect nonrelativistic fluid" Europhys.Lett. **62** 1-7 gr-qc/0211029

2003 Carlos Barcelo, Stefano Liberati, Matt Visser "Towards the observation of Hawking radiation in Bose-Einstein condensates" Int.J.Mod.Phys. **A18** 3735 gr-qc/0408022

2004 Carlos Barcelo, Stefano Liberati, Sebastiano Sonego, Matt Visser "Causal structure of acoustic spacetimes" New J.Phys. **6** 186 gr-qc/0408022

2004 S. Giovanazzi, C. Farrell, T. Kiss, U. Leonhardt "Conditions for one-dimensional supersonic flow of quantum gases" cond-mat/0405007

2004 Tapas K. Das "Analogue Hawking Radiation from Astrophysical Black Hole Accretion" Class.Quant.Grav. **21** 5253-5260 gr-qc/0408081

2005 Samuel Lepe, Joel Saavedra "Quasinormal modes, Superradiance and Area Spectrum for 2+1 Acoustic Black Holes" Phys.Lett. **B617** 174-181 gr-qc/0410074

2005 Carlos Barceló, Stefano Liberati, and Matt Visser "Analogue Gravity" gr-qc/0505065

Effects of relative motion on light (→ 2000)

1728 James Bradley "... an Account of a New Discovered Motion of the Fix'd Stars" Philosophical Transactions of the Royal Society pp.637-660

1859 H. Fizeau "Sur les hypothèses relatives à l'éther lumineux et sur une expérience qui parait démontrer que le mouvement des corps change la vitesse avec laquelle la lumière se propage dans leur intérieur" Ann. de Chim et de Phys. **57** 385-404

1886 A. A. Michelson and E.W. Morley "Influence of motion of the medium on the motion of light" Am. J. Science **31** 377-386

1887 A. A. Michelson and E. W. Morley "On the Relative Motion of the Earth and the Luminiferous Ether" Am. J. Science **34** 333-345

1893 Oliver Lodge, "Aberration Problems - A Discussion concerning the Motion of the Ether near the Earth, and concerning the Connexion between Ether and Gross Matter; with some new Experiments." Philosophical Transactions of the Royal Society **184** sections 56-57 727-804

1895 H.A. Lorentz "Michelson's interference experiment " (1895), translated and reprinted in
   The Principle of Relativity (Dover, NY, 1952) pp.1-7

1909 Oliver Lodge **The Ether of Space** (Harper, London, 1909), Chapter X: "General Theory of Aberration" – *pp.135 aberration shifts*

1914	Pieter Zeeman "Fresnel's coefficient for light of different colours. (First part)" Proc. Kon. Acad. van Weten., **17** 445-451

1914 Pieter Zeeman "Fresnel's coefficient for light of different colours. (First part)" Proc. Kon. Acad. van Weten., **17** 445-451

1915 Pieter Zeeman "Fresnel's coefficient for light of different colours. (Second part)". Proc. Kon. Acad. van Weten. **18** 398-408.

1940 George Gamow, **Mr. Tompkins in Wonderland** (CUP, Cambridge, 1940) pp. 2-5

1955 P. Hickson, R. Bhatia, and A. Iovino, "No relativistic aberration of liquid mirrors," Astron. Astrophys. **303** L37-39

1959 James Terrell "Invisibility of the Lorentz Contraction" Physical Review **116** 1041-1045

1959 Roger Penrose, "The Apparent Shape of a Relativistically Moving Sphere" Proc. Cambridge Phil. Soc. **55** 137-139 "Research Notes" section, effectively a letter, no abstract "received 29 July 1958"

1959 Eugene Feenberg "Doppler Effect and Time Dilatation" Am.J.Phys. **27** 190

1959 E. Dewan and M. Beran "Note on Stress Effects due to Relativistic Contraction" Am.J.Phys. **27** 517-518

1960 V.F. Weisskopf "The visual appearance of rapidly moving objects" Physics Today pp.24-27 (Sept 1960).

1960 Roy Weinstein "Observation of Length by a Single Observer" Am.J.Phys. **28** 607-610

1960 A.A. Evett and D.C. Fried "Speed of Light in Flowing Dispersive Liquids" Am.J.Phys **28** 733-735
– *Fizeau-ey stuff*

1961 George Gamow "Remarks on Lorentz Contraction" Proc. Nat.Acad.Sci. U.S.A. **47** 728-729

1961 Mary L. Boas "Apparent shape of large objects at relativistic speeds" Am. J. Phys. **29** 283-286

1961 W. Rindler "Length Contraction Paradox" Am.J.Phys. **29** 365-366 – *"manhole" or "skateboard" problem*

1961 Willard H. Wells "Length Contraction Paradox" [letter] Am.J.Phys **29** 858
– *Re Rindler paper, criticises notion of rigidity*

1962 Hirsch I. Mandelberg and Louis Witten, "Experimental Verification of the Relativistic Doppler Effect" J.Opt.Soc.Am. **52** 529-536 (1962).

1962 R. Shaw "Length contraction paradox" (letter) Am.J.Phys. **30** 72
– *re: Rindler's paper, a variation with an angled pierced table passing across the rod's path.*

1962 Paul J. Nawrocki "Stress Effects due to Relativistic Contraction" Am.J.Phys. **30** 771-772

1962 Sten Yngström "Observation of moving light-sources and objects" Arkiv för Fysik **23** 367-374

1963 Edmond M. Dewan "Stress effects due to Lorentz contraction" Am. J. Phys. **31** 383-386

1964 Hermann Bondi, **Relativity and Common Sense**, (Heinemann, London, 1964)

1965 G.D. Scott and M.R. Viner "The Geometrical Appearance of Large Objects Moving at Relativistic Speeds" Am. J. Phys. **33** 534-536

1968 N. C. McGill, "The Apparent Shape of Rapidly Moving Objects in Special Relativity" Contemp. Phys. **9** 33-48 – *review and summary of relevant papers*

1970 Donald W. Lang "The Meter Stick in the Match Box" Am.J.Phys. **38** 1181-1184 (1970).

1970 G.D. Scott and H.J. van Driel "Geometrical Appearances at Relativistic Speeds" Am.J.Phys. **38** 971-977 – *aberrated starfields*

1971 Dale C. Ferguson "Stellar Aberration and Apparent Rotation: A Direct Link" Am.J.Phys. **39** 1089-1090

1971 R. V. Jones "Aberration of light in a moving medium" Journal of Physics A (Mathematical and General) **4** L1-L3

1972 G.E. Stedman "The Transverse Contraction Factor of Special Relativity" Am.J.Phys. **40** 782

1972 John A. Winnie "The Twin-Rod Thought Experiment" Am.J.Phys. **40** 1091-1094

1973 Hermann Erlichson "The Rod Contraction-Clock Retardation Ether theory and the Special Theory of Relativity" Am.J.Phys. **41** 1068-1077

1974 H.A. Atwater "Nonsimultaneity in the Aberration of Starlight" Am.J.Phys. **42** 1022-1024
– *"Notes and Discussions" section*

1974 R.C. Jennison and P.A. Davies "Reflection from a transversely moving mirror" Nature **248** 660-661

References

1976 David Hollenbach "Appearance of a rapidly moving sphere: A problem for undergraduates" Am.J.Phys. **44** 91-93

1977 W.J. Thompson "Note on the appearance of rapidly moving objects" Am.J.Phys. **45** 1008

1978 Von H.-J. Treder "Vakuum-Lichtgeschwindigkeit und differentielle Aberration" Annalen der Physik **35** 70-76

1979 P. Stumpff "The Rigorous Treatment of Stellar Aberration and Doppler Shift, and the Barycentric Motion of the Earth" Astron.Astrophys. **78** 229-238

1980 P. Stumpff "On the Relationship between Classical and Relativistic Theory of Stellar Aberration" Astron.Astrophys. **84** 257-259

1987 Asher Peres "Relativistic telemetry" Am.J.Phys. **55** 516-519

1987 G. P. Sastry "Is length contraction really paradoxical?" Am. J. Phys **55** 943-946

1988 Eric Sheldon "The twists and turns of the Terrell Effect" Am. J. Phys. **56** 199-200

1988 H. Blatter "Aberration and Doppler shift: An uncommon way to relativity" Am.J.Phys. **56** 333-338

1988 Kevin G. Suffern "The apparent shape of a rapidly moving sphere" Am.J.Phys. **56** 729-733

1989 James Terrell "The Terrell Effect" Am.J.Phys. **57** 9-10
– *retrospective by Terrell, mentions Bronowski, additional note by the editor*

1989 Eric Sheldon "The Terrell Effect: Eppure si contorce!" Am.J.Phys. **57** 487
– *"Letters to the Editor", rotation vs. shear*

1989 Lawrence Cranberg "Recognition, priority, plagiarism and nomenclature in physics" Am.J.Phys. **57** 488 – *in the "Letters to the Editor" section, on the community's response to the Terrell effect*

1989 Thomas E. Phipps, Jr. "Relativity and aberration" Am.J.Phys. **57** 549-551

1990 Dragan V. Redžić "The Doppler effect and conservation laws revisited" Am.J.Phys. **58** 1205-1208

1990 M. Torres, J.M. Gonzaléz, A. Martin, G. Pastor ... "On the surface charge density of a moving sphere" Am.J.Phys. **58** 73-75

1990 Samuel Mockovcîak "An aberration formula and the angle of Lobachevski" Am.J.Phys. **58** 113-1114

1990 T. Greber and H. Blatter "Aberration and Doppler shift: The cosmic background radiation and its rest frame" Am.J.Phys. **58** 942-945

1990 T.M. Kalotas and A.R. Lee "A 'two-line' derivation of the relativistic longitudinal Doppler formula" Am. J. Phys **58** 187-188

1991 John Robert Burke and Frank J. Strode "Classroom exercises with the Terrell effect" Am.J.Phys. **59** 912-915

1991 Don S. Lemons "Doppler Shift and stellar aberration from conservation laws applied to Compton scattering" Am.J.Phys. **59** 1046-1047

1992 W. Mückenheim "Is Lorentz Contraction Observable?" Annales de la Fondation Louis de Broglie **17** 351-354

1992 E. F. Taylor and J. A. Wheeler, **Spacetime Physics 2nd edition** (Freeman, NY, 1992)

1994 William Moreau "Wave front relativity" Am.J.Phys. **62** 426-429
– *relativistic ellipses as a consequence of the relativity of simultaneity. very visual explanation with nice diagrams*

1995 John R. Graham "Special relativity and length contraction" Am.J.Phys. **63** 637-639

1995 P. Hickson, R. Bhatia and A. Iovino "No relativistic aberration of moving mirrors" Astron Astrophys. **303** L37-L39

1995 Lorentz contraction: "A real change of shape" Am.J.Phys. **63** 413-415

2000 J.H. Field "Two novel special relativistic effects: Space dilatation and time contraction" Am.J.Phys **68** 367-374

2000 S. Carlip "Aberration and the Speed of Gravity" Phys. Lett. **A 267** 81-87

$E=mc^2$

- 1905   Albert Einstein, "Does the Inertia of a Body depend on its Energy-Content?" (1905), translated and reprinted in
  - 📖   **The Principle of Relativity** (Dover, NY, 1952) pp.67-71
- 1935   Albert Einstein "An Elementary Derivation of the Equivalence of Mass and Energy" Bull. Am. Math. Soc. **41** 223-230 (reprinted in new series, **37** (2000) 39-44) www.ams.org/bull/2000-37-01/ , and in
  - 📖   **Out of My Later Years** (Philosophical Library, NY, 1950).
- 1946   Albert Einstein "$E=mc^2$: The most urgent problem of our time" Science Illustrated (April 1946), reprinted in
  - 📖   **Out of My Later Years** (Philosophical Library, NY, 1950)
- 1987   Ø. Grøn "A modification of Einstein's first deduction of the inertia-energy relationship" Eur.J.Phys **8** 24-26
- 1988   Mitchell J. Feigenbaum and N. David Mermin "$E=mc^2$" Am.J.Phys. **56** 18-21
- 1988   W.L. Fadner Did Einstein really discover "$E=mc^2$" Am.J.Phys. **56** 114-122
- 1990   Fritz Rohrlich "An elementary derivation of $E=mc^2$" Am.J.Phys. **58** 348-349
- 1991   Fritz Rohrlich "A response to 'Comment on "An elementary derivation of $E=mc^2$" ' ..." Am.J.Phys. **59** 757
- 1991   Lawrence Ruby and Robert E. Reynolds "Comment on 'An elementary derivation of $E=mc^2$,' by F. Rohrlich" Am.J.Phys. **59** 756

Warp drives (→ 2000)

- 1963   Robert L. Forward "Guidelines to Antigravity" American Journal of Physics **31** 166-170
- 1971   📖 Charles W. Misner, Kip S. Thorne and John Archibald Wheeler (a.k.a. "MTW") ("world line of an observer who has undergone a brief period of acceleration") in **Gravitation** (Freeman, New York, 1971), figure 6.2, pp. 169
- 1972   Robert Wald "Gravitational Spin Interaction" Physical Review **D 6** 406-413
- 1989   📖 Clifford Will, "The renaissance of general relativity" in **The New Physics** ed. Paul Davies (Cambridge University Press, Cambridge, 1989) pp.31-32.
- 1994   Miguel Alcubierre, "The warp drive: hyper-fast travel within general relativity" Class. Quantum Grav. **11** L73-L77
- 1994   Michael Spzir, "Spacetime hypersurfing" American Scientist **82** 422-423 (Sept/Oct 1994)
  - *– commentary on the Alcubierre piece*
- 1996   I.B Khriplovich and A.A. Pomeransky "Gravitational interaction of spinning bodies, center-of-mass coordinate and radiation of compact binary systems" Phys. Lett. **A 216** 7-14
- 1997   Allen E. Everett and Thomas Roman, "Superluminal subway: The Krasnikov tube" Physical Review **D 56** 2100-2108 (August 1997). gr-qc/9702049
- 1997   William A Hiscock, "Quantum effects in the Alcubierre warp-drive spacetime" Class. Quantum Grav. **14** L183-L188 gr-qc/9707024
- 1997   M.J. Pfenning and L.H. Ford The unphysical nature of 'warp drive' Class. Quantum Grav. **14** 1743-1751
  - *– suggests that warp drives would require negative energy densities*
- 1998   D. H. Coule, "No warp drive" Class. Quantum Grav. **15** 2523-2527
- 1998   Ken D. Olum, "Superluminal Travel Requires Negative Energies" Physical Review Letters **81** 3567-3570 gr-qc/9805003
- 1998   S. V. Krasnikov, "Hyperfast travel in general relativity," Physical Review **D 57** 4760-4766 gr-qc/9511068
  - *– arbitrarily-short round-trip transit times ("Krasnikov tube")*
- 1999   Robert J. Low, "Speed Limits in General Relativity", Class. Quantum Grav. **16** 543-549 gr-qc/9812067

GENERAL REFERENCES:

1772   Joseph Priestley, **The History and Present State of discoveries relating to Vision, Light and Colours** – a.k.a. *"History of Optics"*

1904   H.A. Lorentz "Electromagnetic phenomena in a system moving with any velocity less than that of light" – *"Lorentzian electrodynamics"* ... see *(various 1952)*

1905   Albert Einstein "On the electrodynamics of moving bodies" dynamics paper **Band 17** 891-921
– *Einstein's paper on what later became known as special relativity* ... see *(various 1952)*

1905   Albert Einstein "Does the inertia of a body depend on its energy-content?" ...
– *Einstein's "E=mc²" paper* ... see *(various 1952)*

1910   E.T. Whittaker, **A History of the Theories of Aether and Electricity** ... (Dublin University Press)

1911   Albert Einstein, "On the Influence of Gravitation on the Propagation of Light" (1911) ...
– *Einstein's gravity-shift paper* ... see *(various 1952)*

1913   W. deSitter "A proof of the constancy of the velocity of light," Kon. Acad. van Weten. **15** 1297-1298

1913   W. deSitter "On the constancy of the velocity of light," Kon. Acad. van Weten. **16** 395-396

1916   Albert Einstein "The foundation of the general theory of relativity" Annalen der Physik **49** 769-822
– *transl.* *(various 1952)*

1916-1954
     Albert Einstein, **Relativity: The Special and the General Theory** (final ed., Meuthen, 1954)

1916   Arthur Eddington "Gravitation and the Principle of Relativity" Nature **Dec 1916** 328-329

1920   A. Einstein, "Aether and the Theory of Relativity" (1920), translated in
     Sidelights on Relativity (Dover, NY, 1983) pp.1-24.

1921   Albert Einstein "A Brief Outline of the Development of the Theory of Relativity" Nature **106** 82-784

1923   Albert Einstein "The Theory of the Affine Field" Nature **112** 448-449

1930   H.T.H. Piaggio "The Concept of Space" (report of an Einstein lecture), Nature **125** (3163) 897-898

1931   Albert Einstein "Knowledge of Past and Future in Quantum Mechanics" Physical Review **37** 80-781

1931   Albert Einstein "News and Views" (lecture account), Nature **127** 19

1935   Albert Einstein "Can quantum-mechanical description of physical reality be considered complete?" Physical Review **47** 777-780

1935   A. Einstein and N. Rosen "The particle problem in the general theory of relativity" Physical Review **48** 73-77

1936   Albert Einstein "Lens-like action of a star by the deviation of light in the gravitational field" Science **84** 506-507

1937   A. Einstein and N. Rosen "On gravitational waves" Journal of the Franklin Institute **223** 43-54

1937   Herbert E. Ives "The Doppler Effect Considered in Relation to the Michelson-Morley Experiment" J. Opt. Soc. Am. **27** 389-392

1938   Herbert E. Ives and G.R. Stilwell "An Experimental Study of the Rate of a Moving Atomic Clock" J. Opt. Soc. Am. **28** 215-226 – *shift measurements made on a particle-beam at 7 degrees from collinear*

1938   Albert Einstein, L. Infeld, and B. Hoffman "The gravitational equations and the problem of motion" Annals of Mathematics **38** 65-67

1940   Albert Einstein and L. Infeld "The gravitational equations and the problem of motion. II" Annals of Mathematics **41** 455-464

1941   Herbert E. Ives and G.R. Stilwell "An Experimental Study of the Rate of a Moving Atomic Clock. II" J. Opt. Soc. Am. **31** 369-373

1950   Albert Einstein, "On the Generalized Theory of Gravitation" Scientific American **182** 13-17 (April 1950).

RELATIVITY IN CURVED SPACETIME

1952 (various), **The Principle of Relativity** (Dover, NY, 1952)
− *Compilation of translated and republished papers by Einstein, Minkowski, Weil, and others*

1954 A. Einstein "Relativity and the Problem of Space"
− *published as the fifth appendix to Einstein's "popular" relativity book.*

1955 John Archibald Wheeler "Geons" Physical Review **97** 511-536 − *particles as geometrical features*

1955 A. Einstein, **The Meaning of Relativity**, final 1955 editions (6th. ed./Princeton 5th ed.(revised)) (Dover, NY, 1955) − *Einstein's 1921 Princeton lectures, plus some additional material*

1958 C.B Leffert and T.M. Donahue, "Clock Paradox and the Physics of Discontinuous Gravitational Fields" Am. J. Phys. **26** 514-523

1959 C. Møller, "Motion of free particles in discontinuous gravitational fields" Am. J. Phys. **27** 491-493

1959 J.P. Cedarholm and C.H. Townes "A new experimental test of special relativity" Nature **184** 1350-1351

1959 R. V. Pound and G. A. Rebka, Jr. "Gravitational Red-Shift in Nuclear Resonance" Phys. Rev. Lett. **3** 439-441

1960 Asher Peres and Nathan Rosen "Gravitational Radiation Damping of Nongravitational Motion" Annals of Physics **10** 94-99

1960 T.E. Cranshaw, J.P. Schiffer, and A.B. Whitehead "Measurement of the gravitational red shift using the Mössbauer effect in Fe^{57}" Phys.Rev.Lett. **4** 163-164

1960 H.J. Hay, J.P. Schiffer, T.E. Cranshaw and P.A. Egelstaff "Measurement of the red shift in an accelerated system using the Mössbauer effect in Fe^{57}" Phys.Rev.Lett. **4** 165-166
− *uses equivalence principle to apply gravity-shift result to the case of a centrifuged detector*

1960 L.I. Schiff "Possible new experimental test of general relativity theory" Phys.Rev.Lett. **4** 215-217
− *rebuttal" paper*

1960 R.V. Pound and G.A. Rebka, Jr. "Variation with temperature of the energy of recoil-free gamma rays from solids" Phys.Rev.Lett. **4** 274-275

1960 V. F. Weisskopf "The visual appearance of rapidly moving objects" Physics Today (7 Sept 1960) 24-27

1960 Arthur A. Evett and Roald K. Wangness "Note on the Separation of Relativistically Moving Rockets" Am.J.Phys. **28** 566

1960 Gerald Holton "On the Origins of the Special Theory of Relativity" Am.J.Phys. **28** 627-636

1962 Hirsch I. Mandelberg and Louis Witten, "Experimental verification of the relativistic Doppler effect" J. Opt. Soc. Am. **52** 529-536

1963 R. S. Shankland, "Conversations with Albert Einstein," Am. J. Phys. **31** 47-57

1963 D.C. Champeney, G.R. Isaac, and A.M. Khan An "'aether drift' experiment based on the Mössbauer effect" Phys. Lett. **7** 241-243

1963 R. H. Dicke, "Cosmology, Mach's Principle and Relativity," Am. J. Phys. **31** 500-509

1963 Roy P. Kerr "Gravitational field of a spinning masses an example of algebraically special metrics" Phys.Rev.Lett. **11** 237-238

1964 W.G.V. Rosser, **An Introduction to the Theory of Relativity** (Butterworths, London, 1964) section 4.4.7 pp.160.

1964 K.C. Turner and H.A. Hill "New experimental limit on velocity-dependent interactions of clocks and distant matter" Physical Review **134 B** 252

1964 Hermann Bondi, **Relativity and common sense: A new approach to Einstein** (Doubleday, NY, 1964)

1964 R.V. Pound and J.L. Snider "Effect of gravity on nuclear resonance" Phys.Rev.Lett. **13** 539-540

1965 R.V. Pound and J.L. Snider "Effect of gravity on gamma radiation" Phys. Rev. **140 B** 788-803
− *gravity shifts across a ~25m tower*

1966 P. Kustaanheimo "Route dependence of the gravitational red shift" Phys. Lett. **23** 75-77

1967 G.J Whitrow (ed.), **Einstein: The man and his achievement** (BBC, London 1967)
− *Lanczos quote pp. 48-49*

1970 G.R. Isaak "The Mössbauer Effect: Application to relativity" Phys. Bull. **21** 255-257

References

1971 R.G. Newburgh and T.E. Phipps Jr. "Brewster Angle and the Einstein velocity addition theorem" Am.J.Phys **39** 1079-1084

1971 C. W. Misner, K. S. Thorne and J. A. Wheeler (a.k.a. "MTW") **Gravitation** (Freeman, NY, 1971)

1972 Isaac Asimov, **Asimov's Biographical Encyclopedia of Science and Technology** (Pan Books Ltd., London, 1975)

1973 Wolfgang Rindler, **Essential Relativity, revised 2nd ed.** (Springer-Verlag, 1973)
– section 3.1 "Only theories consistent with special relativity need be considered."

1974 D.T. Whiteside (ed), **The Mathematical Papers of Isaac Newton, Volume VI, 1684-1691** (CUP, Cambridge, 1974)

1974 Roger W. Clay, Philip C. Crouch "Possible observation of tachyons associated with extensive air showers" Nature **248** 28-30

1975 John David Jackson, **Classical Electrodynamics**, 2nd ed. (Wiley, NY, 1975) section 11.2(b)

1976 Arthur I. Miller "On Einstein, light quanta, radiation, and relativity in 1905" Am.J.Phys. **44** 912-923

1976 Leigh Hunt Palmer "Simple form of the law of addition of parallel velocities from the special theory of relativity" Am.J.Phys. **44** 702

1976 J.R. Prescott "Tachyons revisited-comments on a search for faster-than-light particles" J.Phys.G: Nucl.Phys. **2** 261-267

1976 S.W. Hawking "Breakdown of predictability in gravitational collapse" Physical Review **D 14** 2460-2473

1976 R.W. Ditchburn, **Light, 3rd ed. Volume 1** (Academic Press, London, 1976) sections 11.31 "Reflection of Light by a Moving Mirror", 11.33 "Experiments with a Moving Medium", 17.23 "Angular momentum associated with circularly polarised light"

1977 Arthur I. Miller "The physics of Einstein's relativity paper of 1905 and the electromagnetic world picture of 1905" Am.J.Phys. **45** 1040-1048

1977 Kenneth Brecher "Is the Speed of Light Independent of the Velocity of the Source?" Phys. Rev. Letters **39** 1051-1054 – *repeat of the de Sitter test*

1977 Reza Mansouri and Roan U. Sexl "A Test Theory of Special Relativity: I. Simultaneity and Clock Synchronisation" General Relativity and Gravitation **8** 497-513

1977 Reza Mansouri and Roman U. Sexl "A Test Theory of Special Relativity: II: First Order Tests" General Relativity and Gravitation **8** 515-524

1977 I. Lerche "The Fizeau effect: Theory, experiment, and Zeeman's measurements" Am.J.Phys. **45** 1154-1163

1977 H.M. Schwartz "Einstein's comprehensive 1907 essay on relativity, part I" Am.J.Phys. **45** 512-517

1977 H.M. Schwartz "Einstein's comprehensive 1907 essay on relativity, part II" Am.J.Phys. **45** 811-817

1977 Steven Weinberg, **The First Three Minutes** (Basic Books, 1977) pp.184-185 '*The Critical Density*'

1978 Joe Kiskis "Disconnected gauge groups and the global violation of charge conservation" Physical Review **17** 3196-3202 – *"Alice" universes*

1979 Clifford M. Will "The confrontation between gravitation theory and experiment" 24-29 in, **"General Relativity - An Einstein Centenary survey"**, S.W. Hawking and W. Israel (eds.) (CUP, Cambridge, 1979)

1979 D. Hasselkamp, E. Mondry and A. Scharmann "Direct observation of the transversal Doppler-shift" Zeitschrift für Physik A **289** 151-155 – *redshift measured at 90 degrees to a particle beam*

1979 Kenneth Brecher "Albert Einstein: A guide for the perplexed" Nature **278** 215-218 – *Einstein quotes*

1981 James T. Cushing "Electromagnetic mass, relativity, and the Kaufmann experiments" Am.J.Phys. **49** 1133-1149

1982 A.S. Shwarz "Field theories with no local conservation of the electric charge" Nucl.Phys. B **208** 141-158 – *"Alice" universes*

1983 Werner Heisenberg, **Encounters with Einstein** (Princeton University Press) *pp.* 10-11, 53, 113-115 – *Originally published as "Tradition in Science"*

1984 Allan Franklin "Forging, cooking, trimming, and riding on the bandwagon" Am.J.Phys. **52** 786-793

1985 P. Juncar, C.R. Bingham, J.A. Bounds, D.J. Pegg, H.K. Carter, R.L. Mlekodaj and J.D. Cole "New Method to Measure the Relativistic Doppler Shift: First Results and a Proposal" Phys.Rev.Lett. **54** 11-13

1985 Matti Kaivola, Ove Poulsen, and Erling Riis "Measurement of the Relativistic Doppler Shift in Neon" Phys.Rev.Lett. **54** 255-258

1985 A.K.A. Maciel and J. Tiomno "Experiments to Detect Possible Weak Violations of Special Relativity" Phys.Rev.Lett. **55** 143-146

1985 F. Wesemael "A comment on Adams' measurement of the gravitational redshift of Sirius B" Roy.Astr.Soc. QJ **26** 273-278

1986 D.W. McArthur "Special relativity: Understanding experimental tests and formulations" Physical Review A **33** 1-5 – *useful review and overview of past experiments*

1988 Clifford M. Will, **Was Einstein Right?: Putting General Relativity to the Test** (Oxford University Press, Oxford, 1988)
 – *Excellent overview of GR-testing for the general reader. The Appendix (pp.245-257) argues the mainstream case for special relativity being considered to be correct "without a shadow of a doubt".*

1988 Clifford M. Will, "Henry Cavendish, Johann von Soldner, and the deflection of light" Am.J.Phys. **56** 413-415

1988 Erling Riis, Lars-Ulrik Aaen Andersen, Nis Bjerre, Ove Poulsen, Siu Au Lee and John S. Hall "Test of the Isotropy of the Speed of Light Using Fast-Beam Laser Spectroscopy" Phys.Rev.Lett. **60** 81-84

1988 Michael S. Morris and Kip S. Thorne "Wormholes in spacetime and their use for interstellar travel: A tool for teaching general relativity" Am.J.Phys. **56** 395-412

1988 Michael S. Morris, Kip S. Thorne, and Ulvi Yurtsever "Wormholes, time machines, and the weak energy condition" Phys.Rev.Lett. **61** 1446-1449

1989 Erling Riis, Lars-Ulrik Anderson, Nis Bjerre, O Poulsen, Siu-Au Lee, and John L. Hall Riis et al "reply [to: Comment on 'Test of the isotropy of the speed of light…']" Phys.Rev.Lett. **62** 842

1989 Matt Visser "Traversable wormholes: Some simple examples" Physical Review D **39** 3182-3184
 – *pp.285-, "non-orientable wormholes"*

1989 Matt Visser "Traversable wormholes from surgically modified Schwarzchild spacetimes" Nucl.Phys. B **328** 203-212

1989 Zoltan Bay and John A. White Comment on "Test of the isotropy of the speed of light…" Phys.Rev.Lett. **62** 841

1989 Clifford Will, "The renaissance of general relativity," in **The New Physics** ed. Paul Davies (Cambridge University Press, Cambridge, 1989)

1990 G.P. Sastry and Tushar R. Ravuri "Modelling some two-dimensional relativistic phenomena using an educational interactive graphics soft" Am.J.Phys. **58** 1066-1073

1990 John Archibald Wheeler, **A journey into gravitation and spacetime**
 – *discussion of Gravity Probe B and the democratic principle as a double-page spread, pp. 232-233*

1990 John Preskill and Lawrence M. Krauss "Local discrete symmetry and quantum-mechanical hair" Nucl.Phys. B **341** 50-100 – *Alice universes and charged Alice handles*

1990 Timothy P. Krisher, Lute Maleki, George F. Lutes, Lori E. Primas, Ronald T. Logan, and John D. Anderson "Test of the isotropy of the one-way speed of light using hydrogen-maser frequency standards" Physical Review D **42** 731-734

1991 Jeremy Bernstein, **Einstein (second edition)** (Fontana, London) (first edition was 1973)

1991 Paul J. Nahin, **Time Machines: Time Travel in Physics, Metaphysics and Science Fiction** (AIP Press, New York)

1991 T.R. Sadin "In defense of relativistic mass" Am.J.Phys. **59** 1032-1036

1992 R. Klein, R. Grieser, L. Hoog, G. Huber et. al. "Measurement of the transverse Doppler shift using a stored relativistic $_7Li^+$ ion beam" Z. Phys. A **342** 455-461
 – *measurement of non-transverse shifts down a straight length of particle accelerator tube*

1993 Clifford M. Will, **Theory and experiment in gravitational physics** (Cambridge University Press, Cambridge, 1993) section 2.2 (iii) – *cites SR-compliance as a prerequisite for a theory to be considered "credible". Hence there are not supposed to be any "credible" theories of relativity that do not reduce to SR, by definition.*

References

1993 David Hochberg and Thomas W. Kephart "Wormhole Cosmology and the Horizon Problem" Phys.Rev.Lett. **70** 2665-2668

1993 Marek Artur Abramowicz "Black Holes and the Centrifugal Force Paradox" Scientific American (March 1993) cover, & 26-31

1993 Don N. Page "Relative Alternatives" Scientific American **(August 1993)** [letter] p.5
— *comment on the March issue's cover-story*

1993 Roger W. McGowan, David M Giltner, Scott J. Sternberg, and Siu Au Lee "New Measurement of the Relativistic Doppler Shift in Neon" Phys.Rev.Lett. **70** 251-254

1993 A. Rupert Hall **All Was Light: An introduction to Newton's Optics** (Clarendon Press, Oxford, 1993)

1993 Jeremy Bernstein **Cranks, Quarks and the Cosmos** (Basic Books, NY 1993)

1993, 1998 Harry Collins and Trevor Pinch, **The Golem: What You Should Know about Science** (Cambridge University Press, second edition 1998)

1994 Wolfgang Rindler, "General relativity before special relativity: An unconventional overview of relativity theory" American Journal of Physics **62** 887-893

1994 Achin Sen "How Galileo could have derived the special theory of relativity" Am.J.Phys. **62** 157-166

1994 R. Anderson, H.A. Bilger and G.E. Stedman "Sagnac effect: A century of Earth-rotated interferometers" Am.J.Phys. **62** 975-985

1994 Richard A. Mould, **Basic Relativity** (Springer-Verlag, NY, 1994) pp.80

1994 William I. McLaughlin, "Resolving Zeno's Paradoxes" Scientific American (Nov. 1994) p.66-71

1995 C. Lagoute and E. Davoust "The interstellar traveller" Am.J.Phys. **63** 221-227

1995 James B. Hartle "Spacetime information" Physical Review **D 51** 1800-1817

1995 Matt Visser "Natural wormholes as gravitational lenses" Physical Review **D 51** 3117-3120

1995 Gérard Clément "Wormhole cosmic strings" Physical Review **D 51** 6803-6809

1995 John G. Cramer, Robert L. Forward, Michael S. Morris, Matt Visser, Gregory Benford, and Geoffrey A. Landis, "Natural wormholes as gravitational lenses," Phys. Rev. **D 51** 3117-3120
`astro-ph/9409051`

1995 L. Sartori "Elementary derivation of the velocity addition law" Am.J.Phys. **63** 81-82

1996 A. G. Kelly, "Reliability of Relativistic Effect Tests on Airborne Clocks" Inst.Engineers.Ireland Monograph No. 3 (February 1996) 1-14 ISBN 1-898012-22-9

1996 Matt Visser, **Lorentzian wormholes: From Einstein to Hawking** (Springer-Verlag, 1996)

1997 Jürgen Renn, Tilman Sauer, John Stachel "The Origin of Gravitational Lensing: A Postscript to Einstein's 1936 'Science' paper" **275** 184-186

1997 Brett Mines "Alice Universes" Class. Quantum Grav. **14** 2527-2538

1997 Simon Singh, **Format's Last Theorem** (Fourth Estate, London, 1997)

2000 J.H. Field "Two novel special relativistic effects: Space dilatation and time contraction" Am.J.Phys. **68** 367-374

2000 Robert V. Pound "Weighing photons" Class. Quantum Grav. **17** 2303-2311

2000 Ignazio Ciufolini "The 1995-99 measurements of the Lense-Thirring effect using laser-ranged satellites" Class. Quantum Grav. **17** 2369-2380

2000 Robert L. Park, **Voodoo Science: The Road from Foolishness to Fraud** (OUP, 2000)

2002 John Waller, **Fabulous Science** (OUP) – *esp. Chapter 3, pp.48-63 – Eddington's expedition*

2005 Alex Harvey and Engelbert Schucking "A Small Puzzle from 1905" Physics Today **58** no.3 34-36
— *clocks at sea level run at the same rate, Alley experiment*

2006 Monwhea Jeng "A selected history of expectation bias in physics" Am. J. Phys. **74**, 578-583
`arXiv:physics/0508199`

RELATIVITY IN CURVED SPACETIME

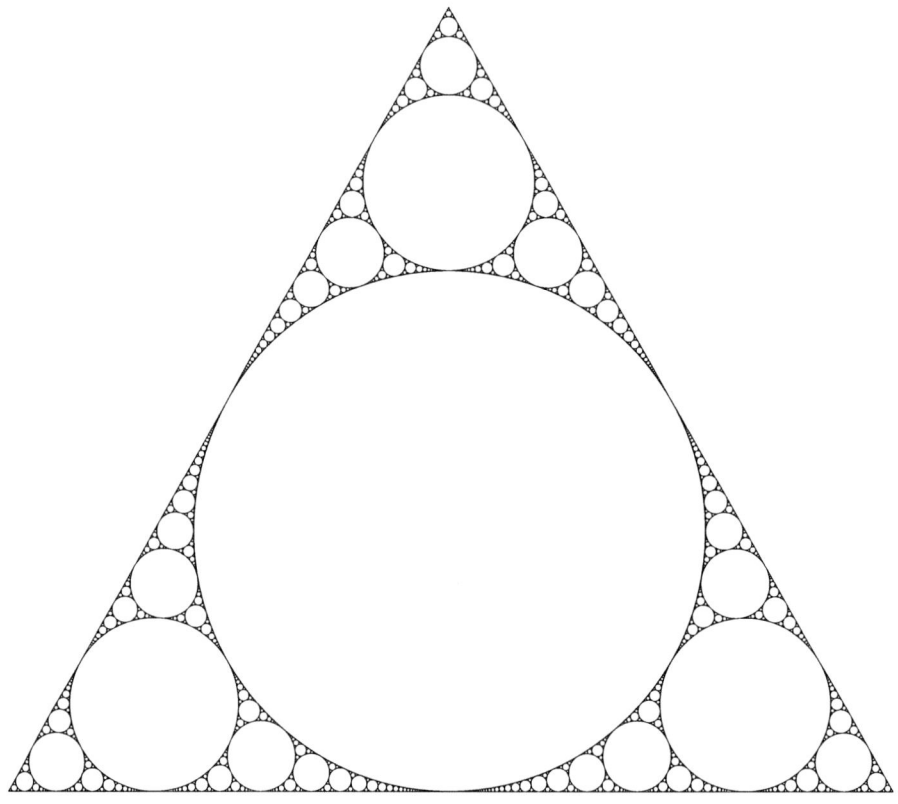

INDEX

A

aberration of angles..................................
... 67, 77, 80, 82, 84, 88, 93-96, 165, 175, 190, 213, 216, 287, 351-353, 362-363
acceleration radiation................................136
accelerator, particle ....
.... 201, 210-211, 215, 218, 221, 368
acoustic
  horizon.. 60, 99, 124, 132, 136, 143, 163, 279, 311, 361
  metric 60, 98, 123, 124, 133-136, 153, 160, 163, 166-167, 170, 180, 203, 221, 249, 280, 296, 311, 327, 339, 359
addition of velocities..................................
.... 91, 191, 211, 216, 367, 369
aether (/ether)
  absolute............................69, 186, 188-189, 212
  Lorentz Ether Theory (LET) 188-189, 212
  wind... 186
Alcubierre metric (warpdrive theory)............
.... 273-274, 290, 364
Alice universe 259, 368
Alley, Carroll 201, 369
amplitude 28-29, 105, 107-108, 138, 179
analogue gravity.. 124
analogues....... 23, 28, 51, 54, 71, 85, 94, 132, 155, 164, 228, 237, 279, 280, 285, 360
angular momentum.................... 129, 285-289
anthropomorphic principle 237
antielectron (positron)...................... 235, 259
antihorizon... 256
antimatter .. 114-115, 238, 242, 257-259, 355
antiparticle 115, 133, 261, 356
antiwormhole.. 259
area-increase theorem (black holes)........ 128
Aristotle... 358
arrow of time...................... 97, 190, 235, 255
artefacts 198, 268

B

baby universe.............................. 137, 241-242
backreaction 186
baseball 8, 50, 96, 121, 163, 257
Bekenstein, Jacob 359-360
Bernstein, Jeremy ....
.... 144, 317, 320, 360, 368-369, 379
Big Bang... 234

black hole 7, 9, 16, 58, 76, 98-99, 115, 119-123, 125-134, 136-144, 152-153, 164, 166, 169-170, 178-180, 203, 221, 227, 241-242, 260, 275, 279-280, 283, 286-287, 289, 296, 298, 314, 339, 356, 359-361
  Holographic Principle................. 141-143, 245
  information paradox.......................... 356, 360
  Kerr 127, 129, 179, 260, 286
  kugelblitz (self-trapped light)...................... 16
  logical... 296, 298-301
  Membrane Paradigm....................................360
blueshift.. 17, 31, 76, 95, 155, 164, 165, 178, 188, 204, 206, 213
Bohr, Niels................................. 3, 107, 293, 358
boomeranging 206, 272
box of frogs..354-356
Bradley, James........................... 77, 358, 361
Brans-Dicke theory.................................... 170
Bruno, Giordano 37, 300, 358
bubble 137, 155, 234, 236, 241, 249, 254-255, 263, 273, 276, 278, 332
bucket 33, 38, 122, 134, 157

C

catastrophe
  Newtonian 51-60
  theory..338
  Ultraviolet 105
causality 138, 163, 247, 277
Cavendish, Henry....................................
.... 59-60, 119, 358-359, 368
centrifugal field ....
.... 157, 260, 269, 274, 283-284, 287
centrifugal forces 33
centrifuge redshift ....
.... 157-158, 201, 205, 222, 342
Cerenkov radiation................................... 217
Cerenkov, Pavel................................. 217, 358
chaos, Chaos Theory 111, 113, 138, 153, 247, 249, 264, 327, 331
Cheshire Charge 259
chirality... 259, 285
classical Hawking radiation 354
Classical Theory (reference for SR)...........
.... 153, 337

Clifford, William Kingdon..............................
........85, 98, 147, 209, 225, 229, 273, 358, 364, 367-369, 379
clock (inertial)... 28
clock hypothesis/ clock postulate (SR).........
.... 158, 205, 218
coffee 41-42, 257, 356
collapse............................... 126, 129, 131, 143, 154, 231, 234, 236, 242, 254-256, 286
complementarity /complementary variables
.... 111
computer......7, 122, 141, 160, 198, 246, 295, 304, 315-316, 328, 329
connectivity 156, 261, 266
conservation
    of energy...154, 236
    of mass... 288
    of momenergy.......................................173, 288
    of momentum..80, 350
conservation laws ....
........22, 80, 155, 236-237, 265, 288, 363
consistency problem (SR) 198
consistent ....
42-43, 105, 107, 110, 114, 139, 156, 158, 170, 173-174, 192-194, 198, 202, 280, 299, 301-302, 310, 312, 334, 337, 367
constant
    Cosmological...................... 154, 208, 231, 320
    Demon (=2) 150, 213, 317, 323, 337, 345
Copenhagen Interpretation (QM).................
.... 107, 109, 138
Copernicus...23, 358
Coriolis field 269, 287
Coriolis, Gaspard de 45, 269, 287, 358
cosmic censorship 127, 227, 324
cosmic ray 217
Cosmological Constant (GR).........................
....154, 208, 231, 320
Coule, D.H. 276, 364
coupling efficiency 215
credible 84, 161-162, 170, 174, 180, 212, 262, 287, 289, 302, 316, 330, 338-369
cresting problem (warp theory) ....
.... 276, 278-279, 280
curvature factor, ©..
.... 174-179, 337, 345-346

D

dark energy 169, 240
dark matter.........................156, 169, 299, 308
dark star........60, 99, 119-123, 125, 132-134, 137, 140, 142-144, 164, 178, 275, 360
decay (particle) 8, 13, 169, 216-218

Democratic Principle.................50, 278, 368
Demon Constant (=2)....................................
.... 150, 213, 317, 323, 337, 345
Dicke, Robert H. 366
domain
    frequency... 96
    of applicability 220
    time... ....96, 165
Doppler effect
    longitudinal...
.... 65-66, 69, 211, 214, 351-353, 363
    transverse 63, 65, 66, 67, 69, 70, 71, 73, 77, 164, 165, 168, 216, 363
Doppler shift........20, 22, 67, 83, 84, 96, 132, 164-165, 167, 191, 212, 214, 337, 348, 350, 352-353, 363
drag.......6, 15, 46, 48, 67-68, 85, 91, 99, 129, 167, 179, 199, 203, 277, 279
duality 59, 103, 180

E

eclipse 219
economics 225, 332
Eddington, Arthur Stanley ....
....317, 358, 365, 369
Einstein, Albert..
.......3, 12-13, 21, 23, 30-32, 40-42, 50-51, 58-60, 63, 69, 85, 98, 103, 107, 117, 125, 127, 144, 147, 149, 150-155, 159, 161, 167-171, 180, 183, 186, 189-192, 194-195, 197, 200-201, 203, 206-207, 209, 211-213, 219-221, 229, 247, 251, 269, 274-275, 290, 296, 302, 317, 320-321, 335, 337, 339, 341, 358-361, 364-369, 379
electric charge...............................87, 114, 367
electron............... 105, 201, 235, 240, 259-260
EM radiation 22
embedding diagram.......... 9, 27, 59, 263-264
emission theory...
.... 21, 68-69, 83, 179, 219, 222, 349, 352
emission theory of light................80, 83, 348
energy conservation 154, 236
entropy 128, 140, 235-236, 239, 249, 256, 356, 361
Eötvös, Baron Roland von..........18, 28, 358
equipotential surface........................ 201, 312
equivalence of mass and energy, $E=mc^2$......
.... 21-22, 80, 95, 167, 173, 179, 205, 211-222, 288, 303, 342, 350-351, 364-365
equivalence principle (inertia/gravitation) ...
....136, 158, 197, 218, 299, 302, 337, 366

error bars 324
escape velocity 119-121, 179
Escher 309, 313
ether .... *see:* aether
event 7, 35, 59-60, 76, 111, 121, 123-125,
    127-128, 130-131, 133, 137-142, 144,
    163-164, 179, 192-193, 202, 227, 229,
    232, 234, 241, 260, 272, 275, 296, 316,
    356, 360
event horizon 7, 60, 76, 121, 123-125,
    127-128, 130-131, 137-142, 144, 164,
    179, 260, 272, 275, 296, 356, 360
Everett, Hugh 110, 240, 364
exotic matter 260, 273-274
expectation bias 325, 369
experiment
    Alley 201, 369
    Fizeau 85, 91, 99, 167, 185, 216, 220-221,
        358, 361-362, 367, 380
    Hafele-Keating 218, 343
    Harwell centrifuge redshift 157
    Hasselkamp (transverse redshift) ....
        .... 213-214, 222, 337, 367
    Ives-Stilwell (longitudinal shifts) ....
        .... 214, 222, 365
    Michelson-Morley ....
        .... 91, 186, 212, 320-321, 358, 361, 365
    Pound-Snider 150, 205, 366
    Shapiro timelag 7, 59, 149, 151
    two-slit 108-109, 138, 240
extinction theorem 221
extrusion 10, 176

F

factor
    curvature, © 174-179, 337, 345-346
    fudge 130, 154
    Lorentz 21, 70, 75, 173-174, 188-190, 192,
        216, 351
faster than light 58, 91, 125, 217, 359
field
    centrifugal ... 157, 260, 269, 274, 283-284, 287
    Coriolis 269, 287
    electric 7-8, 94, 97, 257, 261
    gravitational .... …
        .... 7-9, 11-12, 16, 18-20, 23,
        28, 30, 32, 35, 38, 42-50, 53-55, 57-60,
        88-89, 91, 95-96, 98-100, 115, 120-121,
        129, 136, 143, 149-150, 156-157, 165-
        168, 179, 197, 200, 206, 218, 221, 227,
        239, 249, 254, 256, 265, 272-274, 278,
        280-281, 283-288, 312, 359, 365-366
    inertial 28, 50, 95, 287
    magnetic 7, 97
    probability 91, 110
    quantum 99, 112, 147

fieldline 46-47, 261, 286-287
Fitzgerald, George Francis 188, 194, 358
Fizeau, Armand 85, 91, 99, 167, 185,
    216, 220-221, 358, 361-362, 367, 380
floor/no-floor assumption (gravitation) ....
    .... 15, 18-20, 37, 42, 156, 239, 249, 355
fluid dynamics 185, 187
force
    centrifugal 33
    tidal 28, 115, 121-122, 132, 248, 253, 278
Forward, Robert L. ....
    .... 269, 285, 290, 358, 364, 369
Foucault, Léon 49, 358, 379
fractal 200, 240, 243, 245-246, 249, 252,
    266-267
fractal sponge 252, 266-267
freefall 15, 31-32, 43, 76, 126
frequency ....
    .... 8, 21, 23, 28-31, 60, 63, 65-66, 68, 71,
    73-76, 83-84, 96-97, 105, 108, 111, 155,
    164, 177-179, 203, 205-206, 215,
    348-350, 368
Fresnel, Augustin Jean ....
    .... 85, 91, 100, 185, 216, 358, 362
frog 354-356
fudge factor 130, 154

G

Galileo Galilei ....
    ... 37, 79, 92, 94, 171, 185, 212, 269, 341,
    358, 369
gamma radiation 366
gamma (γ) factor .... *see:* Lorentz Factor
Gauss, (Johann) Karl Friedrich ....
    .... 10, 32, 60, 338, 358
gee, one Earth-gravity 42, 202
geeforce 28, 37-38, 42, 50, 122, 136, 157,
    202, 218, 272, 281
general principle of relativity ....
    . 41, 43-45, 47, 50, 87, 90, 136, 152, 156,
    160-161, 167-168, 170, 177, 180, 197,
    203, 273, 301, 312
general theory of relativity 32, 51, 60, 98,
    100, 158-159, 195, 221, 365
geodesic 263-264
geoid 201
George (Bishop) Berkeley 38, 358
Giordano Bruno 37, 300, 358
global ....
    ..... 35, 40, 50, 68, 70, 100, 136, 194, 197,
    199, 202, 219, 232, 243, 254, 259, 276,
    296, 307, 309, 313-314, 327-328, 338,
    367

373

global geometry 254
GPS (Global Positioning System) 35, 150
GR1915 .... 41, 50, 98-100, 115,
    119, 123-134, 136, 138-144, 147,
    149-161, 163-164, 166-170, 180, 199,
    203, 207-208, 227, 241-242, 260, 275,
    280, 286-287, 290, 296, 298-299,
    301-302, 316-317, 339, 356, 359
gravitational
    braking 92
    floor 156
    lensing 12, 156, 165, 286
    mass 15-18, 20, 28, 43, 45,
        89, 129, 159, 221, 231
    potential 88, 150, 206, 272, 312-313
    slingshot 89, 271
    time dilation 30, 35, 50, 60, 144, 149,
        151-152, 157, 169, 203, 218, 338
gravitational blueshift ....
    .... 29, 76, 95, 164, 178, 272
gravitational collapse 129, 169, 359, 367
gravitational redshift ....
    ... 29, 31, 76, 95, 150, 152, 157, 166-167,
    169, 205, 207, 272, 368
gravitational well 9-10, 12, 24, 28,
    48, 56, 58, 76, 89, 95, 97, 133, 136, 176,
    197, 206, 271, 274, 278
gyroscope 285

H

Hafele-Keating experiment 218, 343
Hasselkamp transverse redshift test ....
    .... 213-214, 222, 337, 367
Hartle-Hawking bubble 236, 247
Hawking radiation ....
    .... 60, 98, 115, 124, 132-134, 136-137,
    140-144, 153, 163, 180, 215, 221, 229,
    278-280, 356, 359-361
Hawking, Stephen ....
    .... 128, 132, 137, 171, 234, 358-359
Heisenberg, Werner .... ~103, 367
Hippasus (root two) 304
Holographic Principle 141-143, 245
horizon
    acoustic 60, 99, 124, 132,
        136, 143, 163, 279, 311, 361
    black hole 7, 60, 76, 121, 123-
        125, 127-128, 130-131, 137-142, 144,
        164, 179, 260, 272, 275, 296, 356, 360
    cosmological 163, 164, 166-167,
        180, 203, 227-229, 249
    velocity 125, 136, 163, 180
Hubble
    expansion 165, 232
    redshift 154, 165-167, 227, 236

time 232
hypercube 230
hypersphere 207, 230, 236

I

ideogram 307
imaginary time coordinates (cosmology) ......
    233
indirect radiation ....
    .... 119, 122, 124-125, 133-134, 215
inertia 8, 13, 25, 28, 33, 38, 41, 50, 92, 95,
    165, 237, 269, 301, 364-365
inertial field 28, 50, 95, 287
inertial mass 8, 15-18, 20, 22, 28,
    38, 43, 45, 159, 215, 221, 233, 237, 287
infinity 126, 177, 178, 215, 237
inflation 241
information ....
    ...... 8, 83, 96-99, 105, 111, 114, 124, 126,
    128, 130-132, 135-139, 141-144, 152-
    153, 170, 180, 191, 229, 237-238, 241,
    245, 264, 271-272, 278-280, 299, 307-
    308, 315, 321, 326-330, 332, 334, 341-
    343, 348, 356, 360, 369
Information Paradox, Black Hole (BHIP)
    .... 138, 164, 178-179, 356, 360
information theory ....
    ....97, 132, 143-144, 153, 327, 332, 356
interference 54, 108-110, 154,
    241, 262, 279, 286, 361
interpretation, of QM
    Copenhagen 107, 109, 138
    Hidden Variable ... 98, 107, 109, 135, 137-138
    Many Worlds 110
intransitive logic 309, 312
intrinsic 36, 50, 157, 159, 163, 342
inverse square law 156
irrational numbers 304-306
Ives-Stilwell experiment 214, 222, 365

J

Julia set 243, 245-246

K

Kaufmann, Walter 320, 367
Kerr black hole 127, 129, 179, 260, 286
Kerr wormhole 260
kinetic 22, 28, 46, 89, 159, 165, 176, 204,
    221, 271, 349
Klein bottle (topology) 258-259

Index

Krasnikov tube (warp theory) ....
.... 274, 277-278, 281, 364
kugelblitz (self-trapped light) 16

L

Laithwaite, Eric 288-289
laPlace, Marquis 359
laser 16, 19, 26, 47, 77, 369
latitude 234
Lemaître, Abbe G.E. 154, 358
lensing 3, 11-12, 95, 156, 288
LET (Lorentz Ether Theory) .. 188-189, 214
light
   deflection of 11-12, 23, 53, 368
   electromagnetic radiation 236
light, faster than 275
lightcone 175, 184, 192-193
lightspeed
   local 56, 130, 275, 278
lightyear 5, 202
linear logic 41, 298
linear momentum 287-288
local ....
   .. 6, 8-9, 19, 28, 32, 35, 40, 49, 50, 55-58,
   76, 79, 97, 100, 122-123, 130, 134,
   136-137, 161, 163, 179, 191, 197, 202,
   218-219, 228-229, 232-233, 237-238,
   243, 259, 264, 274-278, 280, 298, 309,
   311-312, 314, 332, 338, 342, 367
local geometry 134, 163, 309
local physics 9, 32, 79, 136
Lodge, Oliver Joseph 358, 361-362
logic 21, 111, 133, 143,
   158, 188, 195, 197, 200, 268, 275, 279,
   289-290, 293, 295-299, 304, 308,
   310-311, 315-316, 333, 341
logic trap 297
logical black hole 296, 298-299, 300-301
longitude 35, 234
Lorentz
   contraction 175, 189, 362-363
   factor 21, 70, 75, 173-174,
   188-190, 192, 216, 351
Lorentz Ether Theory 188-189, 212, 214
Lorentz, Hendrik Antoon 188, 358
Lorentzlike factor 173-175, 179, 206
lurking variables 123

M

Mach cone 278
Many-Worlds interpretation (QM) 110
market forces 331-332

mass
   gravitational 15-18, 20, 28, 43, 45, 89, 129,
   159, 221, 231
   inertial 8, 15-18, 20, 22, 28, 38, 43, 45, 159,
   215, 221, 233, 237, 287
   rest 351
matter
   dark 156, 169, 299, 308
   exotic 260, 273, 274
Maxwell, James Clerk 7, 185, 358
mechanics
   fluid 187, 280
   Newtonian 20, 36, 38, 40, 80, 93,
   176-177, 179-180, 203, 206, 208, 212-
   213, 215, 282, 339, 346, 350
   quantum 29, 51, 53-54, 59, 91-92,
   97-99, 103, 105-107, 110-111, 114-115,
   124, 131-134, 136-140, 143-144, 153,
   164, 169-170, 179-180, 200, 221, 227,
   240, 247, 265, 278-280, 287, 296, 327,
   340, 355-356, 360
   statistical 106, 110, 328
Mercury (planet) 122, 147, 149, 151-152
meta-theory 106, 189, 280
metric
   acoustic 60, 98, 123-124, 133-136,
   153, 160, 163, 166-167, 170, 180, 203,
   221, 249, 280, 296, 311, 327, 339, 359
   Minkowski 194, 199
metric theories 161
Michell, John ("dark star" pioneer) ....
   .... 29-30, 60, 117, 119-120, 149, 358-359
Michelson, Albert Abraham ....
   .... 91, 186, 212, 320-321, 358, 361, 365
Michelson-Morley experiment 186, 320
microcausality 138, 280
Minkowski spacetime 97-98, 175, 190,
   192-194, 199, 202, 345
Minkowski, Hermann 175, 192, 358
mirage 140
Möbius(/Moebius) strip, surface with one
   side 258
momenergy 289
momentum ....
   ...... 20-22, 28, 37, 54, 80, 89, 91, 98, 114,
   120, 129-130, 165, 167, 175, 178-179,
   198, 212, 215-217, 241, 286, 288-289,
   349-351, 355, 367
   angular 129, 285-289
   conservation of 80, 350
   conversion of 288-289
   exchange of 89, 91
   linear 287-288
Moon (/moons) ....
   .... 5, 18,19, 23, 35, 37, 248, 313
Mössbauer effect 150, 205, 366-367
multiply-connected 253, 260, 311

multiverse 139, 240-242, 246, 249, 346
muon 211, 217

N

negative gravitation 249, 254, 256, 274
Newcombe, Simon 358
Newton, Isaac ....
..3, 13, 23, 37, 51, 53, 319, 358, 367, 379
nonlinearity 98, 117, 124, 279, 283, 286
nonorientable (wormhole) 258-259
nuclear 22, 165, 211, 248, 257, 259, 366
null 160, 186, 221, 315-316, 321
null physics 160
number
    atomic 22
    imaginary 233
    irrational 304, 305, 306
    random 138
    stupid 305-306

O

observer
    perfect 197
    SR 93, 197, 220, 274
observer principle 139
observerspace 107, 127, 135, 139,
140-141, 143, 163, 165, 197
Occam's Razor ....
.... 38-39, 107-108, 241, 259, 264

P

pair production ....
.... 115, 132, 140, 234, 237, 355, 359
panda, giant 308
paradigm
    flat-spacetime 9, 40, 58, 94, 149, 157,
    159-160, 167-168, 174-176, 190, 181-222,
    263, 283, 289
    Membrane 360
    observerspace 107, 127, 135, 139-141,
    143, 163, 165, 197
paradox
    black hole information ....
    .... 138, 164, 178-179, 356, 360
    no-signal 143
    Olbers' 225, 229
    Zeno's 36, 95-96, 358, 369
Parmenides (teacher of Zeno) 36
particle
    accelerator ... 201, 210-211, 215, 218, 221, 368
    anti- 115, 133, 261, 356
    real 97, 135, 160, 220, 264

test 92, 94, 97
virtual 114, 136
visiting 122
pathological 158, 160-161, 297-298, 308,
319, 326, 330, 334, 343
pendulum 32, 49
phonon 279-280, 361
photon 91, 105, 108, 109, 110, 205, 240
Pi (π) 293, 305-307
pilot wave 51, 53, 59
Planck-scale 134, 140, 358
plane of simultaneity 202
Poincaré, Jules Henri 33, 188, 194, 358
Positive Energy Theorem (...................... 274
positron (=antielectron) 235, 259
potato 99
pragmatic 9, 99, 153, 198, 200
Preskill, John 138, 139, 360, 368
principle
    complementarity 111
    democratic 50, 278, 368
    Eötvös 18, 28, 358
    equivalence ....
    ...... 136, 158, 197, 218, 299, 302, 337, 366
    holographic 141-143, 245
    no random numbers 138
    observer 139
    uncertainty (Heisenberg) 103
Principles of Relativity 40
    general (Einstein) ... 41, 43, 44, 45, 47, 50, 87,
    90, 136, 152, 156, 160, 161, 167, 168,
    170, 177, 180, 197, 203, 273, 301, 312
    special 40
prior geometry 156
probability 91, 103, 107-108, 110,
240, 319, 329, 338
probability field 91, 110
proper time 232, 236
pseudowormhole 267-268, 277
public transport 271
Pythagoras 304-305, 358
Pythagorean Brotherhood 304

Q

quantum
    economics 111
    field theory (QFT) 99, 112, 147
    fluctuation 234
    foam 110, 265, 267-268
    fuzz, fuzzball 131, 227
    gravity (QG) ....
    98, 124, 139, 153, 168-169, 180, 279, 280
    mechanics (QM) 29, 51, 53-54,
    59, 91-92, 97-99, 103, 105-107, 110-111,
    114-115, 124, 131-134, 136-140, 143-144,
    153, 164, 169-170, 179, 180, 200, 221,

227, 240, 247, 265, 278-280, 287, 296, 327, 340, 355-356, 360
-tunnel (action) 131, 278-279, 331
sociology 327
photocopier 139
quantum electrodynamics (QED) ....
.... 132, 168, 179, 200, 221
quasi-Euclidean 154, 207

R

r=2M (radius of a black hole or dark star)
..... 121-122, 126-127, 130, 139-140, 143, 359
radiation
accelerational / Unruh 136
gravitational 169
Hawking 60, 98, 115, 124, 132-134, 136-137, 140-144, 153, 163, 180, 215, 221, 229, 278-280, 356, 359-361
indirect/acoustic ....
.... 119, 122, 124-125, 133-134, 215
radiation pressure 19-20, 164, 185, 215
random 84, 98, 109, 111, 138, 143, 235, 246-247, 299, 326, 328, 356
rankings 309
reactionless 289
recoil 205, 214, 366
redshift 17, 67-68, 70, 75-76, 96, 149, 154-155, 158, 165-166, 178, 201, 203-208, 212-213, 227, 297, 320, 337, 345, 349, 352-353, 357, 367
centrifuge/accelerational ....
.... 157-158, 201, 205, 222, 342
cosmological/Hubble ....
.... 155, 163, 166, 169, 207-208, 227
gravitational 29, 31, 76, 95, 150, 152, 157, 166-167, 169, 205, 207, 272, 368
recession 204-205
transverse 68-69, 83, 201, 211-214, 222, 297, 303, 337, 342, 352
reduction to flat spacetime 159
refraction (deflection of signals by a speed gradient) 54-55, 283, 286-287
refractive index 53, 56, 59
relativity of simultaneity 183, 191, 363
rest energy 20
rest mass 351
Riemann, Georg Friedrich 10, 147, 358
route-dependence ....
.... 132, 136, 197, 232, 259, 265, 311, 313

S

sanity testing (logic) ... 58, 198, 201, 298-299
Schrödinger, Erwin 138

Schwarzchild
radius (see also: **r=2M**) 121
Schwarzchild, Karl 121, 358
searchlight effect 80, 211, 216
self-referential .... see: self-referential
semiclassical 153, 215
Shapiro effect 7, 59, 149, 151
shell 38, 44-45, 47-48, 76, 126, 143, 274
singularity
point 131, 169, 227, 234, 238
ring 129, 260
skew 193-194
smell 144
SOD (Second-Order Doppler) effect ....
.... 205, 321
space
absolute 33, 38
relative 33, 37
spacetime, Minkowski ....
.97-98, 175, 190, 192-194, 199, 202, 345
spaghettification 121-122
spectrum 7, 29, 60, 105, 119, 205
spookiness (QM) 107, 111, 138, 153, 280
statistical mechanics 106, 110, 328
stratification of theories 200
string theory 180, 187
superluminal 215-216
superradiance 132
Susskind, Leonard 141, 144, 360-361

T

terminal velocity 121, 165, 178, 206
Terrell, James
.... 71, 191, 197, 342-343, 362, 363
test theory 213, 334, 337
tests of general relativity 149
theorem
extinction 221
positive energy (PET) 274
thermal redshift 204-205, 208
thermodynamics 236
Thorne, Kip 138, 358, 360
thought-experiment 17, 20, 42, 115, 138, 194, 201, 203, 283, 348
tidal forces ....
.... 28, 115, 121-122, 132, 248, 253, 278
tilt ... 11, 56, 80, 95, 114, 192, 193-194, 202, 312
time
absolute 37, 200, 232
axial (T) 232-233, 235-236
cosmological 232-233
arc-parameter (η) 233
Hubble (H) 232

imaginary 233
proper (*t*) 232, 236
radial (*a*) 232-233, 235
time dilation 30, 60, 150, 157-158,
201, 204, 211, 213, 216-218, 285
timelag 44, 73, 87, 90, 151, 191, 262
Titanic Syndrome 327, 339
topology 261, 265-266, 278
topology (effective) 266
torus 253, 255, 269, 282-287, 290, 310
toy model 106, 124
transverse redshift 68-69, 83, 201,
211-214, 222, 297, 303, 337, 342, 352
two-slit experiment 108-109, 138, 240

U

ultrafast (potentially > background *c*) ....
.... 115, 217, 275, 277-289
ultraviolet catastrophe 105
uncertainty 103, 114, 132, 315, 340
unification 53, 94, 165-166, 180, 227
units 22, 202, 218, 285, 289-290
universe
  baby 137, 241-242
  closed 229
  fractal 249
  multiverse 139, 240-242, 246, 249, 346
  reflecting 238
Unruh radiation 136

V

velocity addition formulae
  (SR and earlier theory) ....
  .... 91, 191, 211, 216, 367, 369
virtual particle 114, 136
visiting particle 122

Visser, Matt 124, 253, 257-259, 358,
360-361, 368-369

W

Wald, Robert 140, 285, 290, 359, 364
warp drive ....
......50, 142, 223, 249, 256, 268, 271-272,
274-276, 279, 281-282, 285, 288-290,
364
warp field 256, 274, 276-282
wave
  electromagnetic 7, 138, 176, 185
  gravitational ....
  ....44, 89, 206, 219, 278, 281, 289, 365
wavelength 28-29, 54-55, 65, 83-84,
175-176, 214, 276
wave-particle duality 53, 55
Wheeler, John Archibald ... 16, 50, 147, 209,
261, 265, 269, 272, 317, 358, 363-364,
366-368, 379
worldline ... 97, 127, 165, 184, 190, 192-194,
232, 234, 237-238, 240, 263-264
wormhole 137, 249, 251-268, 277,
286, 360, 368-369
  Kerr 260
  nonorientable (antiwormhole) 258-259
  pseudowormhole 267-268, 277

X

x-ray 7, 105

Z

Zeeman, Pieter 91, 216, 362, 367
Zeno of Elea 36, 95-96, 358, 369

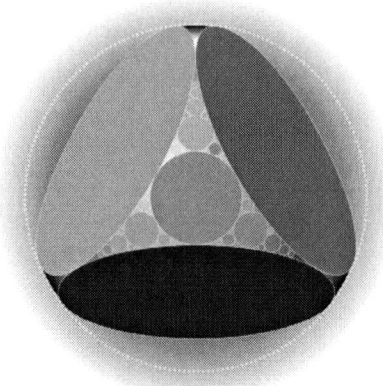